Philosophy of Ecology

Handbook of the Philosophy of Science

General Editors

Dov M. Gabbay
Paul Thagard
John Woods

ELSEVIER

AMSTERDAM • BOSTON • HEIDELBERG • LONDON • NEW YORK • OXFORD
PARIS • SAN DIEGO • SAN FRANCISCO • SINGAPORE • SYDNEY • TOKYO
North Holland is an imprint of Elsevier

N·H

Handbook of the Philosophy of Science

Volume 11

Philosophy of Ecology

Edited by

Kevin deLaplante
Iowa State University,
USA

Bryson Brown
University of Lethbridge,
Canada

Kent A. Peacock
University of Lethbridge,
Canada

ELSEVIER

AMSTERDAM • BOSTON • HEIDELBERG • LONDON • NEW YORK • OXFORD
PARIS • SAN DIEGO • SAN FRANCISCO • SINGAPORE • SYDNEY • TOKYO
North Holland is an imprint of Elsevier

N·H

North Holland is an imprint of Elsevier
The Boulevard, Langford lane, Kidlington, Oxford, OX5 1GB, UK
Radarweg 29, PO Box 211, 1000 AE Amsterdam, The Netherlands
225 Wyman Street, Waltham, MA 02451, USA

First edition 2011

British Library Cataloguing in Publication Data
A catalogue record for this book is available from the British Library

Library of Congress Cataloging-in-Publication Data
A catalog record for this book is available from the Library of Congress

ISBN: 978-0-444-51673-2
ISSN: 0031-8019

For information on all North Holland publications
visit our web site at *elsevierdirect.com*

11 11 10 9 8 7 6 5 4 3 2 1

GENERAL PREFACE

Dov Gabbay, Paul Thagard, and John Woods

Whenever science operates at the cutting edge of what is known, it invariably runs into philosophical issues about the nature of knowledge and reality. Scientific controversies raise such questions as the relation of theory and experiment, the nature of explanation, and the extent to which science can approximate to the truth. Within particular sciences, special concerns arise about what exists and how it can be known, for example in physics about the nature of space and time, and in psychology about the nature of consciousness. Hence the philosophy of science is an essential part of the scientific investigation of the world.

In recent decades, philosophy of science has become an increasingly central part of philosophy in general. Although there are still philosophers who think that theories of knowledge and reality can be developed by pure reflection, much current philosophical work finds it necessary and valuable to take into account relevant scientific findings. For example, the philosophy of mind is now closely tied to empirical psychology, and political theory often intersects with economics. Thus philosophy of science provides a valuable bridge between philosophical and scientific inquiry.

More and more, the philosophy of science concerns itself not just with general issues about the nature and validity of science, but especially with particular issues that arise in specific sciences. Accordingly, we have organized this Handbook into many volumes reflecting the full range of current research in the philosophy of science. We invited volume editors who are fully involved in the specific sciences, and are delighted that they have solicited contributions by scientifically-informed philosophers and (in a few cases) philosophically-informed scientists. The result is the most comprehensive review ever provided of the philosophy of science.

Here are the volumes in the Handbook:

Philosophy of Science: Focal Issues, edited by Theo Kuipers.

Philosophy of Physics, edited by Jeremy Butterfield and John Earman.

Philosophy of Biology, edited by Mohan Matthen and Christopher Stephens.

Philosophy of Mathematics, edited by Andrew Irvine.

Philosophy of Logic, edited by Dale Jacquette.

Philosophy of Chemistry and Pharmacology, edited by Andrea Woodsy, Robin Hendry and Paul Needham.

Philosophy of Statistics, edited by Prasanta S. Bandyopadhyay and Malcolm Forster.

Philosophy of Information, edited by Pieter Adriaans and Johan van Benthem.

Philosophy of Technology and Engineering Sciences, edited by Anthonie Meijers.

Philosophy of Complex Systems, edited by Cliff Hooker.

Philosophy of Ecology, edited by Bryson Brown, Kent A. Peacock and Kevin deLaplante.

Philosophy of Psychology and Cognitive Science, edited by Paul Thagard.

Philosophy of Economics, edited by Uskali Mäki.

Philosophy of Linguistics, edited by Ruth Kempson, Tim Fernando and Nicholas Asher.

Philosophy of Anthropology and Sociology, edited by Stephen Turner and Mark Risjord.

Philosophy of Medicine, edited by Fred Gifford.

Details about the contents and publishing schedule of the volumes can be found at http://www.elsevier.com/wps/find/bookdescription.cws_home/BS_HPHS/description#description

As general editors, we are extremely grateful to the volume editors for arranging such a distinguished array of contributors and for managing their contributions. Production of these volumes has been a huge enterprise, and our warmest thanks go to Jane Spurr and Carol Woods for putting them together. Thanks also to Lauren Schultz and Gavin Becker at Elsevier for their support and direction.

PREFACE

The most pressing problems facing humanity today—over-population, energy shortages, climate change, soil erosion, species extinctions, the risk of epidemic disease, the threat of warfare that could destroy all the hard-won gains of civilization, and even the recent fibrillations of the stock market—are all ecological or have a large ecological component, and it is fitting that philosophers turn their attention to understanding the science of ecology and its huge implications for the human project. Numerous excellent collections on the philosophies of biology, physics, and mathematics have appeared in the past twenty years, but there have been relatively few books actually to have the phrase "philosophy of ecology" in their titles. A notable exception is the fine anthology edited by Keller and Golley [2000], which appeared almost ten years ago. That seems like a long time passing; since then we have had "wars and rumours of wars," the report of the IPCC in 2007, SARS and H1N1, devastating earthquakes and tsunamis, summers when the forests of Europe burned, melting icesheets and a dramatically warming Arctic, and an increase in the human population of nearly another billion hungry mouths. While not all papers in the present volume are directly concerned with the enormous and urgent challenge of environmental remediation, all seek philosophical perspectives on the scientific study of "organisms at home (*oikos*)" in the biophysical world they have built.

The science of ecology directly confronts the huge intellectual challenge posed by our efforts to understand biophysical systems that are immensely rich and complex, and subject to outside influences that can shift and disrupt the patterns of interaction that unify them. Attempts to model complex, open systems cannot be expected to lead to reliable predictions of specific outcomes (regardless of the pressures that practical policy concerns may place on scientists to produce such predictions). However, they can help us identify trends and possible responses (sometimes obvious, sometimes not) to such trends. We can identify important ecological processes and gain more than a glimmering of the various risks posed by changes in ecological systems and their surroundings. The science of ecology is of special philosophical interest because of the synergies between the purely theoretical and the grassroots-practical levels of understanding that it demands. We can't get the application of ecology to policy or other practical concerns right unless we have a clear and disinterested philosophical understanding of ecology

Handbook of the Philosophy of Science. Volume 11: Philosophy of Ecology.
Volume editors: Kevin deLaplante, Bryson Brown and Kent A. Peacock. General editors: Dov M. Gabbay, Paul Thagard and John Woods.

which can help identify the practical lessons that science can teach us. Conversely, the urgent practical demands humanity faces today cannot help but direct scientific and philosophical investigation toward the basis of those ecological challenges that threaten human survival. To adapt a phrase from Dr. Johnson, the prospect of ecological catastrophe focusses the mind wonderfully. We hope that this book will help to fuel the timely renaissance of interest in philosophy of ecology that is now occurring in the philosophical profession.

This volume owes the possibility of its existence to the imagination and initiative of the series editors, John Woods, Dov Gabbay, and Paul Thagard, who conceived of an ambitious, multi-volume set that could present the latest thinking in the philosophies of all the key sciences. Everyone involved in the *Handbook* series is deeply indebted to Elsevier Publishers for making this adventure possible. The editors of this volume enjoyed the almost unprecedented luxury of being able to tell its authors that they had no specific length limits and that this was their chance to write that opinionated review of their field they had always wanted to write. The result is a richly diverse collection of papers. While some have an encyclopedic character, all attempt to synthesize in novel ways, to break ground, and to challenge. This volume is not merely a Handbook (if one conceives of that sort of book as merely a work of reference) but a call to intellectual arms for many of the key issues that will define philosophical thought about ecology in the next decades.

Thanks and acknowledgements are due to many people and organizations. All three editors are very grateful to Jane Spurr and John Woods for their help, good advice, and patience during the long gestation period of this project. K. P. and B. B. thank the Social Sciences and Research Council of Canada for financial support of their research, and the University of Lethbridge for sustenance, financial and otherwise. K. P. is grateful to Richard Delisle, Cody Perrin, and Sharon Simmers for assistance and advice. B. B. thanks Ron Yoshida, for his support and encouragement as co-developer and teacher of our earth and life sciences course, and especially Linde Bruce-Brown for her support and patience with the long process of working on this volume. K. D. offers thanks and gratitude to Iowa State University for support and assistance; to Arnold van der Valk for his partnership as co-instructor of our history and philosophy of ecology course; to Kent Peacock for introducing K. D. to the environmental philosophy literature as a young graduate student; and to Brenda Theoret for her love and endless patience.

BIBLIOGRAPHY

[Keller and Golley, 2000] David R. Keller and Frank B. Golley (eds.), *The philosophy of ecology: from science to synthesis.* University of Georgia Press, 2000.

CONTRIBUTORS

Bryson Brown
University of Lethbridge, Canada.
brown@uleth.ca
J. Baird Callicott
University of North Texas, USA.
johnbaird.callicott@unt.du
John Collier
University of KwaZulu-Natal, South Africa.
collierj@ukzn.ac.za
Mark Colyvan
University of Sydney, Australia.
mcolyvan@usyd.edu.au
Graeme S. Cumming
University of Cape Town, South Africa.
Graeme.Cumming@uct.ac.za
Kevin deLaplante
Iowa State University, USA.
kdelapla@iastate.edu
Christopher Eliot
Hofstra University, USA.
christopher.eliot@hofstra.edu
James Justus
Florida State University, USA and University of Sydney, Australia.
jjustus@fsu.edu
Brendon M. H. Larson
University of Waterloo, Canada.
blarson@uwaterloo.ca
Gregory M. Mikkelson
McGill University, Canada.
gregory.mikkelson@mcgill.ca
Bryan Norton
Georgia Institute of Technology, USA.
bryan.norton@pubpolicy.gatech.edu

Jay Odenbaugh
Lewis and Clark College, USA.
jay@lclark.edu
Kent A. Peacock
University of Lethbridge, Canada.
kent.peacock@uleth.ca
Valentin D. Picasso
University of the Republic, Uruguay.
vpicasso@fargo.edu.uy
Sahotra Sarkar
University of Texas at Austin, USA.
sarkar@mail.utexas.edu
Katie Steele
The London School of Economics and Political Science, UK.
k.steele@lse.ac.uk
Arnold van der Valk
Iowa State University, USA.
valk@iastate.edu

CONTENTS

General Preface v
Dov Gabbay, Paul Thagard, and John Woods

Preface vii
Kevin deLaplante, Bryson Brown and Kent A. Peacock

List of Contributors ix

Introduction

Philosophy of Ecology Today 3
Bryson Brown, Kevin deLaplante and Kent A. Peacock

Part 1. Philosophical Issues in the History and Science of Ecology

Origins and Development of Ecology 25
Arnold van der Valk

The Legend of Order and Chaos: Communities and Early Community Ecology 49
Christopher Eliot

Philosophical Themes in the Work of Robert MacArthur 109
Jay Odenbaugh

Embodied Realism and Invasive Species 129
Brendon M. H. Larson

A Case Study in Concept Determination: Ecological Diversity 147
James Justus

The Biodiversity-Ecosystem Function Debate in Ecology 169
Kevin deLaplante and Valentin Picasso

A Dynamical Approach to Ecosystem Identity 201
John Collier and Graeme S. Cumming

Symbiosis in Ecology and Evolution 219
Kent A. Peacock

Ecology as Historical Science 251
Bryson Brown

Part 2. Philosophical Issues in Applied Ecology and Conservation Science

Environmental Ethics and Decision Theory: Fellow Travellers or Bitter Enemies? 285
Mark Colyvan and Katie Steele

Postmodern Ecological Restoration: Choosing Appropriate Temporal and Spatial Scales 301
J. Baird Callicott

Habitat Reconstruction: Moving Beyond Historical Fidelity 327
Sahotra Sarkar

Modeling Sustainability in Economics and Ecology 363
Bryan G. Norton

Diversity and the Good 399
Gregory M. Mikkelson

Index 417

Introduction

PHILOSOPHY OF ECOLOGY TODAY

Bryson Brown, Kevin deLaplante and Kent A. Peacock

INTRODUCTION

Ecology is a young science, having emerged as a discipline during the latter half of the nineteenth century. It is also contested ground, both because of the richness and complexity of its subject matter, and because of its close ties to important political and economic issues. This volume gathers reflections on the science, its history and its applications to policy-making and ethical choices. We have divided the papers into two groups, the first group focusing on philosophical questions about ecology and its history as a science while the second focuses on applications of ecology to environmental issues.

One theme that makes an appearance in many of the essays, and lies close below the surface for many others, is a sense of deep worry about the state of our world. Aside from the familiar and already troubling damage that we humans continue to wreak on our environment, from deforestation, soil-depletion, desertification to the rapid decline of fisheries due to devastating over-exploitation, it has become increasingly clear over the last decade that we are now conducting one of the most dangerous uncontrolled experiments in history: the increasingly rapid increase of atmospheric levels of greenhouse gases. The implications of this experiment for climate, ocean levels and ocean pH are truly frightening; still more frightening is the possibility that positive feedbacks may become too strong for us to stop these changes from reaching catastrophic levels. Nearly every ecological system in the world (and just about every system that affects our own well-being) is threatened by this possibility. We hope that this experiment can be shut down before disaster ensues, and that the deep interest in scientific, ethical and public policy issues in ecology expressed by all our contributors may inspire in some of our readers the political will to change course.

PART 1: PHILOSOPHICAL ISSUES IN THE HISTORY AND SCIENCE OF ECOLOGY

The first paper here is "Origins and Development of Ecology" by **Arnold van der Valk.** In it, van der Valk explores the origins of ecology and asks, following C. S. Peirce, what new abductions (hypotheses) were at the root of ecology's emergence as a science, and to what extent ecologists have managed to converge on some central hypotheses.

Handbook of the Philosophy of Science. Volume 11: Philosophy of Ecology.
Volume editors: Kevin deLaplante, Bryson Brown and Kent A. Peacock. General editors: Dov M. Gabbay, Paul Thagard and John Woods.

When new abductions that take us outside of the range of hypotheses considered in established sciences arise, scientists can ignore them, expand existing science(s) to include them or begin a new branch of science based on them. In order for the last of these possibilities to occur, both novelty and success or productivity are required. For example, in ecology, we need the new abductions to be useful in a wide range of geographical situations or species. Van der Valk aims to identify the novel abductions, their sources, their influence on ecology's development, and how much convergence on a "consistent and widely accepted" set of hypotheses has occurred subsequently.

Van der Valk describes the origins of ecology as polyphyletic. In its early stages, the field was dominated by scientists trained as botanists and zoologists. Some of these figures focused on terrestrial systems (many among this group were located in the U.S. midwest) while others concentrated on the oceans (many of these were, naturally enough, located in coastal regions). Very different field techniques were involved, and the work of both groups proceeded quite independently. Drawing on early texts, van der Valk identifies the interests and insights of these 'pioneer ecologists'. Among the early ecological topics van der Valk identifies the following: factors limiting growth, the distribution of organisms, communities of organisms, their organization, food chains, and succession.

Van der Valk's investigation reveals three important "initial defining hypotheses" that were seminal for the development of ecology: (i) that adaptations to varying environmental conditions are responsible for the distribution of organisms; (ii) that ecological communities tend toward equilibrium; and (iii) that communities are a type of organism that develop along predictable lines (as in Clementsian succession). All three defining hypotheses resulted in the development of major ecological research agendas in the late 19th and early 20th centuries. Van der Valk notes with some irony that these three hypotheses may in fact be inconsistent, as the first provides the foundation for the reductionistic, evolutionary, population-oriented approaches to ecology that developed later, while the second and third were the foundation for the more holistic approaches in community and ecosystem ecology that emphasize the analogy between community and ecosystem development and the ontogeny of individual organisms.

The essay closes with some worries about convergence. Van der Valk is concerned that the diversity of ecologists has allowed dubious or even refuted ideas to continue in use, thus blocking the development of "a unified ecology with consistent hypotheses". Here van der Valk is less generous than Christopher Eliot in his contribution regarding the possibility of reconciliation between mechanistic, individualistic views and holistic views of ecological phenomena. For van der Valk, the popular analogy between ontogeny of organisms and succession is simply false. More generously, we would recognize that while the analogy can be, and often has been, taken too far (especially in its rhetorical employment) it has also been a useful guide to inference, and served to inspire much further inquiry. It was not a fruitless notion, despite the obvious fact—acknowledged by Clements, as Eliot notes—that organisms have far more systematic and tightly unified responses to disturbance or perturbation, and their organ systems show far less independence than the various components of a climax community do.

Van der Valk's concern for theoretical or conceptual unity in science is still an important one, but it is more crucial to sciences centered around systematic theories than it is to historical sciences focused on understanding a wide range of actual phenomena that don't (yet, at least) have a systematic theoretical treatment. What emerges from the fascinating history that van der Valk reviews here are insights into how our understanding of various kinds of ecological phenomena, their importance, and some of the processes that take place in ecological systems (from populations on up to ecosystems) developed. This kind of work can leave an impression of inconsistency (as inferences are made in one area that would not be reliable in another, and metaphors that are useful guides to inference in one area fail in others). But it can be a fruitful inconsistency.

Our next paper is "The Legend of Order and Chaos: Communities and Early Community Ecology", by **Christopher Eliot**. The paper opens with two striking epigraphs; one assumes concrete, discrete entities with clear extents and boundaries, and strong unifying features, while the other rejects the notion of such discrete ecological units. The paper develops these two conflicting themes in an account of the debate between Clements and Gleason on ecological communities and "plant succession".

Eliot's aim is to link contemporary discussion of the metaphysics of ecological communities by figures including Kristin S. Schrader-Frechette, Jay Odenbaugh, Earl D. McCoy and Kim Sterelny with answers and views from the early 20th century debate that "retain currency" today.

The division here is between two poles of attraction for ecological thinkers. One views the development and final stage or climax of organic communities as parallel to the development and maturity of individual organisms. The other views the different organisms making up a community as independent and "randomly" associated. Parallels with social and political debates (contrasting communism and socialism with individualistic capitalism) played an important rhetorical role here, adding to the perceived significance of the dispute. But for Eliot, this stark, polar opposition account of the debate is not true to Clements' and Gleason's actual views, and the persisting debate in contemporary work has been (mis)shaped by this false dichotomy.

Drawing on work by Moss, Eliot takes a causal approach to sorting things out. For example, Eliot interprets Clements' (1905) notion of 'habitat' as a causal concept. The idea is that we are not just to look at the phenomena of plants making up a community, but also to consider the climate—and soils—that underlie them. An entity's boundaries, on this reading, are fixed by the extent of the causes or causal conditions that it depends on. From this point of view, ecologists must re-arrange mere associations of organisms into formations based on real affinities and underlying causes.

But this optimism for Clements' view as an advance towards more scientific ecology was supplanted by a contradictory narrative deriding it as altogether unscientific. Gleason's rhetoric contrastingly emphasized a Heraclitean flux at all spatial scales, though this assertion needs qualification because it clearly depends on invoking long enough temporal scales. Chance events, including "accidents of dispersal" play a central role for Gleason, who speaks of the "mathematical laws of probability and chance". Yet on this very point, the opposition between Gleason and Clements is far from clear.

Eliot's discussion appeals to the predictive successes and "usefulness" of models and descriptions to ground the view that there really is something valuable in the ecological scale and perspective, a theme echoed in some of the other papers of this volume. We believe this point merits further exploration and development: some inferences involving ecological concepts are indeed reliable (given basic constraints on external goings-on, such as a reasonably constant solar energy flux). Appeals to natural selection have an interestingly similar status: the explanatory and inferential power of natural selection depends on a kind of stability of circumstance and advantage that causally underwrites "patterns of success and failure". Without a surrounding environment that sustains those patterns, this cannot make sense, but as Eliot explains, the invocation of such an environment constitutes a kind of ecological unit. Of course, it remains open that those patterns are really dependent on circumstances all the way up and all the way down, and hence that a kind of holism, in which we must take account of conditions and factors at every scale, is the only sort of full understanding we can hope for.

Eliot's discussion here refines Clements' view of the organism-analogy; plant interactions are clearly seen as indirect, operating causally via various resources—such as the impact on water, shade, etc. of various kinds of plants. The upshot is that strong readings of the analogy overplay Clements' language at the expense of his science: The climax of a "sere" is seen as the "final" cause only in the sense that, once established, it is self-perpetuating (in the absence of disturbance), *i.e.*, further invasion and replacement (by the local candidates for such roles) is blocked by the conditions that characterize the climax vegetation state.

Such a system is "functional" in the sense that it is self-sustaining, and each part explains its presence by what it contributes to and how it persists as part of the stasis of the climax sere. But the issue of boundaries remains. Eliot considers treating this by examining "many, nested scales". Causal connectedness also comes in as a possible marker, with "zonation" used to describe transitions from one to another. But the structure of such transitions is tied to specific causes (pH, salinity, etc.) which give less sharp or uniform boundaries than the literal 'organism' reading would require.

Gleason's disordered or random vision also emerges as more ordered and structured than many imagine from the central metaphor of chance. Early on, Gleason's work was very 'Clementsian', and Eliot urges that his break with Clements, roughly from 1916 on, should not be exaggerated: Gleason does not abandon all his previous views. Gleason still recognizes the 'normal structure of prairie vegetation' and the 'tension zone' between prairie and forest in Illinois. In 1917, Gleason recognizes 'definite units of vegetation' with self-maintaining structure.

But he advocates a non-Clementsian way of interpreting this. Eliot identifies the key differences thus: Gleason rejects similarity of climax systems to organisms, and does not like the developmental picture with its climax sere and the stages that regularly lead up to it. Instead of a temporal sequence, Gleason focused on varying associations of plants in a territory.

Gleason expresses worries about Clements' vocabulary and methodology, saying that Clements defines away exceptions. Still, even Gleason's move to probabilities requires an account of normal conditions producing such probabilities. The key difference is Gleason's emphasis on the lack of direct constraints and interactions between plants and species of plant, but this is already acknowledged by Clements and indirect forms of mutual constraint remain. Gleason sees such interactions as central to 'maintaining the uniformity and the equilibrium...of the association'. Thus Gleason does recognize plant communities as real things, along with the causal interactions (involving water, shade, nutrients, etc.) that connect their members.

Eliot considers several prior treatments of the conflict between Clements and Gleason in the light of this partial reconciliation. Finding them wanting, he proposes an 'error theory' of previous accounts of the conflict. By failing to trace strong parallels in the causal stories related by Clements and Gleason, we have mistaken their differences in rhetoric and emphasis for a far more fundamental sort of incompatibility.

Part of the story here has to invoke interaction—causal interaction is a sine qua non for what Hume called *inference concerning matters of fact*, but it is far from sufficient. But Eliot emphasizes that interaction alone is not enough to ground realism about communities, or the related concerns of environmental ethics and conservation. Instead, Eliot emphasizes dependence, an interaction that supports very substantive inferences. However, Eliot notes that for Clements, dependencies are 'neither necessary nor sufficient'. For example, exclusive competition relations and significant competitive interactions also count. So the criteria come to be relative: we don't in general know what communities are until we know what purposes we want to identify them for. Illustrations and applications of this provocative proposal will have to wait for another occasion, but we anticipate that interesting cases will be found.

Eliot's discussion ends with a critique of historiography grounded on central similes, metaphors and imagery. While these literary devices are important to the rhetoric of science, they can distract from the causal, explanatory and investigative approaches that unify figures like Clements and Gleason.

Jay Odenbaugh gives an historical overview of MacArthur's work with an eye toward this question in "Philosophical Themes in the Work of Robert MacArthur". MacArthur was one of the most influential theoretical ecologists of the post-WWII era. His work ranges widely, but his most important contributions involve the development of mathematical models that bring evolutionary, genetic and biogeographic factors to bear on the explanation of population and community-level patterns of distribution and abundance. Consequently, MacArthur's work has been described as contributing to the "unification" of these various branches of population biology under a common theoretical framework. But is unification the right word to describe MacArthur's achievements?

Odenbaugh's discussion of island biogeography begins with the 'species-area' effect and two factors that drive diversity on islands: the area of the island and its distance from the mainland (diversity on the mainland also comes in here). The most successful model of this effect is due to MacArthur and E. O. Wilson: an equilibrium model balancing immigration with extinctions, according to which the distance and species-richness of the mainland 'source' determine immigration rates, while the area of the island determines extinction rates. Subsequently, Odenbaugh turns to MacArthur's work on 'limiting similarities' and competitive exclusion as a constraint on ecological overlap in resource use between species in a community. The discussion develops the origin and history of that debate as well.

MacArthur's approach to ecological models illustrates a robust understanding of various causal factors (including evolution, competition, immigration and extinction) that must inform ecological thinking, along with clever formal ways of trying to get some useful inferential mileage out of these factors, despite the richness and complexity of the underlying phenomena.

Odenbaugh's conclusion emphasizes the importance of integration rather than unification in MacArthur's work; integration is a broader term (what's unified is integrated, but not necessarily vice versa). We might say that the difference lies in the fact that integration allows a more *piecemeal* approach to arriving at models and inferences that combine elements from different sciences, while unification demands a more theoretically structured, general program for producing models.

The generality or scope of such an integrated perspective is limited, because the wide variety of cases that actually arise generally includes circumstances where the models fail. The integrative approach is open to the existence of circumstances where a model's regularities and explanatory usefulness break down: after all, the claim is not to have given a general recipe for unification, but only local and limited ones whose applicability depends on conditions, though these conditions can often be inferred from the structure of the models themselves.

Paper number four is "Embodied Realism and Invasive Species", by **Brendon Larson.** Larson's paper develops some ideas about the concepts we employ in thinking about invasive species, drawing on work by Lakoff and Johnson on "constitutive metaphors". For Lakoff and Johnson, broadly phenomenalistic sources of 'meaning' are taken to be primary, as illustrated in Larson's discussion of basics like *container*, *path* and *force*. Larson accepts this approach and explores its implications for the notion of an invasive species. Here it's worth remarking that metaphors like that of 'pressure' or 'container' can also be rooted in 'external phenomena', *i.e.*, goings-on in the external world. For example, the parallels linking inferences about pressure (often applying sufficient pressure on one object with another leads to penetration, for example) to observable phenomena (movement of species, their presence in new regions) that occur in the context of invasive species are pretty clear. These inferential parallels reflect real parallels in the phenomena, from familiar physical events (pressure produced by freezing water bursting a pipe in your home) to more metaphorical applications (a newly introduced species spreading wildly and displacing native species). Nevertheless, Larson's treatment of the kinds of connections developed by Lakoff and Johnson illuminates important aspects of our attitudes towards invasive species.

Closely related to the phenomenal content of such notions, the invocation of the 'invasion' metaphor is presented as casting a negative light on invasive species, by implicitly invoking a negative norm about invasion. But this negative aspect of 'invasion' is culturally relative rather than universal (perhaps regrettably so). Consider early and enthusiastic efforts to introduce European species to Australia, generally coupled with a simplistic view of marsupials as 'primitive' and inferior (an attitude reminiscent of similar attitudes towards native Australians and the absurd legal doctrine of *Terra nullius*). One might try to separate the descriptive from the normative aspects of this language, and to separate the scientific content of the metaphor (and its limitations in application to real biological phenomena) from its roots in either phenomenal experience or everyday illustrations of 'real' pressure—here it might be helpful to consider influential work by Mary Hesse and by Wilfrid Sellars on the use of metaphor in science. But Lakoff and Johnson have argued that this kind of separation is not really possible.

Larson's notion of embodied realism is aimed at achieving a balance between the recognition that, on one hand, there really are invasive species (*i.e.*, they are real things in a real, public, culturally-independent physical world), but on the other hand the conceptual scheme we apply to them is human-generated and involves tendencies (and even outright commitments) that may not be borne out in reality and that are, in general, culturally-dependent.

The notion of 'invasive species' is illuminated by this kind of examination. This paper is a revealing examination of some central associations that shape our responses and attitudes to invasive species. The cultural tropes of primitiveness and inferiority were applied during colonial times to other cultures, and also to the flora and fauna of some regions. It is, as Larson's account suggests, not accidental that the end of colonialism (and the increasingly bad reputation of imperialism) has also been accompanied by increasingly negative views on invasive species, though this may also be due in part to the fact that some important examples of invasive species are now invading what we think of as our turf.

In "A Case Study in Concept Determination: Ecological Diversity", **James Justus** considers the long controversy over the best measures and ecological significance of diversity, including measures due to Shannon, Simpson, and others. Species richness and evenness figure prominently in all these measures, but convincingly combining them to produce a satisfying measure of diversity is still challenging. Justus concludes that the widely used measure due to Shannon is not as good as Simpson's, and ends the paper with a discussion of the role of 'diversity' in ecology.

The first part of the paper develops some adequacy criteria for measures of diversity. A simple and highly abstract sort of measure can begin with the proportional species abundance vector, V_p, of a community:

$$V_p = \langle p_1, \ldots, p_i, \ldots, p_n \rangle$$

where n is the number of species and p_i is the proportional abundance of the ith species, as determined either by the number of individuals or its proportional biomass.

A few constraints arise: p_i can be 0 for some values of i, but in such cases the ith species just isn't part of the community, and we will ignore them here (though allowing such vectors can still be useful in cases where we want to represent certain types of change in the community, such as local extinction or complete out-migration). This vector includes the two key elements in diversity noted above: n measures species richness (so long as we don't count species with no members in the community), while the p_i allow us to capture the evenness of the various species' representation in the community. An increase in either of these components (holding the other fixed) intuitively increases a community's diversity. But so far this does not tell us how to measure evenness, so what we mean by 'holding evenness fixed' is still up in the air. One way to proceed is to restrict the principle here very tightly: Evenness must be maximized (for a given n) when there is an equal distribution of numbers of individuals across species.

Some widely discussed (and widely accepted) adequacy conditions emerge from these premises. To extend the constraints beyond these so as to select between measures that survive these conditions, distance metrics are considered as a further source of formal constraints. With a perfectly even distribution defined as one in which $p_i = 1/n$ for all i, a 'Euclidean' metric of distance from $V_p^{\max}$ can be defined. Adding this distance measure to the toolkit, Justus turns to consider Simpson and Shannon's measures of diversity. Simpson's is based on the simple formula,

$$\sum_{i=0}^{n} p_i^2.$$

This represents the sum of the probabilities, for each i, that two randomly selected organisms will both belong to species i (assuming that each individual in the community is equally likely to be selected). This measure is maximized when (intuitively) diversity is minimized by the community having a single species that includes all but $n-1$ individuals. Subtracting this measure from 1 gives us the probability that two randomly selected individuals will belong to *different species*, an intuitively appealing measure of diversity.

Shannon's measure emerges from his work on information theory, in the form:

$$\sum_{i=0}^{n} p_i \ln p_i$$

For Shannon this is a measure of the information contained in a message made up of n characters, each of which has probability of appearing, in any position in the message, of p_i. However, Justus shows that this measure violates his condition A6, which requires that an equal increase in the abundance of a common species and decrease in the abundance of a rare species have the same impact on diversity: Shannon's measure is more sensitive to decreases in rare species than to increases in common ones.

Such abstract measures of diversity have encountered resistance from ecologists, concerned that they simply ignore the relations between different species, along with any differences in their causal roles in the community; taxonomic and functional features are entirely ignored, leaving substantial ecological questions about the significance of diversity underdetermined. Thus the relations between the formal diversity concepts explored here and the causal concepts of, for example, E. P. Odum and MacArthur, which are discussed in the next paper, need examination. Richer measures, or links between these abstract measures and richer, more causal features of ecosystems, may be needed to underwrite the ecological significance of formal notions of diversity. Still, there are very few measures satisfying the abstract criteria set out by Justus, which weakens the case of those who suggest the diversity of formal measures implies there can be no well-determined concept of diversity.

The first paper by one of the editorial team is "The Biodiversity-Ecosystem function Debate in Ecology", co-authored by **Kevin deLaplante** and **Valentin Picasso**. DeLaplante and Picasso begin with a historical review of shifts in the debate over diversity and stability, emphasizing how shifts in the definition of stability affected the course of the debate, and how we've wound up with the present emphasis on measures of ecosystem function and groups of organisms occupying certain roles in the ecosystem rather than populations of species and their variation.

Another theme here is the tension between ecological policy making and the pure science of ecology. This tension has led some ecologists to claim that others (and even an association newsletter) are spinning the evidence for diversity-stability links to support certain political policies on the environment. But the debate seems to turn more on different definitions of stability—from (loosely) bio-functional measures to strictly statistical measures (population size equilibrium or extinction resistance) and then back to (richer or more formal) bio-functional measures. This shift in definitions takes us from early arguments due to Odum and MacArthur, supporting a positive link between diversity and stability, to statistical arguments from May and Pimm supporting a negative link, and back again to a subtler view that recognizes a positive link, with qualifications. The recent work is underwritten by an experimental approach exemplified in Tilman's work, in which diversity leads to greater instability for individual species' population sizes, but also to greater stability in community and ecosystem properties.

Holistic views and equilibrium or balance perspectives on ecology are often accompanied by acceptance of the diversity-stability hypothesis. The hypothesis is also linked with ecological activism, which makes strong rhetorical use of the ideas of balance and the importance of preserving biodiversity. On the other hand, non-equilibrium views are connected to a reductionist perspective on communities and associated with neo-Gleasonian views, currently quite influential in plant ecology. These connections have added fuel to the fire in disputes over diversity and stability, as deLaplante and Picasso note.

The sheer complexity of ecosystems and their components creates serious challenges for attempts to explore connections between diversity and stability. Experimental work depends on empirically applicable measures of both, limiting the range of diversity and stability concepts that can be tested. On the other hand, modelling requires a high level of abstraction, which inevitably raises questions about the significance of model results for actual ecological systems. Finally, the needs of policy makers put pressure on scientists to reach conclusions providing clear and fairly simple grounds for action.

DeLaplante and Picasso explore the connection between work on ecosystem functions and teleological ideas generally associated with holistic, Clementsian views of plant ecology. They present a brief review of ideas about functions in biology, identifying the tendency of ecological holists to appeal to functions in a broader way than reductionists. Work on links between biodiversity and measures of ecosystem functions needs to be examined in the light of these ideas. Evolutionary ecologists (focused on gene- and individual-level selection) tend to be skeptical of such talk, restricting functions to individual-environment interaction and genes' contribution to individual success in a population. But more bare-bones views of functions are generally compatible with the empirical measures employed in various studies, providing some encouragement for regarding the results of empirical work as largely independent of the debate over teleology.

This is a good thing; how to reconcile the avowedly non-teleological world-view of biology today with the widespread appeal to functions and purposes in everyday biological discourse remains one of the deeper puzzles in philosophy of biology. L. Wright's influential account of functions echoes Kant, identifying a function of some feature as something that feature does (or causally contributes to) which explains, in turn, the presence of that feature: here, history grounds attributions of teleology and evaluations of 'malfunction'. But R. Cummins' account, also influential, focuses instead on the (typical) causal role of some feature in a (kind of) system and removes the dependence on history.

DeLaplante and Picasso remark that among ecologists, willingness to attribute functions to features of ecological systems correlates with holistic views of those systems; this makes perfect sense, since such views typically describe these systems in terms that parallel descriptions of individual organisms and their parts, for which teleological description is nigh-irresistible, and underwritten by evolutionary descriptions of features and how they arise under natural selection.

Jax's four part division of the uses of 'function' in the ecosystem literature is also discussed. One important use of function-talk in ecology relies implicitly on 'types' or 'typicality' of the processes occurring within an ecosystem, invoking a form of normativity via an appeal to a standard or 'reference' state of such systems. This raises a methodological concern over whether further norms and values are slipping into our choice of a typical 'functioning' ecosystem, due to a failure to clearly define what the 'types' here are.

Finally, the detailed history presented in section 4 of the paper emphasizes the intimate links between ecology and the challenges of policy making. These links lead to a strong emphasis on a human-centered, utilitarian notion of 'ecosystem function', exemplified by work on *ecosystem services* and their value to human beings (and economies). Such analyses connect ecological studies directly to concerns already understood and valued by policy makers. Funding for work in the area grew dramatically, but vigorous debate over the science undermined its application: scientific debate (and even the illusion of scientific debate, as recent political responses to global warming show) often creates sufficient uncertainty to undermine any policy response beyond support for 'further research'.

Encouragingly, though, deLaplante and Picasso's historical study shows that scientists successfully reconciled many of their differences and also managed to identify avenues of investigation that promise to settle many of the questions that remained open, spawning a second generation of biodiversity studies.

In "A Dynamical Approach to Ecosystem Identity", **John Collier** and **Graeme Cumming** focus on the need for identity criteria for ecosystems. For Collier and Cumming, such a criterion must be *dynamic*, because measurements along with other interactions and interventions in ecosystems are themselves dynamic, and because important accounts of ecosystems focus on processes rather than mere descriptions of their structures or states at various times. An emphasis on the importance of processes is a recurring theme in this volume, reinforced by concerns ranging from the strictly empirical or methodological to the metaphysical concerns of Collier and Cumming. Collier and Cumming back up the importance of a focus on dynamic concepts in our view of ecosystems with remarks on Amazonian deforestation and debates over the wider vs. narrower applicability of models and interventions to protect the forests in different areas; these questions turn on dynamic features of these forests, and thus fit well with a dynamic notion of ecosystem identity.

Collier and Cumming also link their view of ecosystem identity to Michael Ghiselin's and David Hull's view of species as individuals, not natural kinds: species are likewise unified by the dynamic relations, including common descent and a shared gene pool or (in cases of isolation) potential for shared gene pool. Like species, ecosystems are capable of persistence, division and merger, and links and conditions on change over time are crucial here.

The logic of basic metaphysical notions is entirely conventional here: ideas about parthood and unity are drawn from Perry, and identity is used in its strict logical sense, as a strongest equivalence relation. The multi-level nature of ecosystem ontology, in which smaller systems can be parts of larger ones, is allowed for by appeal to the different unifying relations that pick out the parts of systems at different levels. Later discussion also addresses concerns about intermediate cases of ecosystems, paralleling their treatment with the treatment of the same problem for species. One important consequence here for the metaphysics of ecosystems is that the 'same ecosystem' relation is not transitive, just as the 'same species' relation fails to be transitive, both over time and also, sometimes, across geographic ranges, as in ring species.

In connection with the role of geographic boundaries for ecosystems, one might suggest that Collier and Cumming's dismissal of *displacement* of ecosystems is a bit quick. As a matter of conceptual possibility, interactions between species and with non-living features of the environment might actually allow for migration or even a kind of colonization of territory by another ecosystem over time: suppose some invasive species in a particular case tended to alter features of the invaded environment so as to enable companion species from the original ecosystem of the invasive species to in-migrate more easily, gradually establishing a dynamic system very similar to the initial source of the invasion. Migrations of humans along with their domesticates (both crops and animals) and pests (rats, parasites) can be thought of as a deliberate case of displacement. However, the view of ecosystems as historical individuals that Collier and Cumming defend here suggests that such cases would be better interpreted as producing *new* ecosystems of the same (or similar) kind as the original.

Collier and Cumming close their paper with an exploration of what they call *meta-models*, "more general models that incorporate and summarize the findings of many specific models." We think of these as, at least in part, tools or general components for building more specific models of ecosystems and their properties. Collier and Cumming note a wide range of dynamic metamodels for complex adaptive systems, grounded in thermodynamics, agency and the process of adaptation. All of these, they suggest, are sources of insight into system individuation, and illustrate how multiple metamodels can help illuminate ecosystem processes. In particular, we believe that evidence of the *robustness* of system-identity under different metamodels could help to underwrite the idea that there are indeed objective boundaries for ecosystems.

Kent A. Peacock's study of symbiosis, titled "Symbiosis in Ecology and Evolution", begins with a historical discussion of a surprisingly wide range of thinking about symbiosis; Peacock's aim in this introduction is to argue that symbiosis is more important than has yet been appreciated. But the challenge (as it has always been) is to explain how selection for symbiosis arises and is sustained without degenerating into competitive, conflicting or aggressive interactions. Even with a long list of examples like lichen, resistance to the general importance of symbiosis persisted. The success of Margulis' ideas in the 1960s and later transformed the field, finally vindicating her 'serial endosymbiosis' theory of eukaryotic evolution, confirming the insights of predecessors including Watase, Poirier, Wallin, *et al.*

Still, Peacock points out, Margulis' actual definition of symbiosis leaves quite a lot to be desired. The "[l]iving together in physical contact of different species..." is neither precise nor general enough. We need something that allows for more indirect forms of interaction—a *causal* reading, rather than mere contact.

This emphasis on causal interaction allows us at least to consider predator-prey relations as potentially symbiotic (not to mention pollinator to pollen-producer relations): mutual inclusion in life-cycles is the theme here. There is no presumption of a symmetrically beneficial relation. Symbiosis, on this account, can run all the way from straight-out parasitism to obligate mutualism or even fusion into a new single species. Transitions between conditions favouring (selecting for) mutualism and favouring parasitism or competition remain difficult to predict or even to frame in general theoretical terms. However, when mutualism dominates, we can see a shift towards the formation of entirely new units of selection.

Peacock proposes that the difference between mutual and parasitic relations can be expressed thermodynamically, in terms of partners adding to each others' free energy. It may be hard to get into the mutualistic regime, but once there, selection for cooperative interactions can be very strong. Thermodynamically, stable (rich) 'dissipative structures' can be favoured strongly when there is a generous flow of available free energy and physical conditions are 'benign'.

On Peacock's account, there is plenty of room for a real 'genome' in symbiotic systems (even for Gaia) so long as it's recognized to be a distributed genome: the interactions of symbiotic, mutualistic organisms can support the reproduction of each separately, and thus the reproduction of the entire system. This is clearly recognized in the case of some endosymbionts; Peacock also considers the evolution of organ systems in metazoans here. He goes on to suggest that the long-term persistence of life on earth can only be understood as supported by a 'rough-and-ready mutualism', since mere commensalism or parasitism points towards reduced availability of free energy and, ultimately, general extinction. There is lots of room, as Peacock points out, for feedback relations that could provide selective pressure for cooperative behaviour even at one or two removes, although it would be interesting to explore mathematical models that might help us evaluate the likelihood of such indirectly mutualistic associations arising.

Thermodynamically, the ability to capture and store a surplus of free energy is essential, and allows for symbiotic contributions to other organisms that enhance stability and opportunities for the surplus-generator. Even heterotrophs can contribute (indirectly) in this way.

Peacock briefly considers cancer as a vivid example of the breakdown of mutualism: obviously, rogue cells can be selected for, in the local environment that their success eventually destroys. All the necessary capacities for them to exploit and finally destroy the metazoan body that constitutes their ecosystem (reproduction, mobility, ability to recruit blood vessels and other essential services) are built into the repertoire of normal cells already. Here we see a nice example of the balance of selection between mutualism and parasitism.

Ecologically, Peacock turns to a view of humans as parasites on the earth's ecology that is all too convincing. Here, he claims, we need cultural evolution to reshape our behaviour into a form that will allow a sustainable interaction between ourselves and the living environment we depend on. 'Ethics' [Leopold, p. 26] limits self-serving action (both individually and, at the ecological level, for species). Eugene Odum and Grant Whatmough make closely related points about the need for a more balanced interaction between humans and our environment, pointing to

Malthusian and worse-than Malthusian limits on parasitic species expansion and even their persistence, and contrasting those negative results with the artifactual ecologies of Japan and England, which were richer than the wild ecology that predated them. Peacock's essay closes with an emergency room metaphor: at this point things are looking bad, and we have to intervene to support and salvage what we can lest things get much, much worse.

In "Ecology as Historical Science" **Bryson Brown** explores the place of ecology among the sciences, emphasizing methodological and epistemic features that group ecology amongst the historical sciences. Although most ecological work does not include a deep-time perspective like that of geology and paleontology, the kinds of inferences that are made and the epistemic status of the resulting models show many similarities. Among these the role of a long and growing list of basic processes (understood, in general, not in terms of a fundamental theory, but instead in terms of patterns of conditions in which they occur and traces they leave) is central. For example, erosion, transportation, sedimentation and cementation together form a cycle that has been a central explanatory trope in geology from very early days, despite fundamental changes in our understanding of the mechanics and chemistry underlying these processes. The inferences that first shaped our understanding of phenomena like beds of sedimentary rock or eroded flows of lava remain reliable, despite the vast increase and refinement of our current understanding of the processes involved, as well as the much wider range of processes whose workings and effects we can now invoke in studying them.

The resulting explanatory narratives are anchored in mutually confirming patterns of traces that fit the predicted results of the various processes, their order and interaction. Our confidence that processes fitting the descriptions invoked in these narratives have occurred is strongly grounded in straightforward observation of processes at work today and the traces they leave behind. Similarly, many basic ecological processes and interactions, from reproduction and growth to parasitism, predation, the food chain and the contribution of photosynthetic plants to the Earths' atmosphere, are firmly established on the basis of common sense observations, as well as results from chemistry and other sciences. The epistemic status of ecological models is not, in general, simply grounded in predictive success; many of their features are instead a matter of common sense reflection on the processes involved in, for example, populating a newly formed island; the difficulty, of course, lies in combining them into a model that usefully captures some features of what is, in every particular case, a very complex story.

Reduction relations and emergence are also examined, with an eye to relations between the sciences. While metaphysical forms of reductionism, involving ontological mappings from the entities of a reduced theory to collections of entities in a reducing theory, are relatively simple, more substantive reductions, which would require also capturing the observational and inferential uses of the reduced theory within the reducing theory, are clearly beyond us to carry out. In this practical sense, different sciences may be integrated by inferential links (as Odenbaugh's paper also suggests), but they cannot be converted into a single science with a general and unified theoretical structure.

PART 2: PHILOSOPHICAL ISSUES IN APPLIED ECOLOGY AND CONSERVATION SCIENCE

As we've already noted, especially in regard to deLaplante and Picasso's paper, the development of ecology has been greatly influenced by the need for information that could guide rational and ethical action in general, and public policy in particular. In the second part of the book we have gathered papers that address some of the philosophical questions that arise from this close connection between the science of ecology and the various ecological values that inform, or should inform, policy making and personal ethics.

Often decision theory and ethics are depicted as in conflict: they seem to give different advice about certain kinds of compromises. But in their paper, "Environmental Ethics and Decision Theory: Fellow Travelers or Bitter Enemies?" **Mark Colyvan** and **Katie Steele** argue that this is more a matter of the idealizations, budget constraints and time-scales attended to in decision theory-based approaches. There is, they claim, no in-principle conflict, although our practice would need to change to reflect time-scale and dynamics explicitly in order for the two to be reconciled. Specific cases discussed are environmental triage and carbon trading/offsets.

Although an appeal to strong incommensurability or infinite values can add a kind of deontic absolutism to decision theoretic apparatus, the trade-offs under discussion in these cases are clearly not well expressed by these approaches—both undermine any decision's justification, if we continue to believe that our evaluation of the decision's outcome actually matters.

However, there are also problems of relying on the market or on axiological considerations: lack of information undermines both these processes for decision making. There are also reasons to be concerned about the political side of the issue—in general, decision theoretic approaches need probabilities and values as inputs. The more uncertainty there is (or can be politically generated) about these, the easier it is to paralyze the political process. Together with the challenge of future values (economics tends to assume substitutions are always available, so it sets the value of a future forest—say, 40 years hence—so low that we can't economically justify the cost of replanting today), these difficulties make public policy decisions look very difficult indeed.

Despite these practical challenges, which any account of policy making must face, Colyvan and Steele's argument is helpful: the traditional account of how deontological criteria for choice-making conflict with axiological criteria is far too simplistic. As Colyvan and Steele urge, following a deontological rule cannot be *motivated* in any particular case unless following it will, or at least will be likely to, contribute best to bringing about the ends we value. An important corollary to this is the recognition that accepting deontic constraints on our choices makes good sense from the point of view of a *modest* view of our ability to anticipate the effects of our choices, reflecting the value of prudence and the limits of our predictive powers.

In "Postmodern Ecological Restoration: Choosing Appropriate Temporal and Spatial Scales", **J. Baird Callicott** explores a puzzle about ecological restoration. Clements' notion of the climax ecosystem as a 'super-organism' which represents the natural system for that region provides a simple basis for deciding what ecological restoration requires: when human action derails the process of climax-system development following disturbance, that's unnatural, and restoration of the climax ecosystem is restoration of the natural, stable situation for the region.

However, Gleason proposed an individualistic view, in which communities are merely accidental and probabilistic 'assemblages' of organisms adapted to similar conditions; in the last 25 years of the 20th century this individualistic and reductionistic outlook had become standard amongst plant biologists. Having given up the notion of an objective climax ecosystem (for example, conditions prior to European settlement in North America), the question of what ecosystem restoration would require is hard to answer. Restoration to what state? Given that conditions prior to the arrival of Europeans had already been greatly altered by Native Americans, it has been suggested that the post-glaciation/pre-human (Pleistocene-Holocene transition, ca. 30Kya) state, including elephants, cheetahs and lions on the plains, would be appropriate.

Callicott argues instead for a return to the pre-European settlement standard, based on a choice of temporal and spatial scale that draws them from ecosystem scales. On Callicott's view, where a disturbance reaches beyond such scales (as in the effects of European settlement and introduced species in North America) restoration to the status quo ante has a privileged status.

Many major disturbances are 'abnormal' only relative to time and scale as well: on larger scales, forest fire is 'incorporated' as part of the 'system'; on smaller scales, it looks more like an external event. With disturbances incorporated on the right scales, even human disturbances can become 'part of the system'. Callicott's analysis shows how disagreements about what state ecological restoration should aim at are tied to issues of scale, especially temporal scale: a long, macroevolutionary (rather than ecological or historical) time scale favours more radical sorts of restoration.

Callicott's argument is premised on the observation that time scales go with processes—macro-evolution with the arising and extinction of species (millions of years), history with migrations, the rise and fall of countries, governments and civilizations (tens to hundreds or perhaps thousands of years), and ecology with succession, disturbance and recovery processes (tens to hundreds of years). What's important to Callicott here is the fit between ecological process time scales and historical time scales. From a purely biological point of view, the historical time scale is arbitrary—but it matches, broadly, the time scale of ecological processes.

Callicott concludes in favour of restorations that include the effects of typical behaviours of indigenous humans, who have been part of the ecosystems around the world for a long time (save in Antarctica), that exclude organisms that have not been part of the ecosystem in a region for some centuries even if their extinction was originally anthropogenic, and that are *adaptive*, responsive to experience and engaging humans in interaction with the ecology rather than treating it as isolated: there should be something in it for us, as well as for the ecology.

Sahotra Sarkar's "Habitat Reconstruction: Moving Beyond Historical Fidelity" contributes to the same discussion, proposing reconstruction rather than restoration as a goal for environmental policy. Human impact on life and its environmental conditions continues around the world. Anthropogenic changes have had huge impacts, including negative impacts on human living conditions and vulnerability to various natural changes. 'Reservation' approaches to preserving at least some areas in a natural state have been partly successful but insufficient, leading us to consider efforts at *restoration.*

But Sarkar argues that standard accounts of restoration are too restrictive; we need a notion with broader scope. Historical fidelity as the goal of restoration is the main problem (although Sarkar also argues that more than just integrity is needed as a dynamic criterion). He proposes *reconstruction* as a better term and process for our efforts to improve ecological conditions in various areas; reconstruction pursues a different set of natural values, values that are more suitable both to our means and our well-considered ends. What Sarkar aims to do here is to change ideas about theory, foundations and normative reconsideration, not practice; in fact, Sarkar holds, much actual practice already fits the reconstruction view that he espouses here.

Sarkar uses Eric S. Higgs as his primary source for the concern with historicity in 'restoration'. Sarkar's critique of Higgs argues that historicity is more a means than the end in view; that sometimes historicity is at odds with better ends and that the values behind such judgements about ends, not the past itself, must be our main concern. For example, while getting simpler or closer to nature may be goals, historicity needn't be the best way of getting there. Similar considerations apply to the links Higgs asserts between historicity and providing people with narrative links and continuity with their environment. Finally, with respect to the value of rarity, time depth may be statistically related to rarity, but rarity can be valued for itself without needing a conceptual link to historical fidelity.

Another concern of Higgs' is that, without a standard for reconstruction rooted in history, the 'caprices of the present' can distort our ecological goals. But Sarkar argues that there is little room left for caprice after we've considered sustainability, rarity and other natural values. Moreover, Sarkar points out, there's room for caprice in history too, raising a question we've already encountered here: what historical state should be our target? The rewilding proposal for North America, based on lions, cheetahs, elephants, etc., is a telling example.

The upshot, as Sarkar sees it, is that a focus on historical fidelity is expensive, often beyond our ability to measure confidently (records just don't exist), and likely to harm valuable species now present. Finally, despite our best efforts, it will be prevented or rendered unsustainable by coming climate change.

Bryan Norton opens his "Modeling Sustainability in Economics and Ecology" with a discussion of tensions between economics and ecology over what he calls 'the accounting problem' and the 'substitution' problem. The first problem concerns just what values actually get to be considered in decision making; in their efforts to be scientific and thus value-neutral, ecologists have often ceded the debate and accepted too-narrow economic views of what values should guide ecological decision making. The second is based in the common economic assumption that substitutions can always be made to replace resources consumed. More generally, to the economist, ecosystems merely produce some contribution(s) to human welfare, contributions that may vary but are not subject to sudden and massive disruptions, while ecologists see them as capable of varying immensely in what they produce and how.

When it comes to the accounting issue and how impacts on the environment are viewed as economic costs comparable to the depreciation of capital, Norton recognizes that we can quibble about the details of how this is done and whether all environmental services are included properly, but prefers to set these questions aside. Instead, Norton aims to arrive at a larger view of theoretical differences about how to evaluate environmental change.

Issues of scale are also important to Norton, who is worried about the apparent arbitrariness of scale choices, an issue also addressed in Callicott's paper. Questions about the appropriate scales for ecological intervention and concerns about the reversibility of our impacts on the environment are considered, together with their implications for appropriate decision-making rules. However, Norton insists that we still need to decide what spatial and temporal scales we should focus on when we plan to reverse past damage to the environment—this decision must be prior to more concrete policy choices, and sometimes it will be a difficult one.

Norton endorses Callicott's appeal to 'ecological scales' to reject 'rewilding' projects—but he questions if this constraint can do the finer work of helping choose policies that are really on the table, and worries about the 'hyper-realism' of Callicott's assumption that we can really pick out a sufficiently objective ecological scale on which to work. Here, however, the dynamic criteria explored by Collier and Cummings may be promising: if there really is a fair bit of unity in ecological systems, their scales may really help set the scale of our restoration efforts on objective grounds. Here, the pragmatists' concerns may converge with those of a (modest) scientific realist.

Norton continues with a critique of economics and its use of 'preferences' as a starting point for analysis. Norton points out that preferences don't seem to be settled prior to choices, they can be altered, manipulated and even generated by context, as a wide range of studies in cognitive science and psychology have shown. Norton concludes that algorithmic, utility-maximizing approaches to policy evaluation need to be replaced with approaches that acknowledge that the difficult pre-conditions for such evaluation cannot be mutually agreed upon. Instead, we need to focus on 'fair and intelligent processes seeking cooperative solutions'. In connection with this challenge to decision theoretic approaches to environmental problems, Norton appeals to the striking notion of 'wicked' problems, "problems that have no single, uncontested formulation because different individuals and different groups come to the situation with differing values and perspectives."

Norton concludes by proposing *reflexive ecology*: this approach to ecology allows for multiple, partial models, where choice of models is not determined by 'simple observation, on pure ecological theory, or strong forms of realism', involving instead a continual feedback between modeling choices, efforts shaped by them, results and reconsiderations of the model and problem, to guide the process.

There is room for continuing discussion here. In particular, an exchange between Norton, Callicott and Collier and Cumming would be very interesting: the descriptive conditions Collier and Cumming propose for setting the scale and boundaries of 'ecosystems' in terms of the dynamic character of the system and the balance of 'centripetal' and 'centrifugal' influences, and Callicott's appeal to the temporal scale of ecological processes sound like *descriptive* reasons for adopting a scale in ecological thinking and practice. Their practical importance, if these proposals are successful, would presumably lie in the *reliability* of the inferences provided by models that draw the line at these 'joints', and the central role that reliable inferences play in guiding action in the world.

The last paper in the collection is **Gregory Mikkelson's** "Diversity and the Good." This paper also addresses the line between the descriptive and the normative. Mikkelson's concerns here begin with the disproportionate and damaging demands human beings are now placing on the ecological systems we depend upon. In response to this difficult challenge, Mikkelson proposes that we need "an unprecedented integration of science with ethics". Recognizing the role of diminishing returns and higher-level interactions (derived in part from health-care economics) suggests a way of understanding the role of diversity in ecosystem productivity, and a link from the diversity-ecosystem productivity connection to a more general point about value.

Mikkelson's ecological discussion focuses on studies of the productivity of plots populated with varying numbers and proportions of grassland plant species. While assortment according to 'best conditions' for each species within the plot can explain some of the productivity advantage of more diverse plots (these are called *compositional effects*, mediated separately by the conditions best for each species), the productivity advantage is too large to be explained without including positive effects of interactions between species or the positive effects of some species on the growing conditions of others (these are called *contextual effects*, involving same-level interactions between species, and higher-level interactions between species and the ecosystem they belong to).

Mikkelson goes on to link the productivity-diversity relation in ecology to the economic value of income equality. Mikkelson sees equality as related to diversity via an analogy between the number of 'points of view' with the means to develop and express themselves, and the number of species with numbers sufficient to contribute substantially to ecological processes. On the economic side, the idea that economic equality is valuable in general follows from the diminishing utility of money. (Mikkelson acknowledges that the 'diversity' of a group might alternatively be seen as increased by a wider gap between rich and poor, but sets this unattractive reading of diversity aside). Other positive relations between income equality and various social values, from health to trust and the functioning of social institutions are noted, and interpreted as parallel to the contextual benefits of diversity for ecological productivity noted at the outset. Measures of the value of equality in economics and of diversity in ecology provide a basis for measuring the value of these wholes (socio-economic and ecological) in a way that makes it more than the sum of the values of its parts, because part of the total value emerges from positive contextual interactions.

Interestingly, the values of economic equality and ecological diversity are not merely related by this parallel—they are also related by a causal connection: societies that have lower levels of economic equality also perform less well when it comes to protecting ecological diversity. Rich people in general tend to spend more on the preservation of ecological diversity, but they spend proportionately less than poorer people. So greater income equality contributes to higher levels of expenditure on preserving ecological diversity.

Mikkelson's closing discussion draws on a proposal by C. Kelly, advancing this pattern as a general account of the good as *richness* (read as unified variety). The idea seems to be that compositional effects can be captured by variety alone, but the contextual effects of diversity require 'unification', that is, interaction, interdependence and especially mutualism. Such a view of value conflicts directly with attempts to express the value of ecological preservation or restoration in terms of willingness-to-pay or the economic value of 'ecological services', and Mikkelson urges ecologists and others to speak directly in terms of the intrinsic value of rich ecological systems rather than frame their defense of ecological diversity in the terms favoured by our present modes of political discourse.

This is an interesting, ecologically-inspired attempt to link the descriptive and normative in a substantial way. Like any such effort, it invites criticisms and attempts to produce counter-examples. But we hope any such efforts will be more substantial than G. E. Moore's rather facile open-question argument.

Part 1

Philosophical Issues in the History and Science of Ecology

ORIGINS AND DEVELOPMENT OF ECOLOGY

Arnold G. van der Valk

INTRODUCTION

How did ecology develop as a distinct science? What distinguished it from already existing sciences? Why did it develop when it did? Although ecologists seem largely unaware of his work [Loehle, 1987; Krebs, 2006], I will use two concepts developed by Charles S. Peirce (1839–1914) to examine the origins and development of ecology: (1) his concept of abduction, *i.e.*, hypothesis generation; and (2) his concept of convergence. For Peirce, it is the collective judgment of a scientific community that will eventually determine which hypotheses have been sufficiently confirmed by observation and/or experiments to be accepted as beliefs. This is what he meant by convergence. An outline of Peirce's philosophy of science can be found in his two seminal essays, "The fixation of belief" and "How to make our ideas clear," which were published in 1877 and 1878, respectively, in the *Popular Scientific Monthly* [Hartshorne and Weiss, 1931–1935]. For an alternative take on the development of ecology, see Graham and Dayton [2002] who examined the evolution of ecological ideas using Kuhn's concept of paradigm shifts.

Abduction is basically guessing or conjecturing what is responsible for (*i.e.*, explains) an observed pattern. For Peirce, abduction is the only mechanism that produces new knowledge or insight, and he proposes a number of characteristics that make hypotheses plausible, including consistency with already confirmed hypotheses, simplicity, and generality. What were the abductions that resulted in the development of ecology? What was so novel about them that they could not be accommodated by existing sciences?

For Peirce, the scientific community's evaluation of the correspondence between the observations predicted by an abduction (hypothesis) and the actual field or experimental observations resulting from studies designed to test the hypothesis determines if a hypothesis has been confirmed or not. Thus critical scrutiny by the scientific community results in an increasing or decreasing probability that a hypothesis has been confirmed. Hypotheses that have been repeatedly confirmed eventually become beliefs. Ecology, like any other science, is ultimately a set of beliefs. In effect, convergence toward a set of confirmed hypotheses and the elimination of unconfirmed hypotheses are the marks of a mature science. Convergence

Handbook of the Philosophy of Science. Volume 11: Philosophy of Ecology.
Volume editors: Kevin deLaplante, Bryson Brown and Kent A. Peacock. General editors: Dov M. Gabbay, Paul Thagard and John Woods.

is scientific progress for Peirce. Has there been a convergence in ecology? Does it have a well confirmed and universally accepted set of hypotheses?

It is my claim that a new abduction or novel hypothesis that falls outside the scope or form of those in existing sciences can trigger the development of a new branch of science. When novel hypotheses are proposed that do not conform to the kinds of hypotheses that scientists perceive to be relevant to their discipline, they can respond in three basic ways: (1) they ignore such hypotheses, at least in the short term; (2) they expand the boundaries of an existing science in order to accommodate them; or (3) they begin to develop a new branch of science based on them. For a new branch of science to develop, a new abduction must be not only novel but scientifically productive. In the case of ecology, it must be applicable to a wide variety of geographic situations and/or to many kinds of organisms and thus be potentially relevant to many biologists and other scientists. In this chapter, I am specifically concerned with identifying and characterizing those abductions that triggered the development of ecology as a distinct discipline and that established its initial research agenda. I will also briefly examine the role of convergence in ecology.

It is not my claim that *only* novel abductions will lead to the development of a new science. For example, many scientific disciplines that overlap with ecology (forestry, fisheries biology) or subdisciplines of ecology developed because of the common interest of a group of scientists in some organism, *e.g.*, plants (plant ecology) or insects (insect ecology), or some natural system, *e.g.*, lakes (limnology), grasslands (rangeland ecology) or wetlands (wetland ecology). In some cases, these overlapping disciplines, *e.g.*, forestry, became organized prior to ecology.

There were also many hypotheses that were already well established prior to the development of ecology that were simply assimilated by early ecologists because they were relevant to their interests [Park, 1946]. A good example of such a hypothesis is that population sizes of organisms are always limited by predation, disease, starvation, etc. As Charles Darwin (1809–1882) emphasized in *The Origin of Species* [1859], the various factors that control population sizes are responsible for natural selection [Stauffer, 1960]. By the first-half of the nineteenth century, the first mathematical models of human population growth had already been developed such as the geometric growth model of Thomas Robert Malthus (1766–1834) and the logistic growth curve of Pierre-Francois Verhulst (1804–1849); the latter emphasized that there was an upper limit to the size of human populations. The Darwinian hypothesis of natural selection and hypotheses about population regulation were both important hypotheses that were assimilated into ecology, but they were not the novel hypotheses that triggered the development of ecology. In short, early ecologists continued to accept and utilize hypotheses that they had acquired as part of their training in botany or zoology.

Ecology is known to be polyphyletic [McIntosh, 1985]. Most early ecologists were trained primarily as botanists and zoologists. Thus plant- and animal-oriented ecologists were usually hired and housed in different university departments or research institutes [McIntosh, 1985; Kingsland, 2005]. Some pioneer ecologists were primarily interested in terrestrial systems (forests, grasslands) while others were interested in aquatic systems (initially primarily oceans and lakes). The later division is largely due to the different techniques used to sample the dissimilar organisms that dominate terrestrial (vascular plants, mammals, birds, insects, etc.) and aquatic (algae, aquatic invertebrates, fish, etc.) systems. Some of the earliest scientists who are now recognized as proto-ecologists include oceanographers like Edward Forbes (1815–1854) because of his studies on marine benthic invertebrates and limnologists like Francois-Alphonse Forel (1841–1912) because of his studies on Lake Léman (=Lake Geneva) [Acot, 1998a; Elster, 1974; McIntosh, 1985]. Geographic influences also had some bearing on the polyphyletic nature of ecology. Oceanographers were, not surprisingly, found at institutions on or close to a coast while early terrestrial ecologists, at least in the United States, were at institutions in the Midwest (especially in Nebraska and Illinois) about as far from any ocean as it is possible to get in North America. Consequently, the novel ideas that triggered the development of ecology would arise more or less simultaneously in a number of different disciplines and locations.

In this chapter, I will specifically address five questions concerning the origins and development of ecology: (1) What were the novel abductions or hypotheses that set ecology apart from existing sciences? (2) What was the origin or inspiration of these hypotheses? (3) How much have these initial hypotheses affected the subsequent development of ecology? (4) Who exactly constituted the community of pioneer ecologists? (5) How much convergence towards a consistent and widely accepted set of hypotheses has occurred? To keep the task manageable, I will restrict myself to the very earliest stages of the development of ecology in the nineteenth and early twentieth centuries in Europe and North America and to an examination of a limited number of hypotheses concerning topics identified by pioneering ecologists as central to ecology. For more detailed historical accounts of the origins of ecology, see Worster [1977], McIntosh [1985], Acot [1988, 1998a], Cittadino [1990], Golley [1993], Hagen [1992], Kingsland [1985; 2005], and Egerton [2008]. The original works of early ecologists can be easily accessed through compilations such as Kormondy [1965], Egerton [1977], and Acot [1998a].

1 WHAT WERE THE NOVEL ABDUCTIONS OR HYPOTHESES THAT SET ECOLOGY APART?

It is possible to identify the core interests of pioneer ecologists by examining the contents of early ecology texts. Because I am focusing on the origins of American and British ecology, Charles Elton's [1927] *Animal Ecology*, one of the first animal ecology textbooks, and Frederic E. Clements [1907], *Plant Physiology and Ecology*, an early plant ecology text, were selected. Elton [1927] covers topics such as environmental factors limiting growth, distribution of organisms, community organization, food chains, and succession. Clements [1907] covers much the same ground, but he emphasizes the importance of plant adaptations to various environmental conditions for understanding their distribution. This reflects his reliance on late nineteenth-century German physiology and ecology texts for much of the material in his book. Why and how did pioneer ecologists develop an interest in these topics (limiting factors, adaptations and distribution, community organization, food chains, and succession)? I will examine in some detail each topic to determine if an interest in any of them developed because of a novel hypothesis. From here on, any novel hypothesis that triggered the development of ecology will be designated a "defining" hypothesis to distinguish it from other novel and established hypotheses that ecology assimilated more or less unmodified from other scientific disciplines.

1.1 Factors Limiting Growth

Justus von Liebig (1803–1873), whose studies of plant nutrition were made possible by early nineteenth-century advances in chemistry in Germany, proposed what has become known as the law of the minimum in 1840: plant growth is limited by that required nutrient whose supply in the soil is least adequate [Blondel-Mégrelis, 1998, pp. 311–313]. Liebig stressed that the soil has only a limited supply of available nutrients and that nutrients removed from the soil due to the harvesting of plants by man or by domestic animals like cows and sheep can eventual limit the growth of these plants unless these nutrients are replaced. The basic concept of a soil nutrient budget and nutrient cycles were established in the 19th century by Liebig and other European scientists who recognized that the store of soil nutrients is limited, that plants are nutrient pumps that reduce the stocks of nutrients in the soil, that animals obtain their nutrients from plants that they eat, and that the return of nutrients to the soils is through litter decomposition and that this could take a long time. Liebig and other soil scientists of his era were largely concerned with maintaining soil fertility at levels needed for the growth of crops by farmers.

Liebig's novel hypothesis was that plant growth is a function of the amount of the most limited nutrient present in the soil, regardless of what it is. In other words, plant growth can be stunted by an inadequate level of just one nutrient even if all the other nutrients and requirements for growth like water, light and air and soil temperature are at optimal levels. This and his other discoveries put agriculture on a more scientific basis and stimulated the use of fertilizers to restore the fertility of crop fields. Liebig's hypothesis was not so radical, however, that it resulted in the development of ecology. What is striking about Liebig's work and those of his fellow nineteenth-century soil and crop scientists is that early ecologists would ignore their major findings and insights about limits to primary production, nutrient budgets, litter decomposition, and nutrient cycling for nearly a century. Liebig was, in effect, a proto-ecosystem ecologist. There are some echoes of Liebig's ideas about the importance of nutrient limitations and cycles in the later part of the nineteenth and early twentieth century among scientists studying freshwater and marine systems, who in hindsight are considered to be proto-ecologists. For example, Francois-Alphonse Forel describes in some detail the carbon cycle in Lake Léman [Acot, 1998b, pp. 163–164].

The idea that the growth of organisms can be limited by the absence of only one necessary factor was eventually taken up and expanded by animal ecologists to provide a framework for explaining animal distributions. In a 1911 paper and in his 1913 book *Animal Communities in Temperate America*, Victor Shelford proposed a generalization of Liebig's hypothesis in the form of his law of tolerance: an organism can only persist or remain in a given environment, which is characterized by a complex set of physical and chemical factors, if all these factors are within the tolerance range of that organism and, if any one factor exceeds its minimum or maximum tolerance, it will fail in that environment. Shelford's law of tolerance had a profound impact on the development of animal physiological ecology [Feder and Block, 1991]. It is this reformulation of Liebig's hypothesis to explain plant and animal distribution that first found its way into ecology.

Although the law of the minimum was a novel hypothesis when proposed by Liebig and it was eventually taken up by ecologists, it was not a defining hypothesis of ecology. If it had been a defining hypothesis, ecology would have begun to develop many years earlier than it did and it would have been much more focused in the nineteenth century on primary and secondary production and nutrient cycling than it was.

1.2 Adaptations and Distributions

During the 19th century, Germany was the major center for advances in all aspects of botany, not just crop production and nutrition. Major advances were also made in plant anatomy/morphology and many areas of plant physiology such as water relations [Cittadino, 1990]. During this period, German botanists had access to the best and latest technology, especially high resolution microscopes, and German physiologists benefited from rapid advances in analytical chemistry. In reaction to earlier, more speculative botanical theorizing that was largely based on vitalism and idealism, German botanists were the first to begin to apply more rigorous "scientific" or mechanistic approaches to the study of plants. Prior to 1850, botany in Germany, as elsewhere in Europe, had been largely descriptive studies of plant tissues and cells, classificatory studies of plant species, and descriptive studies of plant distribution. Among the last, one of the most influential was Alexander von Humboldt's (1769–1859) *Essai sur la géographie des plantes* [1805–1807] in which he emphasized that botanists should study the physical factors that control plant distribution. This largely descriptive and correlative approach to plant geography reached its ultimate form in August Grisebach's (1814–1879) *Die Vegetation der Erde nach Ihrer Klimatischen Anordnung*, a two volume set published in 1872. (Nicolson [1996] provides a excellent overview of European plant geography during the nineteenth century and the development of several different European schools of vegetation studies.)

Establishing mechanistic relationships between plant adaptations and plant distributions began in the mid-1870s with the anatomist/morphologist Simon Schwendener (1829–1919) and his students as well as other German botanists [Cittadino 1990]. Schwendener's work established that morphological and anatomical adaptations had physiological consequences for plants. Gottlieb Haberlandt (1854–1945), one of Schwendener's students, published his *Physiologische Pflazenanatomie* in 1884. In it, he stressed that to understand plants you had to study the functions of their tissues and organs. Plants have to exploit and cope with their environment. What adaptations do they have to do this? Haberlandt introduced Darwinian thinking about natural selection into botany. In his book and other writings, Haberlandt provided mechanistic explanations, putatively the result of natural selection, for functional adaptations of plants. His emphasis on natural selection and how it produced plant adaptations to environmental conditions stimulated an interest in the study of plants and their environments where they naturally grew. Haberlandt, like many German botanists in the late 19th century, began to travel outside of Germany, and in 1891–1892, he traveled to the Indo-Malaysian tropics to study the adaptations of leaves of tropical plants.

Georg Volkens (1855–1917), who like Haberlandt was a student of Schwendener and who was also interested in the ecological significance of plant adaptations, conducted a study of the anatomical-physiological adaptations of desert plants in Egypt. This study was undertaken because Schwendener had proposed that plant adaptations to environmental conditions could best be studied under extreme climatic conditions. Volkens' book *Flora de ägyptisch-arabischen Wüste* [1887], although more focused on taxonomy than ecology, is among the first scientific works that could be described as ecological. In retrospect, he himself viewed it that way "...[my book] helped found and develop a special discipline of botany, the ecology of plants." (quoted in Citadino [1990, p. 66]. Volkens' book, however, was of minor significance compared to those shortly to be published by Andreas Schimper and Eugenius Warming.

Because of his travels in the Caribbean in 1881–1882 to study epiphytes, Andreas F. W. Schimper (1856–1901) began to recognize that factors other than plant adaptations to environmental conditions like light, temperature, and moisture were responsible for the distribution of epiphytes on Caribbean islands, including distance from the continent, ocean and wind currents, and bird migration patterns. On a later trip to Brazil, he studied the interactions of ants and trees. The ants protect the trees from other herbivores and in turn are supplied with food by the trees in the form of special structures at the base of their leaf petioles. In 1898, Schimper published, *Pflazengeographie auf physiologischer Grundlage*. There is a strong natural selection-adaptationist slant to his book. As the title implies, Schimper stressed the importance of plant physiological adaptations for understanding the distribution of plants. There are 170 pages on environmental factors (temperature, soils) and other factors (animals) as well as 600 pages on the relationships between plants and environmental conditions. In effect it is one of the first ecology textbooks. The first so-called ecology text, however, was written by Eugenius Warming (1841–1924) who is often recognized as the founder of plant ecology. Warming was a Dane who was trained in Germany. In 1895, he published a book in Danish, *Plantesamfund*, which was quickly translated into German as

Lehrbuch der okologischen Pflanzengeographie; eine Einführing in die Kenntniss der Pflanzenvereine [Warming, 1896], and eventually into English much modified as the *Oecology of Plants* [Warming, 1909].

The novel hypothesis developed by Schwendener and his students that plant adaptations to environmental conditions can explain plant distributions was a defining hypothesis of ecology. This hypothesis is the central hypothesis of both Schimper's and Warming's books. Botanists who saw the implications of this hypothesis quickly began to do studies of plant distribution from a physiological perspective all over the world. These were the first ecologists. The rapid adoption of this defining hypothesis in the United States is reflected in the establishment in 1903 of the Desert Laboratory in Tucson, Arizona, by the Carnegie Institution of Washington. The primary focus of studies at the new Desert Laboratory was the physiological basis for the distribution of desert plants [Craig, 2005]. The new field, however, was still trying to decide on an appropriate name [McIntosh, 1985]. Most early plant ecologists viewed what they were doing as an extension of plant physiology [Clements, 1905; 1907]. In the 1890s and early twentieth century, however, ecology began increasingly to be viewed and described as a new field distinct from the established fields of plant physiology and plant geography.

1.3 Community Organization

A second novel hypothesis that became a defining hypothesis for ecology was proposed in a paper on ways to improve oyster cultivation by Karl August Möbius (1825–1908). Over-exploitation had led to a decline in oyster and mussel beds off the German coast, and Möbius was charged with studying the feasibility of promoting oyster and mussel farming. His studies resulted in Möbius proposing that an oyster bank is a biocönose (biocenose) or social community by which he meant "a community of living beings where the sum of species and individuals being mutually limited and selected under average external conditions of life, have, by means of transmission, continued in possession of a certain definite territory." (English translation in 1880, p. 723 [Acot 1998a, p. 228] of Möbius [1877, p. 76]). When environmental factors are altered or new species invade, the composition of the biocenose or community changes and a new equilibrium community develops. He also points out that the over-exploitation of a target species can result in its extinction locally from a community and its replacement by other species. "If in a community of living beings the number of individuals of one species is lessened artificially, then the number of mature individuals of other species will increase." [Möbius, 1880, p. 726]. In fact Möbius had introduced the same idea under a different name, "Lebensgemeinschaft" or "life community", in an earlier publication [Acot, 1998b, p. 156].

Möbius' novel hypothesis is not that organisms are found in communities or that there are species interactions within these communities. These were already well established concepts. His novel hypothesis is that, because of their interactions, species in a community are in dynamic equilibrium with each other and thus form a stable community. The community will remain unchanged as long as nothing disturbs this dynamic equilibrium. This hypothesis in less explicit form, the balance of nature, can be found in earlier natural history writings often more as a theological than an ecological concept [Egerton, 1973] but it is also found in Darwin [1859] as resulting from the "struggle for existence" among organisms. Möbius' claim is that this equilibrium occurs even at the scale of a square meter or less. Another formulation of this hypothesis was proposed in 1887 in the writings of another early ecologist, Stephen Forbes (1844–1930). Forbes [1887, pp. 86–87], in his most famous and influential paper, "The lake as a microcosm", views a hypothetical lake as being a microcosm in equilibrium and that this equilibrium is the result of interactions among the organisms, particularly predator-prey interactions. "The interests of both parties [prey and predator] will therefore be best served by an adjustment of their respective rates of multiplication, such that the species devoured shall furnish an excess of numbers to supply the wants of the devourer, and that of the latter shall confine its appropriations to the excess thus furnished." "We see that there is a close *community of interest* between these seemingly deadly foes." Forbes views this community as being a product of natural selection. As with Liebig's studies of nutrients in soils, the studies of aquatic communities by Forbes and Möbius can also be viewed as precursors of ecosystem ecology. However, their preoccupation with explaining distribution patterns of organisms made pioneering ecologists overlook the functional implications of the Möbius-Forbes hypothesis.

The Möbius-Forbes hypothesis that organisms are found in communities that are in, or tend toward, equilibrium is one of the defining hypotheses in ecology. It introduced into ecology a more holistic perspective that has had a profound impact on its development. Nevertheless, like the German physiological ecologists, who wanted to provide mechanistic explanations for plant growth and plant distribution, Möbius and Forbes seemed to envision that it is mechanistic species interactions that result in a community tending toward equilibrium. Disputes about the nature of communities and community equilibrium, however, would dominate much of plant ecology in the first half of the twentieth century, especially when theories about the development of equilibrium communities (succession) began to be proposed [Worster, 1977; McIntosh, 1985]. See section 1.5.

1.4 Food Chains

The concept of food chains was already established prior to the development of ecology [Egerton, 2007]. Darwin's *The Origin of Species* [1859] described a now-famous food chain in rural England: red clover, bumble bees, field mice and cats. More detailed studies of food chains were done later in the nineteenth century by Stephen Forbes who studied the importance of aquatic invertebrates in fish diets and published a monograph on the topic, *The Food of Illinois Fishes* [1878]. His data were derived primarily from studies of fish stomach contents. He similarly studied the diets of birds in order to discover whether their predation of insects was beneficial to farmers or not [Croker, 2001]. From these studies, Forbes began to understand the central importance of food (energy) in ecology because this was one key link between plants and animals and among animals themselves. However, Forbes and his contemporaries never quantified their studies of food (energy) uptake.

Closely related to the Möbius-Forbes hypothesis, especially Forbes' version of it, was a novel hypothesis about energy losses along food chains proposed by the German zoologist, Carl Gottfried Semper (1832–1893), who worked primarily in the Philippines. Semper [1881, p. 52] hypothesized that inefficiencies in the transfer of energy from one feeding or trophic level (plants, herbivores, carnivores) to another limited the number of organisms at each feeding level or, in other words, limitations in energy transfer were a major factor controlling the types of species and their abundances in communities. Semper proposed a hypothetical community in which there were only 1,000 units of plant food and that only 10% of this food could be transferred to herbivores. This means that this community can only sustain 100 units of herbivores. Assuming the same transfer efficiency from herbivores to carnivores (only 10%), this community could only sustain 10 units of carnivores. It was Semper's novel hypothesis that a community like Forbes' microcosm is structured in large part by energy losses from one trophic level to another and that this limits the number of herbivores and carnivores in any given area. In effect, Semper had interjected thermodynamics into ecology. Although Semper's trophic hypothesis in the form of the pyramid of numbers was popularized by Elton [1927] and it strikingly presages Lindeman's [1942] trophic-dynamic hypothesis, it was not one of the defining hypotheses of ecology. Although it is closely related to the Möbius-Forbes community-equilibrium hypothesis, Sempers' hypothesis had little impact on late nineteenth and early twentieth century ecology. The inefficiencies in energy transfer along food chains that Semper highlighted would not become relevant to ecologists until the latter half of the twentieth century when ecosystem energetics became a major research agenda. Like Liebig, Semper was ahead of his time.

1.5 Succession

Initially ecologists were concerned primarily with trying to explain spatial patterns, *i.e.*, plant and animal geographic distributions. Nevertheless, many observers had noted that temporal changes in vegetation or succession often occurred locally [Clements, 1916; Acot, 1998]. Clements [1916] reviewed the early literature on succession and found numerous descriptions of temporal changes going back to the seventeenth century. (See also Egerton [2009] for a recent review of succession studies.) A couple of examples will illustrate the character of these observations. The French writer Dureau de la Malle (1777–1857) in 1825 published a paper primarily on crop rotation that describes the succession of species in forests and meadows. He concludes that changes in plant species "est une loi générale de la nature" [Acot, 1998, p. 130]. Likewise, Henry David Thoreau (1817–1862) in

1860 gave an address on "The succession of forest trees" in which he describes changes in forest vegetation that he had observed in New England [Spurr, 1952]. Thoreau "recognized the effects of wind-throw and fire in the forests found by the original European settlers, and distinguished between successional trends in small clearings, following cutting, following single fires, and as a results of agricultural use" [Spurr 1952, p. 426]. Such observations, however, had little influence on early ecologists. Although Warming [1895, 1896] had previously described the phenomenon of succession and even postulated some rules that govern it, the studies of succession that most influenced the development of ecology were those of Henry Chandler Cowles (1869–1939). Cowles described succession, more correctly a chronosequence, in the sand dunes along the south shore of Lake Michigan in a series of papers published in 1899. (For a detailed account of Cowles life and works, see Cassidy [2007].) Cowles was able to place the vegetation types observed into a crude chronological sequence because the dunes became older as you moved inland from Lake Michigan. He interpreted this chronosequence as a putative successional sequence from pioneering dune to mature forest vegetation. Cowles describes the various kinds of vegetation found in the dunes in considerable detail, but he does not hypothesize much about the patterns observed beyond noting that physiographic (landscape) position and dune age are correlated with vegetation types. In short, Cowles' study is transitional in that it focuses primarily on the distribution of vegetation types and only secondarily on temporal changes. It was the temporal dimensions of his studies, however, that were to have the most lasting influence on the development of ecology in the twentieth century [McIntosh, 1985; Cassidy, 2007]. Early animal ecologists, most notably, Victor Shelford (1877–1968) quickly picked up the concept of succession first from Cowles and later from Frederic E. Clements [Croker 1991].

In 1917, Frederic E. Clements (1874–1945) published a massive monograph on succession in which he proposed another defining hypothesis of ecology: succession is the development of a climax formation [Clements, 1916]. A climax formation (a vegetation type defined by the growth form of its dominant species, *e.g.*, deciduous trees) was in equilibrium with its climate and thus was able to persist until the climate changed. A formation is for Clements an organism that "arises, grows, matures, and dies." In short, a climax formation has both an ontogeny and phylogeny just like an individual plant. Like the ontogeny of a plant, succession is directional and irreversible (progressive in Clements' words). Nevertheless, Clements also recognized that succession was much more "complex and obscure" than the development of an individual plant and his descriptions of specific vegetation changes are often highly mechanistic. In short, Clements' novel hypothesis is that a climax formation is a "super-organism" and that its ontogeny is the result of succession. Clements makes the claim that there is a strong but not perfect analogy between an individual organism and a formation. Nevertheless, he seems to be making a metaphysical claim that there is a level of biological organization, the climax formation, above the species level and that formations have characteristics, *e.g.*, an ontogeny, similar to those of individual organisms.

In Chapters 1 and 2 of *Bio-Ecology* [Clements and Shelford 1939], one of Clements' last major works, Clements and Shelford review various hypotheses about the nature of communities and defend in considerable detail Clements' hypothesis that communities are "complex" organisms (formerly super-organisms). In this work, Clements and Shelford now call the endpoint of succession a "climax community" rather than a climax formation. "One of the first consequences of regarding succession as the key to vegetation was the realization that the community ... is more than the sum of its parts, that it is indeed an organism of a new order" [Clements and Shelford, 1939, p. 21]. They continue "... it is essential to bear in mind the significance of the word "complex" in this connection, since this expressly takes the community out of the category of organisms as represented by individual plants and animals" [p. 21]. They try to clarify their definition of complex organism again by analogy and state that it bears "something" of the same relation to the individual plant or animal that "each of these does to the one-celled protophyte or protozoan". In other words, the formation is a real entity, but one that is not as integrated as a higher plant or higher animal. Not surprisingly, the exact metaphysical status of Clements' complex or super-organism is still being debated [Eliot, 2007].

According to Clements, ecology is fundamentally a holistic science [Clements 1935]. The Möbius-Forbes hypothesis about communities tending toward equilibrium had holistic overtones, but it did not necessarily imply that communities are metaphysically distinct entities. Clements' critics like Henry Gleason [1917], who saw communities as groups of overlapping populations of species, believed that Clements confused change [*e.g.*, in species composition] with development. Nevertheless, Clements' novel succession/super-organism hypothesis was to be one of the most important defining hypotheses in American ecology in the first half of the twentieth century [Worster, 1977; McIntosh, 1985; Kingsland, 2005].

1.6 Defining Hypotheses

Only three of the topics emphasized in early ecological texts seem to have had their origin in defining hypotheses: the adaptation-distribution, community equilibrium, and succession/super-organism hypotheses. The law-of-the-minimum hypothesis of Liebig and the trophic-limitation hypothesis of Semper were not influential enough in the nineteenth century to require a new discipline, although both would eventually play a major role in shaping ecological thinking and research agendas in the second half of the twentieth century. Many other hypotheses were assimilated unchanged by pioneer ecologists from existing disciplines, *e.g.*, various hypotheses about factors controlling populations sizes in animals.

2 WHAT WERE THE ORIGINS OR INSPIRATIONS OF THESE DEFINING HYPOTHESES?

In general early ecologists acknowledged to only a limited extent the sources or inspirations for their hypotheses. In the case of the adaptation-distribution hypothesis, however, a number of its sources are obvious and are acknowledged by its originators. These include the studies of various plant geographers [Coleman, 1986; Nicolson, 1996] and Darwin's *The Origin of Species*, especially through Darwin's influence on German morphological/anatomical studies, which began to focus on the functional significance of plant adaptations [Cittadino, 1990]. As noted, many of these nineteenth century German scientists were also reacting against the vitalistic and idealistic biology that had dominated German biological thought in the early nineteenth century. Consequently, the adaptation-distribution hypothesis which developed when plant physiology and plant geography began to overlap, was formulated as a mechanistic/reductionistic hypothesis.

Some aspects of the origins of Möbius' concept of the biocoenosis have been examined by Nyhart [1998]. Based primarily on an examination of Möbius earlier writings and his professional activities, Nyhart concludes that the concept of an equilibrium community was shaped in large part by his teaching, previous research on marine fauna, civic experiences, and work with culturing marine organisms in aquaria. What Möbius did in his 1877 monograph on oysters was to propose a Greek neologism, biocönose, to make his hypothesis of a living community in equilibrium appear more significant and profound to the scientific community of his day. What is missing from Nyhart's account of influences on Möbius is an assessment of the general intellectual milieu in which he worked.

Hagen [1992, pp. 4–7] has pointed out that Forbes' concept of the microcosm seems to be based on Hebert Spencer's (1820–1903) evolutionary philosophy. Spencer proposed that all things in the universe are a product of evolution. For Spencer evolution always involves the transformation of the homogenous into the heterogeneous and progress toward heterogeneity is inevitable at all levels of organization from the molecular to cosmological (see Freeman [1974] for more detail about Spencer's philosophy of evolution). However, evolutionary progress need not be continuous. Spencer believed in a "moving equilibrium." For Spencer equilibrium is the result of a temporary balance of the forces of evolution and dissolution. In the case of biological systems, external forces, *e.g.*, changes in environmental conditions, can result in dissolution. Changes in environmental conditions, for example, can have an adverse effect on the production of plants. This in turn will adversely affect the herbivorous animals that depend on these plants for food and this will affect the predators and parasites of the herbivores and so on. Internal evolutionary forces, in this case the development or acquisition of morphological structures that enable the plants to cope with the new environmental conditions, will result in the establishment of a new equilibrium. Although Spencer was interested primarily in human societies, his ideas about the nature and development of human societies are not only reflected in Forbes' community equilibrium hypothesis, but even more so in Clements' succession/super-organism hypothesis.

In a paper entitled "The Social Organism" [1860], Herbert Spencer outlined how human societies developed much like organisms. He admitted that this organic analogy was not exact, but that there were many similarities between the development of organisms and societies. Worster [1977, p. 212] and Tobey [1981, pp. 64–85] point out that Frederic E. Clements was familiar with the works of Spencer. He discussed Spencer's ideas with his colleague Roscoe Pound and Spencer's work is cited in *Bio-Ecology* [Clements and Shelford, 1939, p. 24]. Moreover, Tobey [1981] makes the point that both Herbert Spencer and the pioneer American sociologist, Lester Frank Ward (1841–1913), both of whom conceived of human societies as super-organisms, influenced Clements, but that this conception can also be traced back to German idealistic plant geographers like Oscar Drude (1852–1933). More than any other pioneer ecologist, Clements was conscious, if perhaps only in hindsight, of his intellectual influences. Spencer's ideas of inevitable progress and moving equilibrium seem to be the philosophical underpinnings of Clements' concept of succession. In his *Principles of Biology*, Spencer [1898–1899] states that evolution is responsible for the increasing integration of the plants and animals and their increasing mutual dependence on each other. Spencer's increasingly integrated assemblage as Worster points out bears a strong resemblance to Clements' climax formation. Clements *et al.* [1929, p. 314] quote from Spencer's writings: "Spencer has discussed the concept of the social organism with special clarity, and the student of community development can still turn with great profit to his treatments of this theme [1858, 1864]. It is both interesting and suggestive to find that he anticipated certain axioms of plant succession by the statements 'Societies are not made but grow' and 'Man may disturb, he may retard or he may aid the natural process of organization [development], but the general course of this process is beyond his control.'" Thus it seems that Herbert Spencer and his evolutionary philosophy played a major role in the development of both the equilibrium community and the closely related succession hypotheses.

Prior to the nineteenth century speculations about human societies had been part of philosophy [Tucker, 2002]. Spencer's writings as well as those of other early sociologists like Henri Saint-Simon (1760–1825) and Auguste Comte (1798–1857) provided early social scientists and ecologists with concepts and terms for describing social groups and the development of such groups, particularly the organic analogy between the organization of human societies and organisms [Hagen, 1992]. There were many interactions between early ecologists and sociologists. For example, the sociologist E. A. Ross and Clements were at the University of Nebraska at the same time and, according to Ross' biographer Gross [2002], Clements and Ross became friends at Nebraska and they continued to correspond for several decades after both left Nebraska. Clements' contacts with sociologists were strong enough that he had several papers [*e.g.*, Clements 1935, 1943] published in social science treatises. Clements [1905, p. 16] actually has a short section in his textbook, *Research Methods in Plant Ecology*, on sociology in which he notes that plants and humans are subject to the same "laws of association." In turn,

Clements and other early ecologists influenced the development of some schools of sociology, particularly the "human ecology" of R. D. McKenzie. McKenzie's [1943] *Readings in Human Ecology* has selections from the writings of a number of American and British ecologists: plant communities (W. B. McDougall), animal communities (Charles Elton), competition (Clements, Weaver and Hanson), plant dominance (McDougall), and animal dominance (C. C. Adams). By the mid-1920s, the tables had turned and sociologists were now looking to ecology for inspiration; for example, the development of human societies is now being compared to Clements' succession/super-organism hypothesis [McKenzie, 1924].

3 HOW MUCH HAVE THESE INITIAL HYPOTHESES AFFECTED THE SUBSEQUENT DEVELOPMENT OF ECOLOGY?

All three defining hypotheses to this day continue to shape ecological thought and research agendas as is illustrated in many of the other chapters in this book and numerous books on the history of ecology [Worster. 1977; McIntosh, 1985; Acot, 1988; Hagen, 1992; Golley, 1993; and Kingsland, 1985; 2005]. Both plant [Lambers *et al.*, 1998] and animal [Feder and Block, 1991] physiological ecologists have continued to study the physiological significance of adaptations and their utility for understanding plant and animal distributions. Their techniques and tools have become more sophisticated but the core topics that dominate these fields today would be familiar to their nineteenth and early twentieth century predecessors. The long-lasting impact of this approach can be seen in comparing major monographs on aquatic plants: Agnes Arber's *Water Plants* [published in 1920], C. S. Sculthorpe's *The Biology of Aquatic Vascular Plants* [1967], and Julie Cronk and Siobhan Fennessy's *Wetland Plants* [2001]. Although all three books cover many aspects of the morphology, taxonomy and ecology of aquatic plants, anatomical and morphological features (adaptations) that control their distribution within and among wetlands are a central focus of all three. Because organisms have to cope with more than just their physical environments, by end of the nineteenth century, the inadequacies of the adaptation-distribution hypothesis had already been noted by Schimper. Consequently, topics like chemical defense mechanisms against predators or pathogens and adaptations to disturbances became more prevalent in the twentieth century. This defining hypothesis, however, continues to be influential. This general approach to understanding and predicting the distribution of plants and the composition of plant communities gained new life in the works of J. P. Grime [1979]. He emphasized the importance of three kinds of plant adaptations: to environmental conditions (stress), to periodic disturbances, and to competition.

Whether communities actually are in, or tend to, equilibria as proposed by Möbius and Forbes is still being debated by ecologists. Prior to World War II, this was primarily in the form of a debate about Clements' holistic and Gleason's reductionistic ("individualistic") hypotheses about the nature of plant communities (associations) [McIntosh, 1985, pp. 263–267]. One attempt to resolve this debate was made by Arthur Tansley [1935] who proposed the term "ecosystem" as a less extreme holistic formulation of communities (associations) than that proposed by Clements. Tansley's ecosystem has much in common with Möbius' biocenose. During the middle years of the twentieth century, this debate about the nature of communities sparked a series of field studies by plant ecologists, most notably Robert H. Whittaker (1920–1980) [McIntosh, 1985]. These studies collectively resulted in what Simberloff [1980] called the "materialistic and probabilistic revolution" in ecology that overthrew Clements' succession/super-organism hypothesis, at least among plant community ecologists. However, the debate about the nature of communities did not end. In the 1960s and 1970s, the debate over the equilibrium theory of island biogeography of MacArthur and Wilson [1967] was another version of it, primarily among animal ecologists, and this sparked a secondary debate about the role of competition in structuring communities. Its most recent incarnation has been the debate over biodiversity and community or ecosystem stability [Naeem, 2002]. This version of it began in the 1970s and then re-emerged in the 1990s [McCann, 2000].

Although Clements' succession/super-organism hypothesis was quickly challenged by H. A. Gleason [1917; 1926] and others, Clements' holistic claims about the nature of communities and succession were and continue to be immensely influential in ecology as has been well documented in Worster [1977], Simberloff [1980], McIntosh [1985], Hagen [1992], Golley [1993], and Kingsland [2005]. By the middle of the last century, a new succession theory began to develop among plant ecologists based on the individualistic hypothesis of H. A. Gleason. Glenn-Lewin *et al.* [1992] provide a detailed treatment of post-Clementsian succession theory. Nevertheless, Clements' succession/super-organism hypothesis has not been completely abandoned. In E. P. Odum's paper "The strategy of ecosystem development," he reformulated Clements' hypothesis about succession as a hypothesis about ecosystem development [Odum, 1969]. Among ecosystem ecologists it continues to have traction [Golley 1993] and applied ecologists (see section 5).

4 WHO EXACTLY CONSTITUTED THE COMMUNITY OF PIONEER ECOLOGISTS?

Ecology was proposed as the name for a discipline that was needed but did not exist by Ernst Haekel (1834–1919) in 1866. Haeckel was inspired by the chapters in Darwin's *Origin of the Species* on the Struggle for Existence and Natural Selection to propose that a new science was needed to investigate what regulates population sizes of organisms and allows them to co-exist in nature's economy. Haeckel defined the proposed new science of ecology thus (as translated in Stauffer [1957]): "By ecology, we mean the whole science of the relationship of organism to environment including, in the broad sense, all the 'conditions for existence.'" In reality, the term ecology did not begin to be used until nearly 30 years later after its spelling was Anglicized to ecology at the 1893 meeting of the AAAS in Madison, WI. Initially, the term "oecology" was introduced to distinguish field studies in plant physiology from laboratory studies, only the field studies were designated ecological. Rapidly other types of related field studies, such as studies of community composition and succession, became recognized as ecological studies. Thus the most important common denominator among early ecologists was their field orientation. Ecology was to be the study of nature in nature, *i.e.*, in the field, not the laboratory.

Although Darwin, Haeckel, and even Clements viewed humans as part of nature, Clements and Shelford [1939] pointed out that early ecologists were trained almost exclusively in botany and zoology departments and thus they concentrated on studies of plant and animal species. As consequence, ecology was "generally hostile or indifferent" to the study of man. The lack of interest by ecologists in human societies was also due in large part because sociology and economics were already established academic disciplines when ecology began to "crystallize" [McIntosh, 1985] in the late nineteenth and early twentieth centuries. Thus initially the community of ecologists was a small group primarily of botanists and zoologists with an interest in plant and animal distribution, animal population regulation, native plant and animal community composition, and succession. It was not till the first couple of decades of the twentieth century that there were enough people who viewed themselves as ecologists that they could form their own societies and establish their own journals. The first such society was the British Ecological Society (BES) which was established in 1913. Its inaugural meeting was attended by fewer than 50 people and by 1917 its total membership was only around 100 people [Sheail 1987]. The Ecological Society of America (ESA) was founded in 1917 and had about 300 inaugural members. A decade after its establishment, the Ecological Society of America had around 600 members and the British Ecological Society in 1930 had about 450 members [McIntosh, 1985, p. 161]. Early ecologists were finding it increasingly difficult to get their ecological studies published in existing journals. Consequently, one of the major motivations for establishing (and later for joining) these new societies was that they would publish a journal. The first ecological journal, *Journal of Ecology* (BES), began publication in 1913 and the second, *Ecology* (ESA), in 1920.

In spite of their small numbers, from the beginning ecologists were splintered into many subgroups (plant ecologists, animal ecologists, limnologists, marine ecologists, etc.) and the ecological community overlapped with many already existing scientific communities like foresters, fisheries biologists, geographers, soil scientists, etc. This is well illustrated in a survey of the disciplinary interests of the inaugural members of the Ecological Society of America [Burgess, no date]. Although plant and animal ecology were, not surprisingly, the most common disciplinary areas (57% of its members), nearly 40% of the ESA's inaugural members indicated that their primary disciplinary interest was not ecology (Table 1). Forestry and entomology were fairly common disciplinary interests of non-ecologists, and geology, climatology, soil physics and animal parasitology were the major interest of a few inaugural members. Over time, the number of subgroups in ecology has actually increased dramatically with the proliferation of national ecological societies and increasingly more specialized groups of ecologists focusing on some kind of vegetation (*e.g.*, tropical forest ecology), ecosystem (*e.g.*, wetland ecology) or application (*e.g.*, restoration ecology). Different groups of ecologists often emphasize different hypotheses. In the early twentieth century, for example, plant ecologists tended to focus on studies of plant distribution and succession while animal ecologists focused more on population regulation.

Table 1. The primary disciplinary interests of the inaugural members of the Ecological Society of America, adapted from Burgess [no date].

Discipline	Members	Percent
Plant Ecology	88	29%
Animal Ecology	86	28%
Forestry	43	14%
Entomology	39	13%
Marine Ecology	14	5%
Agriculture	12	4%
Plant Physiology	7	2%
Plant Pathology	4	1%
Climatology	4	1%
Geology	4	1%
Animal Parasitology	3	1%
Soil Physics	3	1%
TOTAL	307	100%

5 HOW MUCH CONVERGENCE TOWARDS A CONSISTENT SET OF HYPOTHESES HAS OCCURRED?

Peirce's philosophy of science relies on a community of scientists to judge how compellingly hypotheses have been confirmed. As has been noted by many authors (*e.g.*, [Peters, 1991]), there has not been in ecology the convergence toward a universally accepted and integrated set of hypotheses along the lines suggested by Peirce. Allen and Hoekstra [1992] made an attempt to unify ecology using hierarchy theory, but they did so by expressly not dealing with the reductionistic and holistic dichotomy among ecological hypotheses: "We will not rely on assertions that any ecological entity is real in an ultimate sense" [Allen and Hoekstra, 1992, p. 14]. There are many possible reasons for the lack of unity in contemporary ecology, including the relatively young age of the field, the historical and geographical contingencies of most aspects of ecology, and the multiple levels of organization (organisms, populations, communities, landscapes, etc.) at which ecologists work. It seems to me, however, that two important and overlooked factors are (1) problems with the formulation of some defining hypotheses and (2) the lack of a uniform community of ecologists.

Peirce stressed the importance of avoiding ambiguous terms in hypotheses. Of the three defining hypotheses, the first, the adaptation-distribution hypothesis, was not ambiguously formulated, but its importance was overstated by pioneer ecologists. The Möbius-Forbes hypothesis of community equilibrium is ambiguous because neither Möbius nor Forbes defined precisely what they meant by "equilibrium" and they were vague about the ontological status of their biocenoses and microcosms. They both suggested, however, that underlying mechanistic interactions among species would result in the development of a community in which all the species would be mutually limited in abundance. They also overstated the importance of their hypothesis because they failed to take into account the importance of ubiquitous disturbances, both abiotic and biotic, on community composition and species abundances that were subsequently documented [Botkin, 1990; Glenn-Lewin *et al.*, 1992; Johnson and Miyanishi, 2007]. Even Clements was somewhat ambiguous about the true nature of plant formations. His detailed accounts in his *magnum opus*, *Plant Succession* [Clements, 1916], of various factors controlling the dispersal, establishment, and growth of plant species during succession are very mechanistic [Tobey, 1981; Eliot 2007]. Clements' main hypothesis is the succession/super-organism hypothesis and he defended it repeatedly [Clements, 1916; 1936].

Because Clements' claim that succession represents the ontogeny of climax formation is based on an organic analogy, it is necessary to examine the logic of analogies in order to evaluate it. An analogy is a proposed correspondence between two things in some respect [*e.g.*, structure, function] that are otherwise dissimilar. All analogies are of the general form: **A** is like **B**; **A** has property **P**; Therefore, **B** has property **P**. A hypothesis derived by analogy is only as reliable as the assigned property (P) on which it is based [Juthe, 2005]. The organic analogy, in which a climax formation (target subject) is said to be comparable in some respect (assigned property) to an individual organism of some kind (analogue) is an example of a different-domain-analogy [Juthe 2005]. For such cases, assigned properties can only be validly projected from the analogue to the target subject if each of the elements of the analogue which determine the assigned property corresponds one-to-one with counterpart elements in the target subject. Assigned properties that meet this requirement are called projectible. When analogies are based on non-projectible assigned properties, *i.e.*, the assigned property of the analogue has no exact counterpart in the target subject; the analogy is false. For example, birds and bats are both group of vertebrates that fly; birds lay eggs; therefore, by analogy bats must also lay eggs. This obviously false analogy is based on a non-projectible property, in this case egg laying. Unlike birds, bats do not have the morphological and physiological means to produce eggs. Hence egg laying is not a projectible property.

Organisms are complex entities whose constituent cells, tissues and organs interact to produce identifiable, self-replicating units. Clements believed that climax formations are also highly complex assemblages in which the constituent species interact in a variety of ways to produce identifiable, self-replicating units (climax formations or communities). Organisms (the analogue) undergo development (ontogeny); they develop in a predictable sequence from fertilized eggs to mature individuals. It is development, defined as ontogeny, a known characteristic of organism, which is the assigned property in the analogy. The climax formation, the target subject, by analogy must also undergo ontogenic development, *i.e.*, succession is an ontogenic process with a defined endpoint. The features of the analogue, however, that are responsible for its development, *i.e.*, primarily its genes, have no exact counterparts in a climax formation. According to Clements the overarching controls of succession are macro-climatic conditions. In other words, the organic analogy is false because development (ontogeny) is not a trait that is projectible from organisms to plant formations. Plant formations can change significantly over time for a variety of internal and external reasons, including disturbances and inter-annual fluctuations in environmental conditions [van der Valk, 1985; 1992], but the responses to these changes are not controlled or limited by some internal feature of the climax formation analogous to genes. Their lack of truly ontogenic development suggests that Clements made a category mistake when he hypothesized that formations are some kind of organism as was first suggested by Gleason [Gleason, 1917].

Meaningless, ambiguous and false hypotheses, however, are expected to occur in any science and according to Peirce will eventually be rejected or modified by the scientific community when these hypotheses fail to be confirmed by observations. This assumes that there is a unified community of ecologists who will determine whether a hypothesis has been confirmed or not. As noted, hypotheses such as Clements' succession/super-organism hypotheses have been investigated and rejected by twentieth-century plant community ecologists because their field observations did not confirm its predictions [Whittaker, 1975; Simberloff, 1980; McIntosh, 1985; Botkin, 1990]. The existence of many subgroups in ecology, however, allows hypotheses like Clements' succession/super-organism hypothesis to persist in other subgroups. For example, Davis and Slobotkin [2004] criticized the Society for Ecological Restoration for their outmoded (read Clementsian) ecological concepts about communities and ecosystem development (succession) in the Society's "Primer for Ecological Restoration". These "outmoded" concepts, however, were immediately defended as valid by leading members of the Society [Winterhalder *et al.*, 2004]. Quine and Ullian [1970, p. 6] point out that "Evidence for belief must be distinguished from causes of belief. Often we gather evidence to defend a belief that we already hold, while the cause of our already having held the belief is forgotten or undiscovered." This unfortunately seems to reflect to some extent the state of ecology today. Ecologists often are more concerned with collecting data to support their hypotheses (beliefs) than with critically evaluating and reconciling their hypotheses with their observations. Consequently, in the short term, the development of a unified ecology with consistent hypotheses has not yet happened as Peirce believed that it would.

6 SUMMARY AND CONCLUSIONS

In the nineteenth and early twentieth century, abduction as defined by C. S. Peirce, produced a number of novel hypotheses that did not fall within the perceived boundaries of existing biological sciences. Some of these novel hypotheses resulted in the development of the new science of ecology. These are called its initial defining hypotheses: (1) adaptations to various environmental conditions are responsible for the distribution of organisms; (2) communities tend toward equilibrium; and (3) communities are a type of organism that develops along predictable lines (succession). Investigating the implications of these hypotheses initiated lines of field research that were different from those in established sciences like botany and zoology. Two other novel hypotheses (Liebig's law of the minimum, Semper's hypothesis about inefficiencies of energy transfer along food chains) that were proposed in the nineteenth century failed to have much impact on ecology until the mid-twentieth century. Many other hypotheses were also assimilated more or less unchanged into ecology from other disciplines.

The stimulus for the development of these defining hypotheses varied. The adaptation-distribution hypothesis developed from the realization by nineteenth-century German botanists that the physiological implications of anatomical and morphological features of plants could be used to explain the distribution of these plants that had been previously documented by plant geographers. Both the community equilibrium and succession hypotheses seem to have been inspired, at least in part, by the evolutionary theories of the influential, nineteenth-century British philosopher, Herbert Spencer.

All three defining hypotheses, but especially the adaptation-distribution and succession hypotheses, resulted in the development of major ecological research agendas in the late nineteenth and early twentieth centuries. These three defining hypotheses, however, are not logically consistent with each other. The adaptation-distribution hypothesis provides a mechanistic/reductionist explanation of the distribution of species and hence implicitly a mechanistic/reductionist explanation of the current composition and future composition of communities. The community equilibrium hypothesis and succession/super-organism hypotheses are more holistic formulations of community composition and community change. The succession/super-organism hypothesis seems to be based on a false analogy that equates the ontogeny of organisms with succession.

Peirce believed that false, ambiguous and inconsistent hypotheses would eventually be eliminated or reformulated when the scientific community compared observations made to test a hypothesis to the predictions made by it. In ecology, hypotheses have been tested and in some instances rejected by some ecological subgroups. Ecology, however, has always been composed of many subgroups and the research programs of these subgroups are focused on different hypotheses and hence have different beliefs. Because subgroups are often intellectually isolated from each other, hypotheses rejected by one subgroup can continue to be held by other subgroups. Peirce predicted that a unified community of scientists would eventually eliminate inconsistent hypotheses by comparing predictions to observations. This elimination or reformulation has not occurred as predicted in ecology because, in large part, ecologists are still not a unified community.

BIBLIOGRAPHY

[Acot, 1988] P. Acot. *Histoire de l'écologie.* Paris, France: Presses Universitaires de France, 1988.

[Acot, 1998a] P. Acot (ed.), *The European Origins of Scientific Ecology.* Amsterdam: Overseas Publishers Association, 1998.

[Acot, 1998b] P. Acot. The structuring of communities. In Pascal Acot (ed.), *The European Origins of Scientific Ecology,* pp. 151–164. Amsterdam: Overseas Publishers Association, 1998.

[Allen and Hoekstra, 1992] T. F. H Allen and T. W. Hoekstra. *Toward a Unified Ecology.* New York, NY: Columbia University Press, 1992.

[Arber, 1920] Agnes Arber. *Water Plants: A Study of Aquatic Angiosperms.* Cambridge, UK: Cambridge University Press, 1920.

[Blondel-Mégrelis, 1998] M. Blondel-Mégrelis. Agrochemistry and bacterial autotrophy. In Pascal Acot (ed.), *The European Origins of Scientific Ecology,* pp. 309–321. Amsterdam: Overseas Publishers Association, 1998.

[Botkin, 1990] D. B. Botkin. *Discordant Harmonies: A New Ecology for the Twenty-first Century.* New York, NY: Oxford University Press, 1990.

[Burgess, no date] R. L. Burgess. *The Ecological Society of America: Historical Data and Some Preliminary Analyses.* Oak Ridge National Laboratory, Oak Ridge, TN, USA. No date. Available on the Ecological Society of America Webside `http://www.esa.org/history/docs/BurgessHistory.pdf`

[Cassidy, 2007] V. M. Cassidy. *Henry Chandler Cowles: Pioneer Ecologist.* Chicago, IL: Kedzie Sigel Press, 2007.

[Cittadino, 1990] E. Cittadino. *Nature as the Laboratory: Darwinian Plant Ecology in the German Empire, 1880–1900.* Cambridge, UK: Cambridge University Press, 1990.

[Clements, 1905] F. E. Clements. *Research Methods in Ecology.* Lincoln, NE: University Publishing Company, 1905.

[Clements, 1907] F. E. Clements. *Plant Physiology and Ecology.* New York, NY: Henry Holt and Company, 1907.

[Clements, 1916] F. E. Clements. *Plant Succession: An Analysis of the Development of Vegetation.* Washington, DC: Carnegie Institution of Washington, 1916.

[Clements, 1935] F. E. Clements. Social origins and processes among plants. In C. Murchison (ed.), *A Handbook of Social Psychology* Vol 1, pp. 22–48. Worcester, MA: Clark University Press, 1935.

[Clements, 1936] F. E. Clements. Nature and structure of the climax. *Journal of Ecology* 24: 252–284, 1936.

[Clements, 1943] F. E. Clements. Succession in the plant community. In R. D. McKenzie (ed.), *Readings in Human Ecology,* pp. 439–440. Ann Arbor, MI: George Wahr, 1943.

[Clements and Shelford, 1939] F. E. Clements and V. E. Shelford. *Bio-Ecology.* New York, NY: Wiley, 1939.

[Clements *et al.*, 1929] F. E. Clements, J. E. Weaver, and H. C. Hanson. *Plant Competition: An analysis of Community Functions.* Publication No. 398, Washington, DC: Carnegie Institution of Washington, 1929.

[Coleman, 1986] W. Coleman. Evolution into ecology? The strategy of Warming's ecological plant ecology. *Journal of the History of Biology* 19: 181–196, 1986.

[Cowles, 1899] H. C. Cowles. The ecological relations of the vegetation of the sand dunes of Lake Michigan. *Botanical Gazette* 27: 95–117; 167–202; 281–308, 1899.

[Craig, 2005] P. Craig. *Centennial History of the Carnegie Institution of Washington. VI. The Department of Plant Biology*. Cambridge, UK: Cambridge University Press, 2005.

[Croker, 1991] R. A. Croker. *The Life and Work of Victor Ernest Shelford 1877–1968*. Washington, DC: Smithsonian Institution Press, 1991.

[Croker, 2001] R. A. Croker. *Stephen Forbes and the Rise of American Ecology*. Washington, DC: Smithsonian Institution Press, 2001.

[Cronk and Fennessy, 2001] J. K. Cronk and M. S. Fennessy. *Wetland Plants: Biology and Ecology*. Boca Raton, FL: Lewis Publihsers, 2001.

[Darwin, 1859] C. R. Darwin. *The Origin of Species by Means of Natural Selection of the Preserved of Favoured Races in the Struggle for Life*. London, UK: John Murray, 1859.

[Davis and Slobotkin, 2004] M. A. Davis and L. B. Slobotkin. The science and values of restoration ecology. *Restoration Ecology* 12: 1–3, 2004.

[Egerton, 1973] F. N. Egerton. Changing concepts of the balance of nature. The *Quarterly Review of Biology* 48: 322–350, 1973.

[Egerton, 1977] F. N. Egerton. History of American Ecology. New York, NY: Arno Press, 1977. (Reprinted by the Ayer Company, Salem, NH, USA, 1984.)

[Egerton, 2007] F. N. Egerton. Understanding food chains and food webs, 1700–1970. *Bulletin of the Ecological Society of America* 88: 50–69, 2007.

[Egerton, 2008] F. N. Egerton. History of Ecology. In S. E. Jorgensen and B. Fath (eds.), *Encyclopedia of Ecology*, pp. 1864–1878. Oxford, UK: Academic Press, 2008.

[Egerton, 2009] F. N. Egerton. Homage to Frederic E. Clements, historian of plant succession studies. Bulletin of the Ecological Society of America 90: 43–79, 2009.

[Eliot, 2007] C. Eliot. Method and metaphysics in Clements's and Gleason's ecological explanations. *Studies in History and Philosophy of Science Part C: Studies in History and Philosophy of Biological and Biomedical Sciences* 38: 85–109, 2007.

[Elster, 1974] H-J. Elster. History of Limnology. *Mitteilungen der Internationale Vereinigung für Theoretische und Angewandte Limnologie* 20: 7–30, 1974.

[Elton, 1927] C. Elton. *Animal Ecology*. London, UK: Sidgwick and Jackson, 1927.

[Feder and Block, 1991] M. E. Feder and B. A. Block. On the future of animal physiological ecology. *Functional Ecology* 5: 136–144, 1991.

[Forbes, 1878] S. A. Forbes. The food of Illinois fishes. *Bulletin of the Illinois State Laboratory of Natural History* 1: 71–89, 1878.

[Forbes, 1887] S. A. Forbes. The lake as a microcosm. *Bulletin of the Science Association*, Peoria, IL, 1887.

[Ford, 2000] E. D. Ford. *Scientific Method for Ecological Research*. Cambridge, UK: Cambridge University Press, 2000.

[Freeman, 1974] D. Freeman. The evolutionary theories of Charles Darwin and Herbert Spencer. *Current Anthropology*. 15: 211–237, 1974.

[Gleason, 1917] H. A. Gleason. The structure and development of the plant association. *Bulletin of the Torrey Botanical Club* 43: 463–481, 1917.

[Gleason, 1926] H. A. Gleason. The individualistic concept of the plant association. *Bulletin of the Torrey Botanical Club* 53: 7–26, 1926.

[Glenn-Lewin *et al.*, 1992] D. C. Glenn-Lewin, R. K. Peet and T. T. Veblen (eds). *Plant Succession: Theory and Prediction*. London, UK: Chapman and Hall, 1992.

[Golley, 1993] F. B. Golley. *A History of the Ecosystem Concept in Ecology: More than the Sum of the Parts*. New Haven, CT: Yale University Press, 1993.

[Graham and Dayton, 2002] Graham M. H. and P. K. Dayton. On the evolution of ecological ideas: Paradigms and scientific progress. *Ecology* 83: 1481–1489, 2002.

[Grime, 1979] J. P. Grime. *Plant Strategies and Vegetation Processes*. Chichester, UK: Wiley, 1979.

[Grisebach, 1872] A. Grisebach. *Die Vegetation der Erde nach Ihrer Klimatischen Anordnung*. 2 volumes, Leipzig, Germany: Engelman, 1872.

[Gross, 2002] M. Gross. When ecology and sociology meet: The contributions of Edward A. Ross. *Journal of the History of the Behavioral Sciences* 38: 27–42, 2002.

[Haberlandt, 1884] G. Haberlandt. *Physiologische Pflanzenanotomie im Gundriss dargestellt*. Leipzig, Germany: Engelman, 1884.

[Hagen, 1992] J. B. Hagen. *An Entangled Bank: The Origins of Ecosystem Ecology*. New Brunswick, NJ: Rutgers University Press, 1992.

[Hartshorne and Weiss, 1931–1935] C. Hartshorne and P. Weiss (eds.), *Collected Papers of Charles Sanders Peirce*, Volume I–VI. Cambridge, MA: Belknap Press, 1931–1935.

[Johnson and Miyanishi, 2007] E. A. Johnson and K. Miyanishi. *Plant Disturbance Ecology. The Process and the Reponse.* Burlington, MA: Academic Press, 2007.

[Juthe, 2005] A. Juthe. Argument by analogy. *Argumentation* 9: 1–27, 2005.

[Kingsland, 1985] S. E. Kingsland. *Modeling Nature: Episodes in the History of Population Ecology.* Chicago: University of Chicago Press, 1985.

[Kingsland, 2005] S. E. Kingsland. *The Evolution of American Ecology, 1890–2000.* Baltimore, PA: Johns Hopkins University Press, 2005.

[Kormody, 1965] E. J. Kormody (ed.), *Readings in Ecology.* Englewood Cliffs, NJ: Prentice-Hall, 1965.

[Krebs, 2006] C. J. Krebs. Ecology after 100 years: Progress and pseudo-progress. *New Zealand Journal of Ecology* 30: 3–11, 2006.

[Lambers *et al.*, 1998] H. Lambers, F. S. Chapin III, and T. L. Pons. *Plant Physiological Ecology.* New York, NY: Springer, 1998.

[Lindeman, 1942] R. L. Lindeman. The trophic-dynamic aspect of ecology. *Ecology* 23: 399–418, 1942.

[Loehle, 1987] C. Loehle. Hypothesis testing in ecology: Psychological aspects and the importance of theory maturation. *Quarterly Review of Biology* 62: 397–409, 1987.

[MacArthur and Wilson, 1967] R. H. MacArthur and E. O. Wilson. *The Theory of Island Biogeography.* Princeton, NJ: Princeton University Press, 1967.

[McCann, 2000] K. S. McCann. The diversity-stability debate. *Nature* 405: 228–233, 2000.

[McIntosh, 1985] R. P. McIntosh. *The Background of Ecology: Concept and Theory.* Cambridge, UK: Cambridge University Press, 1985.

[McKenzie, 1924] R. D. McKenzie. The ecological approach to the study of the human community. *American Journal of Sociology* 30: 287–301, 1924.

[McKenzie, 1943] R. D. McKenzie. *Readings in Human Ecology.* Ann Arbor, MI: George Wahr, 1943.

[Möbius, 1877] K. Möbius. *Die Auster und die Austernwirthschaft, 10. Eine Austernbank ist eine Biocönose oder Lebensgemeinde.* Berlin, Germany: Wiegandt, Hempel and Parey, 1877.

[Möbius, 1880] K. Möbius. *The oyster and oyster-culture. 10. An oyster-bank is a biocönose, or a social community.* (Translated by H. J. Rice). Fish and Fisheries Annual Report of the Commission for the year 1880, Vol 3. appendix H, 1880.

[Naeem, 2002] S. Naeem. Ecosystem consequences of biodiversity loss: The evolution of a paradigm. *Ecology* 83: 1537–1552, 2002.

[Nicholson, 1996] M. Nicholson. Humboldtian plant geography after Humboldt: The link to ecology. *British Journal for the History of Science* 29: 289–310, 1996.

[Nyhart, 1998] L. K. Nyhart. Civic and economic zoology in nineteenth century Germany: The "living communities" of Karl Moebius. *Isis* 89: 605–630, 1998.

[Odum, 1969] E. P. Odum. The strategy of ecosystem development. *Science* 164-262-270, 1969.

[Park, 1946] T. Park. Some observations on the history and scope of population ecology. *Ecological Monographs* 16: 313–320, 1946.

[Peters, 1991] R. K. Peters. *A Critique for Ecology.* Cambridge, UK: Cambridge University Press, 1991.

[Quine and Ullian, 1970] W. V. Quine and J. S. Ullian. *The Web of Belief.* New York, NY: Random House, 1970.

[Schimper, 1898] A. F. W. Schimper. *Pflanzengeographie auf physiologischer Grundlage.* Fischer, Jena, Germany, 1898.

[Sculthorpe, 1967] C. S. Sculthorpe. *The Biology of Aquatic Vascular Plants.* London, UK: Edward Arnold, 1967.

[Semper, 1881] K. Semper. *Animal Life as Affected by the Natural Conditions of Existence.* New York, NY: Appleton, 1881.

[Sheail, 1987] J. Sheail. *Seventy-five Years in Ecology: The British Ecological Society.* Oxford, UK: Blackwell, 1987.

[Shelford, 1911] V. E. Shelford. Physiographical animal geography. *Journal of Morphology* 22: 551–618, 1911.

[Shelford, 1913] V. E. Shelford. *Animal Communities in Temperate America.* Chicago, IL: University of Chicago Press, USA, 1913.

[Simberloff, 1980] Daniel Simberloff. A succession of paradigms in ecology: essentialism to materialism to probabilism. *Synthese* 43: 3–29, 1980.

[Spencer, 1860] H. Spencer. The social organism. Originally published in the *Westminster Review*, January, 1860. Reprinted in H. Spencer. *Essays: Scientific, Political and Speculative*, 3 volumes. New York, NY: Appleton, 1891.

[Spencer, 1898–1899] H. Spencer. *The Principles of Biology*, 2nd Edition. London, UK: Williams and Norgate, 1898–1899.

[Spurr, 1952] S. H. Spurr. Origin of the concept of succession. *Ecology* 33: 426–427, 1952.

[Stauffer, 1957] R. C. Stauffer. Haeckel, Darwin and ecology. *Quarterly Review of Biology* 32: 138–144, 1957.

[Stauffer, 1960] R. C. Stauffer. Ecology in the long manuscript version of Darwin's Origin of Species and Linneaus' Oeconomy of Nature. *Proceedings of the American Philosophical Society* 104: 235–241, 1960.

[Tansley, 1935] A. G. Tansley. The use and abuse of vegetational concepts and terms. *Ecology* 16: 284–307, 1935.

[Tobey, 1981] R. C. Tobey. *Saving the Prairies: The Life Cycle of the Founding School of American Plant Ecology, 1895–1955.* Berkeley, CA: University of California Press, 1981.

[Tucker, 2002] K. H. Tucker, Jr. *Classical Social Theory.* Malden, MA: Blackwell Publishers, 2002.

[van der Valk, 1985] A. G. van der Valk. Vegetation dynamics of prairie glacial marshes. In J. White (ed.), *Population Structure of Vegetation*, pp. 293–312. The Hague: Junk, 1985.

[van der Valk, 1992] A. G. van der Valk. Establishment, colonization and persistence. In D. C. Glenn-Lewin, R. K. Peet and T. T. Veblen (eds.), *Plant Succession: Theory and Prediction*, pp. 60–102. London, UK: Chapman and Hall, 1992.

[Volkens, 1887] G. Volkens. *Die Flora der ägyptisch-arabischen Wüste auf Grundlage anatomische-physiologischer Forschungen dargestellt.* Berlin, Germany: Gebrüder Borntraeger, 1887.

[von Humboldt, 1805–1807] A. von Humboldt. *Essai sur la Géographie des Plantes, accompagné d'un Tableau physique des Régions équinoxiale.* Paris, France: Schoell, 1805–1807.

[Warming, 1896] E. Warming. *Lehrbuch der ökologischen Pflanzengeographie: Eine Einführing in die Kenntnis der Pflanzenvereine.* Berlin, Germany: Gebrüder Borntraeger, 1896.

[Warming, 1909] E. Warming. *Oecology of Plants: An introduction to the study of plant communities.* Translated and edited by P. Groom and I. B. Balfour. Oxford, UK: Clarendon Press, 1909.

[Whittaker, 1975] R. H. Whittaker. *Communities and Ecosystems*, 2nd Edition. New York, NY: McMillan, 1975.

[Winterhalder *et al.*, 2004] K. Winterhalder, A. F. Clewell and J. Aronson. Values and science in ecological restoration—A response to Davis and Slobotkin. *Restoration Ecology* 12: 4–7, 2004.

[Worster, 1977] D. Worster. *Nature's Economy: A History of Ecological Ideas.* San Francisco, CA: Sierra Club Books, 1977. 2nd Edition, Cambridge, UK: Cambridge University Press, 1994.

THE LEGEND OF ORDER AND CHAOS: COMMUNITIES AND EARLY COMMUNITY ECOLOGY

Christopher Eliot

> "We must admit that a *stand* of vegetation is a concrete entity." "Extent, boundary, uniformity: these are the *sine qua non* of every community." Henry A. Gleason (1936)

> "No serious student of succession (a process) has ever claimed that a succession is made up of 'discrete units.'" E. Lucy Braun (1958)

A community, for ecologists, is a unit for discussing collections of organisms. It refers to collections of populations, which consist (by definition) of individuals of a single species. This is straightforward. But communities are unusual kinds of objects, if they are objects at all. They are collections consisting of other diverse, scattered, partly-autonomous, dynamic entities (that is, animals, plants, and other organisms). They often lack obvious boundaries or stable memberships, as their constituent populations not only change but also move in and out of areas, and in and out of relationships with other populations. Communities are consequently interesting to philosophers interested in ontology—in what kinds of things exist—as unusual scientific objects. But others with interests in communities, including ecologists, conservationists, policy-makers, land-managers, environmental philosophers, and philosophers of science, have an interest in whether these unusual features make communities unreal. Familiar objects have identifiable boundaries, for example, and if communities do not, maybe they are not objects. Maybe they do not exist at all. The question this possibility suggests, of what criteria there might be for identifying communities, and for determining whether such communities exist at all, has long been discussed by ecologists. This essay addresses this question as it has recently been taken up by philosophers of science [Shrader-Frechette and McCoy, 1993; Shrader-Frechette and McCoy, 1994; Sterelny, 2006; Odenbaugh, 2007], by examining answers to it which appeared a century ago and which have framed the continuing discussion.

Plant ecologists struggled openly and vigorously through the early twentieth century with the definitions, and then with the legitimacy, of their basic units. Though this discussion continues, a conversation about the discipline's foundations prospered from the 1910s to 1950s with a rough continuity of participants

Handbook of the Philosophy of Science. Volume 11: Philosophy of Ecology.
Volume editors: Kevin deLaplante, Bryson Brown and Kent A. Peacock. General editors: Dov M. Gabbay, Paul Thagard and John Woods.

and issues. As those decades advanced, plant ecologists paid increasing attention to the roles of animals in ecological communities, but animal ecology developed largely independently (as described by Gregg Mitman, for instance [Mitman, 1988; Mitman, 1992]). Focusing on vegetation as the basis for defining ecological units even when animals were integrated, plant ecologists asked how portions of it can be demarcated from their surroundings for analysis. Current ecology generally calls distinguishable, multi-species groupings "communities," though in the early twentieth century, depending on their scales, on the more-specific properties attributed to them, and on the preferences of individual scientists, they were variously called associations, societies, facies, and formations. Except where noted, I will use "community" in the general, contemporary sense which includes all these multi-species groupings.

Disregarding for now the particular connotations of these various kinds of groupings, whether any groupings exist at all in such a way that they can be substantively differentiated from their surroundings as communities is of interest for several projects in philosophy of science. First, working towards understanding the ecological usage of "community" parallels philosophical attempts to define "chance," "time," "species," and "gene" in other sciences. That is, some philosophers have hoped to make a contribution to science itself by clarifying terms or pointing to their ambiguities. Examining communities (sometimes independently of the term "community") contributes to science insofar as biology texts sometimes muddle this part of their field's development, potentially confusing current discussion—see section 5, below. Second, some philosophers have used scientific fields' terminology to assess them. Scientific realists in philosophy of science have linked the question of whether theoretical progress has occurred in a particular scientific field to the question of whether the entities postulated by its theories really exist [Leplin, 1984]. Ian Hacking worried, for instance, that because gravitational lenses postulated by theoretical astrophysics had never been observed, that field might therefore not be on the right track [Hacking, 1989]; a few years later, the predicted gravitational lenses were observed, and their existence bolstered confidence that astrophysics had made progress towards revealing nature. Whether or not entities' status is the best standard for progress in ecology, it is useful to ask whether ecology meets it, and how the discipline has fared against it over the course of its development. Third, others have recommended that philosophy and science can work together to produce a richer understanding of nature than either one can generate independently, a project Peter Godfrey-Smith has recently called developing a "philosophy of nature" [Godfrey-Smith, 2009].

But also, beyond philosophy, policy-makers, land managers, and conservationists have an interest in whether communities of some sort can be reliably distinguished from their surroundings. And so, environmental philosophers consequently inherit the question of communities' existence as they examine conservation's and policy's foundations, and environmental ethicists face it (whether they recognize it or not) as they consider what kinds of things humans have duties towards, or duties to preserve. If some areas of vegetation can persist autonomously as units better than other groupings can, or are more stable or real or integrated than others, or consist of parts which are especially fragile when dissociated, they may be better candidates for protection than other areas. Casually speaking, they might be more ecologically-sound or ecologically-significant than areas of remnant vegetation in landscapes modified by humans. Conversely, if ecology *cannot* recognize any robust groupings larger than single-species populations—where "robust" means that they can be reliably identified on the basis of well-grounded criteria—attempts to protect them come to seem misguided.

Both epistemological and environmental projects have an interest in the status of ecological communities, but environmental philosophy has special reason for concern, in that if calling a group of organisms a community is akin to *arbitrarily* drawing a line around them on a map, a number of its projects are jeopardized. So, without entirely resolving it in this essay, I consider the recognition of and challenge to communities as they appeared in the early twentieth-century debate where they were first influentially asserted. Then I assess the implications of that debate for our current philosophical discussion of ecology. Specifically, I discuss how two scientists' theories have come to frame a debate about communities widely reported by biology textbooks and histories, and reproduced by philosophers. I argue that accounts of this dispute—the Clements/Gleason debate—have inaccurately radicalized the views of the scientists whose names are attached to it. Worse, these fanciful, radicalized positions are untenable in themselves in a way that infects their enduring currency in debates, whether their authors are treated as allies or antagonists.

Philosophical discussions taking impoverished concepts as starting points risk being unproductive at best and muddled at worst, while enjoying a superficial patina of biological respectability. So, after unyoking the scientists' positions from the standard versions of them, I argue that a class of arguments against community-preservation is thereby undermined. Then, after stepping back to consider how and why inaccurate versions of scientific positions could have such longevity and cachet, I consider how their untenability affects current philosophy of ecology. I argue that, as an example, Jay Odenbaugh's recent appeal to Clements and Gleason in arguing for realism about communities is diminished by his problematic versions of them. Yet, observing the significant common ground between them clarifies the way forward.

1 CLEMENTS, GLEASON, AND PRESERVABILITY

Among episodes in the history of ecology that have drawn attention from beyond that discipline, the debate about plant community structure during the early and middle twentieth century has attracted a remarkably diverse and persistent congregation of commenters. Much of the credit for this goes to the episode itself, for its distinctive features. First, it was set off by what was arguably the first body of general theory in ecology, a theory of plant succession—of the development of vegetation through time—advanced by Frederic E. Clements and others. Second, this theory touched off especially vigorous criticism, even condemnation, immediately on publication. The main justifications offered for its rejection did not arise over a long series of papers generating accumulating anomalies, but instead were already in print within a year of the theory's kernel publication, Clements's comprehensive 1916 volume *Plant Succession: An Analysis of the Development of Vegetation* [Clements, 1916]. By the following year, October 1917, ecologist Henry A. Gleason had published "The Structure and Development of the Plant Association," which introduced what would remain for some time the core objections to Clements's theory [Gleason, 1917]. Third, these criticisms, both Gleason's and others', have appeared to some to have produced one of ecology's most visible paradigm shifts, one close to, though not exactly fitting, the mold developed in Thomas Kuhn's *Structure of Scientific Revolutions* [Kuhn, 1962]. Fourth, this conspicuous shift of opinion away from Clements's theory has seemed to others less-attracted to Kuhn's flirtations with incommensurability and relativism to be a laudable instance of progress already appearing in the early decades of a young, diverse science with few unequivocal instances of theoretical progress. Accounts of it often serve the point: 'see how wrong ecology used to be, and how much it has since learned?' Fifth, and of most interest to this essay, the scientific debate (including the position eclipsed by opposition) set up terms for subsequent debates in community ecology, terms persisting here and there to the present. Though there certainly have been other motivations, these features especially have attracted commentators and retellings.

To situate the two ecologists who are the episode's protagonists: Clements and Gleason each performed his formative research in the Midwestern United States, near the historical boundary between eastern forests and Midwestern plains. Clements began to develop his ideas about plant ecology while studying at the University of Nebraska as both an undergraduate and graduate student under Charles E. Bessey, leader of the discipline-shaping Botanical Seminar. After earning his PhD in 1898, he remained on the faculty there until 1907, when he left to become the chairman of the Botany Department at the University of Minnesota. After completing his most influential theoretical work, he was employed by the Carnegie Institution, traveling and researching around the United States, until 1942. Gleason left the Midwest after undergraduate and masters work at the University of Illinois, to earn his PhD at Columbia University in New York in 1906, but then held faculty positions at Illinois and the University of Michigan until 1919. He finally returned to the East Coast to spend the rest of his career at the New York Botanical Garden studying plant taxonomy more than ecology. Histories of these figures and their scientific and cultural contexts have been produced from a wide variety of perspectives by both biologists [Phillips, 1931; McIntosh, 1975; Tobey, 1981; Hagen, 1988; Nicolson, 1990; Hagen, 1992; Worster, 1994; Barbour, 1995; Nicolson and McIntosh, 2002] and historians [Malin, 1947; Hagen, 1988; Hagen, 1992; Worster, 1994; Kingsland, 2005].

For philosophy of ecology, one noteworthy aspect of these ecologists' historical situation—and a theme well-developed in the works just mentioned—is that the context of the putative waning of Clementsian ecology and the ecological ascendance of Gleasonian ecology was the Dust Bowl disaster in the Great Plains of the American Midwest. As crops on land recently converted from prairie to cultivation were obliterated by drought and wind in the early 1930s, it became easier to interpret vegetation as directed more by disturbances than by orderly processes. The variability of habitat stood out more vividly than its stability, in a way that cast doubt on ecology which seemed to presuppose long-term stability of habitats. The Dust Bowl context itself has been offered as at least as plausible an explanation of the shift of favor as the theories' predictive and explanatory strengths and weaknesses. Whatever the appropriate sociological explanation, understanding the changes in ecological science through this period requires realizing that this was not an episode of theory-change in a constant context. Circumstances Clements's theories were designed to help understand themselves shifted in a way which made the theory appear extraneous or false. Since the scientific concepts developed during this period retain a life in philosophy of science, we should consider how the concepts' scientific careers were influenced by context.[1] The ecological severity of the circumstances in which Clements's theory was partly abandoned by ecologists and the degree to which its central concepts have endured in the ecological literature[2] and in natural history together suggest that the theory did not wane in popularity solely on account of its wrongness *per se*. That external rather than strictly internal (or evidential) factors promoted its demise has contributed to its enduring relevance as something other than a discarded falsehood.

Also noteworthy in this historical situation of the two ecologists is that early in the debate, Gleason moved to an institution where his research became only indirectly ecological, so that the ascendance of Gleasonianism included only limited ascendance of Gleason as a professional ecologist. Moreover and at first glance surprisingly, Clements never published a response to his nominal antagonist. For both these reasons, to the extent there was a Clements-Gleason debate, it was only fractionally a debate between Clements and Gleason.

Yet as their positions have been taken to frame a debate, the positions have been offered as contraries, in the following way. Clements's theory has traditionally been tied to two related claims: (1) that vegetation develops in any given area in a way comparable to, or literally identical with, the development of an individual organism; and (2) that the development of vegetation in an area necessarily culminates in a particular type of vegetation, called that area's "climax," which is determined by its climate. Gleason and his ecological theory have correspondingly been associated with the rejection of these claims, and identified with an alternative he called the "Individualistic Concept of Ecology." The positive core of the

[1]One need not accept much of Kuhn's framework for scientific change, for instance, to agree with him that the acceptance and rejection of scientific theories can be and has been influenced by external circumstances. That influence has been accepted even by accounts of theory-change trying to defend its potential rationality much more than Kuhn did [Kitcher, 1993, for instance].

[2]To mention scattered well-cited examples of their endurance: identifying climax communities in intestinal fauna and ocean-floor cyanobacteria, and determining spider abundance at various forest-succession stages [Bultman *et al.*, 1982; Reid *et al.*, 2000; Hooper, 2004].

individualistic theory has usually been summarized as the view that individual plants disperse and establish themselves independently of others, so that plant communities are merely unstructured aggregates of independent plants. Such unstructured Gleasonian communities would have dynamics rather unlike those of the integrated collectives attributed to Clements which develop as units towards definable climaxes.

As I remarked, an attraction of this episode for commentators has been the tension between these two apparently-opposite theories. Their neat opposition supports a framework for a narrative of conflict and theory-replacement in early plant ecology. But even more beguiling have been their vivid images, their provocative similes and metaphors. Tossing up fodder for poetic imagination, the debate sets the idea of a collective social organism against the idea of disaggregated, dissociated individuals—images which evoked in the subsequent decades the antagonisms of communism and capitalism, totalitarianism and democracy. Through the Cold War period of the middle twentieth century, such similes resonated with popular ambitions and popular fears burgeoning beyond the scientific discussion [Mitman, 1995]. Within scientific ecology and at its boundaries, these images set up a pointed question about the nature of nature. They frame the possibilities for what can be an object of ecological inquiry. Further, the opposing icons of collectives of organisms and dissociated, free individuals crystallize opposing answers to the question of whether communities are structured, organized entities or whether they are randomly-assembled aggregates of individuals. They have provoked curiosity within science and beyond it about whether nature and the nominal objects of ecology are essentially functions of order or of chaos.

Yet, the possibility that nature at the ecological scale might be chaotic has serious practical implications. When conservationists, politicians, and ethicists aim to preserve communities or endorse preserving them, they assume that communities are more than arbitrarily-identified fictions. If ecology demonstrates that communities are mere fictions, a *modus tollens* inference is licensed, concluding that community-preservation efforts are futile or misguided. This inference depends on an underlying conditional claim:

> Communities can be preserved only if they are real, orderly entities and not chaotic (not, that is, unreal fictions imposed on real chaos).

At this level of generality, the conditional claim expresses a sensible view. A thing cannot be preserved as such if it does not exist. Nor can it be meaningfully preserved if it has arbitrary boundaries and negligible structure. Insofar as communities are collections, in this case *some things* might be preserved, but not *something*. Imagine, as an image of a worst-case scenario, trying to preserve a liter of the sea *in situ*. Tracking the individual components, one finds they rapidly dissipate and mix with others; tracking the location, one finds that it rapidly changes as particles arrive and depart. In the absence of ecological communities, one could identify the organisms in an area and attempt to keep them there or within some dynamic limits. Or, one could avoid interfering with an area, come what may. But those projects are not what those seeking to preserve communities normally understand themselves as doing. They take themselves to be preserving *something*. So, as it relates to conservation practice and advocacy, the claim is sensible at this level of generality.

Problems begin when this reasonable claim is made more specific by affixing it to scientific positions, to draw further conclusions. Arguments employing the claim have instantiated the positions it mentions with Clements's and Gleason's theories, so that the claim becomes:

> Communities can be preserved only if they are Clementsian, not Gleasonian.

Drawing inferences via this conditional requires that the ecologists' assertions of order and chaos deny each other—that their ecological theories have communities either existing in an orderly way or being fictions imposed on chaos, and that these claims negate one another, or are mutually exclusive. Read in a straightforward way, the concepts do negate one another: clearly-bounded functionally-structured super-organisms in which individuals are controlled by their systems are not unbounded, unstructured collections of causally-unrelated individuals. As Jay Odenbaugh suggests, some purposes may be served by engaging with these concepts in abstraction, whether or not they reflect the views of any scientists [Odenbaugh, 2007, p. 629]. I will argue for the inaccuracy of aligning these polar concepts with those scientists, but also that more is at stake than historical accuracy in determining the scientific legitimacy of these concepts.

What is at stake appears when we notice how a conditional claim employing the concepts is used to draw further conclusions. Donald Worster uses it to complain about contemporary ecology, lamenting its inadequacies for supporting conservation efforts [Worster, 1990]. When ecologists approach nature assuming it is Gleasonian, he worries, they fail to produce science which can support conservation.[3] J. Baird Callicott identifies "residual traces of the early twentieth-century Clementsian super-organism paradigm" in Aldo Leopold's defense of his land ethic, and finds a broader commitment to community stability in the environmental ethics tradition following him [Callicott, 1996, p. 358]. Aligning "the insidious challenge to nature conservation posed by poststructuralists," with a scientific challenge to communities, as revealed by Worster's "exposé" of a "deconstructive siege of nature," Callicott worries about the consequences for community-preservation if nature is "chaotic, changing unpredictably, and disturbance ('perturbation') by wind, flood, fire, pestilence, not freedom from disruption is nature's normal state" [pp. 353–355]. Leopold's commitment to community-preservation is undermined by normalizing disturbance, and Worster and Callicott are each alarmed by the threat this poses to conservation. Then, with opposite sympathies, Allan Fitzsimmons argues against the existence of ecosystems in a broad sense incorporating

[3]Worster is actually concerned, in this well-cited article, with two kinds of disordered ecology: Gleasonian and systems theory on the model of Odum. Only the first, community-ecology branch is at issue here.

the equilibrium assumptions of Clementsian "organicism" [Fitzsimmons, 1999, p. 143]. He reasons that because there are no ecosystems (in a broad sense incorporating communities),[4] we should not attempt to preserve any such thing. Treating ecosystems-science as a Kuhnian paradigm (with the associated implications for anti-realism), he offers that "we know that the boundaries of ecosystems are guesswork and rarely represent real features on the landscape. We know that the landscape is in constant flux so that the ecosystems depicted by researchers constantly change in space and time in poorly understood ways, turning ideas such as ecosystem stability and sustainability into oxymorons" [p. 161]. If so, he reasons, conservation policies are flawed because they attempt to preserve illusions glommed onto real chaos from questionable motives.

However, while the first conditional is reasonable, the second is mistaken. A mythology has grown up around the early ecologists and their concepts, mistakenly aligning them with order and chaos.[5] If, as I will argue, this is a mistake, the second claim is not an instance of the first. The correctness or coherence of Clements's or Gleason's ecology in particular do not have the general implications for conservation they have been taken to have. Obviously, one upshot of this argument is that these challenges to community-preservation from classical ecology do not go through so straightforwardly. Towards arguing against the second conditional claim and reject accepting it as an instance of the first, I begin by positioning the historical debate between these theories, and then observe how the debate has been reconstructed, comparing that to the scientists' research.

2 THE PROSPECT OF SCIENTIFIC ECOLOGY

Already in the first decade of the twentieth century, C. E. Moss describes the terminology for classifying vegetation inherited from the nineteenth as being in disarray: "the subject of ecological plant geography has suffered and still suffers very considerably from a lack of uniformity in the use of its principal terms" [Moss, 1910, p. 18]. His 1910 survey focuses specifically on discrepancies in the usages of "formation," "association," and "society"—three terms for different kinds of groupings of plants which would be, in our current usage, different kinds of communities. Moss observes the variety of usages among German botanists through the nineteenth century (Schouw, Griesbach, Hult, Kerner, Drude, Flahault, Schimper, Warm-

[4]He defends this claim, for instance, by using Shrader-Frechette and McCoy's argument against community stability to undermine ecosystem stability. If his general argument runs that *ecosystems* can be preserved only if they are *Tansleyan*, but in fact they are not, still many of his sub-arguments apply equally well to communities.

[5]An earlier essay [Eliot, 2007] also argued that the Clements/Gleason debate should be understood differently than it often has been. That article focused on the assumptions shaping science-interpretation, and specifically on whether it has been reasonable to interpret the Clements/Gleason debate as having been about the assertion of an ecological law of succession and the denial of that law. The analysis here works alongside that one by focusing instead on the metaphysics of Clements's and Gleason's and other ecologists' positions, which is to say, on their arguments concerning communities as things.

ing), following them up to the adoption of similar terminology by British and American ecologists at the beginning of the twentieth (Cowles, Moss, Clements). His verdict is that ecologists have not settled on a shared set of terms, but moreover that their problem is not just finding the right words; it is that there are substantive disagreements about what the words should refer to, and about how these terms should be related to one another. Moss indicates optimism for improvements in the uniformity of usage. But, concerned that terms for subdivisions of plant associations ("plant societies" and "facies") have already been used in multiple senses, he concedes that "in fact, so many terms have been used by ecologists and plant geographers with so many different significations that it would appear to be impossible to find any term to which the above objection does not apply" [p. 48].

Moss's solution for terminological disorder is causal investigation. Though in 1910 he can be aware only of Frederic Clements's earliest work produced by that time (1899–1907), his optimism lands on a concrete project in Clements's approach. More adamantly than others, Clements had begun to argue that formations, the largest-scale groupings of species, should not be identified in the field primarily by physiognomic criteria—that is, on the basis of the *appearances* of vegetation—but recognized instead on the basis of common *habitats*. Extending back at least to the German *Naturphilosophie* tradition in botany, physiognomic approaches to vegetation sought to identify the character of landscapes just as one might identify people's characters by visually scanning their faces. As Moss describes, Clements responded to this tradition by arguing in his early work that areas of vegetation should be differentiated by their differing causes. "Habitat" thus becomes a way of referring to these causes. This approach can work at the various scales at which there are common conditions. "Formations" are the units of vegetation at the largest scale in space and time, and consequently reflect the widest range of habitat conditions, the most inclusive scope of similarity. Their appearance at any time is normally accordingly diverse. Formations are comprised of "associations" which may be recognized empirically, and aligned with more temporary causes. At both levels, units' boundaries are determined by the extent of action of causes. Moss endorses this initiative to investigate causes as a way out of the conceptual morass.

Crucially in Moss's estimation, Clements's causal analysis would render ecological classification more *scientific* by aligning ecological units with identifiable causes that can be investigated experimentally. The result would be concepts and corresponding units reflecting reality better than do the subjective impressions of naturalists:

> Although many earlier writers regarded the formation and the habitat as vitally connected, it is to Clements (1905) that ecologists owe the most emphatic expression of this view. Clements (1905:292) stated unequivocally that 'the connection between formation and habitat is so close that any application of the term to a division greater or smaller than the habitat is both illogical and unfortunate. As effect and cause,

> it is inevitable that the unit of the vegetative covering, the formation, should correspond to the unit of the earth's surface, the habitat.' This view, as has been shown, was by no means new; but no one had previously stated with sufficient emphasis and in general terms what must be regarded as the foundation of the modern treatment of vegetation. The concept is much more stimulating and much more scientific than a merely physiognomical view of the formation; and this latter view, useful enough in the early days of plant geography, has now been quite outgrown. It is no longer possible to regard a forest as a 'formation,' nor even a coniferous forest. Such complex pieces of vegetation must be resolved into separate associations, and the latter rearranged into formations on a basis which shall commend itself to those who search after real affinities and underlying causes. The rearrangement of associations into formations will not be accomplished at once, except in the case of well-marked habitats. Where the habitats are less sharply defined, much exact and quantitative experimental work remains to be done; and here again Clements, in his *Research Methods in Ecology*, has performed useful and pioneer work. Until much work of this character has been performed, until certain habitats have been more closely investigated, ecologists and plant-geographers must be content to refer certain communities simply to their associations, rather than hastily build up formations on flimsy foundations. [Moss, 1910, p. 33]

What is most notable here is why Clements's theory seems progressive to Moss in 1910. It seems to him an advance because it attempts to ground the differentiation of communities on their distinct causal backgrounds rather than on common appearances, and to identify causal backgrounds through "exact and quantitative experimental work." Investigating causes this way is "more scientific," and yields more accurate units. It offers a more scientific response to an existing demand: botanists and geographers had announced the need for sound nomenclature at international congresses because their fields require discussing areas of vegetation, and they want to establish their work as scientific at a time when many other fields of biology had been professionalizing. Clements's contribution is simply to offer a means for remodeling botany as scientific, by pushing its investigation of causes.

It is problematically artificial to claim a sharp boundary between natural history and science, either in the historical development of ecological knowledge or in the practice of contemporary researchers, but even so, descriptive natural history adds something recognizably different as it engages with the investigation of causes. Causal investigation can help us understand unlike things as instances of single kinds for substantive reasons, which is to say, understand how they participate in the causal structure of the world. Recognizing the participation of individual things in the causal structure of the world is in turn a large part of what it means to gain scientific understanding. So, if descriptive classification can also be scientific, causal investigation unites it with theory in a way that helps it further understanding. Just as in systematics, where understanding of evolution by natural selection (and later genetics) restructured how organisms are arranged into species and higher taxa, so in ecology classificatory optimism arose about the same time.

In fact, this parallel is drawn explicitly by F. F. Blackman and Arthur Tansley in their early review of Clements's *Research Methods* [Blackman and Tansley, 1905; Clements, 1905]. Taxonomists at the beginning of the twentieth century had faced the same kind of conceptual disarray ecologists now were facing. And they, too, had earned some optimism about gradually revealing objective edges for their groupings through a combination of new theory and emerging experimental techniques developed to assess theories' applicability to nature. Beginning with the problem of identifying entity-boundaries in ecology, Blackman and Tansley land alongside Moss on Clements's work as a basis for optimism. Here they discuss how to probe the boundaries of communities as a problem paralleling one in taxonomy. In both cases, units grade into one another. Yet, in both disciplines it is possible through investigation to discern the actions of different causes which objectively distinguish the units. Referring first to vegetation, they write:

> If you get a gradual and continuous change of one or more factors in passing away from a given spot characterised by a definite assemblage of plant-forms you may pass through a region which shows a continuous change in vegetation structure and composition till you arrive at quite another definite assemblage. At what point is "the final test" to be applied? The difficulty here seems to be fairly comparable with the difficulty of delimiting species in taxonomy. Critical study will in very many cases enable us satisfactorily to delimit formations which at first present bewildering difficulties. The same is true of species. There may be cases in which the difficulties are so great that there is still room, after the best investigation we can give, for difference of opinion as to whether the assemblages in dispute shall be "split" or "lumped"; which means that the subjective element cannot at present be entirely eliminated. The same is true of species. Meanwhile we are convinced that both species and formations have a real objective existence, though widespread doubt exists in both cases, especially among those who have not given attention to their actual study. The real differentiating factors in the two cases are probably of entirely different nature and in both cases we are far from having explored them to the bottom. Nevertheless we have full confidence that finality in these provinces will be reached in the course of future work. The work of Jordan, of De Vries, and of the Mendelians seems to furnish a beginning in one province, while Dr. Clements's researches constitute an important advance in the other. [Blackman and Tansley, 1905, pp. 250–251]

Noticing Blackman and Tansley's early optimism about defending communities' objective existence invites curiosity about what happened to ecology's optimism for this new "more scientific" approach. One kind of answer to this question is defended in Ronald Tobey's history. In *Saving the Prairies*, he describes how a Clementsian paradigm (in Kuhn's sense) eroded and collapsed though a combination of scientific, contextual, and internal sociological factors. Along with Tobey, a number of historians and historically-oriented biologists have analyzed this shift in similar ways, including those discussed below in section 5.

For philosophy of ecology, however, a further question emerges when we observe that the eclipse of Clements's ecology involved not only a theory-shift away from a theory deemed worse or less-true, but also decades of claims that Clementsian ecology is *unscientific*. Ecologist Daniel Botkin, for instance, remarks that the Clementsian approach to communities will soon seem "silly" as an explanation of nature, and that it "by the 1940s had been completely dismissed in the United States, where it remained a historical curiosity, useful in explaining to students of ecology why it is an inappropriate perception." For Botkin, Clements's account of communities was not merely mistaken, but moreover "quickly dismissed when proposed in the scientific age," because it was not only wrong, but also deficient as science [Botkin, 1990, pp. 98–99]. It is one thing to be wrong; it is another to be unscientific. How did conventional wisdom switch from treating Clementsian ecology as scientifically progressive to unscientific? Then, how does its ontology—the units and entities it employs—contribute to or impede its success as science?

From Moss's early review all the way through to the present, one finds arguments both (a) that Clements's theory was scientifically progressive, while Gleason's undermined ecology's goals, and (b) that Clements's theory was unscientific, while Gleason's was scientifically progressive. I suggest this feature of the debate is not unrelated to the other feature I noted, the prominence of its similes in discussions of both theories from their early reception to the present. Commentators have sometimes taken the scientists' similes to *be* their positions, and even where commentary has been more sophisticated, it has usually understood the positions as what one *should* be committed to if one is committed to a certain simile, rather than what the scientists themselves actually did commit to.

3 ORDER AND CHAOS

Donald Worster [1990], positioning Clements's and Gleason's theories as opposites in an influential article, has labeled them "the ecology of order and chaos." On the order side is of course Clements's structured climax sere (the sequence of vegetation leading up to a climax state). Most of Clements's 1936 essay, his most widely reprinted piece of writing, is devoted to the structure of the climax sere, and this emphasis on its structure has suggested that the sere itself imposes a causal structure on the organisms within it. That impression derives from an abundance of terminology. Though I will not explain its full structure here, the climax sere has numerous parts, fostering the impression that these parts, and consequently vegetation itself, are structured by an overarching organizing-principle. In discussing later developments of terminology, Clements writes that "the climax group now comprises the following units, viz. association, consociation, faciation, lociation, society, and clan," and these are only some of the components of the climax [Clements, 1916, p. 272].[6] As Clements worked out how to explain the complexities of vegetation, his theory became increasingly laden with vocabulary he deployed to handle variation. It contains names for vegetation at multiple scales, from whole regions to tiny clusters of a few plants, and names especially for describing variation due to particular classes of causes. Clements's prodigious collection of terms, whose coinage from Latin and Greek roots he displays special relish in detailing, has struck few readers, scientific or lay, as palatable. For most published respondents, the unsavory volume of the coined vocabulary has led to a sense of the whole theory as an undigestible effort to force an imaginary order onto disorderly nature. Botanist Neil Stevens, for example, approvingly cites this damning review of Clements and Shelford's *Bioecology* in 1950:

> This book is a fine example of an important and already difficult subject discussed in an abstruse, involved, pompous and thoroughly tiresome manner. Simple things are made complex, and complex things made well-nigh incomprehensible.... Nor is the mounting use of coined words helpful in elucidating the text. One is led almost to believe that ecology, as understood in the Clements-Shelford biome, is the occupation of thinking up new names for old things. [Stevens, 1950, p. 112][7]

Beyond the sheer volume of terms *per se*, Clements's coinages have led to a number of further commitments being attributed to him, typically as part of repudiating them. Most commonly, the baroque order the terms and their relations suggest have often been thought to imply deterministic orderliness. They have been thought to cumulatively describe a system in which areas' climates are tied to their climax vegetation in lawful associations—despite the *prima facie* implausibility of doing so—as if Clements were naming the mechanical parts of a landscape-factory. Or, they have been thought to correspond to the myriad anatomical parts of complex organisms, with the connotation of governed, purpose-driven development. Furthermore, the vocabulary has been considered ontologically overeager, in that each of the newly-coined vegetation terms have been thought to reflect discrete units of vegetation, so that Clements is thought to recognize entities everywhere, including where they do not exist. Though this assumption of existence applies also to his various concepts like subclimax, disclimax, and serclimax, the impression of Clements's ontological naïvete has especially derived from the impression that his stages of plant succession are supposed to be temporally discrete, and that his areas of vegetation are supposed to have sharp boundaries. Nascent suspicion of ontological naïveté is clinched by Clements's attraction to the organism-simile,

[6]Confusingly, towards the end of his career, Clements came to use this term to refer not only to the end-point of seral sequences but also to vegetation units or formations as a whole, in order to identify the units with their best-adapted species.

[7]I have been unable to identify the original source of this review, and it may be unpublished except in quotation by Stevens.

which seems to treat vegetation units as existing as discretely as individual organisms.

Gleason's terms and phrases—here metaphors—are as evocative of chaos as Clements's "organisms" are of order. To mention one later example:

> Suffice it then to repeat that on every spot of ground the environment is continually in a state of flux, and that the time-period in which a certain environmental complex is operative is seized on by the particular kinds of places which can use it. The vegetation of every spot of ground is therefore also continually in a state of flux, showing constant variations in the kinds of species present, in the number of individuals of each, and in the vigor and reproductive capacity of the plants. [Gleason, 1939, p. 99]

Gleason's description of both environments and the vegetation occupying them as in continuous flux, like his remarks elsewhere that the co-occurrence of organisms in a particular place is a matter of mere "coincidence," evoke the opposite of Clementsian orderliness. Where Clements appears holistic, Gleason appears reductionistic—at least where holism means that vegetation can only be understood as whole entities clearly distinguishable from their surroundings, while reductionism means that no such whole exists as a real thing, and that only the plants which are its supposed components do. Consequently, while Clements laid out explanatory theory and a research program for ecology, Gleason's ideas may entail that community ecology is impossible. Communities seem not to exist beyond the dynamics of individual organisms, and those dynamics are themselves chaotic or random [Gleason, 1926, p. 16]. Gleason writes that "the distribution of species is primarily a matter of chance, depending on the accidents of dispersal" [Gleason, 1925, p. 74]. And a year later he offers "that careful quantitative study of certain associations from 1911 to 1923 produced the unexpected information that the distribution of species and individuals within a community followed the mathematical laws of probability and chance" [Gleason, 1926, p. 16]. This phrase, "follow[ing] the mathematical laws of probability and chance," demands clarification about what the laws of chance are and what they might apply to, since it cannot mean that plants have equal likelihoods of appearing anywhere.[8] But if vegetation is best described by randomness, that outcome leaves precious few research avenues for community ecologists.

Beyond its political resonance, this language of order and organization found in Clements and the opposing flux and continuum found in Gleason set up this scientific debate as an iteration of a very old philosophical discussion. From the beginning of Western philosophy, the fragmentary remains of presocratic philosopher Heraclitus's writings offer provocative images of nature as in flux, in part or wholly. Heraclitus's remark that "upon those that step in the same rivers, different

[8]Obviously, in the Sonoran Desert, one's odds of finding a saguaro are rather different than one's odds of finding water lilies, for both contingent reasons having to do with past dispersal and necessary reasons having to do with the dynamics of physiologies and environments.

and different waters flow," or colloquially, "you can't step in the same river twice," offers the best-known image of a portion of nature being dynamic to such a degree that its identity is compromised [Kirk *et al.*, 1983, p. 195]. Its implicit suggestion is that dynamic entities, those with shifting properties, not only become different in shifting their features from moment to moment, but also through these shifts become different things entirely. Yet this image, like Descartes's famous example of a ball of wax which retains its identity despite radical changes in all its properties, depicts a problem different than that which ecologists face in recognizing communities. Ecologists' problem with communities is closer to that hinted at by another of Heraclitus's fragments about flux. This fragment proposes an implication of bringing various things together in a group, as communities do by definition: "Things taken together are wholes and not wholes, something which is being brought together and brought apart, which is in tune and out of tune; out of all things there comes a unity, and out of a unity all things" [Kirk *et al.*, 1983, p. 190]. That is, once one starts uniting one thing with another, one finds that all its neighbors can be united with it, and one is left with a unity which excludes nothing. Once one starts thinking of one's siblings and parents as united with oneself to form a family, one finds that "family" can include third cousins and in-laws' siblings' spouses, too.

Yet, the relatedness and consequent potential unity of everything with everything else, as illustrated by Heraclitus's fragments, is not normally a problem. It does not derail our ordinary thought or everyday activities. One can easily delineate one's immediate family for whom taxes must be paid from the larger portion of one's family one wants to invite to a reunion, and this latter group from the whole of humanity (who are in the end all family, too). Similarly, differentiating in our perception of the world around us the signals reflecting the presence of entities from those signals reflecting the space around them—being able, that is, to recognize the edges of ordinary objects—is not normally a practical problem. It is a philosophically-loaded problem for neuropsychology [Marr, 1982, for instance]. But, at least when watching ourselves, we do not set coffee mugs on thin air instead of our desktops, despite that the information our senses gather about the world around us is continuous and constantly fluctuating. We are not fooled by the changing colors and shapes of things into seeing the world as an undifferentiated mishmash, like a pointillist painting viewed from too close. William James evocatively imagines that a baby, confronted with its own new sensations, "assailed by eyes, ears, nose, skin, and entrails at once, feels it all as one great blooming, buzzing confusion" [James, 1890, p. 488]. But James makes this famous remark only on the way to analyzing how children learn *not* to experience the world as an undifferentiated continuum. They learn quickly how to individuate things from their surroundings—chairs from floors and food from spoons. Kuhn revives this phrase from James to present an image of what our experience of the world would be like in the absence of a paradigm that structures that experience for us [Kuhn, 1962, p. 113]. But even if learning strongly influences how we perceive, our ordinary perception can succeed when supported by a store of previous experiences amounting to considerably less than a scientific theory or theoretical paradigm [Richeimer, 2000, pp. 388–391].

Ecological objects are different. We do not normally learn as children to distinguish ecological communities from their surroundings, because their boundaries are rarely obvious, and may never appear at all to laypeople. That is, differentiating communities from their surroundings often requires expertise, and perhaps also a theoretical framework like Kuhn has in mind. But if individual plants are in flux, and if when they appear together it is merely a matter of coincidence, theories permitting groups to be distinguished from one another as units are *prima facie* suspicious. We should mistrust what passes for expertise about their dynamics. We should suspect that supposed experts perceive order where it does not exist.

Wondering whether expertise about ecological communities is possible, we can find Plato's response to Heraclitus handy. While making a broader argument in *Theaetetus* against the idea that to perceive something is to know it, Socrates urges that if one adopts Heraclitus's position that everything is in flux, one can no longer count on one's language. If not only the properties but also the boundaries and identities of certain things are unstable and consequently indeterminate, those are not the sort of things which can anchor the meanings of our words. For two beings to communicate (which is implicitly to say, communicate *meaningfully*) about desks, there must be some actual or possible thing which is distinguishable from its surroundings as a desk [Burnyeat, 1990]. Since the expression of the position itself that everything is in flux depends on the usability of the words it is couched in, this is damning. Relatedly, in *Sophist*, Plato's Visitor from Elea raises the alternative specter, the problem which arises if words pick out only unique things (and perhaps only as they exist statically at a single point in time). Ignoring the relatedness of things, we find that "to dissociate each thing from everything else is to destroy totally everything there is to say" [Plato, 1993, *259e*]. That is, thought and speech generally, and scientific description and explanation in particular, require that the things they refer to be able to be disentangled fairly reliably from their spatiotemporal surroundings as repeated instances of the same kind of thing.

Tailoring Plato's point to fit ecological objects: as far science goes, chaos, in the sense of complete disorder, is a nonstarter. Science cannot gain a foothold in a completely disordered world. One might object that scientists have recognized and developed a mathematical account of chaos. But such "chaos theory" discusses a different kind of chaos than disorder. That other chaos is what Stephen Kellert in a philosophical account defines as "the qualitative study of unstable aperiodic behavior in deterministic nonlinear dynamical systems" [Kellert, 1993, p. 2]. Like any other science, applying chaos theory requires, trivially, that there be systems to which it may be appropriately applied. Whether systems may be singled out for substantive description, and what kinds of systems those are, are precisely what is in question in the ecological discussion. Mathematical chaos moreover explains how there might be order in apparently disordered systems. It is therefore not what is at issue in Worster's usage of "chaos," for instance. Far from it being a nonstarter, ecologists have found mathematical chaos useful in ecological analysis, for instance of population cycling.[9]

So, for present purposes I mean by "chaotic," for a system, that it is entirely disordered. If a system is entirely disordered, any predictive science concerning it is stymied, except that one could banally predict disorder. But more significantly, if attributing this sort of disorder entails that every arbitrarily-bounded unit works equally well for characterization at some level of description, then there are no entities a scientific discipline can call its own at that level of description. Without a substantive level of description of its own, a discipline can be folded into others which do have unique domains about which substantive claims may be made, without any loss of information or understanding. Applying this idea to ecology, if ecology is to be a partly-autonomous discipline in the way that naming it and working on it implies, research in ecology requires employing *some* units. It requires them insofar as it needs to discuss parts of the universe in isolation from other parts. Furthermore, it needs for there to be some units which are better suited for theorizing than others are. Otherwise, ecology cannot uniquely contribute to understanding.

In a nominally-Gleasonian spirit one could offer individual organisms as the units for ecology and reject any units larger than individuals. But this move is not independently open even to the ecological reductionist: if a reductionist wishes to understand the features and dynamics of an ecological system as a function of its components, she still presupposes that a *system* may be identified in some way or other. This system could in principle be the entire world, or the collection of all living things. But for pragmatic reasons, her understanding anything requires that some parts of nature be substantively isolatable from the rest of the universe as systems (even if those systems' dynamics are still entirely a function of their parts). A first pragmatic reason is that sometimes we want explanations of the dynamics of particular systems like lakes and forests. If all ecology can tell us is that the dynamics of a forest depend on everything else in the world or universe, ecology is not worth pursuing. That is because we can rarely acquire very much data about any given system (especially without enormous expense), so that if our understanding of particular dynamics depends on data about all phenomena, we will not reach understanding of those particulars. A second pragmatic reason is that if the entire world or universe could only be understood as such, we would be unlikely to reach such understanding without understanding something about the dynamics of particular parts first. So, even if systems' dynamics are entirely a function of their parts, ecology needs systems as units.

Yet, this argument risks begging the question. Perhaps there are no parts of the living world which may be distinguished from their surroundings enough that they support better predictions or explanations than other parts. Perhaps ecology is not a science because it cannot be. There are at least two good reasons not to accept this alternative. First, ecology has predictive successes which are demonstrably not just lucky guesses, and some of its models and descriptions have proved useful.

[9]See discussion in [Pool, 1989], and, as an example, [Tilman and Wedin, 1991]).

Second, what understanding we have of the features of organisms from elsewhere in biology is largely due to our recognizing the action (or process or mechanism) of natural selection. That natural selection works, that it has successfully produced complex forms, entails that there have been patterns of success and failure in the struggle for survival among individuals and populations. Those patterns are ecological patterns. So while ecological prediction and even description may be difficult, its domain is not patternless, not entirely random. And significantly, the action of natural selection suggests that some of those patterns are ecological in that they involve relationships among populations, not just individuals.

Accordingly, when ecologist Michael Barbour writes, criticizing Clementsian ecology, "if one wishes to recognize associations, perhaps on the basis of the presence of certain dominant species, one can do so and even draw lines on maps; but this activity must be recognized as arbitrary, subjective, and a gross simplification of nature," we should not read his argument as rebutting *all* ecological entities as they appear in scientific discussion [Barbour, 1995, p. 237].[10] Even a gross simplification is not necessarily a falsehood when it serves a descriptive function. In examining Barbour's own research one immediately recognizes it as not at all defeatist about the whole enterprise of ecology, and moreover open to the possibility that ecology might offer substantive claims about parts of nature in isolation from the entire universe. That Barbour discusses systems does not prove that communities exist. Rather, it deflects criticisms like Barbour's from the position that community ecology and its systems stand or fall together, where that ecology's systems stand means that it has ability to discuss real patterns in partly-isolatable portions of nature. Natural selection suggests that there are real patterns, and empirical successes suggest that ecology, however nascent, sometimes stands.

If all is not lost to chaos, then, the difficult remaining problem becomes how to draw lines which are not entirely "arbitrary, subjective, and a gross simplification of nature." What should guide delimiting systems, and what degree of confidence should we place in the entities they delimit? What kind of characteristics, that is, does a community or other ecological system need to have in order to persuade us of the reasonableness of describing it in at least provisional, partial isolation? If figuring out how this works for ordinary objects provides puzzles for philosophers of language but poses no problem for ordinary thought, disentangling ecological objects from their surroundings in this way produces a real, practical problem for ecologists, and raises interesting philosophical questions about ontology, too.

I have mentioned the workaday and even moral consequences of this practical problem. In Donald Worster's inference, the shift from a Clementsian to a Gleasonian ecology produced a crisis for anyone wanting or needing to employ ecological units, most of all environmentalists and conservationists. In "The Ecology of Order and Chaos," Worster's chief concern is with the idea that during the latter half of the twentieth century, ecology experienced a paradigm shift from a

[10]Barbour's objection here can be read as tailored to association-types rather than community tokens, but types are not bounded by "lines on maps"—community tokens are.

Clementsian framework to a Gleasonian one, a shift away from finding order in nature towards regarding nature as in flux.[11] The pathology of this shift lies in its crippling conservation or preservation; if scientists do not treat communities as real things, the justification for conserving them is injured:

> There is a clear reason for that outcome, I will argue, and it has to do with drastic changes in the ideas that ecologists hold about the structure and function of the natural world. In [mid-twentieth-century environmentalist Paul] Sears's day ecology was basically a study of equilibrium, harmony, and order; it had been so from its beginnings. Today, however, in many circles of scientific research, it has become a study of disturbance, disharmony, and chaos, and coincidentally or not, conservation is often not even a remote concern.[12] [Worster, 1990, p. 3]

One might justifiably balk at criticizing science for not offering up an ontology or account of nature supporting applying one's own value-system to the world. Even if one accepts, with Helen Longino for instance, that values from the context of science can constrain scientific reasoning, it is wrongheaded to criticize scientific results for not fortifying our values. Yet, if the concept of order or chaos is assumed by theories pre-theoretically or pre-empirically, it becomes reasonable to criticize it, and even for laypeople to do so. Longino argues that this even shores up science's objectivity [Longino, 2004]. So, there is a reasonable interpretation of Worster's complaint.

Though Worster ends up resigning himself in the essay's final sentence to accepting the theoretical complexity needed to describe nature, it is only after lamenting at length that ecology has lost something if it moves away from equilibrium assumptions. Confidence in order has been lost, and Worster understands "order" to have several components. Orderly systems, he believes, have equilibria, are "perfectly predictable" [pp. 13–14], and have holistic dynamics, especially "emergent collectivity" [p. 8]. Worster takes Gleason to have aimed to demolish ecology's confidence in meeting all three of these standards. For Gleason, indicates Worster,

> there is no such thing ... as balance or equilibrium or steady-state. Each and every plant association is nothing but a temporary gathering of strangers, a clustering of species unrelated to one another, here for a brief while today, on their way somewhere else tomorrow. [pp. 8–9]

Yet, equilibrium, emergent collectivity, and predictability are quite different features of systems. Why should one suppose that orderliness of communities involves

[11] Worster treats Eugene Odum's early systems ecology along with Clements on the side of order, but since systems ecology, shifting focus to abiotic components and ignoring species boundaries raises distinct issues, it can be understood historically as a separate tradition.

[12] "Coincidentally or not" offers Worster a rhetorical hedge here; the essay overall suggests that he does not regard this as coincidental, and the essay would not be worth discussing absent its implication of that relationship.

these particular commitments? Must it? And what is the degree of orderliness one needs to suppose in order to understand the dynamics of systems? If one recognizes with Plato that to discuss nature one needs to suppose it orderly to some degree, the question becomes whether equilibrium, emergent collectivity, and perfect predictability need to be supposed to discuss systems, or something else, perhaps something less. What motivates treating deterministic holism as the main alternative to disorder?

To be discussable, a thing minimally needs to be at least in principle reliably identifiable and roughly distinguishable from its context. Meeting that minimal criterion means already resisting Heraclitus's image of the un-ignorable unity of all things. At least some communities, like those of terrestrial organisms on very remote islands or aquatic organisms in very isolated ponds, meet it easily. No ecologist I am aware of, including Gleason, denies that some communities meet this criterion of orderliness—being roughly distinguishable in such a way that a layperson could find their boundaries. But then, to be the sort of thing scientists can successfully theorize about, a thing's dynamics also need to be patterned or regular or organized to some degree—there must be repetition of phenomena for instance. As I argue in the next section, even Gleason agrees that communities have this sort of order. (If regularity is denied by part of ecology, it is not in Gleasonian ecology, that is.) So, a first step away from chaos requires some minimal distinguishability and regularity. But a thing may be orderly in this way, having regular internal dynamics, and yet be very difficult to engage predictively, much less predict perfectly. Even perfect predictability, though is a long step from having holistic dynamics.

So, preservationists face the question of what kind of order communities need to have to be preservable, beyond just being describable. If communities have real equilibria or emergent properties, those would contribute significantly to the case that they are appropriate objects for conservation. But need they have? That is less clear. Whenever natural selection is operating in communities, at least that force opposes their equilibria. Large-scale climate shifts may work against equilibria, too. But, we do preserve communities with degrees of success whether or not they are in fact equilibrial or have emergent properties. We should determine what and how they need to be for this preservation to be realistic. With both questions in mind—what is required for recognizing and then for preserving communities—I return to the nominally opposite views to find their common ground on this point, along the way considering how they came to be regarded as opposites, and then how that matters to what communities are.

4 CLEMENTS'S AND GLEASON'S ONTOLOGIES

The legend of order and chaos, as expressed by Worster and repeated by others, is supported by the legend of Clements and Gleason, and the former unravels with the latter. By "the legend of Clements and Gleason" I mean (a) the two claims traditionally attributed to Clements—that communities are like organisms, and they develop according to a simple, deterministic law, (b) Gleason's rejection of these claims in his "individualistic view," and (c) the broader narratives their scientific claims have been embedded in, which further the impression of that (a) and (b) are polarized claims. Having discussed what is at state in order and chaos, (c), I turn now to comparing the specific content of Clements's and Gleason's views with the legend about (a) and (b).

Clements's putative law

A foundation of the legend is the law which is supposed to be the backbone of Clementsian order. Eliot [2007] argued that in his mature theory Clements does not assert a deterministic law of vegetation, that "climates beget climaxes." Even so, laws of vegetation *have* been asserted. In 1825, Adolphe Dureau de la Malle claims that the improved success of crops when they are rotated reflects a general law of nature:

> The alternance or alternative succession in the reproduction of plants, especially when one forces them to live in societies, is a general law of nature, a condition essential to their conservation and development. This law applies equally to trees, shrubs, and undershrubs, controls the vegetation of social plants, of artificial and natural prairies, of annual, biennial, or perennial species living socially or even isolated. This theory, the basis of all good agriculture, and reduced to a fact by the proved success of the rotation of crops, is a fundamental law imposed upon vegetation. [Dureau de la Malle, 1825; Clements, 1916]

This "law" is a statement of the idea that often plants modify external conditions in a way that makes those conditions more suitable for plants other than themselves. In other words, Dureau de la Malle has identified in a very general way the chief mechanism driving successional change, the dynamic between plants' physiologies and their habitats. This "law" does not claim that certain plants are everywhere followed by certain other plants. While identifying a facilitation relationship, it does not ground precise predictions. Clements, too, in the appendix of his earliest book, lists several "laws of succession" [Clements, 1904]. But none of them express constant conjunction, much less nomic expectability or necessitation or counterfactual dependence in the way philosophers of science have typically expected scientific laws to. "Law" is being used in a different sense here than any of the usual philosophical ones, to refer to a combination of local mechanisms which do not operate in isolation, and to some non-predictive, non-universal regularities.

So while Clements occasionally makes leading claims which seem to indicate that a simple law is about to be unveiled—like that the relation between habitat and plant "is precisely the relation that exists between cause and effect," and "the essential connection between the habitat and the plant is seen to be absolute"—such comments are significantly tempered by reminders that, for instance, "the habitat is the sum of all the forces or factors present in a given area" [Clements,

1905, pp. 17,18]. Consequently, while one might usefully remark that it is true, in a law-like way, that habitat controls vegetation, this is not yet to commit to anything Gleason would disagree with.

The question of whether Clements's theory can be distinguished from Gleason's through his commitment to ecological laws turns on how Clementsian explanations of vegetation actually proceed, and what they invoke. This most-general law-like association suggests a methodological starting-point rather than (assuming that particular habitats and particular vegetation-types are filled in) the autonomous explanatory apparatus it has been made out to be. It suggests an idealized sequence of development of idealized vegetation best adapted to an area's climate. But it does so in order to help explain actual instances of vegetation (which, he recognizes rarely match idealized conditions) by appeal to departures from idealized conditions. What seems to be a law is the basis of a framework for incorporating diverse local causal factors and using them to explain. Indeed, a 1906 *American Naturalist* review of Clements's 1905 *Research Methods* emphasizes that its significance lay in shifting vegetation science *away* from naïve generalizations and towards investigation of specific causal factors: "This work should do much towards establishing ecology and experimental plant evolution upon a firmer basis by pointing out the need and the method of making absolute determinations of factors, instead of the inaccurate generalizations so often recorded" [Allen, 1906, p. 805].

Later, in Clements's major theoretical work of 1916, *Plant Succession*, the phrase "law of succession" appears twice, with two different meanings. In one instance, Clements writes, "to this fact," that in open associations immigration is inhibited by present occupation, "may be traced the fundamental law of succession that the number of stages is determined largely by the increasing difficulty of invasion as the area becomes stabilized" [Clements, 1916, pp. 77–78]. This is to say that increasing occupation of an area makes invasion by new plants increasingly difficult and that this impediment to invasion affects how many immigrants one actually detects. The statement is a causal generalization, but as what it explains is the stages of vegetation in idealized sequences, it does not yet explain what the theory is designed to explain—what vegetation appears in an area. It is certainly not a climate-begets-climax law.

At another point, while speculating about the history of vegetation, Clements mentions "the basic law of succession that life-forms mark the concomitant development of the habitat and formation stage by stage, and that this development is reflected in the structure of the vegetation" [p. 494]. This is closer to a climate-begets-climax law, but importantly different; it suggests that changes in vegetation track *shifts* in habitat, and it opens a discussion of a range of sources of habitat dynamics. Recognizing this association between changing habitat and changing vegetation helps interpret this next statement, which is closest of all to causal law:

> The habitat is the basic cause, and the community, with its species or floristic, and its phyads and ecads, or physiognomy, the effect. But the effect in its turn modifies the cause, which then produces new effects,

> and so on until the climax formation is reached. A study of the whole process is indispensable to a complete understanding of formations. [p. 123]

An obvious, reasonable interpretation of such claims is that they express a law like Newton's $force = mass \times acceleration$, where if one provides a set of circumstances, one can derive an outcome, and thereby predict and explain that outcome. That is, one could construe these statements as functioning as laws in a hypothetico-deductive system, in Carl Hempel's sense [Hempel, 1966]. Philosophers have produced various other interpretations of scientific laws, too (see, *e.g.*, discussion in [Weinert, 1995; Psillos, 2002]). But whatever conception of laws philosophers favor themselves—empirical regularities, or casual regularities, or inference rules, or unifying axioms of deductive systems or something else—the usual motivation for identifying laws is that they license a *modus ponens* form of explanatory reasoning, deriving some y from some 'if x then y' and some information x. The initial acceleration of a baseball, for instance, can be derived from its mass and the force applied to it, using the $F = ma$ law. Whatever particular conception of law is in use, invoking a law to produce explanations typically requires that what is explained corresponds at least roughly to some such derived y, and that the circumstances employed to explain correspond at least roughly to some such x or x's.

Unlike law-based theories in this most general sense, Clements's explanatory framework associates climates and climaxes through a kind of idealized association *not* normally realized. The association between climate and climax is that the climax is comprised of the species which most successfully outcompete other species in an area over the course of long-term competition, while the climate is the area's long-term average habitat characteristics. The connection between the two lies in the physiological adaptation of the former to the latter, as a consequence of evolution. Yet this framework explains the *changes* in vegetation in terms of *changes* in habitat. It does so by appealing to the ways actual habitat characteristics *deviate* from average conditions, dynamically. Because habitats, as they affect the survival of particular kinds of plants, depart from average conditions in extremely many ways, the idealized association between average conditions and climax vegetation offers a methodological starting point for empirically investigating the effects of certain kinds of deviations *as* deviations. Conceiving habitat conditions as deviations from an idealization is what allows their incorporation into explanatory structure at all.

Unlike a simple climate-begets-climax law, Clements's theory thus attempts to offer explanations by drawing on a whole variety of causes rather than a single one like "climate." It enumerates four general classes of causes as producing vegetation: initial causes, ecesic causes, reactions, and stabilizing causes. Initial causes are those instigating succession, such as a clear-cut or forest fire; they create the possibility of succession by eliminating current vegetation. The particular character of the initial causes affects which plants are able to establish by determining initial habitat features. Ecesic causes are the characteristics of plants affecting their establishment and growth, including the ranges of conditions in which they survive and their adaptations for dispersal and immigration. Reactive causes are the ways in which plants themselves affect habitats for other plants. Stabilizing causes are the features or activities of plants adjusting habitat characteristics in such a way that habitats become unfavorable to new immigrants of other species (*e.g.*, by reducing nutrients or light). Reactive and stabilizing causes are distinguished by what they benefit or harm. If a plant's changing its habitat benefits the plant itself, this change is called stabilizing; otherwise, it is reactive. After initial causes instigate succession, its dynamics are a function of reactive and ecesic causes, until—in stable habitats, anyway—stabilizing causes explain the persistence of particular species.

The myriad causes in these four classes are non-synonymous with 'climate' and can give rise to a range of outcomes, depending on the particulars of habitat and available species. Which is to say, Clements's theory is neither a 'monoclimax' theory, expecting a single outcome of succession, nor a monocausal theory treating vegetation as arising from a single cause. Whether or not 'climate begets climax' should count as a law is partly a function of how inclusive we are willing to be with the term 'law.' But Clements's explanatory strategy is enough different from the way laws are employed, I suggest, that we miss something about its approach to explanation when we call this part of the theory a law in the usual sense connecting one kind of cause to one kind of outcome. Indeed, the accounts recounted above demonstrate that treating the generalization this way has contributed to misunderstandings of the theory as a whole as deterministic.[13]

Clements's loose organism

If, comparing the explanatory and predictive resources employed by Clementsian ecology with those appearing in standard accounts of its simple law, we find them richer than that caricature, attention to the Clementsian account of communities reveals a comparably more complex treatment of them, as well. Fostering confusion, Clements makes claims nurturing the conclusion that he believes in ecological communities every bit as discretely-bounded and functionally-integrated as human bodies. The best-known is the provocative sentence, "as an organism, the formation arises, grows, matures, and dies" [Clements, 1916, p. 16]. Odenbaugh has taken this to suggest a close resemblance to a multicellular organism:

> a community may be a tightly integrated group of species that bear various causal relations among their component species. The community forms an individual, as if it were a multicellular organism. This is a Clementsian community: a group of species that strongly interact with one another [Odenbaugh, 2006, p. 217]

How strongly, then, does Clements intend the comparison to organisms? We can answer this by appraising what work he puts it to. The comparison to organisms

[13]And see [Clements, 1916; Eliot, 2007] for further details.

encompasses four similarities which are of at least heuristic value in analyzing units of vegetation. First, it suggests that unlike vague entities such as seas or clouds, vegetation units have boundaries. Clearly, not all organisms have clean boundaries,[14] but it is a general characteristic of organisms that they exhibit edges between themselves and their surroundings. Second, organisms exhibit patterns of growth and change which are predictable at least in a general way. Third, these roughly predictable patterns can be explained by evolutionary adaptation, or at least by descent with modification. Finally, their component parts demonstrate interdependence acquired through historical adaptation or accommodation to one another.

But Clements never suggests that these similarities between ecological units and individual organisms rise to the level of homologies. Where historical adaptation has produced similar features in unrelated entities (sometimes even by response to similar environments), like producing wings separately in birds, bats, and insects, these are analogous features. Clements's claims to similarity are never stronger than this. "As an organism, the formation arises, grows, matures, and dies" might be read as suggesting that plant formations are themselves organisms. The "as" here can be read as suggesting that formations *are* organisms (that is, it can be read in the "*qua*" sense of "as," meaning, "with respect to its being ...") or as marking a comparative simile. Evidence appears already four sentences later for the latter interpretation, when Clements writes that "the life history of a formation" (here sounding literal) "is a complex but definite process, comparable in its chief features with the life-history of an individual plant." If formations *compare* to individuals in a few chief features, they clearly are not *literally* individuals themselves. And in the next paragraph, Clements remarks that vegetation units differ from individuals in that they are capable of altering their habitats, whereas individuals acting alone cannot do much to alter their habitats (at least at the scale of influencing other populations) [Clements, 1916, pp. 124–125].

Moreover, there are other features Clements attributes to units of vegetation which are not even analogous, much less homologous, to individual organisms. Foremost is the climax concept—the idea that units develop towards a mature state which has the capacity to endure indefinitely unless interfered with. Individual organisms cannot do so, insofar as they always, unless terminated sooner, advance through developmental stages to senescence and death. Though some can persist a very long time, none can endure stably for indefinitely long. Nor in many cases are their later features predictable from before they are born or produced. Given these disanalogies, it is then even more noteworthy that appeal to the organismic analogy offers no resources for explaining community dynamics other than that they can be understood as consequences of adaptive histories. This assertion that evolutionary processes produced each of them hardly rises to the level of homology; it is exactly the relationship shared by evolutionary analogs!

So, if Clements does not intend the organismic metaphor in a strong or literal sense (as I suggest he cannot, given what he attributes to it), why does he open

[14] See the interesting discussion of this point, at length, in [Wilson, 1999].

his discussion of vegetation with it in his 1916 book, and remark again on its significance elsewhere? Part of the answer lies in what he holds a comparison to organisms involves. Joel Hagen, perhaps uniquely attentive to Clements's use of the simile, points out that Clements must have had in mind for comparison a quite simple organism:

> It most certainly was not an organism in the same sense as a vertebrate animal or even a higher plant. What Clements seemed to have in mind as models for the community-organism were much simpler plants and animals, perhaps what we would refer to today as protists. [Hagen, 1992, pp. 22–23]

Suggesting that Clements's implicit object of comparison is simple living structures, Hagen further remarks that such a comparison would have been "quite unremarkable" to biologists at the beginning of the twentieth century. It flags some basic similarities: continuity and growth in size through time, relationships among components, and a physiology, in the sense that the whole adapts to changing circumstances by modifying its components.[15] I depart from Hagen in asserting that while Clements extends physiology to populations, his wholes do not themselves have causal agency. Causal agency explicitly, exclusively lies with plants and habitats. But, following Henry Cowles's research on the Indiana Dunes, Clements attempted to represent vegetation as dynamic, rather than static. Treating vegetation as a consequence of the physiological interactions of constituent plants encouraged him to extend the concept of physiology to cover populations, and then the causal interactions among populations as their competition for resources changes the overall shape of the whole over time. While I do not take physiology to connote holism, nonetheless, as Hagen explains, the simile would have conveyed to his contemporaries that Clements was examining vegetation physiologically and as a dynamic unit able to shift in response to circumstances.

Still, while accepting that the comparison to organisms is not literal and that it is not to complex organisms, one might respond that Clements expresses certain other commitments by invoking it. Even protists distinguish themselves from disordered aggregates in two ways: functional integration and reasonably clear boundaries. Their functional integration is a matter of their exhibiting repeated, complex internal dynamics, and parts whose presence and structure can be explained by their functions in serving the fitness of their organisms (and the structures from which they have evolved). Does not Clements mean by invoking the organism simile that each individual species serves the development of the community towards climax in just this way? Then, while they don't exhibit perfectly sharp boundaries, every organism exchanges matter and energy with its surroundings. Both of these features have been attributed to Clementsian communities.

A first reason to conclude that Clements does not attribute this kind of thick functional integration to the relationship between individuals and communities (or, "seres," specifically, which is to say, communities developing over time) is that

[15]See [Hagen, 1992] for further, useful discussion.

he does not believe that communities in particular areas necessarily have certain species-components. Each area, as defined by environmental conditions that are similar to some degree, has species which have evolved to be adapted to its normal conditions, and these species often alter local conditions (*e.g.*, by changing the nutrient profile of soil and creating shade) in ways facilitating the entrenchment of other species well adapted to those conditions. But this process is in no way inevitable: " In the case of invasion," Clements writes, "it is obvious that the failure of the dominants of a particular stage to reach the area would produce striking disturbances in development. Likewise, the appearance of alien dominants or potential climax species would profoundly affect the usual life-history" [Clements, 1916, p. 33]. Lest one imagine the theory assumes that invasions and disturbances are rare in nature, consider Clements's comment that "unlikeness and variation are universally present in vegetation" [Clements, 1907, p. 289], and further that a primary reason Clements coined his much-lamented, expansive collection of terms for different kinds of vegetation is that each of them reflects variations on idealized, normal vegetation. Each such variation reflects the action of a different kind of disturbing cause, and thus departure from idealized normal vegetation.

There is a second reason to conclude that Clements understands vegetation to have rather less functional integration than even rather simple organisms like protists have. Animals exhibit centralized control, with a central nervous system issuing signals to bodies' components and receiving signals back from them, while seres clearly do not exhibit anything comparable. But even in much simpler organisms, components directly serve one another. In paramecia, vacuoles transport nutrients to lysosomes, fusing with them to accomplish digestion. These components of the organism interact directly. In plant successions, on Clements's view, *all* interactions among plants are indirect. All of them are accomplished through an intermediary, usually a resource. It is by adjusting the water content of soil or the amount of sunlight other organisms encounter in those other media that one plant affects another. Clements calls this driving force of succession "reaction on" an external medium. But since that is the nature of interactions among component species, any two organisms or species are completely intersubstitutable if they can survive in the same conditions and react on conditions in the same way. Expressed casually, a shade-loving fern does not care whether its shade is produced by an oak or a jacaranda, and Clements asserts no other way in which any one plant cares, so to speak, about other plants' identities beyond how they affect habitat. This is not to deny that there are much more specialized and fragile interactions in nature between mutually-adapted species, only to note that Clements's theory never asserts them among plants. His strongest statements of causal relationships do not therefore connote them. It is easy to imagine, if one starts with the organism simile, that Clements *would* assert such direct interactions, but that is to be misled by the metaphor's possible connotations. For him, internal community dynamics are exclusively indirect and identity-independent. This commitment is entirely missed by Richard Levins and Richard Lewontin, for instance, who assert that for Clements, "the behavior of the parts [of a plant community] is wholly subordinated to this abstract principle," and that "Clementsian idealism sees the community as the only causal reality, with the behaviors of individual species populations as the direct consequence of the community's mysterious organizing forces" [Levins and Lewontin, 1985, p. 135].

Yet, the organism comparison implies a degree of functional integration, even if it is loose. I suggest that understanding areas of vegetation as functionally integrated serves as a cornerstone of Clements's strategy for explaining their dynamics. Its implication is that the idealized sere has a defined endpoint, the climax, and that successions are end-directed in this sense. The emergence and entrenchment of climax vegetation require a causal, developmental sequence leading up to it. One can reason towards the idealized developmental sequence for an area only by considering the causal influence plants have on one another through reaction. Since the idealized sequences are essential to this explanatory approach, and identifying causal connections among plants is essential for establishing sequences, explaining requires identifying actual or potential causal connections. As they produce sequences, they are also trivially ends-directed, and functional in that sense. As discussed, Clementsian explanations frequently employ these idealized sequences to explain vegetation departing from them, and in these cases, too, they invoke the causal contributions of plants themselves alongside the causal contributions of environments. In sum, using this approach to explaining vegetation trivially requires discussing causal relationships among plants, and in the idealized sequences with defined end-points, vegetation is treated as aiming towards an abstract end-point. So, vegetation explained this way is treated as trivially functionally integrated. The functional integration is trivial in the sense that while it is necessary for constructing explanations this way, it does not involve the claims that vegetation itself has these functions or end-points, that idealized endpoints ought to emerge, that they necessarily do emerge, that communities are causes, or that causation is top-down.

However, recognizing the functional or organizational looseness of the Clementsian community (and seral sequence) does not help with, and even amplifies, the problem of communities' boundaries. On this point, Clements can seem elusive. On the one hand, he sometimes characterizes formations as potentially "continental in extent" [Clements, 1936, p. 253], but elsewhere, especially in his work on indicators, he isolates areas of developing vegetation as small as "the north side of a rock" [Weaver and Clements, 1938, p. 373]. In practice, at any rate, he identifies areas of vegetation at many, nested scales, not taking their boundaries to be fixed, rigid, or definite. While some organisms are nested, as our intestinal fauna are nested inside us, that communities can be identified at such a range of scales suggests a further disanalogy with organisms. Individual plants participate in communities to the degree that they causally contribute to communities, or are produced by common causes. The strength of plants' and habitats' causal contributions can be used to identify better and worse boundaries for dividing up communities.

Building on this idea, Clements navigates the problem of community-boundaries by rejecting that communities have sharp boundaries, even as they are bounded. He adopts a particular term for the phenomenon of gradation between communities at their edges: "zonation." Zonation is the blending of one population into another at its edges, a phenomenon often taken to have been used by Robert H. Whittaker to refute Clements after he observed it in the Great Smoky Mountains and Siskyou Mountains [Ricklefs, 1997, pp. 507–510, for a textbook example]. What keeps community boundaries from becoming arbitrary where blending occurs is that zonation aligns with particular causes as multiple causes work simultaneously. For instance, soil salinity-levels may correlate with the abundance of one species along one gradient, while soil acidity is correlated with the abundance of another, along a different gradient [Weaver and Clements, 1938, pp. 226–233]. There is order along these boundaries, but it is order tied to underlying causes, not the order of discrete objects and sharp edges. It is considerably different than the sort of order imagined by those starting with higher animals or plants as models for the organism simile. Further inferences about sharp boundaries drawn from the simile are false, and Clementsian ecology in no way denies the kind of visual disorderliness one observes in overlapping populations like Whittaker identifies.

Gleason's order

If Clements's communities turn out, on examination, to be compatible with a much greater degree of observed disorder than is imagined by commentators inferring his theory from his simile, Gleason's explanatory strategies similarly involve a much greater degree of order than is usually imagined. I argue this point in three steps. First, prior to his move away from views resembling Clements's, Gleason's early work supposes a great deal of order in plant succession. Much of the terminology he uses is Clementsian, and his explanatory strategies align with Clements's. As a second step, a few years later in 1917 and 1926, reacting to Clements's major publication of 1916, Gleason offers two sets of criticisms of Clements's approach. But in examining these objections while keeping an eye on the obvious, earlier affinity with Clements they arise from, we find Gleason objecting to fewer Clementsian ideas than is usually supposed. We should not automatically assume that Gleason's views shifted where there is no evidence of him abandoning his earlier views. This recognition attributes a burden of proof to anyone claiming that Gleason moved towards radical views opposite Clements's—a burden which I do not think any evidence bears. As a third step, I note the degree of community order and organization Gleason assumes in his later ecological writings, which further undermines the attribution to him of radical views of the sort inferred from his individualistic concept alone.

Early in his career, three years after his 1906 PhD, Gleason's analysis of vegetation conveys all the orderliness of mature Clementsian ecology. Here, he analyzes the ecology of the Midwestern prairie as a Clementsian sere:

> Within every complex of related plant associations, there are one or

> more definite orders of succession, leading from pioneer to climax associations. The steps in the succession follow each other in a regular series and constitute what may be called a normal succession. [Gleason, 1909, pp. 269–270]

At this early stage of Gleason's career, he treats succession as having a normal order, proceeding from associations of colonizing, pioneer species to associations of entrenched climax species. Associations are structured and bounded. Prairie remnants "still existing along our railroad tracks give only a faint idea of the normal structure of the prairie vegetation" [p. 269]. Not only do these remnant communities have structure, they are recognized as resembling the normal structure for vegetation in that area. Then, as he observes the prairie and the forest butting up against each other in Illinois, he remarks on a "tension zone between the two associations" [p. 270]. That is, the associations are identifiable, and competing. But of course, this sort of analysis precedes Gleason's famous reaction eight years later to Clements's *Plant Succession.*

Even in 1917, right alongside his famous criticisms, Gleason makes some fairly strong gestures towards the nominally-Clementsian pole. For instance, he announces a commitment to the "actual existence," as a matter of observable fact, "of definite units of vegetation" with self-maintaining structure:

> Of the actual existence of definite units of vegetation there is no doubt. That these units have describable structure, that they appear, maintain themselves, and eventually disappear are observable facts. That to each of these phenomena a definite or apparent cause may be assigned is evidenced by almost any piece of recent ecological literature. But the great mass of ecological facts revealed by observation and experiment may be classified in different ways, and from them general principles may be derived which differ widely in their meaning or even in their intelligibility. [Gleason, 1917, p. 464]

The qualification accompanying this declaration of allegiance to units is that there are different ways to make sense of ecological observations. And this idea points towards the further conclusion that Clementsian ecology offers just one way of doing so. Gleason's particular resistance to Clements's way of assembling ecological facts into ecological units crystalizes in two objections:

1. the units of vegetation are dissimilar to organisms;
2. Clements should not enlarge the unit of vegetation to include a climax and the stages leading up to it. [p. 463]

These dismissals do not necessarily reject the causal story about what produces vegetation, but focus rather on the language used to describe it. Of course, the content we assign to these complaints turns on the degree and kind of difference Gleason actually takes there to be between vegetation and organisms, considered below. The second objection explicitly concerns terminology; units of vegetation are not exactly like organisms, and the seres Gleason recognizes in his early work should not be treated as fundamental units for description. The primary unit should be the association (or community, in our terms), rather than the sequence of associations called a sere. Gleason's third and fourth objections function similarly:

3. new terminology like Clements's is not needed for describing succession;
4. Clements excludes apparent exceptions to his generalizations by definition. [Gleason, 1917, p. 463]

The third objection is obviously terminological, decrying Clements's terminological enthusiasm noted earlier. Gleason would later refer in passing to Clements as "an enterprising classicist," and he came to dislike Clements's approach of understanding associations in relation to normal types, where the various kinds of departure were what drew Clements to classifications requiring nomenclature [Gleason, 1936b, p. 41]. The fourth objection concerns Clements's putative law, unpacked above. It reflects a methodological departure, resisting Clements's use of an unfalsifiable idealization as part of his explanations. Neither of these are yet objections to communities' organization, boundaries, or functional integration.

In a second critical essay of 1926, Gleason restates the first of these objections more adamantly, and adds two further criticisms directed at Clements's strategy of associating particular environments with particular vegetation types. In short, he claims that,

5. similar and homogeneous environments exhibit varied vegetation;
6. associations under same name occur in different environments. [Gleason, 1926, p. 17]

Taking the first of these objections first, Gleason in 1917 argues that even if units of vegetation are comparable to organisms, they are merely *like* organisms in these respects. They are not themselves literally organisms. So, the question is, What kinds of similarities are there?

> Various analogies may easily be drawn between a unit of vegetation and an organism, but these analogies are always more apparent than real, and never rise to the rank of homologies. For example, it is obvious that an association may appear on a new area, develop to maturity, and finally disappear, but these phenomena are nowise comparable to the life history of an individual. A spore of *Rhizopus*, for example, given the proper environment, will grow to maturity and reproduce without the presence of any living organism. The first pioneer species of an association, on the other hand, will merely reproduce themselves, and maturity of the association will never be reached unless its other species are also present in a neighboring area. Similar exceptions may be taken to all other analogies between the individual and the association, designed to demonstrate the organic entity of the latter. [p. 465]

Gleason's argument here is that while there are similarities between vegetation units and individual organisms, there are also dissimilarities, like the possibility of developing to maturity in the absence of other individuals. Plant associations require immigration of their components, while individuals can develop on their own, apart from other organisms. Consequently, associations cannot be organismic. This is indeed a disanalogy, though not a deep one. Notably, an individual plant cannot mature in the absence of components it can assimilate, either. The difference is just that, as a matter of scale, these components need not be organisms themselves in the case of the individual.

Gleason's use of this particular dissimilarity to object to Clements's analogy rises to the level of rejecting Clements's theory only if (a) Clements intends vegetation units to *be* organisms, rather than be comparable to them, and (b) treating units of vegetation as organisms makes a difference to explanations and predictions for vegetation. Otherwise, it works not so much a rejection of Clements's theory as a reining in of Clements's excesses in comparing seres to organisms. Yet, this move risks begging the question of how rhetoric can be distinguished from theories. Insofar as the simile is offered as part of its author's conceptualization and presentation of the theory, it is questionable to disentangle them. But, Clements's comparison to organisms does not appear among his theory's resources for explanation and prediction, and it does not determine the properties of ecological units to which it has been applied. Consequently, I propose that Gleason be read, at least on this one point, as meaningfully criticizing the usefulness of the comparison. But to the degree this comparison can be disentangled from explanatory and predictive resources and ontological commitments, it is not yet in itself a rejection of the *theory*.

Much less attention has been devoted in the literature to the other five of these objections Gleason raises to Clements. Beyond the rejection of the climax, objection (2) concerns the enlargement of the unit of vegetation. Gleason's other objections, as I enumerated them, include a denial of the need for new terms to describe succession, arguing that Clements excludes some objections by definition, and complaining that Clements's theory attends inadequately to the consequences of variation. Each of these differences, like Clements's statements which suggest that climaxes obtain deterministically, arise from methodological differences between them over how best to represent causes and disturbing conditions. Clements needs more terms for describing succession than does Gleason, for instance, in part because he enlarges the unit of vegetation; and he does this latter as part of a quite different strategy for handling the complexity which *both* of them recognize in successional systems. As Gleason comments, "the great mass of ecological facts revealed by observation and experiment may be classified in different ways," and Clements adopts a physiologically-normal sequence as starting point for general theory organizing them, while Gleason leans towards probabilism (without developing general theory very far beyond leading suggestions) [Gleason, 1917, p. 464]. Among the observations differently organized into explanatory theory are the content of Gleason's objections (5) and (6). Each ecologist has a way of making sense of these observations that depends on how he assigns normalcy and variation to particular states of affairs. In Clements's case, a physiologically-adapted sequence of vegetation is normal and variations from it are explained as departures due to environmental variation from average. In the probabilistic theory Gleason gestures towards (but does not develop), any probabilities would similarly need to be assigned to habitats chosen as normal. A move away from climax sequences does not alleviate the problem of explaining variation, even for a theory assuming radical individualism. The difficulty of handling variation is merely relegated to the project he does not explore—assigning probabilities. Probabilities can only be assigned theoretically in relation to ranges of conditions treated as normal.

To now reach the third step of the case, the extent of order recognized by the later Gleason, who by this point was increasingly working on taxonomy rather than ecology, is apparent in his ecological papers of 1936. In them he reveals an acceptance of the causal basis of Clementsian order. The strongest basis for Clementsian order lies in reactive causes. Recall that what structure a community has for Clements is produced by the action of four kinds of causes—initial, ecesic, reaction, and stabilizing. Initial and ecesic causes refer simply to environmental conditions creating openings for plants to establish, and the physiological characteristics of plants permitting them to establish, respectively. The action of initial and ecesic causes involves nothing not captured by Gleason's comment, for instance, that "all phenomena of succession depend on the ability of the individual plant to maintain itself and to reproduce its kind" [Gleason, 1927, p. 325]. Reactive causes are those covering the relationships among individual plants—Gleason's supposed denial of which offers the basis for the view that considers vegetation less orderly. Reactions are specifically, for Clements, those influences of individual plants on their environments that change their environments and thus the environments of other resident or immigrant plants. (In the special case where reactive causes favor the plants responsible themselves, like when a plant's offspring survive well in its own shade, reactive causes are called stabilizing causes, the fourth class.) Reaction is the force knitting the Clementsian community together as a unit. It is the class of causation supposedly missing in the Gleasonian Individualistic Community, or it is supposed to have vanishingly weak effects therein. Yet, Gleason points to the same phenomenon:

> Nevertheless, these plants [of different species] have definitely an influence on each other. To select perfectly obvious examples, it is clear that the larger plant affects the light and, though its leaf-fall, the soil environment of the smaller, while the latter intercepts rainwater and reduces the light for seedings of the larger one. The two plants have intersecting spheres of influence; each interferes with the environment of the other.... Intensifying the influence of either plant within its sphere has a direct effect on the life and well-being of the other. It may act either favorably or unfavorably. [Gleason, 1936a, pp. 444–445]

That is the essence of reaction, of the sum of causal influence Clements holds is exerted among plants in an association or community. And indeed, Gleason uses even the terminology of reaction in a different paper from the same year:

> It is probable that every species of plant, no matter whether its individuals are large or small, abundant or few, reacts on its environment in a manner peculiar to itself.... It also seems probable that the joint reaction of the whole population is one of the most important factors in maintaining the uniformity and the equilibrium, and therefore the identity of the association. [Gleason, 1936b, pp. 44–45]

Moreover, here the reactions are part of the causal structure contributing to uniformities and equilibria of vegetation. As reactions function this way, all the classes of causation recognized by Clements are thereby asserted by Gleason, along with their putative effects. If so, the ecologists' differences on communities do not lie in their understandings of causal structure.

To understand the relationship between their theories, consider how you and I might analyze the disappearance of a sand castle on the beach.[16] Looking at one sand castle, we each notice it lightly eroded by the breeze, and then, because it was built especially close to the rising tide, the moment when its foundations are first degraded by a trickle lapping its base, before a large wave suddenly reduces it to a undifferentiated lump and it steadily thereafter declines to flatness under light foamy washes. Imagine that, asked separately by children what happened to their creation, we tell nearly identical stories, narratives involving the same series of causes and changes: roughly, breeze, location, base-erosion, splash, fade. Now further imagine that we offer similarly identical narratives to the children who built their castle up closer to the dunes, which lost its towers almost immediately to the high winds there, but otherwise remained intact surprisingly long into the evening before being trampled by teenagers. Imagine that if our particular stories about these particular castles differ slightly, they do not differ much. We recognize the same series of major causes and effects. But now imagine that we are asked for our theories of sand-castle disappearance. Our approaches for producing general theory at this level are underdetermined by our understandings of particular cases. Our sharing causal understanding is consistent with our adopting very different methods for assembling those classes of causes into generally-applicable theories. You propose to model disappearance as a function of location on the beach relative to water down low and wind up high. I build my model on the idea that disappearance has a range of probabilities aligned with time scales.[17] These approaches are starting points. We each still have to figure out what to say about teenagers. You may regard my approach as hopeless, and vice-versa. But to consider our understandings as opposite, to imagine that we disagree about how sand castles disappear, is to grossly underestimate our shared familiarity with those systems.

[16]This example is original, but a similar one is used to different effect by [Jackson and Pettit, 1992].

[17]These modeling strategies are not supposed to capture features of Clements's and Gleason's, but simply contribute to the point that such strategies are underdetermined by causal understanding.

Yet, one might object that when Clements's and Gleason's theories are described as opposite, what is meant is not differences in the analysis brought particular cases, but differences at the level of generalized theory. Clements's and Gleason's general models of communities are indeed very different. Gleason himself pointed to the theories' relationship in remarking how the "great mass of ecological facts ... may be classified in different ways" [Gleason, 1917, p. 464]. But though they raise occasional differences in emphasis like the degree of influence of water, the ecologists' differences do not lie in causal understanding. Nor do they lie in how they treat the nature of communities as things. Consider the robustness of communities in Gleason's account, and the resemblance to Clements's loose organisms:

> Since the first recognition of the plant community, irrespective of the name applied to it, its cause, or its scope, and continuing to the present day, the individual plant community has always been a geographic unit. It occupies space and has boundaries. Moreover, it exhibits uniformity of structure within the area. Extent, boundary, uniformity: these are the *sine qua non* of every community. [Gleason, 1936a, p. 447]

This combination of factors—extent, boundaries, uniformity—makes the community, to the degree it possesses them, an appropriate entity for a single explanation. "We must admit," Gleason writes, "that a *stand* of vegetation is a concrete entity" [Gleason, 1936a, p. 450]. His understanding of vegetation has it very far from chaos.[18]

5 POLARIZING NARRATIVES

This recognition produces a mystery, however: if Clements's and Gleason's understandings of the causal structure of communities are quite close, then, how did the legend of order and chaos become affixed to them? The purpose of this section is to develop an error theory, as J. L. Mackie has used the term, to mean an explanation for the persistence of mistaken ideas [Mackie, 1977]. We typically acquire our understanding of these scientists' theories from historical accounts written by historians and ecologists. My main error theory is that those have tended to emphasize intellectual inheritance of concepts and usage of terminology, at the expense of considering practices or methodologies, especially representational, predictive, and explanatory strategies. While absorption in language is no sign of bankrupt historiography *per se*, in this case alternate modes of historiography focusing on

[18]Malcolm Nicolson [Nicolson, 1990] and Nicolson and McIntosh [Nicolson and McIntosh, 2002] have developed the best existing analyses of Gleason's views and methodology, especially of Gleason's use of mathematics and how he understood the chanciness he occasionally invoked. That they do not revisit Clements leaves them repeating a stronger opposition between the two than I have argued exists. But Nicolson's is the one account I am familiar with that understands Gleason as not having been committed to radical, disordered individualism. I discovered this work on Gleason after developing the account presented here, and it is a welcome complement, as useful for further inquiry into Gleason as Hagen's articles [Hagen, 1988; Hagen, 1992] are for further inquiry into Clements, despite the differences I have with each author's account.

practice or theory-structure have been rare enough to distort our understanding. Linguistic opposition has become theoretical opposition, beyond mere difference. To observe this, I begin with Worster's own history of ecology volume, and then remark on two other historical accounts and on textbooks. (After developing this error theory, I resume the main line of argument in section 6.)

Worster's moral and political narrative

Worster develops the historical narrative behind the ecologies of order and chaos in *Nature's Economy: A History of Ecological Ideas* [1977, and in a revised edition as of 1994], which lends significant portions of two chapters to Clements and his ideas, and to Gleason only a brief appearance, on a few pages. In his account, Worster refers to Clements's science as "dynamic ecology," and associates dynamic ecology with the thesis that "the climax or adult stage [of a plant association] is the direct offspring of the climate"[19] [Worster, 1994, p. 295]. Following tradition, Worster takes Clements's association of climate with successional development to be coupled with a faith in a fairly strong determinism by climate or the character of a climax community in any place. And he cements that association by pointing to Clements's organismic simile:

> Just as physical maturation into adulthood is programmed into the genes of the child or seed, so the climax community marches toward an automatic, predetermined fate. Only in freakish circumstances does the process bog down at a subclimax level of development, a kind of arrested adolescence. [Worster, 1994, pp. 211–212]

Read as an assertion about communities more generally, this sentence has Worster assigning to Clements's theory of vegetation the view that only in exceptional circumstances does a vegetation fail to achieve climax, and likening the development of a community to the physical growth of a child, furthermore understanding the results of each as inevitable. For Worster, the view that climaxes are determined follows from treating plant communities as complex organisms. Worster writes, "undoubtedly the explanation for Clements' emphasis on the sere and its climax lies in his underlying, almost metaphysical faith that the development of vegetation must resemble the growth process of an individual plant or animal organism" [p. 211]. Worster traces this metaphysical faith back to Clements's interest in Herbert Spencer's *Principles of Biology*, which treats society as a "social organism."

In Worster's estimation, Gleason's criticism is thus aimed at Clements's metaphysical faith in social organisms. It takes specific form in objecting to Clements's

[19]This, however, represents a category mistake, in that while Clements refers to his *branch of science* or approach to ecology as "dynamic ecology," meaning the study of ecological dynamics (and in his case, of changes in vegetation), this branch is distinguishable from any particular thesis advanced as part of his work in that area. McIntosh, for example, describes appropriate usage [McIntosh, 1985, pp. 76–77]. Of course, Clements also appears to have had the ambition to synthesize previous work in this area into a foundational theory, and so he does consider his ideas foundational to this branch of inquiry. But Gleason's work is, if anything is, dynamic ecology, too.

organismic analogy, and then by extension, to his assertion that successions culminate in climaxes. Against Clements's organismic concept, Worster suggests Gleason offered the "'individualistic' view of nature" (beyond, apparently a thesis about vegetation). He rejected "rigidity" and a "formal concept of ecological dynamics" and "precise succession" for a looser account; "organized being[s]" for "haphazard, imperfect, and shifting organization;" and "carefully orchestrated" succession for "accidental groupings." "More important," Worster maintains, "Gleason's 'individualistic' view of nature suggested that the climax was a haphazard, imperfect, and shifting organization—one that man need not worry overly much about disturbing" [Worster, 1994, pp. 238–239].

In Worster's broader narrative, the backers of order and chaos are motivated by their politics, by their views of conservation and technology more than vegetation *per se*. He reaches to find motivations for their views in divided politics, so that their positions not only have implications for conservation, but also have roots in different degrees of enthusiasm for it. Worster treats Gleason's central claims as rejections of the claims he understands as the core of Clements's theory—that plant associations advance towards climaxes, and do so as a feature of resembling organisms. Then, treating Gleason primarily as a critic, he ascribes to him the motive of rejecting what Worster thinks are "the anti-technology implications in the climax ideal":

> There were a number of scientists, too, who found the anti-technology implications in the climax ideal hard to accept. From this objection, as much as from any purely scientific quarrel with Clements, there emerged in the thirties an 'anti-climax' party. Earliest to join issue with Clements on this point was Henry Gleason of the University of Michigan. [Worster, 1994, p. 238]

This first inspiration for objecting stands two steps removed from Clements's theory of succession. It is one step removed because it arises from treating the climax sere as an ideal state, where "ideal" means something close to 'desirable from a human perspective,' as opposed to 'physiologically ideal,' which is clearly Clements's usage [Clements, 1907; Eliot, 2007]. In so far as Clements expresses the implications of his theory for conservation, the achievement of a climax sere in any given area becomes a normative state of affairs—the appropriate and best outcome of vegetative development. In the theory of vegetative succession, however, the climax sere is merely a normal ending-point, in the same way protein-folding has a normal end-point, where the normative connotations borne by "normal" do not involve a claim about what ought to be, and consequently fail to bear in any direct way on technology.

Worster's account of Gleason's motive is further removed from Clements's descriptive theory in that it concerns a potential *implication* of that additional claim about climax seres. Worster treats the climax sere as a normatively-ideal state of vegetation (a state, that is, which *ought* to arise), by appealing to Clements's discussions of conservationism in non-scientific writing. He remarks that this understanding of "Clements's doctrine of the climax as a natural ideal was by now firmly lodged in the national imagination"[Worster, 1994, p. 237]. From this assumption, Worster argues that the Dust Bowl of 1934 came to be understood "in the American mind" as a negative consequence of permitting technology (tractors, combines, plows, etc.) to interfere with the climax communities of the prairie. Gleason is then supposed to have advanced his theory of vegetative succession as a reaction to this inferred consequence of an apparent implication of Clements's theory read (problematically) as normative [pp. 237–239].

However, Gleason's objections to Clements's theory do not mention this distant implication Worster considers his central stimulus. For support, Worster reaches to cite poet and journalist Archibald MacLeish as drawing these implications, in an article for *Fortune* on the American grasslands, though MacLeish refers there to neither scientist [Worster, 1994, p. 454, fn. 23]. MacLeish does conclude that agriculture must conform to local environmental conditions—especially soil—or it cannot succeed, landing on the poetic synecdoche that the plow can produce disaster [MacLeish, 1935]. But if nothing in Gleason contradicts this idea, and it does not appear in the articles Worster cites when referring to anti-technology implications, Gleason also mentions no objections to Clements other than to his descriptive theory of vegetation. While very scientific disagreement is embedded in the human world of scientists' concerns with reputation and ambition, support for the terms of this quarrel being not "purely scientific" is missing.

Tobey's intellectual-inheritance narrative

While Worster sketches the Clements and Gleason debate as a dispute over the politics of conservation and technology played out as a scientific debate over the metaphysics of vegetation, historian Ronald Tobey's *Saving the Prairies: The Life Cycle of the Founding School of American Plant Ecology, 1895–1955* treats the rise and fall of Clementsian vegetation-theory as a Kuhnian "microparadigm." Worster, too, renders Clements and Gleason central players in a paradigm shift, but Tobey works out the Kuhnian dynamics in much greater detail [Worster, 1990, p. 11]. He remarks that he takes a Kuhnian approach out of a desire to avoid the "embarrassing methodological fallacies" of conventional intellectual history, which has "isolated the major ideas of the Clementsian ecological theory and followed their development in Clements's published writings, prefaced by a reconstruction of their precursors and suffixed by their denouement in the hands of his critics" [Tobey, 1981, p. 6]. Though Tobey proposes that his alternative method of attending more to how many times scientists cite one another and to "the relationship between ideas and the social and material structures," can move his analysis beyond the "debilitating flaws in the history of ideas" into "the bracing wind of rigor and procedure," he directs relatively little attention to Clements's published writings, to the detriment of his account of them [p. 6].

Tobey locates Clements's work as essentially in competition not with Gleasonian individualism but with ideas from the University of Chicago school of H. C.

Cowles, and (after denouncing conventional intellectual history) at the intersection of various intellectual traditions:

> By the end of the nineteenth century, two distinct approaches of explanation for vegetational change competed for advocates in the United States. One approach, which was to lead to Frederic Clements's mature work, *Plant Succession* (1916), was centered at the University of Nebraska and was the result of a formalization of the experience of Bessey's students by the theory of Oscar Drude and Clements's reading in sociology. [Tobey, 1981, p. 108]

Tobey understands Clements's account of vegetation as essentially an inheritance of diverse influences from his most important teacher at Nebraska, Charles E. Bessey. One of these legacies was Bessey's "pragmatism," closely resembling C. S. Peirce's and emphasizing direct experience with what is described over starting from known categories and assigning names. Such an approach arose in opposition to the German *Naturphilosophie* tradition unfolding from Goethe to Julian Sachs. But the origin of Clements's "organistic metaphors" is, for Tobey, Clements's possible reading of Herbert Spencer and Bessey's affinity with sociologist Lester F. Ward's liberalism. Clements apparently also drew from Bessey the "idealistic tradition" in botany giving ontological status to vegetative formations, inherited ultimately from the European Floristics tradition of Alexander von Humboldt, via Oscar Drude's plant geography. Meanwhile, he was also influenced by its opposite, a reductionist "mechanistic tradition" which denied the reality of vegetative units and was derived primarily from Darwin. Beyond these intellectual bequests from Bessey, Clements's major contemporary influence lay in competition with Cowles, whose "model was built upon a philosophical approach to vegetation quite distinct from that of the Nebraska scientists"—one derived from Danish botanist Eugenius Warming [p. 111]. While Clements "explored [the grasslands] in terms of climatic formations," the Warming-Cowles approach did so "in terms of topographical and biological habitats" [p. 110].

Yet, Clements's explanatory framework incorporates *both* the climatic formation and the topographical habitat. It attempts to assert an ontological framework to hang vegetation on, but does so in order to connect it, in a reductionist manner, with local, individual causes. That is, one can trace these themes in Clements's writing, but they offer little insight into scientific strategies. Tobey presents Clements's understanding of the vegetative formation as a combination of "two conceptions ...—organism and population—[which] implied quite contradictory conceptions of development." The "organismic" concept "implied that development was caused by a major external cause"—climate—while the "population" concept "implied strong habitat influence on the development of vegetation," making climate "of secondary importance" [Tobey, 1981, p. 80]. If Clements's theory can be read as bearing traces of these ideas, this usage of "habitat" (in opposition to climate, rather than incorporating climate) bears no resemblance to Clements's own. As Tobey writes, "according to Clements, eight major physical factors controlled habitat conditions: water content of the soil, humidity of the air, light intensity, air and soil temperature, precipitation, wind, soil class ... and ground physiography" [Tobey, 1981, p. 72]. At least the majority of these factors are functions of climate; all are features of climate at least on some scales.

With respect to the development of the formation through time—which is to say, succession—Tobey again discovers intersecting concepts at the heart of Clements's approach: "Clements's vision was fundamentally ambiguous, expressing both an idealistic interpretation of growth and its contrary, a mechanistic model of change." But then also, "Clements's conception of change was typological ... by providing a schematic growth in terms of distinct categories or types of being, with succession as the change of one type into another" [p. 80]. Clements may again be read as heir to these traditions, but to do so means focusing misleadingly on his language. His "vision" is ambiguous only because rendered as a conflict of inherited concepts. Absent this story of inheritance, the confusion falls away.

So, beyond just mentioning the concepts Clements incorporates, Tobey also attributes some empirical claims to his theory by calling it a "monocausal (climatic) theory of formations," and imparting to this one cause deterministic influence over vegetative change [Tobey, 1981, p. 103]. He writes, "in Clements's *Structure* ... a formation *had* to develop as the terminal climax of succession" [p. 105], and that "development of vegetation towards its terminal climatic formations was always progressive. It could not permanently regress or stall short of the final form" [p. 82]. So, "an essential principle of Clementsian theory was that every succession was headed toward a climatically caused monoclimax" [p. 104]. Neither the term nor its associated concept, "monoclimax," belong to Clements. "Monoclimax" has been used subsequently by others to mean the idea that one, and only one, particular climax-type is determined in some area. And yet, Clements never asserts this.[20]

So with respect to the metaphysics of communities, Tobey, like Worster, treats Clements's organismic simile as central to his explanation: "adoption of the organismic model was not a matter of heuristic convenience for Clements, as it would become for A. G. Tansley, who in 1931 referred to the 'quasi-organism.' For Clements, the formation was ontologically as real as the individual plant or animal" [Tobey, 1981, p. 81]. This is a *non sequitur*; however much the moon might be "as ontologically real as" cheese, this does not entail literalism in their comparison. Whether the organismic simile is literal or not, the passages from Clements which Tobey immediately cites as revealing Clements's realistic (as opposed to pragmatic) usage of "organism" in simile and metaphor do not indicate one usage or the other. Tobey offers:

> Hence: '[Succession] is the basic organic process of vegetation, which results in the adult or final form of this complex organism.' And: 'All the stages which precede the climax are stages of growth.' As he stated in the second sentence of *Plant Succession*, 'As an organism the

[20]See the argument against this misattribution in [Eliot, 2007, pp. 94–97].

> formation arises, grows, matures, and dies.' [Tobey, 1981, pp. 81–82]

Yet, the first and second quotations employ "organism" and associated terms metaphorically, and the third employs it as simile. Leaving the question open, neither wears on its sleeve the strength with which its comparison is intended.

Like Worster, Tobey finds little role for Gleason's alternative views in his account of the rise and fall of Clements's. Partly, this is because he considers Oxford ecologist Tansley Clements's more significant critic [Tobey, 1981, pp. 155–190]. But mostly, Tobey believes that the Clementsian "microparadigm" fell not to criticism, but to the "theoretical exhaustion of [its] intended paradigm examples," in the mold of Wolfgang Stegmüller's account of theoretical decline [pp. 216–219]. Specifically, Clements and Clementsian ecologists relied most strongly on the North American grasslands as examples of succession, but the apparent climax community of the grasslands was replaced by other organisms like *Opuntia* cacti after facing the Great Drought of 1933–1941. Without access to this paradigm example, newer ecologists were not convinced of the theory built from it, and the ascendant "range management literature" of 1947–1955 ceased to cite Clements and his allies.

Tobey does briefly treat Gleason's 1917 and 1926 essays, and considers the "key proposition" of the 1917 article to be the rejection of the plant association's status as organic entity, which Tobey analyzes as untenable. But then, he attributes a radical ontology to Gleason:

> In Gleason's universe, therefore, there were only individual organisms (and, presumably, physical objects). This position was philosophically untenable, as any nineteenth-century idealistic philosopher could quickly have shown, but Gleason was no more a professional philosopher than Clements or Tansley, and whistled his tune, oblivious to the cemetery of buried doctrines similar to his. [pp. 170–171]

If this is an argument that an eliminativist metaphysics is untenable, it depends on one's assent to core ideas of German Idealism, ideas neither then nor now standard issue for scientists. Gleason *could* maintain that there are only individual objects and organisms; the more interesting question is what kind of ecology can be done on that basis. Tobey takes Gleason's theory to suffer from other metaphysical difficulties, suggesting intriguingly that Gleason's later commitment to the existence of species shares ontological problems identical to those threatening Clements and Tansley. But we do not learn more of the substance of Gleason's ecology, probably because Tobey considers it internal to the Clementsian microparadigm, if nonetheless critical of it.[21] So, Clements appears in Tobey as essentially sponsor of an organismic representation of vegetative processes; Gleason appears as essentially its co-paradigmatic detractor. Gleason's three other central criticisms from 1917 are disregarded.

[21] "Although one internal critic, H. A. Gleason was highly cited ..." [Tobey, 1981, p. 140].

Tobey's construction of the Clements/Gleason opposition thus arises from several quirks of his attention to it. He attends primarily to the authors' inheritances of terminology without heeding the terms' specific meanings and roles in their new theoretical contexts. Clements and Gleason patently employ contrasting language, but that state of affairs is something quite different from their having opposite causal understandings of the systems they both study. Tobey reveals little of their understandings, though one is left with the impression of familiarity with the theories via their language.

McIntosh's conflicting-concepts narrative

Ecologist Robert P. McIntosh also characterizes the theoretical differences between Clements and Gleason in his retrospective discussion of Gleason's career, "H. A. Gleason—'Individualistic Ecologist' 1882–1975: His contributions to ecological theory." Though Gleason, rather than Clements, is therefore his central subject, McIntosh characterizes their relationship along much the same lines as Worster and Tobey do.

McIntosh treats Clements as primarily endorsing "the rather extreme position that the successional development of a community is comparable to the development of an individual organism" [McIntosh, 1975, p. 259]. And again, the development of vegetation within a vegetative community is deterministic, with an inescapable terminus fixed by climate: "A key element of Clements' concept of vegetation was that succession was always progressive to a single climax association under the control of the regional climate." In contrast, Gleason is treated as dissenting from this rigid determinism. McIntosh writes, "Gleason, along with W. S. Cooper and others, dissented from the rigid Clementsian concepts of succession.... Thus, he clearly came out against the monoclimax concept proposed by Clements and endorsed a much less rigid view" [p. 255]. The insight behind Gleason's rejection of Clements's account of vegetation was that he "was, more than most of his contemporaries, impressed by the heterogeneity and variation of vegetation both in space and time" [p. 261]. Clements, endorsing the view that vegetation must always progress towards a single climax type, along a determined path, was less sensitive to the reality of change in nature:

> Gleason wrote that as early as 1908 he became convinced that succession could be retrogressive, and that the Clementsian concept of succession, as an irreversible trend leading to the climax, was untenable. He, of course, allowed that succession was influenced by climatic change, while Clements presumed stable climatic conditions. [p. 255]

This is much the same account of their differences that we find in Worster and Tobey, offering undefended attributions.

Even in this retrospective account of his career contributions, Gleason's theorizing appears mostly as criticism of a Clementsian ecology committed to universal, fixed determinism: "Gleason contributed little in the way of detailed studies of succession, but his consideration of succession effectively resisted too rigid a formalization, and his early ecological instincts appear sound, even conventional, by today's hindsight" [McIntosh, 1975, p. 256]. Again, the criticism is of rigid formalization. The positive thesis McIntosh finds among Gleason's "avowedly heretical ideas" amounts to an individualistic explanation of vegetation [p. 261]. McIntosh claims, "Gleason's most significant and most lasting contribution to ecology was his 'individualistic concept.' It persists in the current research literature and recent textbooks of ecology as one of the basic tenets of modern ecology, although it earned him little credit when he propounded it" [p. 258]. The individualistic concept suggests that variable conditions, and varying sets of organisms will produce differing developmental sequences for different areas of vegetation. The individuality of each such process, arising as a consequence of the idiosyncratic activities of its individual components, creates variation in possible outcomes beyond what Clements recognized. McIntosh writes, "each area, he said, is a resultant of a unique mixture of migrants, environment, and historical sequence, and there are no grounds for recognizing one as normal and typical" [p. 261]. This attributes, I believe, two theses to Gleason—that no association is normal to any area, and that associations do not have recognizable identities.

McIntosh does, however, offer the more novel insight that Clements's and Gleason's accounts of vegetation reflect their different modes of thinking, and that this is even perhaps their key difference. So, quoting A. O. Lovejoy's comment from a different context, he summarizes, "there are not many differences in mental habit more significant than that between the habit of thinking in discrete, well-defined class concepts and that of thinking in terms of continuity ..." [McIntosh, 1975, p. 270]. I think that, after characterizing Gleason as more sensitive to variation than Clements, McIntosh emphasizes their differences more in terms of habits of thought than causal or ontological structure, because he has recognized earlier in the article, if more quietly, the closer relationship than his account overall admits between Clements's and Gleason's views. For instance, he observes that Gleason, while emphasizing variation, accounts for uniformity of vegetation (where it appears) as a consequence of the actions of similar causes:

> Under the individualistic concept, [Gleason] said, the fundamental idea is 'the visible expression, through the juxtaposition of individuals, of the same or different species and either with or without mutual influence, of the result of causes in continuous operation.' He noted that similar juxtaposition of plants is simply due to the similarity in contributing causes. [pp. 255–256]

But if so, the difference between Clements and Gleason becomes much more a matter of emphasis than McIntosh concludes. And further, Gleason's account is shown to treat one and the same phenomena under different terms mostly as a consequence of constructing the boundaries of the successional process differently:

> Gleason's view of succession between vegetational provinces had its counterpart in Clements's concept of the clisere; and he differed from

> Clements in including interformational sequences as successional, whereas Clements regarded succession as proceeding to a climax determined by stable climatic conditions. [pp. 255–256]

Clements at least in this instance appears to be equally aware of variation in vegetation, but to have accounted for that variation differently. And this, again, reveals weaker difference between their accounts than McIntosh's conclusions indicate.

McIntosh presents another account of Clements a decade later, in *The Background of Ecology: Concept and Theory* [1985]. Here his depiction of Clementsian ecology is more sophisticated, to the degree that it avoids explicitly attributing "monoclimax" determinism, and treats Clements as primarily concerned with measuring and representing change. Rather than emphasizing a deterministic association of climate with climax, he notes Clements's statement of a "'universal law,' that 'all bare places give rise to new communities except those which present the most extreme conditions of water, temperature, light or soil.'" McIntosh treats this generalization not as a deterministic law but as a *ceteris paribus* law: "In either case, 'except' could be followed by the phrase 'where it does not' with equal validity" [McIntosh, 1985, p. 79]. This is at least an account of the generalization closer to Clements than we find elsewhere.

But note that this generalization is not the same one that others attribute as deterministic to Clements—that being a deterministic connection between climate and climax—as it predicts for bare places *some* new community, and not any particular one (*e.g.*, the expected climax). On this second matter, McIntosh describes the cause and effect relationship Clements asserts for habitat and climax, but immediately suggests that for Clements, "the 'historical fact'" is also explanatorily significant, remarking even that "Clements was the most explicitly philosophical and historical thinker of the early plant ecologists" [McIntosh, 1985, p. 78]. McIntosh includes more about Clements, some of which follows Tobey, but to mention just one of his other observations, he emphasizes like others that "the essence of Clementsian theory of vegetation was that the plant formation was a 'complex organism' and, like an individual organism, it changed not in haphazard ways but by progressive development" [p. 80]. And McIntosh thus fills out the organismic concept as an assertion of deterministic development, remarking that "later ecologists sometimes seized on *development* to avoid the presumably rigid deterministic connotations of *succession*" [p. 80]. Interestingly, McIntosh also believes that "the entire premise of Clements's dynamic ecology was that the 'seral stages' of a series of populations or groups of populations followed in sequence," an idea which appears in textbooks as Clements's central testable claim [p. 82].

McIntosh's characterization as a whole (which includes more than I have included here), is closer to Clements than most others except Hagen [1992]. More than others, however, it is not so much an integrated account of his theory as, in the style of a scientist surveying literature, a catalog of various ideas attributed to Clements. Accordingly, McIntosh does not illustrate any reasoning logically connecting the idea that seral stages following in sequence is "the entire premise" of Clementsian ecology, for instance, with the others he attributes to Clements.

He does not, for instance, connect it to the suggestion that the formation is an organism, or to Clements's assertion of climax, or with the treatment of his quadrat method, or with other points. So while McIntosh's account is somewhat more accurate than others, this may be in part because it does not attempt to provide a unified characterization of Clements's approach to representing vegetation so much as a few disparate observations of it. In summary, then, while McIntosh observes that Clements includes a significant historical element in his explanations, he still describes Clementsian succession as essentially deterministic, and this as a function of its organismic structure. What is emphasized is Clements's unusual language, and speculation about what metaphysics it invokes; what is missing is Clements's reasoning behind it, such as would help make sense of the terminology's role.

Textbook narratives

Ecology textbooks have offered similar narratives. Michael Begon *et al.* treat Clements and Gleason under the heading "The Problem of Boundaries in Community Ecology," and characterize Clementsian ecology with the organismic simile: "Clements (1916) conceived of the community as a sort of *superorganism* whose member species were tightly bound together both now and in their common evolutionary history. Thus individuals, populations and communities bore a relationship to each other which resembled that between cells, tissues, and organisms." Gleason's contribution is thus of course identified in contrast as "the *individualistic* concept," remarking that he "saw the relationship of coexisting species as simply the results of similarities in their requirements and tolerances (and partly the result of chance)." They attribute to the individualistic concept the implications (rather than the justification) that community boundaries may not be distinct, and that "associations of species would be much less predictable than one would expect from the superorganism concept" [Begon *et al.*, 1990, p. 627]. Clements is mentioned in one other segment, in association with his "rather extreme *monoclimax* theory." They write that "Clements argued that there was only one true climax in any given climatic region," and characterize it as an extreme view by virtue of both endorsing the existence of the climax and considering a single type strongly determined [Begon *et al.*, 1990, p. 646, italics original].

In another popular ecology textbook, Robert Ricklefs lends significantly more space to each ecologist. He devotes a section to "the holistic concept" and "the individualistic concept" of community structure, where the former suggests that parts cannot be understood independently of the whole, and that "a community is much more than the sum of its individual parts" [Ricklefs, 1997, p. 500]. But interestingly, he leaves these views unattributed, though Clements and Gleason are the first scientists mentioned afterwards, three pages later. There, Ricklefs treats Clements as "the most influential advocate of the organismal viewpoint," understanding communities "as discrete units with sharp boundaries and a unique organization [sic]." Gleason, he notes, rejects these claims and suggests that a community is not "a distinct unit like an organism." Ricklefs ties, then, the organismal view and its rejection to the open and closed views of community organization; the organismic view suggests that "the ecological limits of distribution of each species will coincide with the distribution of the community as a whole," which is to say, closed community structure. Gleason's view suggests the opposite [p. 503]. Ricklefs also treats Clements in association with "the concept of climax as an organism," even quoting the first paragraph of Clements (1916) which mentions both the organismic simile and the climax formation, before tying Clements to "the monoclimax theory" [pp. 529–530, 538]. Finally, as I mentioned earlier, he sets up R. H. Whittaker's recognition of communities grading into one another against Clements's account of communities as having discrete boundaries.

If the positions comprising this episode have been inaccurately radicalized to polar opposition, despite Gleason supposing a kind of order and Clements recognizing a degree of disorder, I suggest the polarization has been a function of the prevailing historiographic approach taken towards the episode. This approach has paid attention to the theories' language at the most general level, then interpreted their explanatory approaches by speculating about the language's connotations and linking it to intellectual traditions which employ similar language and concepts. It has inferred explanatory approach from similes rather than causal claims. Insofar as metaphors can contribute to understanding, nothing is intrinsically wrong with such historiography, but it can seriously mislead when employed to make inferences about explanatory strategy where actual strategy is ignored. This is a general danger for science studies, one reflected in egregious conclusions drawn elsewhere, too.

But it is not just historians' attractions to language and metaphor which can foster such distortion. Philosophers approaching scientific episodes have frequently approached them with an eagerness for conceptual analysis, foregrounding terminology. If philosophers have occasionally made contributions to science by clarifying concepts and revealing confusions, their disposition for linguistic analysis has also contributed to concepts themselves being centerpieces of theories alongside laws [Hempel, 1966, is just one of many endorsing such foregrounding]. Where philosophers have repeated the polarizing narrative, it may be in part a function of this habit.

Because scientists themselves have also treated these ecologists as opposites, the preceding discussion of narratives supports an error theory. Another contributor to the mistake besides that historiographic tendency is the typical rhetorical polarization of debates in the theoretical disciplines, as repeatedly illustrated by Sharon Kingsland in *Modeling Nature* [Kingsland, 1995]. But if rhetorical and linguistic polarization in the debate has contributed to the opposition legend, it also obscures the full range of intermediate positions held during the period. Though Tobey characterizes the alliances among ecologists in terms of adherence to paradigm, ecologists continued through the early twentieth century to lament the diversity of nomenclature which reflected diverse understandings, like Moss did at its opening.

6 MULTIPLE COMMUNITIES

Now, with the error theory in hand, I turn to considering the implications for philosophy of ecology of recognizing the significant common ground between Clements and Gleason. This essay has not argued for or against their views, nor for nor against the existence of communities, but has sought to contribute to the discussion of *criteria* for community-recognition. At varying distances in the background has been a motivation for developing such criteria: the question of whether communities are preservable. While philosopher Kim Sterelny has discussed community criteria independently of the preservation question, Jay Odenbaugh and Kristin Shrader-Frechette and Earl McCoy, at least, have brought preservation to the foreground [Shrader-Frechette and McCoy, 1993; Shrader-Frechette and McCoy, 1994; Sterelny, 2006; Odenbaugh, 2007].

Recall Worster's Clements-derived features of communities discussed in section 3: having equilibrium, being perfectly predictable, and having holistic dynamics including emergent collectivity.[22] Callicott similarly points to Leopold's Clementsian vision of communities as stable super-organisms. If they were to exist, Clementsian communities, in the various senses of the term, would be preservable. But what has been packed into being "Clementsian" for a community in the accounts just described makes them more complex than we have evidence for any community being, and more than almost any ecologist asserts.[23] In the line used as this essay's epigraph, E. Lucy Braun, prominent American botanist and president of the Ecological Society of America during the 1950s, remarked: "no serious student of succession (a process) has ever claimed that a succession is made up of 'discrete units'" [Braun, 1958].[24] Even the strongest proponent of communities as holistically-organized units, South African ecologist John Phillips, neglects to argue for sharp boundaries in space or time [Phillips, 1935b; Phillips, 1935a; Phillips, 1931]. In a series of discussions of the organism concept, Phillips argues for—or at least describes and asserts—emergent holism, a view that communities are wholes with properties independent of their parts. His argument for this lies in the unpredictability of properties of wholes from the properties of their parts:

> Very briefly and generally stated, it is the view of the authors and disciples of this concept that there is a creative synthesis and emergence of properties, structures, forms, stages or levels; such newness, springing from the interaction, interrelation, integration and organisation

[22] Relatedly, but much more concretely and modestly, Sterelny offers as a criterion possession of "causally salient, functional properties," specifically top-down causal dynamics, such communities themselves "play a role in determining the presence, abundance, and fate of the populations out of which they are composed" [Sterelny, 2006, pp. 216, 217].

[23] That is, there may or may not be communities with internal dynamics of the sort I have attributed to Clements's communities; it is unlikely that there are communities with the internal dynamics typically attributed to Clements's communities.

[24] One should be concerned that this not be a tautology instantiating the No True Scotsman fallacy, ruling out a class of scientists believing in discrete units as "unserious students," but it is clear in the article cited that Braun regards Clements, at least, as, though wrong, very much a "serious student of succession."

> for qualities—whether these be inorganic, organic, or psychic—could not be predicted from the *sum* of the particular qualities or kinds of qualities concerned; integration of the qualities thus results in the development of a whole different from, unpredictable from, their mere summation. [Phillips, 1935b, pp. 489–490]

Yet, oddly Phillips leaves the key premise that there are such unpredictable properties undefended. His three-part article is concerned with "an analysis of concepts," and examination "of the views of certain workers" rather than contributing significantly to their defense [p. 494]. In the analysis of concepts, however, what does become clear is that this emergence is a function of holism. That is, it is not the parts of a community which work together to produce emergent phenomena; instead, wholes themselves are causally efficacious (or, at a further level of abstraction which makes sense of an explicit reference to Plato, the *holism* itself is): *e.g.*, "it should be plain that [emergence, holism, and the complex organism] are inherently related: holism the causal factor: emergence arising from this factor: the complex organism an integration of emergents, of wholes of potential development, to a yet more efficient whole" [Phillips, 1935b, p. 494]. Here is Phillips's most vivid difference with Clements. For the latter, wholes are not the cause of successions. Instead, the four classes of causes linked to plants themselves, as discussed above—initial, ecesic, reactive, and stabilizing—produce successions.

Significantly, Phillips does not attempt to defend these views from empirical results, nor from their being needed to make sense of empirical results. Nor does he attempt to employ them to explain biological systems. Instead of being motivated by research, Phillips's endorsements appear to have been motivated by an antecedently adopted philosophical holism. Noting the philosophical influence on Phillips of South African politician General Jan Smuts, historian Peder Anker describes Phillips as "having fashioned himself as a follower of Smuts's philosophy of holism," and then having "transferred Smuts's theory of the evolution of personalities and wholes into the natural world" [Anker, 2001, p. 134]. Phillips described his "holistic attitude" as revealing in a "spiritual experience" "the greater truth that ecology is an attitude towards facts and their meaning" [Anker, 2001, p. 146]. The terminology does not appear accidental when Phillips refers above to advocates of community-holism as "disciples of this concept." That is, the one prominent example of an ecologist endorsing the sort of view typically attributed to Clements does not involve arguments from evidence (in contrast to the substantive methodological program Clements pursued).[25]

[25]If it complicates the wedge just driven between Phillips and Clements somewhat to note that Clements expresses enthusiasm for Phillips's discussion of climax vegetation and the complex organism concept—"This characterization has recently been annotated and confirmed by Phillips' masterly discussion of climax and complex organism, as cited above, a treatise that should be read and digested by everyone interested in the field of dynamic ecology and its wide applications" [Clements, 1936, p. 262]—that wedge is robust so far as Phillips and Clements are bound by shared philosophical orientation rather than anything resembling shared causal and explanatory theory. Phillips developed none, and his holism never seriously influenced explanation in community ecology, even if Clements expressed enthusiasm for Phillips as an ally.

Consequently, if ecologists, or at least the mainstream of ecologists working towards causal theories of vegetation, did not even in this primitive period of ecology understand communities as discrete units governed from the top down, it is foolish to establish a criterion for conservation to meet which demands that. Such a criterion would extend past what even the scientists describing communities have thought of them as meeting. In trying to understand what parts of nature are preservable, we should not establish demanding requirements for ecological entities without good arguments for those requirements. As "community" in ecology's sense is a term developed by scientists, the question of whether one of them is preservable turns on fixing what one is or would be, for scientists.[26]

In his own conservation writings of the late 1930s, Clements notably does not accept that conservation can be undertaken only if or because communities are discrete, holistic entities [Clements, 1949a; Clements, 1949c; Clements, 1949b]. He repeatedly stresses that the relevance of his theory to conservation lies instead in its predictive power. While the theory he develops in *Plant Succession* does not make falsifiable predictions about what plants will appear where (because it recognizes the commonness of disturbance at a variety of scales and the possibility that best-adapted plants are not present), it provides a foundation for the rarely-discussed work in *Plant Indicators*. That volume develops strategies for drawing inferences from known relationships between climate/habitat and physiology, together with observed vegetation [Clements, 1920]. This work suggests how one can infer predictions about past vegetation and climatic conditions from present observations. One can use this data to make future predictions, too. For example, sowing seeds of desirable plants on dry fields will often not restore those plants to an area, as when they are late-successional species they will tend to be out-competed by early-successional plants attempting to colonize them. One can learn, on his view, by observing both disturbed and less-disturbed instances of vegetation what *sequences* of species are capable of thriving in the area, physiologically. And this investigation will not always arrive at particular native species of plants, either. For landscape restoration, non-native species may even be useful for aesthetic reasons or for creating the conditions under which desirable species (sometimes non-native) grow best [Clements, 1949c, p. 276]. Through his preference for native species, Clements remarks that

> while a natural treatment presupposes the use of species and communities in the regional association or faciation, it also permits modification and enhancement consistent within its limits. The process of succession by which nature reclothes bare areas is to be utilized as the chief tool in landscaping, but the process is often to be hastened or telescoped to secure more rapid and varied results (Plate 70). [p. 276]

The adjacent photographic Plate 70 brings the point home by depicting non-native

[26]There is more to this project than I take up here, as I have focused on early community ecologists' understanding of these groupings. While their concepts have enduring currency, as I take up below, this analysis should complement further investigation of contemporary usage.

"tamarisks planted along highway for ornamentation and shade."

None of this supposes that communities are holistic entities. Highway roadsides are prime areas for ecological restoration in Clements's view, for instance, despite amounting to artificial swaths through landscapes. These human-planted, arbitrarily located and bound spaces invite conservation, that is, subject just to the lesson that plants which cannot outcompete others in local conditions will not remain there long, and will frustrate a restorer. The sense in which communities are preservable or restorable in these spaces (as opposed to component species *per se* being preservable) arises from the facilitative and inhibitory dynamics which take place among species living in the same area. Further, under most natural circumstances late-successional species, for instance, cannot thrive without being preceded by earlier-successional organisms, so that many plants can only be preserved under conditions normally contributed by other plants. This requirement is not due to magical, holistic connections among native species: non-native species are entirely substitutable for native ones so long as they are sufficiently physiologically similar to change abiotic habitat conditions for their neighbors in similar ways. That is, for Clements, plants can be restored or preserved only when and because their dynamics are preserved. Other features traditionally attributed to holistic communities are not presumed in his conservation writings, though he employs that terminology in them at times to indicate interactions.

Stripped of scientific respectability, causal holism recedes in plausibility as a necessary criterion for community-recognition. At least, a burden of justification is thus imposed on defending that criterion. Absent such holism, we are left with causal interactions among plants and their environments as a neutral starting-point. This returns us to the question of criteria for communities. Jay Odenbaugh has recently defended realism about communities by appealing to them [Odenbaugh, 2007]. He sets out three community concepts of increasing strength, assigning them in turn to the three ecologists, Gleason, Hutchinson, and Clements. In his scheme, a Gleasonian community is "a group of species in a particular area and time" [p. 631]. It is distinguished from stronger community concepts by not requiring that to form a community, constituent populations need to interact causally. Immediately on introducing it, however, Odenbaugh argues that the Gleasonian community concept faces an "$(n+1)$th problem." That is, under this concept it is indeterminate for any collection of n populations (of more than one species) in an area, whether an additional population is part of the community. "Particular area and time" is loosely defined enough to include any given area and time that can be specified, and the concept offers no further criterion determining whether any additional population is part of the community. The $(n+1)$th problem becomes a difficulty for the Gleasonian community concept because it permits any given assemblage of populations to count as a community [p. 632]. It is too inclusive.

Odenbaugh recommends that the $(n+1)$th problem is avoided by rejecting the Gleasonian concept and adopting one of two stronger community concepts. The weaker of the two alternatives is his Hutchinsonian community concept, named for ecologist G. Evelyn Hutchinson. Odenbaugh's Hutchinsonian community is "a group of species that at least weakly interact with one another and not others at a time and through time" [Odenbaugh, 2007, p. 633]. It adds weak interaction to the Gleasonian community. Then, Odenbaugh's stronger alternative is the Clementsian community. This is "a group of species that strongly interact with one another at a time and through time" [p. 633]. Odenbaugh's Clementsian community differs from his Hutchinsonian community just in requiring "strong" rather than "weak" interactions. Unfortunately, strong and weak interactions are not further characterized. Surely causal influences come in degrees, and some are stronger than others. But since for Clements all interactions among plants are indirect and intersubstitutable, there is little conceptual space to articulate a Hutchinsonian concept weaker than that. If one takes that to be the Hutchinsonian concept, and makes Clements's stronger, one is invoking a criterion without basis in predictive and explanatory ecological theory.

Whether via the Hutchinsonian or Clementsian community, Odenbaugh takes interactions among populations (whether they are weak or strong interactions) to answer the $(n+1)$th problem. However, interactions on their own do not solve it. Few organisms on earth, if any, live without becoming benefactors or beneficiaries of habitat modification for or by organisms of other species. Thus, interactions alone do not offer a basis for differentiating particular communities from the global community of all organisms (or nearly all organisms, if there are some extremophile species, for instance, which live and die independently of others). An appeal to interactions does prevent a population having no causal interactions with others in a community from being part of it, from being a "+1" for it. Yet this appeal allows the addition of any population causally connected in space and time, and thus licenses treating as communities any of the full series generated by successively adding populations to any population, up to the global community of all organisms. That is, interactions answer the $(n + 1)$th problem, but only at the cost of embracing extreme promiscuity about community-identification. A more significant cost is that such communities are not robustly preservable. Removing a single species-population from a community (assuming the community can endure for some time with the same species-mix in the absence of that species) would often leave one with a community rather than something relatively impoverished. For nearly any community of n population, its $(n-1)$ will still be a community.[27] Thus, one could continue removing populations from the community and still be left at each step with a community. This is not to suggest that single populations *can* always be removed without losing other populations, like when a community loses one of the dozens of species of migratory birds eating its small flies, leaving everything else intact. Rather, the point is that "interactions" are insufficient for

[27]This works unless we have picked a population which is the only bridge between two others. Since interactions include anything affecting the living conditions of other species, interactions are so ubiquitous that that is the exception rather than the rule. This is an empirical claim which could turn out false and weaken the point. But this exception applies only when there is no population which can be individually removed from a community leaving the others intact.

solving the $(n+1)$th problem in a way that helps in contexts like conservation. After all, it is considering contexts like conservation that gives the $(n+1)$th problem its bite.

Though he does not adopt either one of them in particular, Odenbaugh's own view of communities does not add criteria to his Hutchinsonian or Clementsian concepts. He offers: "species populations form an ecological community just in case they exhibit community interactions, or put differently, they possess a community-level property" [p. 636]. Then, he defines a "community-level property" entirely in terms of causal interactions, so that it becomes just another way of referring to them: a community-level property is "any causal biotic relation between two or more species" [p. 636]. "Biotic" might do some work to restrict the class of causal interactions, but since interactions among plants are typically mediated by abiotic resources like nutrients and water, and these are included among causal interactions Odenbaugh recognizes, "biotic" adds no further restriction on causal interactions.

It is sensible to take causal interactions as a starting point for community definition, as Odenbaugh does; we attribute a status to ordinary objects like desks that we do not attribute to smoke rings, because of the strength and persistence of causal interactions among the molecules of the former. But, once we recognize that the denial of causal interactions is a position not advanced in ecology, a view or "concept" merely asserting their presence becomes banal. Responding to the point that the Clementsian and Gleasonian concepts might not have been articulated as such, Odenbaugh offers that whether or not these concepts are historically accurate, "critically engaging the stereotype serves a valuable purpose" [Odenbaugh, 2007, p. 629]. But recognizing interactions alone does not contribute substantially to defending robust realism for ecology or providing an ontology for ethics and conservation. Resisting unrestricted mereological composition, Odenbaugh asks, "do we really want to be ontologically committed to the existence of an object composed just of my left foot, Lewis and Clark College, and Sevilla, Spain?" [p. 631]. No, indeed, there are not apparently any purposes for which that is an interesting object. Yet, the bare existence of interactions, without any further characterization of what kind they are, does not reveal why that object is in distinctly worse shape than the collection of, say, Drake Passage Wandering Albatrosses, anchovy populations off the coast of Peru, and the population of occasional patrons of DiFara Pizza in Brooklyn, New York (some of whom eat anchovies). These populations are joined by causal biotic interactions, and thus should form a community on Odenbaugh's standard, but the same rhetorical question applies: Do we really want to be ontologically committed to the existence of an object composed of Wandering Albatrosses, Peruvian anchovies, and DiFara Pizza patrons? The considerations invoked as weighing against the former by Odenbaugh's rhetorical question weigh equally against the latter being an object, and therefore against it counting as a community (and rightly so). Moreover, even if one were to disagree that the same considerations weigh against each, the existence of my distributed object would not be sufficient for the main purposes for which some philosophers have sought a community concept, namely defending ecology's success in relation to scientific realism or providing a suitable ontology for environmental ethics and conservation. That is, causal, biotic interactions alone do not get us there; they are not enough to establish communities.

Particular kinds of causal interactions, however, *are* important to efforts to preserve biological units. Dependencies are basic facts of life. It is easy to forget that we human beings do not survive long without environmental oxygen, or outside a narrow range of temperatures. Consequently, dependence relations are significant to conservation. The degree to which species depend on one another has been a motivation for discussing preservation of communities rather than just species (as the US Endangered Species Act of 1973 has been employed). For example, keystone predators are those on whom the persistence of a number of other species depends. The loss of wolves in an area can cause a trophic cascade in which species composition of an area is radically altered [Hebblewhite *et al.*, 2005, for instance]. So, conservationists need to attend to this particular kind of causal interaction—dependence—and may refer to organisms connected by such interactions as "communities" in this sense. Dependence relations among populations are sufficient for their becoming potential targets of conservation interest, because such dependence relations causally affect the outcomes of trying to preserve some collections of organisms rather than others.[28] Groups exhibiting obligate, non-intersubstitutable dependencies are sufficient to a further kind of conservation interest. And, if we add the desideratum that we are interested in preserving groupings which can only exist as such, dependence relations become necessary, too.

On the other hand, if our interest is, just to take an example near at hand, a Clementsian approach to long-term vegetation explanations or forecasts,[29] dependence relations are neither necessary nor sufficient for identifying communities. We must instead identify all the populations in an area with significant competitive interactions. A strategy for identifying communities that misses populations engaged in these interactions will fail to capture causally significant influences, and thus fail to capture the relevant kind of communities. Attending to dependence relations among living populations is not sufficient for picking out communities for this purpose. Furthermore, three populations competing for a common abiotic resource with no dependence relations among them count as a community under Odenbaugh's definition. Odenbaugh rightly lists exclusively-competitive interactions among the appropriate relations binding populations together with others. Dependence relations are not necessary for picking out communities for the purpose of long-term forecasting, just as they are not sufficient. So, what is necessary and sufficient for one kind of community is neither necessary nor sufficient for another kind.

---

[28]That is, so far, not to say that anyone should be interested in preserving such groups, only that they now have properties making them something other than random assemblages, and are thus sufficient to a certain interests in groups.

[29]This applies just as well to interest in a number of contemporary approaches like David Tilman's resource competition theory [Tilman, 1982; Tilman, 1988].

The upshot of comparing these two sample projects for which one might employ community-criteria is that while communities may trivially require interactions among their constituent populations—Odenbaugh's criterion—that is not enough to support realism about communities *per se*. So, what is enough? That depends on what we want to identify communities for. 'What are the criteria for communities?' is thus not a productive question for philosophy at that level of generality. Various further interests (conservation, realism, prediction) determine different particular criteria for communities such as can fulfill those interests. Whether collections of populations comprise communities will vary depending on what we need communities to be. Especially as philosophers investigate communities in order to address these further interests, we should thus not attempt to identify what communities are *per se*.

Importantly, this is a different result than the claim that communities are fictions. Every ecologist, including Gleason, recognizes interactions among organisms, including that some require others, to survive. But how collections of interacting populations are rightly attributed a further status depends on both what kinds of interactions are in nature and what that status is. This is also a different result than that there is nothing more for philosophy to say about communities. Particular interests in communities produce puzzles about kinds and strengths of causal interactions and how best to recognize and talk about them. There is a live discussion of the possibility of top-down causation, for instance [Mikkelson, 2004; Mikkelson, in press; Sterelny, 2006], and whether any kind of holism makes sense. But these questions about the features of communities are where the action is, not the general question of whether communities exist.

7 CONCLUSION

This essay has sought to puncture the legend of Clements and Gleason, and along with it another legend it has supported, one which has in turn motivated keeping it alive—the legend of order and chaos. Clements did not treat vegetation as developing like birds and mammals do, nor as sharing many structural commonalities with organisms, like holistic, functional integration or discrete boundaries. Gleason did not treat vegetation as composed of individuals unaffected by their neighbors or difficult to group into robust collections. The ecologists agree that plants are affected by their environments and affect one another indirectly, and that those are the only kinds of causal relationships on which further theory can be built. As they recognize quite similar interactions, they depart from one another at the stage of trying to assemble the various kinds of causes into portable general theory. So, there is not a scientific basis in this debate, where it is usually located, for setting up against one another ecologies of order or chaos.

The ecologies of order and chaos, as they live in narratives, have had, as scientific episodes go, unusual power to inspire outrage and condescension. Why? I noted that one of the most interesting features of the episode is ecologists' shift from treating Clements's causal investigation as the way to render ecology more scientific to treating it as hopelessly unscientific, beyond just right or wrong. Studying the legends of Clements and Gleason and of order and chaos reveals something interesting about historiography—about understanding the history of science. It reveals that focusing on similes, metaphors, imagery, and their potential connotations can seriously mislead us in trying to understand others' understandings. Images help scientists create theories and communicate them, but scientific investigation, understanding, and explanation have other components; investigative and explanatory methodology are especially easy to overlook. In this episode, commenters drawn to the imagery have paid little attention to methodology and causal understanding, aspects which, whether the theories are wildly mistaken or not, are straightforwardly scientific. Historiography itself has made them unscientific.

So, putting the focus instead on how they assign causes to vegetation, that we find Clements and Gleason both recognizing causal interactions provides a basis for rejecting the claim that communities can be preserved only if they are Clementsian not Gleasonian. That in turn provides a basis for rejecting the claim that communities can be preserved as such only if they have some exotic kind of order or structure. If there are reasons certain groupings cannot be preserved as such, we do not learn about them by examining the concepts advanced in this debate.

Turning to implications for philosophy of ecology's discussion of communities, that Gleason did not deny causal interactions supports Odenbaugh's strategy of identifying a modest community concept based on the sort of causal interaction every ecologist recognizes. But the problem for Odenbaugh's approach to defining community criteria is that additional criteria are needed for the particular projects which have motivated seeking them. However, community concepts with additional features making them robust enough to support certain further interests include too much to serve other interests, and vice-versa. The richest challenge for philosophical discussions of communities is therefore not disorder, but multiplicity.

ACKNOWLEDGEMENTS

Many colleagues assisted this work, and I especially thank John Beatty, Ken Waters, Patti Ross, Karin Matchett, Jay Odenbaugh, Kevin deLaplante, Aidan Lyon, and audiences at the AAP and ISHPSSB for productive conversations at various stages. Part of this research was supported by a research fellowship from The Sydney Centre for the Foundations of Science at University of Sydney, which provided excellent conditions for writing.

BIBLIOGRAPHY

[Allen, 1906] G. M. Allen. Review: *Research Methods in Ecology*. *The American Naturalist* 40(479): 804–805, 1906.

[Allred and Clements, 1949] B. W. Allred and Edith S. Clements (eds.), *Dynamics of Vegetation: Selections from the Writings of Frederic E. Clements, Ph.D.* New York: H.W. Wilson Company, 1949.

[Anker, 2001] Peder Anker. *Imperial Ecology: Environmental Order in the British Empire, 1895–1945*. Cambridge, MA: Harvard University Press, 2001.

[Barbour, 1995] Michael G. Barbour. Ecological fragmentation in the fifties. In William Cronon (ed.), *Uncommon Ground: Rethinking the Human Place in Nature*, pp. 233–255. New York: W.W. Norton & Co., 1995.

[Begon *et al.*, 1990] Michael Begon, John L. Harper, and Colin R. Townsend. *Ecology: Individuals, Populations, and Communities*. Boston:Blackwell, 1990.

[Blackman and Tansley, 1905] F. F. Blackman and A. G. Tansley. Review: Ecology in its physiological and phytotopographical aspects. *New Phytologist* 4(10): 232–253, 1905.

[Botkin, 1990] Daniel B. Botkin. *Discordant Harmonies: A New Ecology for the Twenty-first Century*. New York: Oxford University Press, 1990.

[Braun, 1958] E. Lucy Braun. The development of association and climax concepts, their use in interpretation of the deciduous forest. In William Campbell Steere (ed.), *Fifty Years of Botany*. New York: McGraw-Hill, 1958.

[Bultman *et al.*, 1982] T. L. Bultman, G. W. Uetz, and A. R. Brady. A comparison of cursorial spider communities along a successional gradient. *Journal of Arachnology* 10: 23–33, 1982.

[Burnyeat, 1990] Myles Burnyeat. *The Theaetetus of Plato*. Hackett, 1990.

[Callicott, 1996] J. Baird Callicott. Do deconstructive ecology and sociobiology undermine Leopold's land ethic? *Environmental Ethics* 18: 353–372, 1996.

[Clements, 1904] Frederic E. Clements. *The Development and Structure of Vegetation*. Botanical Survey of Nebraska: Studies in the Vegetation of the State. The Botanical Seminar, 1904.

[Clements, 1905] Frederic E. Clements. *Research Methods in Ecology*. Lincoln: University Publishing Company, 1905.

[Clements, 1907] Frederic E. Clements. *Plant Physiology and Ecology*. New York: Henry Holt & Co., 1907.

[Clements, 1916] Frederic E. Clements. *Plant Succession: an Analysis of the Development of Vegetation*. Washington DC: Carnegie Institution of Washington, 1916.

[Clements, 1920] Frederic E. Clements. *Plant Indicators*. Publication 290. Washington DC: Carnegie Institution of Washington, 1920.

[Clements, 1936] Frederic E. Clements. Nature and structure of the climax. *Journal of Ecology* 24: 252–284, 1936.

[Clements, 1949a] Frederic E. Clements. Climatic cycles and human populations in the Great Plains. In Allred and Clements [1949].

[Clements, 1949b] Frederic E. Clements. Climaxes, succession, and conservation. In Allred and Clements [1949].

[Clements, 1949c] Frederic E. Clements. Ecology in the public service. In Allred and Clements [1949].

[Dureau de la Malle, 1825] A. J. C. A. Dureau de la Malle. Mémoire sur l'alternance ou sur ce problème: la succession alternative dans la reproduction des especes végétales vivant en société, est-elle une loi général de la nature? *Ann. Nat. Sci.* 1(5): 353, 1825.

[Eliot, 2007] Christopher Eliot. Method and metaphysics in Clements's and Gleason's ecological explanations. *Studies in History and Philosophy of Biological and Biomedical Sciences* 38(1): 85–109, 2007.

[Fitzsimmons, 1999] Allan K. Fitzsimmons. *Defending Illusions: Federal Protection of Ecosystems*. Lanham, MD: Rowman & Littlefield Publishers, 1999.

[Gleason, 1909] Henry A. Gleason. Some unsolved problems of the prairies. *Bulletin of the Torrey Botanical Club* 36(5): 265–271, 1909.

[Gleason, 1917] Henry A. Gleason. The structure and development of the plant association. *Bulletin of the Torrey Botanical Club* 43: 463–481, 1917.

[Gleason, 1925] Henry A. Gleason. Species and area. *Ecology* 6(1): 66–74, 1925.

[Gleason, 1926] Henry A. Gleason. The individualistic concept of the plant association. *Bulletin of the Torrey Botanical Club* 53, 1926.

[Gleason, 1927] Henry A. Gleason. Further views on the succession-concept. *Ecology* 8(3): 299–326, 1927.

[Gleason, 1936a] Henry A. Gleason. Is the synusia an association? *Ecology* 17(3): 444–451, 1936.

[Gleason, 1936b] Henry A. Gleason. Twenty-five years of ecology, 1910–1935. *Brooklyn Botanical Garden Memoirs* 4: 41–49, 1936.

[Gleason, 1939] Henry A. Gleason. The individualistic concept of the plant association. *American Midland Naturalist* 21(1): 92–110, 1939.

[Godfrey-Smith, 2009] Peter Godfrey-Smith. *Darwinian Populations and Natural Selection.* Oxford University Press, 2009.

[Hacking, 1989] Ian Hacking. Extragalactic reality: The case of gravitational lensing. *Philosophy of Science* 56: 555–581, 1989.

[Hagen, 1988] Joel B. Hagen. Organism and environment: Frederic Clements's vision of a unified physiological ecology. In Ronald Rainger, Keith R. Benson, and Jane Maienschein (eds.), *The American Development of Biology.* New Brunswick: Rutgers University Press, 1988.

[Hagen, 1992] Joel B. Hagen. *An Entangled Bank: The Origins of Ecosystem Ecology.* New Brunswick: Rutgers University Press, 1992.

[Hebblewhite *et al.*, 2005] Mark Hebblewhite, Clifford A. White, Clifford G. Nietvelt, John A. McKenzie, Tomas E. Hurd, John M. Fryxell, Suzanne E. Bayley, and Paul C. Paquet. Human activity mediates a trophic cascade caused by wolves. *Ecology* 86(8): 2135–2144, 2005.

[Hempel, 1966] Carl G. Hempel. *Philosophy of Natural Science.* Englewood Cliffs: Prentice-Hall, 1966.

[Hooper, 2004] Lora V. Hooper. Bacterial contributions to mammalian gut development. *Trends in Microbiology* 12(3): 129–134, 2004.

[Jackson and Pettit, 1992] Frank Jackson and Philip Pettit. In defense of explanatory ecumenism. *Economics and Philosophy* 8: 1–21, 1992.

[James, 1890] William James. *The Principles of Psychology.* New York: Henry Holt & Co., 1890.

[Kellert, 1993] Stephen H. Kellert. *In the Wake of Chaos: Unpredictable Order in Dynamical Systems.* Chicago: University of Chicago Press, 1993.

[Kingsland, 1995] Sharon E. Kingsland. *Modeling Nature: Episodes in the History of Population Ecology,* 2nd Edition. Chicago: University of Chicago Press, 1995.

[Kingsland, 2005] Sharon E. Kingsland. *The Evolution of American Ecology, 1890–2000.* Baltimore: Johns Hopkins University Press, 2005.

[Kirk *et al.*, 1983] G. S. Kirk, J. E. Raven, and M. Schofield. *The Presocratic Philosophers,* 2nd Edition. Cambridge, UK: Cambridge University Press, 1983.

[Kitcher, 1993] Philip Kitcher. *The Advancement of Science: Science without Legend, Objectivity without Illusions.* New York: Oxford University Press, 1993.

[Kuhn, 1962] Thomas S. Kuhn. *The Structure of Scientific Revolutions.* University of Chicago Press, 1962.

[Leplin, 1984] Jarrett Leplin (ed.), *Scientific realism.* Berkeley: University of California Press, 1984.

[Levins and Lewontin, 1985] Richard Levins and Richard C. Lewontin. Dialectics and reductionism in ecology. In Richard Levins and Richard C. Lewontin (eds.), *The Dialectical Biologist,* pp. 132–160. Cambridge: Harvard University Press, 1985.

[Longino, 2004] Helen E. Longino. How values can be good for science. In Peter Machamer and Gereon Wolters (eds.), *Science, Values, and Objectivity,* pp. 127–142. Pittsburgh: University of Pittsburgh Press, 2004.

[Mackie, 1977] J. L. Mackie. *Ethics: Inventing Right and Wrong.* Harmondsworth: Penguin, 1977.

[MacLeish, 1935] Archibald MacLeish. The grasslands. *Fortune* 12: 59ff., 1935.

[Malin, 1947] James C. Malin. *The Grassland of North America.*, 4th Edition Lawrence, Kansas: Edwards Brothers, 1947.

[Marr, 1982] David Marr. *Vision: a Computational Investigation into the Human Representation and Processing of Visual Information.* San Francisco: W.H. Freeman, 1982.

[McIntosh, 1975] Robert P. McIntosh. H. A. Gleason—"individualistic ecologist" 1882–1975: His contributions to ecological theory. *Bulletin of the Torrey Botanical Club* 102(5): 253–273, 1975.

[McIntosh, 1985] Robert P. McIntosh. *The Background of Ecology: Concept and Theory.* Cambridge: Cambridge University Press, 1985.

[Mikkelson, 2004] Gregory M. Mikkelson. Biological diversity, ecological stability, and downward causation. In Markku Oksanen and Juhani Pietarinen (eds.), *Philosophy and Biodiversity.* Cambridge University Press, 2004.

[Mikkelson, in press] Gregory M. Mikkelson. Part-whole relationships and the unity of ecology. In Jr. Skipper, Robert A., Colin Allen, Rachel A. Ankeny, Carl F. Craver, Lindley Darden, Greg Mikkelson, and Robert Richardson (eds.), *Philosophy and the Life Sciences: A Reader.* Cambridge, MA: MIT Press, in press.

[Mitman, 1988] Gregg Mitman. From the population to society: The cooperative metaphors of W. C. Allee and A. E. Emerson. *Journal of the History of Biology* 21: 173–194, 1988.

[Mitman, 1992] Gregg Mitman. *The State of Nature: Ecology, Community, and American Social Thought, 1900–1950.* University of Chicago Press, 1992.

[Mitman, 1995] Gregg Mitman. Defining the organism in the welfare state: The politics of individuality in American culture, 1890–1950. In Sabine Maasen, Everett Mendelsohn, and Peter Weingart (eds.), *Biology as Society, Society as Biology: Metaphors*, pp. 249–278. Dordrecht: Kluwer Academic, 1995.

[Moss, 1910] C. E. Moss. The fundamental units of vegetation: Historical development of the concepts of the plant association and the plant formation. *New Phytologist* 9(1/2): 18–53, Jan 1910.

[Nicolson and McIntosh, 2002] Malcolm Nicolson and Robert P. McIntosh. H. A. Gleason and the individualistic hypothesis revisited. *Bulletin of the Ecological Society of America* 83: 133–142, 2002.

[Nicolson, 1990] Malcolm Nicolson. Henry A. Gleason and the individualistic hypothesis: The structure of a botanist's career. *Botanical Review* 56: 97–161, 1990.

[Odenbaugh, 2006] Jay Odenbaugh. Ecology. In Sahotra Sarkar and Jessica Pfeifer (eds.), *Philosophy of Science: an Encyclopedia.* New York: Routledge, 2006.

[Odenbaugh, 2007] Jay Odenbaugh. Seeing the forest *and* the trees: Realism about communities and ecosystems. *Philosophy of Science* 74(5): 628–641, 2007.

[Phillips, 1931] John Phillips. The biotic community. *Journal of Ecology* 19(1): 1–24, 1931.

[Phillips, 1935a] John Phillips. Succession, development, the climax, and the complex organism: An analysis of concepts: Part II. development and the climax. *Journal of Ecology* 23(1): 210–246, 1935.

[Phillips, 1935b] John Phillips. Succession, development, the climax, and the complex organism: An analysis of concepts: Part III. the complex organism: Conclusions. *Journal of Ecology* 23(2): 488–508, 1935.

[Plato, 1993] Plato. *Sophist.* Hackett, 1993. translated by Nicholas P. White.

[Pool, 1989] Robert Pool. Ecologists flirt with chaos. *Science* 243(4889): 310–313, 1989.

[Psillos, 2002] Stathis Psillos. *Causation & Explanation.* Montreal: McGill-Queen's University Press, 2002.

[Reid *et al.*, 2000] R. P. Reid, P. T. Visscher, A. W. Decho, J. F. Stolz, B. M. Bebout, C. Dupraz, I. G. Macintyre, H. W. Paerl, J. L. Pinckney, L. Prufert-Bebout, T. F. Steppe, and D. J. DesMarais. The role of microbes in accretion, lamination and early lithification of modern marine stromatolites. *Nature* 406(6799): 989–992, 2000.

[Richeimer, 2000] Joel Richeimer. How philosophy lost perceptual expertise. *Synthese* 124(3): 385–406, 2000.

[Ricklefs, 1997] Robert E. Ricklefs. *The Economy of Nature: A Textbook in Basic Ecology*, 4th Edition. New York: W.H. Freeman, 1997.

[Shrader-Frechette and McCoy, 1993] Kristin S. Shrader-Frechette and Earl D. McCoy. *Method in Ecology.* Cambridge: Cambridge University Press, 1993.

[Shrader-Frechette and McCoy, 1994] Kristin S. Shrader-Frechette and Earl D. McCoy. The community concept in community ecology. *Perspectives on Science* 2(4): 455–475, 1994.

[Sterelny, 2006] Kim Sterelny. Local ecological communities. *Philosophy of Science* 73: 215–231, April 2006.

[Stevens, 1950] Neil E. Stevens. The moral obligation to be intelligible. *The Scientific Monthly* 70(2): 111–115, 1950.

[Tilman and Wedin, 1991] David Tilman and David Wedin. Oscillations and chaos in the dynamics of a perennial grass. *Nature* 353(6345): 653–655, 1991.

[Tilman, 1982] David Tilman. *Resource Competition and Community Structure.* Princeton: Princeton University Press, 1982.

[Tilman, 1988] David Tilman. *Plant strategies and the dynamics and structure of plant communities.* Princeton University Press, Princeton, 1988. 87025833 David Tilman. Monographs in population biology; 26. Bibliography: pp. 335–351.

[Tobey, 1981] Ronald C. Tobey. *Saving the Prairies: the Life Cycle of the Founding School of American Plant Ecology, 1895–1955.* Berkeley: University of California Press, 1981.

[Weaver and Clements, 1938] John E. Weaver and Frederic E. Clements. *Plant Ecology.* New York: McGraw-Hill, 1938.

[Weinert, 1995] Friedel Weinert (ed.), *Laws of Nature: Essays on the Philosophical, Scientific, and Historical Dimensions.* Berlin: De Gruyter, 1995.

[Wilson, 1999] Jack Wilson. *Biological Individuality: The Identity and Persistence of Living Entities.* Cambridge: Cambridge University Press, 1999.

[Worster, 1990] Donald Worster. The ecology of order and chaos. *Environmental History Review* 14: 1–18, 1990.

[Worster, 1994] Donald Worster. *Nature's Economy: A History of Ecological Ideas*, 2nd Edition. Cambridge: Cambridge University Press, 1994.

PHILOSOPHICAL THEMES IN THE WORK OF ROBERT H. MACARTHUR

Jay Odenbaugh

1 INTRODUCTION

In this essay, I first provide an introductory sketch of Robert H. MacArthur's academic life and the core elements of his research program in theoretical and community ecology. Second, we consider a tale of two models. Specifically, we consider MacArthur's collaborative work in island biogeography and limiting similarity both to illuminate his research but also to examine some of the successes and controversies that this work inspired. Finally, we consider one of the philosophical debates surrounding his work—the role of unification in ecology and population biology more specifically.

2 A BIOGRAPHICAL SKETCH AND SKETCH OF MACARTHUR'S RESEARCH PROGRAM[1]

Robert Helmer MacArthur was born April 7, 1930 and died November 1, 1972. MacArthur was the youngest son of John Wood MacArthur, a biologist at the University of Toronto who later moved to Marlboro College in Marlboro, Vermont. MacArthur himself received his undergraduate degree there and subsequently went on to receive a master's degree in mathematics at Brown University. In 1957, he entered the PhD program at Yale University under George Evelyn Hutchinson, a preeminent ecologist and limnologist. During 1957–1958, MacArthur worked at Oxford University under the acclaimed ornithologist David Lack in order to build up his background in field ornithology.[2] From 1958–1965, he went from assistant professor to full professor at the University of Pennsylvania and then finally moved to Princeton University where he was the Henry Fairfield Osborn Professor of Biology until his death.

[1] In this section, the biographical material comes from [Wilson, 1993; Wilson and Hutchinson, 1982; Kingsland, 1995].

[2] Hutchinson and Lack shaped MacArthur's views about many things but specifically over the importance of interspecific competition in structuring ecological communities. Hutchinson had articulated the basic template for understanding the role of competition in his [1958]. Lack had been a forceful proponent of the density-dependent position in the population regulation debates.

Handbook of the Philosophy of Science. Volume 11: Philosophy of Ecology.
Volume editors: Kevin deLaplante, Bryson Brown and Kent A. Peacock. General editors: Dov M. Gabbay, Paul Thagard and John Woods.

MacArthur is one of the most important ecologists ever to work in the discipline. Moreover, at that time, he was surrounded by and worked with some of the luminaries of evolutionary biology and ecology including Egbert Leigh, Richard Lewontin, Richard Levins, Leigh van Valen and E. O. Wilson.[3] As important as he was and is, his work has also been exceptionally controversial. Ecologists recognize he did exceedingly important theoretical and empirical investigations; however, many also believe that he took ecology down methodologically and theoretically problematic paths as we shall see. Still, MacArthur's work serves as an interesting case study for philosophers of science. For example, MacArthur prized generality in science and strove to connect ecology with other areas of biology including evolutionary biology and biogeography. As an example of the philosophical import of his work, collaborator E. O. Wilson and mentor G. E. Hutchinson wrote the following of MacArthur after his death in 1972.

> [He] will be remembered as one of the founders of evolutionary ecology. It is his distinction to have brought population and community ecology within the reach of genetics. By reformulating many of the parameters of ecology, biogeography, and genetics into a common framework of fundamental theory, MacArthur—more than any other person who worked during the decisive decade of the 1960s—set the stage for the unification of population biology. [1982, p. 319]

Did MacArthur and his collaborators "unify" evolution, ecology, and biogeography? I will examine this and other issues in this essay.

In order to assess MacArthur's accomplishment, we must understand the components of the program he and others articulated. Here are some of the salient elements.[4] First, MacArthur typically formulated general, simple deterministic mathematical models which lacked a certain degree of precision. In the terms of Richard Levins' [1966] account of model building, precision was sacrificed for generality and realism.[5] Second, MacArthur also emphasized the ecological process of interspecific competition as a mechanism structuring ecological communities. This is evident in his work on limiting similarity and species distributions (*i.e.*, the "broken stick" model). This is not to say that he did not work on other types of processes like predation [MacArthur, 1955]; rather it is that interspecific competition played a predominate role in his thinking. Third, MacArthur rarely evaluated models with sophisticated statistical measures of goodness-of-fit. There are of course exceptions to this rule but mostly he and his colleagues evaluated

[3] For a "feel" of the work these individuals were doing, see [Leigh, 1971; Levins, 1968; Lewontin, 1974].

[4] For discussions of the elements of the "MacArthur school" see [Weins, 1992; Horn and Pianka, 2005].

[5] This is not to say that MacArthur modeled ecological systems realistically; rather, the desiderata of interest were generality and realism and precision less so. As an example of MacArthur's "realism," he devised a mechanistic consumer-resource model with two consumers and two resources and showed how the more phenomenological Lotka-Volterra interspecific competition could be derived from it [MacArthur 1972].

their models by looking for corresponding dynamical patterns such as stable equilibria and various types of cycles. This too would be extremely controversial as we shall see with the debates over "null hypotheses" and the contingencies of history. Finally, he was a master at presenting complex mathematical results with graphical representations [MacArthur and Wilson, 1967]. Specifically, MacArthur used isocline analysis to not only present theory in pedagogically useful ways but also to draw interesting and unobvious implications [Rosenzweig and MacArthur, 1963]. We will see some of this with his work on density-dependent selection.

In the rest of this essay, I present two case studies of MacArthur's own work. First, we will consider what many would consider a moderate success—that of the equilibrium model of island biogeography. Here the model certainly was controversial but was used to stimulate further research on reserve design, other mathematical structures like metapopulation models, and even helped to produce a new discipline: conservation biology. Though the model was confirmed and disconfirmed in different respects it was an important step forward. Second, I consider his work on interspecific competition and specifically his modeling of "limiting similarity." Clearly, the research trajectory he and others moved forward was an obvious answer to the questions ecologists had been asking over the principle of competitive exclusion; nevertheless, these theoretical excursions found themselves mired in controversies. Finally, I consider one of the philosophical issues raised by MacArthur's work and his commentators—did MacArthur unify the various disciplines in which he worked? I will argue that he did not unify those areas though he did integrate them in very important ways.

3 A TALE OF TWO MODELS

Let us begin with MacArthur's work in island biogeography. Obviously enough, island biogeography is the area of biology concerned with the distribution and abundance of species on islands. A chief concern with the theory has been the "species-area effect." One important question that biologists have been attempting to answer is, why is it that larger islands support more species than smaller islands? There appears to be a monotonic relationship between the area and species richness (the number of species in a community). The species-area effect has been decomposed into two distinct effects. First, there is the *area effect* which is simply that more species are expected on larger islands than on smaller islands and secondly, the *distance effect* which is that islands closer to the mainland are more likely to have a greater number of species than islands farther away. Thus, in a somewhat simplistic fashion, we can think of the theory of island biogeography as an attempt at explaining the species-area effect by explaining the area and distance effects. The most successful and controversial attempt to do this is the equilibrium model devised by ecologist Robert H. MacArthur and myrmecologist

E. O. Wilson [1963; 1967].[6] The basic assumptions of the model are as follows:

1. The species richness on an island has a stable equilibrium.
2. The stable equilibrium is the result of a balance between the immigration rate from the mainland and extinction rate on the island.
3. The distance of the island from the mainland alone determines the immigration rate.
4. The extinction rate is determined only by the size of the island.
5. The stable equilibrium is a dynamic equilibrium where the number of species on the island is constant but the identity of the species is constantly changing which is the rate of species turnover.

In mathematical dress, the MacArthur-Wilson equilibrium model appears as follows (see [Wilson and Bossert, 1971, pp. 166–184, Gotelli, 1995, pp. 159–173] for details). Let P represent the number of species in a "pool," that is, the number found in all the surrounding areas which provide immigrants. P is a parameter and takes a constant value. Let us define the total immigration rate λ_S as the number of new species colonizing the island per unit time. The total extinction rate μ_S is the number of species among those already present on the island going extinct per unit time. The rate of change in species richness on the island is dS/dt which equals the difference between the immigration and extinctions rates, or

$$(1)\quad \frac{dS}{dt} = \lambda_S - \mu_S$$

When $\lambda_S = \mu_S$, then $dS/dt = 0$ and the number of species will be at equilibrium. We can now ask what is the total immigration rate in number of species per unit time when S species are present? We must take the average immigration rate of new species per species when S species are present on the island. This we will call λ_A. The total immigration rate then is $\lambda_A(P - S)$. Similarly, what is the total extinction rate in species per unit time? It is the average extinction rate per species μ_A when S species are present multiplied by the number species already on the island, or $\mu_A S$.

MacArthur and Wilson also postulated that the number of species on an island has a stable equilibrium given the rates of immigration and extinction. So, if we suppose $dS/dt = 0$, then we have,

$$(2)\quad \frac{dS}{dt} = \lambda_S - \mu_S = \lambda_A (P - S) - \mu_A S$$

At equilibrium, $dS/dt = 0$ and so,

[6] The rudiments of the equilibrium model were discussed by ecologists prior to MacArthur and Wilson though they took those initial insights much further than anyone heretofore.

(3) $$\left.\frac{dS}{dt}\right|_{S=S^*} = \lambda_S - \mu_S = \lambda_A (P - S) - \mu_A S = 0$$

This is equivalent to

(4) $$S^* = \frac{\lambda_A P}{\lambda_A + \mu_A}$$

If we integrate the differential equation above, we have the following expression,

(5) $$S = \frac{\lambda_A P}{\lambda_A + \mu_A} \left(1 - e^{-(\lambda_A + \mu_A)t}\right)$$

As t becomes very large, then $e^{-(\lambda_A+\mu_A)t}$ will approach zero and therefore S approaches $\lambda_A P/(\lambda_A + \mu_A)$. We can now determine the rate of species turnover. Even though MacArthur and Wilson argued that species richness was at equilibrium on any given island (assuming there has been a sufficiently long period of time after a disturbance), the composition of species would be continually changing. In other words, S^* is a *dynamic equilibrium*. Let us suppose we are interested in some fraction of the equilibrium number of species S^*, for example, 90%. So, we multiply both sides by 0.9.

(6) $$0.9S^* = \frac{\lambda_A P}{\lambda_A + \mu_A} \times 0.9$$

Applying our equations, we then have

(7) $$S = 0.9S^* = \frac{\lambda_A P}{\lambda_A + \mu_A}(1 - e^{-(\lambda_A+\mu_A)t_{0.9}})$$

Since $S^* = \frac{\lambda_A P}{\lambda_A+\mu_A}$ then it follows that

(8) $$(1 - e^{-(\lambda_A+\mu_A)t_{0.9}}) = 0.9$$

If we rearrange and take the natural logarithms of this equation, we find that the species turnover rate is

(9) $$t_{0.9} = \frac{2.3}{\lambda_A + \mu_A}$$

So far, we have constructed the basic elements of the MacArthur-Wilson equilibrium model, but we have not accounted for the species-area effect. In order to do this, let us make two further assumptions. First, let's assume that the total population size for each species S_i is proportional to the island's area. That is, the number of individuals of a species per unit area, or population density, is the same on differently sized islands. Second, let us assume that the probability that a species goes extinct increases as the size of an island gets smaller. Hence, the probability that a species goes extinct decreases with increasing island size. From our model and these assumptions we have the following implications. Let S_{FS}, S_{NS}, S_{FL}, and S_{NL} be different equilibrium numbers of species (where F stands for far, N stands for near, S stands for small, and L stands for large). It follows that the immigration rate of the near island is greater than the immigration rate of the far island. Likewise, the extinction rate of the small island is greater than the extinction rate of the large island. Therefore, the equilibrium number of species on the near, large island, S_{NL}, is the largest equilibrium number of species. Similarly, the island with the greatest turnover rate is the near, small island with equilibrium number of species S_{NS} (where dS/dt is greatest). Thus, we have a purported explanation of the species-area effect.

If we were to test the MacArthur-Wilson equilibrium model, then we would try to establish a fit between the model and the pattern of interest determining whether the laws and assumptions are true or at least empirically accurate of actual island communities. The MacArthur-Wilson model has had mixed success when it has been tested. First, the MacArthur-Wilson model has performed respectably when biologists have evaluated its predictive success. One extremely important experimental study was performed by Wilson and Simberloff [1969] which made general predictions about the relation between island size and area. In the Florida Keys, there are thousands of small mangrove islands with 20–50 species per island each of different areas and distances from the mainland. So, they hired some fumigators and applied to methyl bromide to 6 islands killing the insects [1969; 1970]. Their general conclusions were these. First, species abundance returned to their previous number. Second, species richness was a function of island size and distance. Third, there was substantial species turnover. Likewise, the mangrove islands did apparently have an equilibrium number of species.

Unfortunately, Wilson and Simberloff had great difficulty in estimating colonization and extinction rates. However, they did find that the small, distant islands had fewer species than larger, nearby islands. They could only periodically check the islands for species richness counts and thus could never get an accurate idea of how many colonists were arriving, where they were arriving from, and so on. There has also been quite a controversy over the rate of species turnover. Initially, the data seemed consistent with the model until Simberloff [1976] realized that they had counted transient species as species going extinct and thus in fact the predicted rate of species turnover was far too high. Originally, they estimated that there was a turnover rate of 0.67 per day. In 1976, Simberloff reanalyzed the data eliminating transient individuals who did not stay on the island and reproduce ("false" extinctions). His corrected estimate was that the turnover rate was only 1.5 extinctions per year.

The MacArthur-Wilson model also has many important idealizations. Here are a few.

- We have assumed that the immigration and extinction curves are linear.
- The distance of the island from the mainland alone determines the immigration rate and the extinction rate is determined only by the size of the island.

- Extinction of a local population is independent of species composition on the island [Gotelli, 1995, pp. 181–187].

The first idealization is not terribly problematic since MacArthur and Wilson recognized that changing the immigration and extinction curves to nonlinear ones would not appreciably change their conclusions.

> This linear condition is not so stringent as it may seem, for any transformation of the ordinate is permissible, although not all immigration and extinction curves can be simultaneously straightened. If the immigration and extinction curves are mirror images, then both may be straightened simultaneously by distorting the ordinate; otherwise our results will apply only where immigration and extinction curves are relatively straight. [1967, p. 28]

However, the other assumptions are not so easily discharged. The second assumption is problematic since biologists recognized what has been called the "rescue effect" and the "target effect." The rescue effect is that the smaller the distance of the island from the mainland the lower the rate of extinction since other members of that species will probably be on the island. The target effect is the larger the island the more likely a species is to successfully "intercept" that island (see [Gotelli 1995, pp. 183–5]). Finally, Daniel Simberloff has documented how predatory and competitive relations decrease the equilibrium number of species on an island after a period of time, which results in an "assortative equilibrium" [Simberloff, 1976]. These considerations also show how easily the third assumption is violated. Species interactions on the island, not just the size of the population, affect the extinction rate.

Model building provides important conceptual resources for scientists. In modeling, one must pick out what the salient factors are that bring about a particular phenomenon. Suppose a model is not an accurate representation of a natural system or systems in the laws of succession/coexistence or entities it postulates. Nonetheless, the model can give scientists concepts by which to classify various ecological structures and can help them pose new questions. These concepts highlight kinds of phenomena that are worth researching. The MacArthur-Wilson model provided a set of concepts that allowed biologists, even those who strongly criticized the model, to further investigate island populations. Thus, the concepts of a *species equilibrium*, *turnover rate*, and others enriched the conceptual tool-kit of biologists, leading them to ask new questions and investigate islands in new ways. Two extremely important ideas concerning life history strategies emerged from their models, r and K selection, which have enriched ecological theory. There have been important studies which have documented the possibility of nature reserve design with the help of the MacArthur-Wilson model since fragmented landscapes which have been scarred by development often function as isolated islands (see [Harris, 1984]). So for example, we can ask on the basis of concepts gleaned from the MacArthur-Wilson model questions like, should reserves be a single, large area or several, small areas (*SLOSS*)? Should corridors connect them and what shape should the reserves have? What are the effects of habitat edges on species richness and evenness? Critics of the equilibrium model often point out that the model may not correctly answer these questions (or even provide answers for that matter) [Shrader-Frechette and McCoy, 1993]. However, the equilibrium model can still provide a conceptual framework for such investigations even if it itself is not the means by which to answer such questions.

The MacArthur-Wilson model also has drawn attention to developing related mathematical structures in trying to come to terms with spatial heterogeneity such as metapopulation models. Spatial ecology bears a direct link to the work of MacArthur and Wilson, even though the approaches are very different.[7] Thus, the MacArthur-Wilson model clearly has provided conceptual resources for investigating new questions and looking for new patterns. Models provide much needed conceptual frameworks or heuristics that structure the ignorance of a discipline. This can be so even when they are explanatorily or predictively inadequate in some respects, as MacArthur and Wilson recognized themselves.

> A great deal of faith in the feasibility of a general theory is still required. We do not seriously believe that the particular formulations advanced in the chapters to follow will fit for very long the exacting results of future empirical investigation. We hope instead that they will contribute to the stimulation of new forms of theoretical and empirical studies, which will lead in turn to a stronger general theory and, as R. A. Fisher once put it, "a tradition of mathematical researches upon which a mathematical physicist can draw in the resolution of species difficulties." [1967, p. v]

Let us now turn to a different model, that of "limiting similarity". Ecologists have long argued that species that are similar in morphology and thereby have similar diets will be more similar in allopatry than in sympatry. That is, species that are sympatric will often segregate in body size or foraging behavior. So, this pattern—if it is a pattern—cries out for an explanation. One possible explanation is that species that have the same resource requirements compete and in order to minimize this competition will over time differ in their resource utilization. Famously, G. E. Hutchinson [1959] noted that three European species *Corixa affinis*, *Corixas macrocephala*, and *Corixa punctata* had segregated distributions such that the largest corixid *C. punctata* occurred with either *C. affinis* or *C. marocephala* but the smaller two do not co-occur. He suggested that insofar as species differ in size or other life history characteristics they may also differ in their resource use so as to avoid excluding one another. By examining other taxa, he found that coexisting species tended to differ in some aspect of size by approximately 1.3.

[7]Metapopulation models take as their state variables the frequency of patches occupied whereas many spatial ecological models are diffusion equations describing the spread of individuals of a species through time.

In this work, Hutchinson was attempting to more precisely articulate the competitive exclusion principle: if two species share the same "niche," then they cannot coexist. Hutchinson in his 1957 "Concluding Remarks" reformulated the principle with tools from set theory. Suppose that we represent each independent factor that affects the reproductive output of a species by a variable and suppose there is an n-dimensional space composed of those n variables. This hypervolume will contain a non-empty area which are values of the variables under which the species can persist—$dN/dt \geq 0$ in the continuous case or $N_{t+1}/N_t \geq 1$ in the discrete case. This is what Hutchinson termed the "fundamental niche" of a species. Likewise, there is a region of this space which represents the values of the variables which a species "occupies" as the result of interactions with other species—this is what he called the "realized niche" of the species. Hutchinson reformulated the principle of competitive exclusion as the claim that realized niches of different species do not intersect.

Hutchinson also suggested that if the principle of competitive exclusion was false then one important taxonomic group to examine would be territorial birds [Kingsland, 1995]. If the availability of territory regulated population size and not competitive interactions, then the principle of competitive exclusion would be violated. In his classic 1958 study, MacArthur, having done work with David Lack, examined just such a group. He studied five warbler species in Maine. It did appear that nesting differences in habitat did reduce competition. However, in those areas of feeding where competition would be most likely to occur, he found an amazing degree of niche specificity, though all the species were roughly of the same size and shape and would feed in the foliage of trees. For example, MacArthur found that some species fed high in the trees and others spent their time on the forest floor. Some species fed near tree trunks and other would forage on the branches of the tree. As just two examples, consider the foraging behavior of the Myrtle and the Black-throated green warbler. MacArthur observed and recorded the amount of time spent by each species in various positions in the forests and found that they differed in where they fed in the tree. Thus, the principle of competitive exclusion appeared to be dramatically confirmed by the way in which the warblers subdivided their resources and behavior.

As we saw above, according to the competitive exclusion principle, if several species occupy the same niche, then they cannot coexist. That is, if their resource requirements are exactly the same, then at least one species must go extinct. However, very few species share exactly the same niche; hence, they should be able to coexist. So, how much "niche overlap" is compatible with their coexistence—are there limits to their overall similarity? MacArthur and his colleagues Richard Levins [1967] and Henry Horn [1972] spilt much ink on these questions and we will look at the most sophisticated attempt to answer this question. Robert May and Robert MacArthur [1972] provide an answer to this question within the framework of traditional mathematical ecology. Consider a one-dimensional "resource spectrum," let us say that it concerns food size. Basically, we are considering food size as a variable, the values it can take, and how a given set of species utilizes each value of food size. For a given species, let's suppose it has a mean food size and dispersion about the mean. Let d be the distance between adjacent means, with w the dispersion about the mean for a given species and K the amount of food consumed as a function of food size. May and MacArthur suggest that the ratio d/w then is measure of niche overlap.

We can model a system of species $N_i(t)$ with a set of first-order differential equations.

$$(10)\quad \frac{dN_i(t)}{dt} = N_i(t)(k_i - \sum_{j=1}^{n} \alpha_{ij}N_j(t))$$

The k_i are integrals over the products of the resource spectrum and utilization function of the i-th species and are assumed constant. The competition coefficients α_{ij} are convolution integrals over the utilization functions of a species—a function of niche overlap, *i.e.*, the ratio d/w. After analyzing their model, May and MacArthur arrive at the following results. In the deterministic case where the parameters are constant, there is no limit to the number of species that can be packed along the resource spectrum. In the stochastic case where k_i fluctuate randomly, there is a limit to the number of species that can be packed along the resource spectrum. They write,

> We observe that the species packing parameter d indeed goes to zero when the environmental variance becomes strictly zero, but that for any finite environmental variance, d remains roughly equal to the utilization function width, w. [1972, p. 1109]

To determine stability, we find the equilibrium species densities N_i^* (where $dN_i^*/dt = 0$) and determine the behavior of arbitrarily small perturbations around the equilibrium; $n_i = N_i - N_i^*$. The stability of the community is determined by the dynamics of $d\mathbf{n}/dt = \mathbf{An}$ where $\mathbf{n}$ is a vector of n_i and $\mathbf{A}$ is a matrix of competition coefficients α. If all the eigenvalues λ of $\mathbf{A}$ are positive, then the system is locally stable. For all assumed values of α in our deterministic model, the eigenvalues are positive. Hence, there is no limit to the number of species that can "packed" along the resource spectrum consistent with long-term local stability. However, if we assume that $k_i = \bar{k} + \gamma_i(t)$ where $\gamma_i(t)$ is Gaussian "white noise", then the dynamics change. The probability of any given species going extinct is small if $\lambda_{\text{min}} > \sigma^2/\bar{k}$; that is, the smallest eigenvalue is greater than the variance relative to the mean value of the resource spectrum. The closest niche overlap d/w consistent with long-term community stability in a randomly varying environment whose fluctuations are characterized by a variance relative to the mean. Note that $d/w \approx 1$ for many values of n. So, in a fluctuating environment where the community is stable, adjacent species on a resource spectrum must be separated by $d/w \approx 1$. Crucially, for our purposes, May and MacArthur argue that their results are robust. That is, this result would hold even when the idealizations are "relaxed" by more realistic assumptions. In their [1972, p. 1112] essay in the section entitled "How robust are our results?", they suggest their results hold under the following alterations:

- They could replace Gaussian utilitization functions with others and retain the same results.
- Width and separation are chosen to be constant; however, if they vary but are proportional, then the results are the same.
- It was assumed that at equilibrium, the populations along the resource spectrum are equal. The results remain the same so long as each species is relatively large.
- The Lotka-Volterra equations are the first-term in a Taylor expansion of a larger set of equations; hence, they could have used more complex equations with approximately the same results.
- Gaussian white noise could have been replaced by some other type of random variation so long as fluctuations are correlated at best over short time scales relative to the scale of the system.
- Many bird communities are organized along one-dimension like food size; hence, it is not a completely unreasonable assumption.

Moreover, they argue that there is empirical evidence of $d/w \approx 1$ [May and MacArthur, 1972, p. 1112]. Specifically, they cite several examples. The work of John Terbough suggests that five species of tropical antbird segregate by forest height where $d/w \approx 1$. They mention MacArthur's own analysis of data on food weight distribution of three species of hawks which have a $d/w \approx 1$. Finally, they suggest Diamond's work on tropical birds and their weight sorts out to be between $d/w \approx 0.6 - d/w \approx 0.1$.

In summary, May and MacArthur's work on niche overlap in fluctuating environments suggests that by examining multiple models their prediction is robust; niche overlap $d/w \approx 1$. They argue that it holds over a variety of Lotka-Volterra stochastic mathematical models (and not in the deterministic case). Finally, they suggest that there is some empirical evidence in favor of the robust theorem. Their essay ends in the following way:

> In brief, the basic conclusion that emerges in a non-obvious but robust way from our mathematical model, namely that there is a limit to niche overlap in the natural world and that this limit is not significantly dependent on the degree of environmental fluctuation (unless it be severe, as in the arctic), seems to be in harmony with such facts as are known about real ecosystems. [1972, p. 1112]

Nevertheless, there are several remaining questions which suggest that it was not successful. First, is $d/w \approx 1$ genuinely robust or only apparently so, as some theoreticians latter argued [Abrams, 1983]? Second, is there genuine empirical evidence for $d/w \approx 1$?

First, Peter Abrams [1983] has argued that the result was not robust. For example, consider the formula for determining the competition coefficients in the discrete case where p_{ik} represents the proportion of i's total resource utilization given by resources of type k,

$$(11) \quad \alpha_{ij} = \frac{\sum_k p_{ij} p_{jk}}{\sum_k {p_{ik}}^2}$$

and in the continuous case where $f_i(x)$ and $f_j(x)$ are utilization functions of species i and j,

$$(12) \quad \alpha_{ij} = \int f_i(x) f_j(x)$$

These formulae can be derived from ecological theory; however, they are derived from consumer-resource equations on the assumption that the functional response of the consumer is linear and the resource grows logistically, and neither of these assumptions is realistic. Likewise, the model assumes that the environmental variability which affects different species is uncorrelated, which is also unrealistic, and the size of the environmental variability can have drastic effects on the degree of niche overlap.

Second, many ecologists were suspicious that the pattern of limiting similarity was simply an "artifact." For example, Henry Horn and Robert May [1977] would later argue that there was nothing necessarily biological about such patterns since many objects exhibit a 1.3 difference in ratios. For example, they noted that musical instruments in an orchestra do so. Thus, before theoreticians begin to build models to explain some pattern, some ecologists believed that we should determine that there is a "genuine" pattern to begin with. One way in which these suspicions manifested themselves was in one of the most spirited debates in community ecology, the "null hypotheses" [Gotelli and Graves, 1997].

These debates appeared largely because of those who stressed that interspecific competition was fundamental to structuring properties like body size and resource use of organisms. Still, the disagreements also pointed to more general issues surrounding how ecological models should be tested. In 1975, Jared Diamond—a collaborator of MacArthur's—published a study on the distribution of species of birds among approximately fifty islands in the Bismarck Archipelago near New Guinea [Diamond, 1975]. He noted that certain combinations of species have never been found together in the archipelago. One such example was two species of cuckoo-dove, *Macropygia nigriostris* and *M. mackinlayi*, which occurred on six and fourteen islands respectively though they never co-occurred on any island. This "checkerboard pattern", or what is termed 'complementary distribution', suggested that interspecific competition was at work through differentiation of the species' niches. In the late 1970s, Edward Connor and Daniel Simberloff [1979] argued that Diamond's work was deeply flawed from a statistical point of view. They argued checkerboard distributions could just as easily appear from random colonization as opposed to competition. Methodologically, they constructed "null models" of assemblages retaining certain properties of the communities such as the number of species per island, the relative abundances of species, and their incidence functions (the probability of a species occurring on an island given the total number of species on that island) but reassembled the rest of the salient properties at random attempting to remove the competitive effects. If the actual data differ in statistically significant ways from the null hypothesis, then the null is rejected and the interaction is strongly suggested. Simberloff and his colleagues claimed that null hypotheses were "simpler" and in some sense "logically prior" to competition hypotheses. Interestingly, Simberloff and Connor adopted a Popperian methodology attempting to falsify the competition hypotheses whereas Diamond looked for confirming evidence as opposed to first refuting a null hypothesis.

The work of Simberloff and his group has been criticized. First, in traditional Neyman-Pearson testing as advocated by Simberloff and his colleagues, one formulates two mutually exclusive and exhaustive hypotheses, the null and the alternative. However, the null hypotheses articulated by the Florida group were only sometimes logically inconsistent with competition hypotheses, as argued by Michael Gilpin and Diamond. Key features of the null models—the species pools, dispersal abilities of species, and "incidence functions" of species could have been affected by competition in the past [Gilpin and Diamond 1983]. Hence, the "ghost of competition past" might be a "hidden structure" built into the null model. Second, Connor and Simberloff performed their analyses using groups of species that were not restricted to groups of species that utilize similar resources in similar ways (or what is termed a "guild"). Competition is to be expected between two species if and only if they occupy the same guild. One could thus obscure the competitive effects in a morass of irrelevant data [Gilpin and Diamond 1983].

It should be noted that Connor and Simberloff argued even if one could delineate guilds with good evidence and still had the "checkerboard pattern," one still could not conclude that competition had occurred. Similarly, they claimed that Gilpin and Diamond had not provided independent evidence for their "hidden structure" claims (see Strong *et al.* 1984 for the details). The null model controversy continued in many essays; however, the debate pushed ecologists into discussions as to how ecological theory should be evaluated. As cantankerous as the debate was, the effects it has had on hypothesis testing and model selection in ecology have been good.

4 DID MACARTHUR UNIFY POPULATION BIOLOGY?

Finally, let's turn to more issues of intertheoretic relations or unification. Famously, or infamously, MacArthur wrote,

> To do science is to search for general patterns. Not all naturalists want

> to do science; many take refuge in nature's complexity as a justification to oppose any search for patterns. This book is addressed to those who do wish to do science. [1972, p. 1]

MacArthur was interested in general patterns and finding models which would explain and predict features of those patterns. MacArthur and his colleagues produced a variety of different models involving environmental heterogeneity, density-dependent selection, optimal foraging, limiting similarity, and equilibrium island biogeography. However, they realized that many of the patterns of interest were not ecological *per se* but involved evolutionary and biogeographical factors as well. Thus, somehow ecological, evolutionary, and biogeographical processes had to be jointly modeled. As an example of this sort of work, let us consider his modeling of density-dependent selection [1962]. This is a case where MacArthur attempts to integrate ecological and evolutionary concepts.

In most evolutionary models, according to MacArthur, population geneticists use r, the intrinsic rate of increase of a population, as a measure of fitness. He writes,

> For populations expanding with constant birth and death rates, r, or some equivalent measure (Fisher used r; Haldane and Wright used e^r which Wright called W) is then an appropriate definition of fitness. [1962, p. 146]

However, as MacArthur notes, present values of r may not be reliable predictors of the number of descendants a group of individuals will have since r is an accurate measure of fitness only if the environment is relatively stable. One way in which the environment may be unstable is if population density affects fitness. In fact, MacArthur writes, "[t]o the ecologist, the most natural way to define fitness in a crowded population is by the carrying capacity of the environment, K, ..." [1962, p. 146].

MacArthur devises the following mathematical model. Let n_1 and n_2 represent populations of alleles 1 and 2, respectively, and suppose they are governed by the following equations

(13) $dn_1/dt = f(n_1, n_2)$

(14) $dn_2/dt = g(n_1, n_2)$

Suppose we have a phase space where the x-axis represents the population of allele 1 n_1 and the y-axis represents the population of allele 2 n_2. A point in the space then represents the joint abundances of population n_1 and n_2. Suppose there is a set of values of n_1 and n_2 such that $f(n_1, n_2) = 0$, or equivalently, $dn_1/dt = 0$ for those values of n_1 and n_2. If the population of n_1 is to the left of the f-isocline, it will increase. Likewise, if the population of n_1 is to the right of the f-isocline, it will decrease. Let us further suppose that there are a set of values of n_1 and n_2 such that $g(n_1, n_2) = 0$, or equivalently, $dn_2/dt = 0$ for those values of n_1 and n_2.

If the population of n_2 is below the g-isocline, it will increase. Likewise, if the n_2 population is above the g-isocline, it will decrease. There are four different ways the two isoclines can relate to one another; either allele 1 will outcompete allele 2, allele 2 will outcompete allele 1, there is a stable equilibrium between allele 1 and 2, or finally, whichever allele is more frequent at the outset will outcompete the other.

We can now understand how this model represents both ecological and evolutionary factors. The f-isocline intersects the axis at K_{11}. In this circumstance, the population consists only of allele 1 and K_{11} represents the number of allele 1 homozygotes that can maintain themselves in this environment. In other words, K_{11} is the carrying capacity of the allele 1 homozygotes. Likewise, the f-isocline intersects the axis at K_{12}. K_{12} is the number of allele 2 that can keep allele 1 from increasing and represents the carrying capacity of the environment for heterozygotes expressed in units of allele 1. We can similarly denote the end of points of the g-isocline as K_{22} and K_{21}. MacArthur concludes, "We have now replaced the classical population genetics of expanding populations, where fitness was r, as measured in an uncrowded environment, by an analogous population genetics of crowded populations where fitness is K" [1962, p. 149].

Let us now consider what in fact MacArthur accomplished theoretically in this and other examples. First, let me define the notion of a unifying theory.

> A *unifying theory* applies a single theoretical framework (for example, common state variables and parameters) to a variety of different phenomena.

Often philosophers of science consider a theory to be a *unifying theory* under just the conditions defined above (see [Friedman, 1974; Kitcher, 1989; Morrison, 2000] for discussion and debate). As Margaret Morrison writes of Newtonian mechanics and Maxwell's electrodynamics,

> The feature common to both is that each encompasses phenomena from different domains under the umbrella of a single overarching theory. Theories that do this are typically thought to have "unifying power"; they unify, under a single framework, laws, phenomena or classes of facts originally thought to be theoretically independent of one another [2000, p. 2].[8]

If unification with respect to scientific theories or models minimally consists in a "single overarching theory" accounting for a variety of phenomena, then it appears that MacArthur's framework could not have unified population biology. If one examines the various models that MacArthur devised—equilibrium biogeography, limiting similarity, and density-dependent selection for example—they form an extremely diverse group. The state variables and parameters are rarely the same across models; they are rarely even representing the same phenomena. The

[8]Here the term 'unify 'does not specify whether a theory explains, is true of, or is empirically adequate with regard to a diverse set of phenomena.

state variables of the limiting similarity models are population abundances and the parameters are intrinsic rates of growth, carrying capacities and interaction coefficients, whereas in the equilibrium models of island biogeography the state variable is species richness and the parameters are rates of immigration and extinction. Likewise, in the density-dependent selection model presented above, the state variables are populations of alleles and the parameters are carrying capacities. There is no common overarching structural framework to unify population genetics, population and community ecology, and biogeography.[9] This methodology certainly does not generate a theory like Newtonian mechanics which consists in a small set of schematic equations concerning the motion of objects. Hence, MacArthur provided no theoretical framework of the sort needed to unify population biology.

Nonetheless, MacArthur did show how one could represent ecological, evolutionary and biogeographical factors at different scales in mathematical models. These different areas of population biology had largely proceeded independently of one another. However, if evolutionary and ecological processes are commensurate, then it was increasingly important to theoretically integrate these different processes at work in biological systems. It surely is correct that MacArthur "brought population and community ecology within the reaches of genetics" as claimed by Wilson and Hutchinson. However, he did not do so by "reformulating many of the parameters of ecology, biogeography, and genetics into a common framework of fundamental theory". We can now see how MacArthur approached the relations between theories. Here is another definition.

> An *integrating theory* takes a variety of theories (different state variables and parameters) and combines them in their application to a variety of phenomena.

He supplied a variety of models that incorporated many different evolutionary and ecological state variables and parameters and devised equations representing their interactions thus taking a first step toward integrating population biology.

Integrated models demonstrate how a variety of causal ecological, evolutionary, and biogeographical factors may interact. Whether any such general models will be explanatorily or predictively accurate depends on there being patterns of interest. Put differently, here is "MacArthur's bet":

> If general ecological modeling is to be successful, then there must be discoverable general ecological patterns.

Some ecologists and philosophers have argued that in fact there are few if any discoverable general patterns and thus general ecological modeling and the success it

[9]This is not to say that there is nothing that these models have in common of course. However, the common ingredients are usually that the models represent equilibrium behavior, make important optimality assumptions, and are represented with deterministic equations. Nonetheless, that which is at equilibrium is sometimes population abundances, species numbers, or population of alleles. Likewise, what is considered optimal is sometimes phenotypes, sometimes genes and sometimes the numbers of species and their abundances in a community.

can achieve must be rethought. As ecologists Arthur Dunham and Steven Beaupre write,

> Ecologists have also established that very few general principles apply to all ecological systems and remain valid irrespective of spatial, temporal, or organismal scales.... However, most processes or principles that ecologists use to understand the patterns they study are not general because they are valid over a restricted range of spatial, temporal, or organismal scales (= the domain of generality of a given process or principle). [1998, p. 28]

Kim Sterenly not only notes the debate over general patterns but offers one reason why they might be absent when he writes,

> The worry posed by extreme versions of the contingency hypothesis is that there are no patterns at all. The thought here is that membership and abundance within a community is sensitive to so many causal factors that we cannot project from one community to another. [2001, pp. 158–159]

The "contingency hypothesis" can be understood as the claim that ecological systems are sensitively-dependent on their prior states in the following sense: if the system's state at time t had been at all different, then the system at $t + \Delta t$ would be significantly different. As an example, in 1883 several volcanic explosions removed all of the biota from the island of Krakatoa. The reassembled island biota were the product of its area, the distance from mainland, and the species pool. However, the order of arrival of species was an important factor in determining who survives; and if things had been different, so would species identity, richness, and evenness.[10]

MacArthur's own response to the "contingency hypothesis" was to claim essentially that not all ecological systems are equally contingent. He writes concerning the spatiotemporal patterns of species,

> Ecological patterns, about which we construct theories, are only interesting if they are repeated. They may be repeated in space or time, and they may be repeated from species to species. A pattern which has all of these kinds of repetition is of special interest because of its generality, and yet these very general events are only seen by ecologists with rather blurred vision. The very sharp-sighted always find discrepancies and are able to say that there is no generality, only a spectrum of special cases. [MacArthur, 1968, p. 159]

In other words, some patterns are general though they will have exceptions. Nevertheless, they are to be explained by models which depict those causal processes

[10]MacArthur recognized sensitive-dependency in ecological systems. For example, in the Lotka-Volterra model of interspecific competition, there are circumstances where one of two species will exclude the other based on their initial abundances, which are contingent on a variety of factors.

which generate those patterns. The most direct way to respond to those who worry about contingency is to offer existing general patterns. Here are some general patterns that ecologists have discovered.

- The number of species in most groups of organisms increases along a gradient from the temperate zone to the tropics.
- There is the species-area effect that we saw above discussed in island biogeography.
- There is the phenomena of morphological convergence among unrelated taxa (say African and Neotropic rain forest mammals) which suggests these species occupy similar niches or functional roles.
- There are a variety of allometric relationships in ecology. As one example from physiological ecology, there is an allometric relationship between resting metabolic rate measured by the amount of oxygen consumed per hour and body mass in mammals.

These are not the only patterns available but they are famous and have stood the test of time. MacArthur also notes,

> The theme running through this book is that the structure of the environment, the morphology of the species, the economics of species behavior, and the dynamics of population changes are the four essential ingredients of all interesting biogeographic patterns. [1972, p. 1]

Here he himself is arguing that there are constraints of general models explaining patterns; specifically, if the environmental structure, species' morphology, its "economics" and dynamics are similar enough, then we should expect common explanations of patterns. However, if they differ, then so should their explanations.

5 CONCLUSION

In this essay, we have explored the work of Robert H. MacArthur, one of the most important ecologists of the twentieth century. After a brief biographical sketch and introduction to the elements of the "MacArthur School", we have seen a tale of two models—the equilibrium model of island biogeography and that of limiting similarity. Both garnered support but I have argued that the former did much better than the latter both in the empirical details and in the subsequent work that it spawned. Finally, we have considered the role of generality in MacArthur's work, provided a philosophical defense of the claim that MacArthur integrated rather than unified population biology, and attempted to rebut the charge that general model explanations required non-existent general patterns.

BIBLIOGRAPHY

[Abrams, 1983] P. Abrams. The Theory of Limiting Similarity. *Annual Review of Ecology and Systematics* 4: 359–76, 1983.

[Connor and Simberloff, 1979] E. F. Connor and D. Simberloff. The Assembly of Species Communities: Chance or Competition? *Ecology* 60: 1132–1140, 1979.

[Diamond, 1975] J. Diamond. Assembly of Species Communities, in M. L. Cody and J. M. Diamond (eds.), *Ecology and Evolution of Communities*. Cambridge, MA: Belknap Press of Harvard University Press, 1975.

[Dunham and Beaupre, 1998] A. Dunman and S. Beaupre, in Resetarits, W. J., and Bernardo, J. (eds.), *Experimental ecology: Issues and perspectives*. New York: Oxford University Press, 1998, pp. 27–49.

[Friedman, 1974] M. Friedman. Explanation and Scientific Understanding. *The Journal of Philosophy* 71(1): 5–19, 1974.

[Gilpin and Diamond, 1983] M. Gilpin and J. Diamond. Are Species Co-occurrences on Islands Non-Random, and Are Null Hypotheses Useful in Community Ecology? in Donald Strong, Daniel Simberloff, L. G. Abele, and A. B. Thistle (eds.), *Ecological Communities: Conceptual Issues and the Evidence*. Princeton, NJ: Princeton University Press, 1983.

[Gotelli, 1995] N. Gotelli. *A Primer of Ecology*. Sunderland: Sinauer Associates, 1995.

[Gotelli and Graves, 1996] N. Gotelli and G. Graves. *Null Hypotheses in Ecology*. Smithsonian, 1996.

[Horn and May, 1977] H. Horn and R. May. Limits to similarity among coexisting competitors. *Nature* 270: 660–661, 1977.

[Horn and Pianka, 2005] H. Horn and E. Pianka. Ecology's legacy from Robert MacArthur, in B. Beisner and K. Cuddington (eds.), *Ecological Paradigms Lost: Routes of Theory Change*, pp. 212–232. Elsevier, Amsterdam: Academic Press, 2005.

[Hutchinson, 1957] G. E. Hutchinson. Concluding Remarks. *Population Biology Animal Ecology and Demography, Cold Spring Harbor Symposia on Quantitative Biology* 22: 415–427, 1957.

[Hutchinson, 1958] G. E. Hitchinson. Homage to Santa Rosa, or Why are There So Many Kinds of Species? *The American Naturalist* 93: 145–159, 1958.

[Kingsland, 1995] S. Kingsland. *Modeling Nature*, 2nd Edition. Chicago: University of Chicago Press, 1995.

[Kitcher, 1989] P. Kitcher. Explanatory Unification and the Causal Structure of the World, in P. Kitcher and W. Salmon (eds.), *Scientific Explanation. Minnesota Studies in the Philosophy of Science*, vol. 13, pp. 410–505. Minneapolis: University of Minnesota Press, 1989.

[Leight, 1971] E. Leigh. *Adaptation and Diversity: Natural History and the Mathematics of Evolution*. Freeman, 1971.

[Levins, 1966] R. Levins. The Strategy of Model Building in Population Biology, *American Scientist* 54: 421–431, 1966.

[Levins, 1968] R. Levins. *Evolution in Changing Environments*. Princeton: Princeton University Press, 1968.

[Lewontin, 1974] R. Lewontin. *The Genetic Basis of Evolutionary Change*. New York: Columbia University Press, 1974.

[MacArthur, 1957] R. MacArthur. On the relative abundance of bird species, *Proceedings of the National Academy of Sciences USA* 43: 293–295, 1957.

[MacArthur, 1958] R. MacArthur. Population ecology of some warblers of northeastern coniferous forests. *Ecology* 39: 599–619, 1958.

[MacArthur, 1960] R. MacArthur. On the relative abundance of species. *American Naturalist* 94: 25–36, 1960.

[MacArthur, 1961] R. MacArthur. Population effects of natural selection, *American Naturalist* 95: 195–199, 1961.

[MacArthur, 1962] R. MacArthur. Some generalized theorems of natural selection, *Proceedings of the National Academy of Sciences USA* 48: 1893–1897, 1962.

[MacArthur, 1968] R. MacArthur. The Theory of the Niche, in Richard Lewontin (ed.), *Population Biology and Evolution*, Syracuse: Syracuse University Press, 1968.

[MacArthur, 1970] R. MacArthur. Graphical analysis of ecological systems, in *Some Mathematical Questions in Biology*, pp. 61–72. American Mathematical Society, 1970.

[MacArthur, 1972] R. MacArthur. *Geographical Ecology*. Princeton University Press, 1972.

[MacArthur and Horn, 1972] R. MacArthur. and H. Horn. Competition among fugitive species in a harlequin environment. *Ecology* 53: 749–752, 1972.

[MacArthur and Levins, 1964] R. MacArthur and R. Levins. Competition habitat selection and character displacement in a patchy environment. *Proceedings of the National Academy of Sciences, USA* 51: 1207–1210, 1964.

[MacArthur and Levins, 1967] R. MacArthur and R. Levins. The limiting similarity, convergence and divergence of coexisting species. *American Naturalist* 101: 377–385, 1967.

[MacArthur and May, 1972] R. MacArthur and R. May. Niche overlap as a function of environmental variability. *Proceedings of the National Academy of Sciences, USA* 69: 1109–1113, 1972.

[MacArthur and Pianka, 1966] R. MacArthur and E. Pianka. On optimal use of a patchy environment, *American Naturalist* 100: 603–609, 1966.

[MacArthur and Wilson, 1963] R. MacArthur and E. O. Wilson. An equilibrium theory of insular zoogeography. *Evolution* 17: 373–387, 1963.

[MacArthur and Wilson, 1967] R. MacArthur and E. O. Wilson. *The Theory of Island Biogeography.* Princeton University Press, 1967.

[Morrison, 2000] M. Morrison. *Unifying Scientific Theories: Physical Concepts and Mathematical Structures.* Cambridge University Press, 2000.

[Rosenzweig and MacArthur, 1963] M. Rosenzweig and R. MacArthur. Graphical representation of stability conditions of predator-prey interactions. *American Naturalist* 97: 209–223, 1963.

[Shrader-Frechette and McCoy, 1993] K. S. Shrader-Frechette and E. D. McCoy. *Method in Ecology.* Cambridge: Cambridge University Press, 1993.

[Simberloff, 1974] D. Simberloff. Equilibrium Theory of Island Biogeography, *Annual Review of Ecology and Systematics* 5: 161–82, 1974.

[Simberloff, 1976] D. Simberloff. Species Turnover and Equilibrium Island Biogeography. *Science* 194: 572–8, 1976.

[Simberloff and Wilson, 1969] D. Simberloff and E. O. Wilson. Experimental Zoogeography of Islands: The Colonization of Empty Islands. *Ecology* 50: 278–96, 1969.

[Sterelny, 2001] K. Sterelny. Darwin's Tangled Bank, in *The Evolution of Agency and Other Essays.* Cambridge: Cambridge University Press, 2001.

[Wiens, 1992] J. Wiens. *The Ecology of Bird Communities.* Cambridge: Cambridge University Press, 1992.

[Wilson, 1971] E. O. Wilson. *A Primer of Population Biology.* Sinauer Associates, 1971.

[Wilson, 1993] E. O. Wilson. *Naturalist.* Harper Collins, 1993.

[Wilson and Hutchinson, 1982] E. O. Wilson and G. E. Hutchinson. Robert Helmer MacArthur 1930–1972, A Biographical Memoir. *National Academy of the Sciences*, 1982.

EMBODIED REALISM AND INVASIVE SPECIES

Brendon M. H. Larson

1 INTRODUCTION

In summer 2002, a new species of beetle, the emerald ash borer (EAB), was detected on ash trees in southwestern Ontario, Canada and in adjacent Michigan, USA. It was new to this region, at least, having recently arrived in solid wood packing material from Asia. Given that ash species were a dominant component of regional forests and that EAB soon demonstrated its ability to spread and to kill most adult trees, the potential economic impact of its spread in the United States alone was estimated at $282 billion [Poland and McCullough, 2006], not including tremendous aesthetic and ecological changes to both rural and urban landscapes. Consequently, government agencies enacted a number of measures to prevent its spread. In Essex County, Ontario, healthy ash trees in the vicinity of an infestation were cut down and burned. Furthermore, all ash trees within a 10-km wide "firewall" zone along the eastern edge of the county were cut down to help slow the eastward spread of EAB. The entire county was quarantined by regulations that prevented people from moving ash wood out of the region.

To conservation biologists and ecologists, narratives such as this one are by now familiar, with the EAB simply the latest in a long list of "invasive species" (for review, see [Mack *et al.*, 2000]). For biologists, invasive species provide opportunities to study diverse questions in ecology, evolutionary biology, and related fields [Sax *et al.*, 2005]. In the case of EAB, for example, there has been extensive research into how it spreads and how we might manage it (*e.g.*, [Muirhead *et al.*, 2006]). There has been much less consideration of our conception of this situation, and of invasive species in general. It simply seems commonsensical that there is a border between areas that have been invaded and those that have not, between an inside and an outside. It also seems commonsensical to think of these species as moving across this boundary, exerting effects on species on the other side. Hence, we erect firewalls to prevent their spread. It is this commonsense characterization, however, that I wish to investigate here.[1]

[1]Ecologists also investigate the process of invasion in contexts other than that of invasive species, such as how tree seedlings invade an old field. While the issues considered herein can be generalized to such cases, I will focus on invasive species because "invasive" in this context has a more normative overtone that helps to demonstrate the dualities discussed below.

Handbook of the Philosophy of Science. Volume 11: Philosophy of Ecology.
Volume editors: Kevin deLaplante, Bryson Brown and Kent A. Peacock. General editors: Dov M. Gabbay, Paul Thagard and John Woods.

I will first introduce two general ways that invasive species have been defined, and then consider how these echo the long-standing debate between realist and constructivist philosophies of nature. This debate tends to reinforce conceptual dualities, and I thus propose embodied realism as a means to navigate between the extremes. I specifically examine how the metaphors of invasion biology exemplify embodied realism, with the core of the chapter investigating the image schemata of the field. Finally, I conclude by revisiting the extent to which embodied realism helps us to understand the conceptual underpinnings of invasion biology.

1.1 Two definitions of invasive species and their implications

There is extensive debate about the concept of an invasive species (reviewed in [Colautti and MacIsaac, 2004]). A key element of the usual definition is that they are non-indigenous (also called non-native or alien); that is, people have introduced them to a new region, either intentionally or unintentionally.[2] Either way, the definition then dichotomizes [Lodge *et al.*, 2006; Ricciardi and Cohen, 2007]. To many ecologists, invasive species are defined as non-indigenous species that spread and tend to become abundant in the new region. In contrast, policy papers, legislation, and some ecologists tend to append an additional component to this definition: invasive species are not just invasive, but they *also* cause some form of ecological or economic harm.

Both definitions have in common the concept of invasion, but they differ in terms of their emphasis on impact. This reflects a broader debate about the moral implications of invasive species. Brown and Sax [2004], for example, "plead for more scientific objectivity and less emotional xenophobia" [p. 531] in the study of invasive species, since they are simply "unintentional experiments" that provide a novel opportunity for obtaining insight into how the natural world functions. A key basis of this perspective is that species have always spread around the globe, and that the current biotic interchange is no different. Implicitly, or sometimes explicitly, such views tend to contribute to a less policy-oriented science (see [Larson, 2007a]). Some might fear that they lead to apathy by reducing the incentive to protect "natural" systems.

In contrast, other ecologists emphasize the distinctiveness of modern invasions relative to historic ones, and thus the detrimental effect of some invasive species on native communities and species (*e.g.*, [Cassey *et al.*]). They feel compelled to actively defend these landscapes, and tend to more actively advocate for policy regarding invasive species. They might also feel betrayed by the first camp; Simberloff [2006], for example, denounces the work of some critics as "a rearguard action to convince biologists and the lay public that the ecological threat from introduced species is overblown" [p. 915]. By implication, the scientific questions

[2]Normally, native species are not included in definitions of invasive species, though some ecologists would broaden the definition of invasive species to include native as well as non-native species (*e.g.*, [Houlahan and Findlay, 2004]). The recent literature has begun to refer to superabundant species, either native or non-native ones that have invasive tendencies.

addressed by this group tend to emphasize more applied dimensions of how to reduce the impacts of invasive species.

1.2 Constructivism and realism in invasion biology

This debate can be put in the broader context of the long-standing debate between realist and constructivist philosophies of nature. A realist view of the natural world assumes that it is "real and knowable" and that "facts are not just made-up things ... but rather are claims about the real world that are true to the extent that they correspond to this reality" [Proctor, 2001, p. 231]. It thus makes claims to universality. Whereas many environmental and ecological thinkers have taken for granted this scientific view of the world, some have begun to question the extent to which the resulting understanding is "constructed" [Evernden, 1992; Cronon, 1995]. The perspective of social constructivism serves to remind us

> that any descriptive or normative pronouncement people make on nature is never innocent of its human origins ... we cannot say anything more about [nature] without relying on human modes of perception, invoking human conceptual apparatus, involving human needs and desires—in short, when we speak of nature we speak of culture as well. [Proctor, 2001, p. 229]

In its extreme form, however, critics have argued that this view denies what is actually significant in nature, thus stealing the thunder of those intent on protecting it from plundering humankind [Soulé and Lease, 1995; Crist, 2004].[3] Unfortunately, something has been lost in the extremism and smoke-and-mirrors of this debate. On the one hand, many scientists acknowledge a basic constructivism in human knowing, and on the other, few constructivists deny there is a "reality out there."

Nonetheless, this discussion relates to some of the recent debate about invasive species cited earlier. A geographer has concluded, for example, that "the native/alien polarity is a subset of the discredited nature/culture duality, [so] its conceptual foundations seem irredeemably fractured," [Warren, 2007, p. 427] or, as another put it, "The status and identification of any species as an invader, weed, or exotic are conditioned by cultural and political circumstances" [Robbins, 2004, p. 139].[4] From the other side, an outspoken ecologist denounces a particular conception of invasion biology because it is

> essentially a version of the strong program of social construction of the science, an example of an approach by a small minority of sociologists who construe developments in the sciences as reflecting social

[3]This debate was part of the more general "science wars" between those standing behind the authenticity of scientific truth claims and those arguing for their constructedness, evinced in the polemic of the Sokal [1996] hoax.

[4]Note, however, that the emphasis here is on "conditioning" rather than causation, a marker that this form of construction does not deny material reality.

> factors and the psychology of its practitioners rather than advances in understanding the workings of the universe. [Simberloff, 2006, p. 916]

Such exchanges typify the misunderstanding between the two parties in realist-constructivist debates, with each side presenting a caricature of the other. I wish to skirt simplistic and extremist versions of this discussion that were propounded during the science wars by recognizing that environmental problems are both real *and* constructed. The main challenge lies in explicating what that might mean [Proctor, 2001]. As a starting point, we might recognize that there is truth to both sides. We cannot deny that there are species moving here, and scientists can certainly study that phenomenon as they have any other. Nonetheless, the narrative of invasive species depends on particular notions of space, time and human agency that have seldom been explicated. Thus, we certainly need to question how the category of invasive species has been created here, and even more so, the political and social elements of its construction. In particular, why and how has the leap from "is" to "ought" across the naturalistic fallacy occurred? Numerous philosophers have explored such issues in relation to invasive species [Botkin, 2001; Shrader-Frechette, 2001; Woods and Moriarty, 2001; Lodge and Shrader-Frechette, 2003] yet "the ambiguities [they have uncovered] may perhaps have been neglected or glossed over in the haste to sound the alarm of a crisis" [Foster and Sandberg, 2004, p. 180].

To further clarify social constructivism, we can turn to Hacking [1999]. He explains that it simply interrogates the status quo by demonstrating that X, being what is constructed, "appears to be inevitable" [p. 12], yet "need not have existed, or need not be at all as it is. X ... is not determined by the nature of things; it is not inevitable" [p. 6]. In the current context, we largely take invasive species for granted as a phenomenon because they certainly seem real enough as "objects" [*sensu* Hacking, 1999]. It is our "idea" of invasive species, our way of thinking about them, that is more questionable, and which I will thus examine here.[5] Complex facts such as this are likely to be profitably examined from a constructivist perspective.

Realist and constructivist philosophies related to invasive species particularly appear in terms of the nature of ecological communities. If we perceive native communities as natural kinds, as static entities, then we are more likely to adamantly defend them against invasive species. In the case of EAB, for example, we seek to prevent its impact on "native ash forests" [Muirhead *et al.*, 2006, p. 76]. More generally, we might be realists about invasive species if we perceive them as unproblematically *there*, as a real ongoing phenomenon out there in the world. The solution thus becomes a scientific one. Furthermore, realism about facts typically correlates with realism about values—in this case, it helps support the view that we

[5] Hacking [1999] further notes that constructionist arguments are often extended in a normative and activist direction that devalues X and thus seeks to transform it by "raising consciousness." While part of my motivation for examining invasive species derives from concern about whether our framing of them is appropriate or helpful [Larson, 2005; 2007b], I will only briefly consider this normative aspect here.

ought to prevent invasive species from spreading. While this perspective allows us to scientifically examine the phenomenon, it tends to overlook the theory-ladenness of our observations, the extent to which the observer is always present [Larson, 2007c]. In contrast, if we perceive native communities as temporary assemblages of individualistic species, we might be less concerned about invasive species [Soule, 1990; Botkin, 2001; Brown and Sax, 2005; Vermeij, 2005]. They become less convincing as a category, instead being mere instances of the propensity for all species to move around. Their significance is thus constructed. Though these perspectives are quite oppositional they both contain important elements that engage in our conception of these species.

1.3 Constitutive metaphors and embodied realism

The concept of invasion underlies the field of invasion biology, as exemplified by its role in both of the definitions given above. It is thus a critical concept to examine in terms of the realist-constructivist debate over invasive species because it is taken for granted. And given its metaphorical basis, we first need to consider the role of metaphor in science. Many scientists historically denigrated metaphors such as this as deviant, rhetorical embellishment. However, they are no longer considered mere rhetoric because myriad historical, philosophical and sociological studies have demonstrated that they are integral to scientific practice (*e.g.*, [Kuhn, 1979; Bono, 1990; Brown, 2003]). Cognitive linguists have also shown that metaphor is not just a matter of words, but of thought as well [Lakoff and Johnson, 1980; Johnson, 1987]. In particular, many "dead" metaphors—which have become so ossified that we consider them literal—are "metaphors we live by." Two classic examples include the conceptual metaphors Time is Money (*e.g.*, "How will you spend your weekend?") and Argument is War (*e.g.*, "His criticisms were right on target"), metaphors that structure our normal understanding of time and argument, respectively. Metaphors such as these are neither deviant nor expendable. Rather, they may constitute our interpretation of the world, thereby undermining standard conceptions of the distinction between literal and metaphoric and even of "truth" itself. The concept of invasion in the field of invasion biology is a case in point.

Furthermore, metaphors such as these may in fact be constitutive, defined by Boyd [1979] as those which form "an irreplaceable part of the linguistic machinery of a scientific theory; cases in which there are metaphors ... for which no adequate literal paraphrase is known" [p. 360]. It is often debatable whether particular scientific metaphors are actually constitutive or merely heuristic. Two ecologists, for instance, recently accented the importance of metaphors yet limited them to an heuristic role: "Metaphors ... are crucial stimuli to synthesis and innovation ... [but] in a mature science, the metaphorical assumptions must be stripped from the core definition" [Pickett and Cadenasso, 2002, pp. 6, 8]. Unfortunately, the conceptual foundations of our language may neutralize this intent. For example, while Simberloff [2006], citing Boyd, states that "constitutive metaphors are invitations to future research, including research into the degree of analogy between the developing concept and the referent of the metaphor" [p. 917], that is only true to the extent that we remain conscious of the power of the metaphors that already hold us. I will show herein that metaphors such as invasion may not be that accessible.

The accessibility of constitutive metaphors may relate to their scale. Metaphors may range from cognitive metaphor [Lakoff and Johnson, 1980], through discourse metaphor (*e.g.*, metaphor used at the level of everyday conversation to promote ecological ideas, [Zinken *et al.*, 2008]), and on to root metaphor (such as large-scale underlying meta-tropes, including those at the scale of "system" or "holism," [Taylor, 1988]). Each of these may influence the way we conceptualize and thus approach ecological systems. These levels may also reinforce one another, thus making particular metaphors more naturalized and, hence, less open to change. Take invasion for example. Otis [1999] magisterially analyzed the constitutive role of invasion metaphors in nineteenth-century medicine and demonstrated how they interwove literature, politics and science. In the case of invasion biology more specifically, an historian has claimed that the use of the invasion metaphor in this field derives from political geography [Moore, 2005], and some ecologists have opined that initial concerns about invasive species arose from related concerns about Nazi invasion [Davis *et al.*, 2001]. In short, such large-scale cultural factors may instantiate a metaphor and simultaneously make it relatively inaccessible and intransigent. At the same time, they may be reinforced by cognitive factors that are the focus of this chapter.

Specifically, constitutive metaphors may provide an avenue to what Hayles [1991] calls constrained constructivism. This perspective seeks a middle ground for environmentalist concern between the extremes of scientific realism and social constructivism. Her main claim is that access to the "unmediated flux" is mediated by the form of our embodiment, our particular ways of interacting with the world from specific positions within it. Thus, the key insight here is that there is no view from nowhere. However, constrained constructivism *also* recognizes that scientific inquiry puts constraints on possibility, so it is not totally deconstructive. Science can reject some possibilities. Constrained constructivism simply acknowledges the limitations on our knowing and thus that other possibilities exist.

A related view has been developed in the field of cognitive linguistics, which investigates the specific ways in which our embodiment influences our interaction with the world. Some of these ideas reflect Kantian precursors, specifically the notion of embodied or experiential realism that describes how metaphors reflect the cognitive structure of our way of being in the world [Lakoff and Johnson, 1980; Brown, 2003]. As explained by Brown [2003], embodied realism "makes a case that we know the world only in terms of perceptions, categorizations, and reasoning, both conscious and unconscious, grounded in our bodily capacities and life experiences and inherently limited by them" [p. 187]. Surprisingly, this cognitive linguistic perspective has seldom been considered within the philosophy of ecology, except for occasional blanket citation of Lakoff and Johnson [1980] yet without up-dated analysis that reflects how the field of cognitive linguistics has developed over nearly three intervening decades) *e.g.*, [Lakoff and Johnson, 1999; Frank *et al.*, 2008]).

It may be that philosophers of biology resist metaphor analysis because it is "just linguistics" and seems to be a throw-back to the antiquated view that language is representational [Bono, 2003]. As we well know, however, even semantic analysis can contribute to a science such as invasion biology. For example, Ricciardi and Cohen [2007] provides evidence that "invasive species" are not necessarily harmful ones, so it might bring clarity if the term referred to species that tend to spread rather than confounding this tendency with their impact. While such analyses are important, this chapter instead examines how we conceptualize the process of invasion, specifically in terms of the boundaries we invoke. I also aim to show that rather than just being a representational issue, such metaphoric constitution contributes to action and practice—metaphors are performative. As explained by Bono [2003], this means that "the work of metaphor ... is not so much to represent features of the world, as to invite us to *act upon* the world *as if* it were configured in a specific way *like* that of some already known entity or process" [p. 227]. Let us now examine how this might operate in the field of invasion biology.

2 THE IMAGE SCHEMATA OF INVASION BIOLOGY

This chapter focuses on image schemata as an instantiation of embodied realism in our conception of invasive species. Image schema are "dynamic analog representations of spatial relations and movements in space" [Gibbs, 1999, p. 354], and they are studied extensively in the field of cognitive linguistics. They are crucial to human understanding because they organize our cognition in fundamental, preconceptual ways, as will be demonstrated below.[6] Image schemata are metaphoric in that they derive from our bodily experiences and are projected so that we can understand "external" phenomena. In invasion biology, three key schemata are CONTAINER[7] and PATH (which together give rise to the inside-outside duality and motion implicit in invasion), and FORCE dynamics (giving rise to how we think about the pressure exerted by invasive species). Here, I will follow cognitive linguistic tradition by providing textual examples that reveal these underlying schemata and patterns of thought to provide a better understanding of our conception of invasive species.

---

[6]Mandler [2006] reviews some of the empirical evidence that psychologists have found for the existence of image schemata, and Johnson [1987] and Lakoff and Johnson [1999] examine their existence more generally.

[7]I will use small CAPITALS to indicate image schema, following the convention in cognitive linguistics.

2.1 The CONTAINER *and* PATH *image schemata*

The concept of invasion relies on a particular way of understanding and relating to the world around us. A key component is the CONTAINER image schema, which envisions one's relation with the world in terms of a container, with the human inside and the rest of the world outside. This distinction between *inside* and *outside* can be projected onto the world as a means to structure and understand it, a process which derives in part from the familiar human experience of having a boundary, the skin [Johnson, 1987; Rohrer, 1995; Chilton, 1996; Lakoff and Johnson, 1999]. Given extensive bodily experience of separate inside and outside, we project this CONTAINER image schema in various ways to understand the world. Lakoff and Johnson [1999] give the example of a bee in a garden:

> When we understand a bee as being *in* the garden, we are imposing an imaginative container structure on the garden, with the bee *inside* the container. The cognitive structure imposed on the garden is called *the container image schema.* That cognitive structure plays a *causal role* in bringing about an understanding—a conceptualization of the bee as being *in* something. [p. 117]

The CONTAINER image schema is also implicit in how we think about invasive species [Larson, 2008a]. The container exists around a pre-existing native community (or at larger scales, all the way to biogeographic regions and nations), the one present before a novel species arrives. When this species arrives, it crosses the boundary defining the container by entering the native community; in normal parlance, it invades it.

This notion of invading a pre-existent container depends on yet another schema, the PATH schema. This schema derives from our everyday experience of purposeful movement from a source to a goal or objective, and it is projected onto our understanding of the trajectory or path through space taken by invasive species. As explained by Lakoff and Johnson [1999, p. 33], this trajectory is conceptualized "as a linelike 'trail' left by an object as it moves and projected forward in the direction of motion." We thus conceptualize invasive species moving in such a manner, from a source somewhere else towards extant, integrated communities. At this point, our conception resonates strongly with our conception of political invasion and the invasion of our bodies by disease (see [Larson, 2008]). It is in part for this reason, though more intimately because of the nature of the PATH schema, that invasive species are often portrayed as having a negative *intent* or purpose.

At a relatively unconscious level, the CONTAINER and PATH image schemata together construct how we conceptualize the process of invasion. They thus constitute the field of invasion biology, as revealed by the very fact that our name for the field that studies *invasive* species is *invasion* biology. As with metaphors in general, however, these two schemata highlight some aspects of a relation while at the same time hiding others. A major implication of the CONTAINER schema, for example, is that we can define *inside* here, that is, that there is an enduring inner state that is "native" and which can be contrasted with an external "non-native" state. We see here the persistent, yet problematic boundary between nature and culture that often appears in invasion biology [Milton, 2000; Robbins, 2001]. In contrast, we might recognize that the pre-existing community already contains non-native elements because of the effects humans have had almost everywhere on the planet [Larson, 2007c]. Furthermore, this schema assumes that we can draw a boundary around an integrated community, despite the fact that few ecologists subscribe to the idea that communities are integrated wholes. An integrated community is also one that is implicitly in *balance*.[8]

The performativity of the CONTAINER image schema is instantiated by barrier zones (*e.g.*, the "firewall" of the introduction). A barrier zone can be defined as an area at the front of a population where eradication (or suppression) activity is performed in order to prevent or to slow population spread [Sharov and Liebhold, 1998]. As an example, managers cut ash trees in a 10-km wide by 30-km long barrier zone to prevent the spread of EAB, as discussed in the introduction [Muirhead *et al.*, 2006]. The establishment of this barrier zone plays out on the land the boundary between what pre-exists, inside the boundary, and that which impedes on it from the outside. Here we see humans reinforcing the capacity of natural systems to resist the spread of an invasive species, a force we explore next.

2.2 The FORCE *dynamic schema*

To further develop our understanding of the conceptualization of invasive species, next consider the title of a recent paper published in *Ecology*: "Ecological resistance to biological invasion overwhelmed by propagule pressure" [Von Holle and Simberloff, 2005]. There are a number of conceptual elements operating here, and cognitive linguistic analysis of FORCE dynamics helps to unpack them. But first, we need to expand and define some background terms. Communities can resist invasion in two main ways: environmentally, in terms of those abiotic factors affecting the establishment of an invading species, and biotically, defined as "ways in which the resident species repel invaders" [Von Holle and Simberloff, 2005, p. 3212]. Communities with low invasibility are able to exert enough pressure to prevent invaders from entering. From the other side, we have propagule pressure, which may be defined as "a composite measure of the number of individuals released into a region to which they are not native ... [which] incorporates estimates of the absolute number of individuals involved in any one release event (propagule size) and the number of discrete release events (propagule number)" [Lockwood *et al.*, 2005, p. 223].

[8] "Balance" here derives from yet another image schema, one that cognitive linguists attribute to our familiar sense of bodily balance [Gibbs, 1994]. While ecologists generally consider "balance of nature" an out-dated popularization, it has been argued that it still constitutes ecological theories and that it is "much more than an imprecise precursor of the theoretical concept of mathematical equilibrium" [Cuddington, 2001, p. 465]. Its bodily basis may be one reason for its entrenchment.

Even though the father of invasion biology, Charles Elton, is credited with the concept of biotic resistance to invaders, the joint concepts of biotic resistance and propagule pressure have only recently arisen in the literature. According to a search of ISI Web of Knowledge, the first occurrence of the term "propagule pressure" was Williamson and Fitter [1996]. Since then, it has rapidly grown in prominence (Figure 1), with a total of 138 citations from 1996–2007 (and 88% of them over the four years 2004–2007). Similarly, the phrase "biotic resistance" first occurred in Lake and Odowd [1991], but it has since grown exponentially (Figure 1), with a total of 97 relevant citations from 1991–2007 (and 77% of them over the three years 2005–2007).

Where do these notions come from? According to cognitive linguists, these embodied conceptions derive from our everyday experience of pressures and resistances as we abut against other objects and people [Johnson, 1987; Lakoff and Johnson, 1999; Talmy, 2000]. Again, the basic idea is that as embodied beings we daily experience pressure, and that this familiar experience may ground metaphorical understanding. Evolutionary biologist Richard Dawkins [cited by Gould, 1997], for example, characterizes natural selection as "the pressure that drives evolution up the slopes of Mount Improbable. Pressure really is a good metaphor. We speak of 'selection pressure' and you can almost feel it pushing a species to evolve, shoving it up the gradients of the mountain" [p. 1022]. Although Gould wryly comments, "Surely, we can do better" in response to this metaphor, biologists often conceptualize evolution in terms of selection pressure, and pressure recurs throughout biological conceptions [Young, 1993].

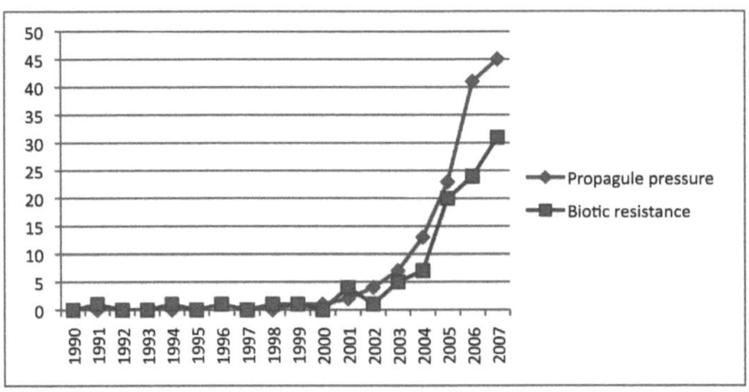

Figure 1. The rise in citations of propagule pressure and biotic resistance, 1990–2007. Figure shows results from search of ISI Web of Knowledge using the keywords (i) "invasi*" and "propagule pressure" and (ii) "invasi*" and "biotic resistance." Only records pertinent to invasion biology have been included in the data presented here, and related terms such as invasion resistance and ecological resistance have not been included.

In the case at hand, we have invasive species exerting propagule pressure on communities from the outside (presumably pressing on their boundary, in the sense discussed above), as represented in Figure 2. The more propagules there are, the greater the external pressure threatening to decompose the pre-existent "native something." From Newton's third law (and daily experience), we know that forces always occur in action-reaction pairs, and we do not have far to look for that oppositional force. Unified native communities exert a force outwards, biotic resistance, which opposes this propagule pressure (Figure 2). We can see such conceptions at play most markedly where the two metaphors occur in concert with one another, as in Von Holle and Simberloff [2005], but also in numerous other papers in the recent literature (*e.g.*, [D'Antonio *et al.*, 2001; Martin and Marks, 2006; Hollebone and Hay, 2007; Perelman *et al.*, 2007]). Regardless of whether these metaphors are made explicit, however, the pressure-resistance pairing is evident in how a wide range of biologists and environmentalists conceptualize and thus respond to invasive species.

As a further example, consider the phenomenon of invasional meltdown, where "interspecific facilitation leads to an accelerating increase in the number of introduced species and their impact" [Simberloff, 2006, p. 912]. Invasional meltdown is now "routinely considered in various explanations by ecologists, conservation biologists, and invasion biologists [and] it has entered the lay literature" [p. 916]. But what is it that is melting-down in an invasional meltdown? I would proffer that the meltdown is a loss of pressure within the pre-existing community and hence of its ability to resist invaders. Once the community loses this resistance pressure entirely, as the pressure of the introduced species and their impacts accelerate, it ceases to exist. It is deflated and over-run. Note the performativity of this metaphorical extension of pressure-and-resistance, despite the fact that "a full 'invasional meltdown' ... has yet to be conclusively demonstrated" [p. 912].

Interestingly, ISI Web of Knowledge also located 2,576 records for "invasive species" through the end of 2007, of which the first was in 1986.[9] We have already seen that invasion may be thought of as constitutive within invasion biology. The terms "biotic resistance" and "propagule pressure" first occurred five and ten years later, respectively, and appear to be an outgrowth of thinking in terms of invasion. For once you have invaders impinging on a container, from the outside, it becomes normal to then think of them in terms of the force they exert, and how this might be opposed by the ecological resistance of the existing community. The coincident rise of these two metaphors occurs in part because thinking in terms of one performatively leads to thinking in terms of the other because of FORCE dynamic conceptualization.

[9]This total may be a slight over-estimate since the data have not been combed for hits in ISI Web of Knowledge that do not correspond to invasive species in the sense pertinent to this discussion.

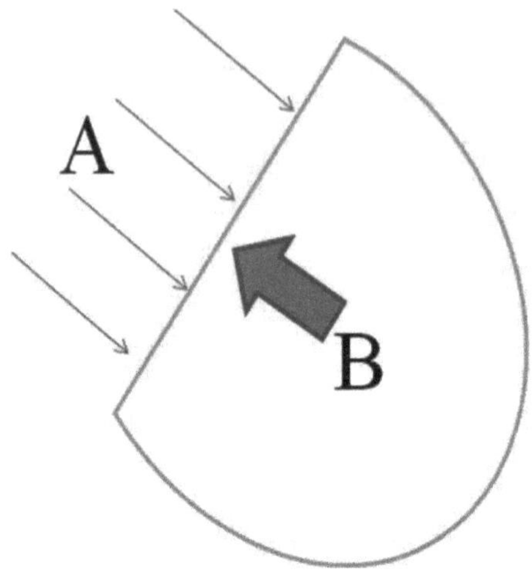

Figure 2. A FORCE dynamic representation of propagule pressure and biotic resistance. The four arrows (A) represent the propagule pressure exerted by a novel invasive species on a pre-existing community, represented by the polygon. The community exerts a force (B), biotic resistance, which opposes this external pressure.

3 DISCUSSION

I began this chapter by inquiring into our conception of invasive species. I then demonstrated that our conception of them depends on our embodiment, specifically in terms of cognitive schemata such as the CONTAINER and PATH image schemata and FORCE dynamics. Accordingly, the reality of invasive species can only be assessed in terms of how humans understand; it is not unembodied truth. Invasive species may nonetheless appear inevitable in the sense outlined by Hacking [1999] earlier, for we cannot conventionally conceptualize a world without boundaries and pressures. However, is our conception actually "inevitable" and in "the nature of things?" We might approach this question in two ways. First, is this conception universal? Is it common to people regardless of their culture? Second, is it the only option? Is this the only way to conceptualize these species? If we answer both of these questions in the affirmative, we would have a more firm basis for concluding that our conception is effectively inevitable since we cannot realistically know the world in any other way.

Let's begin with the first question: Is this conception universal? While I have presented the cognitive schemata of invasion biology as embodied, we have not yet considered the extent to which they may vary depending on cultural context. The importance of this context is contested within cognitive linguistics, but many scholars conclude that image schemata are developmentally and culturally conditioned rather than innate and individualistic [Gibbs, R. W., Jr., 1999; Bono, 2003;

Frank *et al.*, 2008]. Thus, it is by no means clear that such schemata are universal, even if we might assume they are relatively constant in the case of Western scientists studying invasion. That said, I have already shown that biologists exhibit variation with regard to their views of invasive species, perhaps *in part* related to variation in the expression of particular image schemata.

The CONTAINER image schema, for example, relates to our conception of self versus other. While we may perceive an obvious boundary distinguishing ourselves from our surroundings—the skin, the question is whether there is any reason to emphasize this boundary rather than the tremendous flux across it (see [Brown and Toadvine, 2007]). While those of us raised in Western societies lean towards the former view, anthropologist Geertz [1979] observed that

> [t]he Western conception of the person as a bounded, unique, more or less integrated motivational and cognitive universe, a dynamic center of awareness, emotion, judgement, and action organized into a distinctive whole and set contrastively both against other such wholes and against a social and natural background is, however incorrigible it may seem to us, a rather peculiar idea within the context of the world's cultures. [p. 59]

In some of these cultures, invasive species may be perceived differently, with less expectation that what is inside should remain isolated and fixed over time. Bono [2003], for example, contrasts containment in Chinese and Western thought and claims that "the boundaries between what is inside and what is outside are differently drawn and, at its most extreme ... the very notion of a 'boundary' itself is differently constituted in the two cultures" [p. 221].[10] A key boundary here is the nature-culture boundary implicit within invasion biology, a boundary that tends to be less marked in many non-Western cultures. Thus, while this question is by no means closed, it appears that other cultures may place less emphasis on the crossing of boundaries by invasive species.

Second, we can ask whether there are alternatives to the standard conception of *invasive* species. These alternatives may be hard to find since "invasion" is a constitutive, entrenched metaphor. It leads us to conceptualize these species in a particular way, as entities that cross boundaries and exert pressure on intact communities or ecosystems that oppose them. Nonetheless, this boundary-laden Newtonian conception may be inadequate for understanding open biological systems whose boundaries may in fact be permeable and interactive. This metaphor has become self-fulfilling, however, in part because it is also performative. It also contributes to an oppositional response to these species, one predicated in part by FORCE dynamics as described above.

Nonetheless, at a trivial level this conception is not inevitable because every metaphor and schema highlights one aspect of a relation while hiding others. As

[10] On a related note, it has been found that people from Asian cultures tend to conceptualize individuality in terms of interdependence more than Americans (*e.g.*, [Markus and Kitayama, 1991]).

an example, we typically conceptualize arguments as war, as shown by everyday expressions such as "He attacked every weak point in my argument" and "I've never won an argument with him" [Lakoff and Johnson, 1980, p. 4]. It is possible, however, to view them instead as a dance between two partners, which emphasizes the mutual, perhaps positive exchange of ideas and opinions that can occur, dialogically, as two individuals seek consensus, rather than as a confrontational winner-take-all situation.

We might similarly benefit from the broader perspective provided by reframing invasive species. There are certainly alternatives to our current way of conceiving them (see also [Larson, 2007b; Keulartz and van der Weele, 2008]). For example, our typical conception reifies "inside" and "outside" rather than emphasizing simple movement of species. Hence, the discussion typically focuses on which particular species are present in a community (*composition*), and on maintaining it in a native/natural state. An alternative is to emphasize questions about whether desired/important *functions* are maintained [Callicott *et al.*, 1999; Hull, 2006], and whether some of these might be maintained regardless of nativity. Other conceptions might weaken the form of spatiality inherent in the CONTAINER image schema, such as ones that better acknowledge the connection between the spread of these species and human globalization and cosmopolitanism.

For alternatives, we may also turn to the two definitions introduced earlier. One of them accents the tendency of these species to spread, to become superabundant. The other accents the harm they cause in relation to human interests. Neither of these so strongly invoke the inside-outside duality implicit in the concept of "invasion." Similarly, there is no particular reason that either has to attach to non-native as opposed to native species. Recognizing these two dimensions, we might begin to refer to these species as either superabundant or harmful ones, depending on context and intent. Both of these would serve to partially break down the nature-culture, inside-outside dichotomy that remains as long as we continue to conceptualize these species as invasive ones.

Turning to the broader issues raised in this paper, embodied realism allows us to accept that invasive species are real. We can in that sense be realists about them. At the same time, however, we can see that our conception of them derives from the structured way in which humans have evolved to relate to the world. They are thus constructed. So it seems sensible to accept both views, which may help to further weaken any remaining duality between realism and constructivism.[11] Furthermore, whether or not image schemata are universal,[12] biologists (and the general population, more generally) certainly hold varied views on invasive species and their impacts. Embodied realism provides one avenue for a more rich under-

[11]Proctor [2001] sees the relation between these views as one that is fundamentally paradoxical, meaning that the usual attempts to resolve the situation only serve to leave out part of the whole picture.

[12]And this point gives rise to the as yet unmentioned challenge for any deconstructive enterprise: that any statements it makes are no more foundational than any other. Applied here, the science of cognitive linguistics is new enough that I would not want to claim that my hypotheses here about the image schemata of invasion biology are founded in stone.

standing of the reasoning that may underlie these varied perspectives. Ultimately, it thus encourages us to engage in open dialogue about the state of our planet and how we will choose to relate to issues such as invasive species in particular contexts.

In that context, this chapter has focused on the constitution of our *idea* of invasive species, rather than on the related *issue*. The "invasion" metaphor is also performative in terms of the fear-based resonance that it has, which is one reason that people become so adamant about defending landscapes against invasive species. Propagule pressure and biotic resistance are performative too, for when we think in this way, we tend to want to defend the systems that are there. We perhaps relate to the pressure these species exert on intact systems as we feel pressure on our skin. We oppose such unpleasant external pressures on our intact, reified selves; we oppose invasion of our nations as well as the invasion of our bodies by disease. We are thus led to oppose the spread of these species by what I have elsewhere called metaphoric resonance [Larson, 2006]. This resonance by-passes the naturalistic fallacy. By this means, the idea of invasive species inevitably turns into an issue, while we often neglect how this issue rests on a certain value-set that is not necessarily internally consistent and which is by no means acceptable to all [Rawles, 2004]. This is all the more reason for ongoing reflection and discussion about the circumstances under which we wish to enforce boundaries against invasive species.

ACKNOWLEDGEMENTS

I appreciate comments and suggestions from Paul Chilton, Roslyn Frank, Cor van der Weele, and especially Kevin deLaplante.

BIBLIOGRAPHY

[Bono, 1990] J. J. Bono. Science, discourse, and literature: The role/rule of metaphor in science. In S. Peterfreund (ed.), *Literature and Science: Theory and Practice*, pp. 59–89. Boston: Northeastern University Press, 1990.

[Bono, 2003] J. J. Bono. Why metaphor? Toward a metaphorics of scientific practice. In S. Maasen and M. Winterhager (eds.), *Science Studies: Probing the Dynamics of Scientific Knowledge*, pp. 215–233. Transcript Verlag, 2003.

[Botkin, 2001] D. B. Botkin. The naturalness of biological invasions. *Western North American Naturalist* 61: 261–266, 2001.

[Boyd, 1979] R. Boyd. Metaphor and theory change: What is "metaphor" a metaphor for? In A. Ortony (ed.), *Metaphor and Thought*, pp. 356–408. Cambridge: Cambridge University Press, 1979.

[Brown and Toadvine, 2007] C. S. Brown and T. Toadvine (eds.), *Nature's Edge: Boundary Explorations in Ecological Theory and Practice*. Albany, NY: State University of New York Press, 2007.

[Brown and Sax, 2004] J. H. Brown and D. F. Sax. An essay on some topics concerning invasive species. *Austral Ecology* 29: 530–536, 2004.

[Brown and Sax, 2005] J. H. Brown and D. F. Sax. Biological invasions and scientific objectivity: Reply to Cassey *et al.* (2005). *Austral Ecology* 30: 481–483, 2005.

[Brown, 2003] T. L. Brown. *Making Truth: Metaphor in Science.* Chicago: University of Illinois Press, 2003.

[Callicott *et al.*, 1999] J. B. Callicott, L. B. Crowder, and K. Mumford. Current normative concepts in conservation. *Conservation Biology* 13: 22–35, 1999.

[Cassey *et al.*, 2005] P. Cassey, T. M. Blackburn, R. P. Duncan, and S. L. Chown. Concerning invasive species: Reply to Brown and Sax. *Austral Ecology* 30: 475–480, 2005.

[Chilton, 1996] P. A. Chilton. The meaning of security. In F. A. Beer and R. Hariman (eds.), *Post-Realism: The Rhetorical Turn in International Relations*, pp. 193–216. East Lansing: Michigan State University Press, 1996.

[Colautti and MacIsaac, 2004] R. I. Colautti and H. J. MacIsaac. A neutral terminology to define 'invasive' species. *Diversity and Distributions* 10: 135–141, 2004.

[Crist, 2004] E. Crist. Against the social construction of nature and wilderness. *Environmental Ethics* 26: 5–24, 2004.

[Cronon, 1995] W. Cronon (ed.), *Uncommon Ground: Toward Reinventing Nature.* New York: W.W. Norton, 1995.

[Cuddington, 2001] K. Cuddington. The 'balance of nature' metaphor and equilibrium in population ecology. *Biology and Philosophy* 16: 463–479, 2001.

[D'Antonio *et al.*, 2001] C. M. D'Antonio, J. Levine, and M. Thomsen. Propagule supply and resistance to invasion: A California botanical perspective. *Journal of Mediterranean Ecology* 2: 233–245, 2001.

[Davis *et al.*, 2001] M. A. Davis, K. Thompson, and J. P. Grime. Charles S. Elton and the dissociation of invasion ecology from the rest of ecology. *Diversity and Distributions* 7: 97–102, 2001.

[Evernden, 1992] N. Evernden. *The Social Creation of Nature.* Baltimore: Johns Hopkins University Press, 1992.

[Foster and Sandberg, 2004] J. Foster and L. A. Sandberg. Friends or foe? Invasive species and public green space in Toronto. *The Geographical Review* 94: 178–198, 2004.

[Frank *et al.*, 2008] R. M. Frank, R. Dirven, T. Ziemke, and E. Bernárdez (eds.), *Body, Language and Mind. Volume 2: Sociocultural Situatedness.* New York: Mouton de Gruyter, 2008.

[Geertz, 1979] C. Geertz. *Local Knowledge.* New York: Basic, 1979.

[Gibbs, 1994] Gibbs, R. W., Jr. *The Poetics of Mind: Figurative Thought, Language, and Understanding.* Cambridge: Cambridge University Press, 1994.

[Gibbs, 1999] Gibbs, R. W., Jr. Taking metaphor out of our heads and putting it into the cultural world. In Gibbs, Raymond W., Jr. and G. J. Steen (eds.), *Metaphor in Cognitive Linguistics: Selected Papers from the Fifth International Cognitive Linguistics Conference*, pp. 145–166. Philadelphia: John Benjamins, 1999.

[Gould, 1997] S. J. Gould. Self-help for a hedgehog stuck on a molehill. *Evolution* 51: 1020–1023, 1997.

[Hacking, 1999] I. Hacking. *The Social Construction of What?* Cambridge: Harvard University Press, 1999.

[Hayles, 1991] N. K. Hayles. Constrained constructivism: Locating scientific inquiry in the theater of representation. *New Orleans Review* 18: 76–85, 1991.

[Hollebone and Hay, 2007] A. L. Hollebone and M. E. Hay. Propagule pressure of an invasive crab overwhelms native biotic resistance. *Marine Ecology-Progress Series* 342: 191–196, 2007.

[Houlahan and Findlay, 2004] J. E. Houlahan and C. S. Findlay. Effect of invasive plant species on temperate wetland plant diversity. *Conservation Biology* 18: 1132–1138, 2004.

[Hull, 2006] R. B. Hull. *Infinite Nature.* Chicago: University of Chicago Press, 2006.

[Johnson, 1987] M. Johnson. *The Body in the Mind: The Bodily Basis of Meaning, Imagination, and Reason.* Chicago: Chicago University Press, 1987.

[Keulartz and van der Weele, 2008] J. Keulartz and C. van der Weele. Framing and reframing in invasion biology. *Configurations* 16: 93–115, 2008.

[Kuhn, 1979] T. S. Kuhn. Metaphor in science. In A. Ortony (ed.), *Metaphor and Thought*, pp. 409–419. Cambridge: Cambridge University Press, 1979.

[Lake and Odowd, 1991] P. S. Lake and D. J. Odowd. Red crabs in rain-forest, Christmas Island—Biotic resistance to invasion by an exotic snail. *Oikos* 62: 25–29, 1991.

[Lakoff and Johnson, 1980] G. Lakoff and M. Johnson. *Metaphors We Live By.* Chicago: University of Chicago Press, 1980.

[Lakoff and Johnson, 1999] G. Lakoff and M. Johnson. *Philosophy in the Flesh: The Embodied Mind and its Challenge to Western Thought.* New York: Basic Books, 1999.

[Larson, 2005] B. M. H. Larson. The war of the roses: Demilitarizing invasion biology. *Frontiers in Ecology and the Environment* 3: 495–500, 2005.

[Larson, 2006] B. M. H. Larson. The social resonance of competitive and progressive evolutionary metaphors. *BioScience* 56: 997–1004, 2006.

[Larson, 2007a] B. M. H. Larson. An alien approach to invasive species: Objectivity and society in invasion biology. *Biological Invasions* 9: 947–956, 2007.

[Larson, 2007b] B. M. H. Larson. Thirteen ways of looking at invasive species. In D. R. Clements and S. Darbyshire (eds.), *Invasive Plants: Inventories, Strategies, and Action*, pp. 131–156. *Canadian Weed Science Society/Société Canadienne de Malherbologie, Sainte Anne de Bellevue, Quebec*, 2007.

[Larson, 2007c] B. M. H. Larson. Who's invading what? Systems thinking about invasive species. *Canadian Journal of Plant Sciences* 87: 993–999, 2008.

[Larson, 2008] B. M. H. Larson. Entangled biological, cultural and linguistic origins of the war on invasive species. In R. M. Frank, R. Dirven, T. Ziemke and E. Bernárdez (eds.), *Body, Language and Mind. Volume 2: Sociocultural Situatedness*, pp. 169–195. New York, NY: Mouton de Gruyter, 2008.

[Lockwood *et al.*, 2005] J. L. Lockwood, P. Cassey, and T. Blackburn. The role of propagule pressure in explaining species invasions. *Trends in Ecology & Evolution* 20: 223–228, 2005.

[Lodge *et al.*, 2006] D. M. Lodge, *et al.* Biological invasions: Recommendations for US policy and management. *Ecological Applications* 16: 2035–2054, 2006.

[Lodge and Shrader-Frechette, 2003] D. M. Lodge and K. Shrader-Frechette. Nonindigenous species: Ecological explanation, environmental ethics, and public policy. *Conservation Biology* 17: 31–37, 2003.

[Mack *et al.*, 2000] R. N. Mack, D. Simberloff, W. M. Lonsdale, H. Evans, M. Clout, and F. A. Bazzaz. Biotic invasions: Causes, epidemiology, global consequences, and control. *Ecological Applications* 10: 689–710, 2000.

[Mandler, 2006] J. M. Mandler. *Foundations of Mind: Origins of Conceptual Thought.* New York: Oxford University Press, 2006.

[Markus and Kitayama, 1991] H. R. Markus and S. Kitayama. Culture and the self: Implications for cognition, emotion, and motivation. *Psychological Review* 98: 224–253, 1991.

[Martin and Marks, 2006] P. H. Martin and P. L. Marks. Intact forests provide only weak resistance to a shade-tolerant invasive Norway maple (*Acer platanoides* L.). *Journal of Ecology* 94: 1070–1079, 2006.

[Milton, 2000] K. Milton. Ducks out of water: Nature conservation as boundary maintenance. In J. Knight (ed.), *Natural Enemies: People-Wildlife Conflicts in Anthropological Perspective*, pp. 229–246. New York: Routledge, 2000.

[Moore, 2005] J. Moore. Revolution of the space invaders: Darwin and Wallace on the geography of life. In D. N. Livingstone and C. W. J. Withers (eds.), *Geography and Revolution*, pp. 106–132. Chicago: University of Chicago Press, 2005.

[Muirhead *et al.*, 2006] J. Muirhead, B. Leung, C. van Overdijk, D. Kelly, K. Nandakumar, K. Marchant, and H. MacIsaac. Modelling local and long-distance dispersal of invasive emerald ash borer *Agrilus planipennis* (Coleoptera) in North America. *Diversity and Distributions* 12: 71–79, 2006.

[Otis, 1999] L. Otis. *Membranes: Metaphors of Invasion in Nineteenth-Century Literature, Science, and Politics.* Baltimore: Johns Hopkins University Press, 1999.

[Perelman *et al.*, 2007] S. B. Perelman, E. J. Chaneton, W. B. Batista, S. E. Burkart, and R. J. C. Leon. Habitat stress, species pool size and biotic resistance influence exotic plant richness in the Flooding Pampa grasslands. *Journal of Ecology* 95: 662–673, 2007.

[Pickett and Cadenasso, 2002] S. T. A. Pickett and M. L. Cadenasso. The ecosystem as a multidimensional concept: Meaning, model, and metaphor. *Ecosystems* 5: 1–10, 2002.

[Poland and McCullough, 2006] T. Poland and D. McCullough. Emerald ash borer: Invasion of the urban forest and the threat to North America's ash resource. *Journal of Forestry* 104: 118–124, 2006.

[Proctor, 2001] J. D. Proctor. Solid rock and shifting sands: The moral paradox of saving a socially constructed nature. In N. Castree and C. S. Brown (eds.), *Social Nature*, pp. 225–239. New York: Wiley-Blackwell, 2001.

[Rawles, 2004] K. Rawles. Biological diversity and conservation policy. In M. Oksanen and J. Pietarinen (eds.), *Philosophy and Biodiversity*, pp. 199–216. New York: Cambridge University Press, 2004.

[Ricciardi and Cohen, 2007] A. Ricciardi and J. Cohen. The invasiveness of an introduced species does not predict its impact. *Biological Invasions* 9: 309–315, 2007.

[Robbins, 2004] P. Robbins. Comparing invasive networks: Cultural and political biographies of invasive species. *The Geographical Review* 94: 139–156, 2004.

[Robbins, 2001] P. Robbins. Tracking invasive land covers in India, or why our landscapes have never been modern. *Annals of the Association of American Geographers* 91: 637–659, 2001.

[Rohrer, 1995] T. Rohrer. The metaphorical logic of (political) rape: The new wor(l)d order. *Metaphor and Symbolic Activity* 10: 115–137, 1995.

[Sax *et al.*, 2005] D. F. Sax, J. J. Stachowicz, and S. D. Gaines (eds.), *Species Invasions: Insights into Ecology, Evolution, and Biogeography.* Sunderland: Sinauer Associates, 2005.

[Sharov and Liebhold, 1998] A. A. Sharov and A. M. Liebhold. Bioeconomics of managing the spread of exotic pest species with barrier zones. *Ecological Applications* 8: 833–845, 1998.

[Shrader-Frechette, 2001] K. Shrader-Frechette. Non-indigenous species and ecological explanation. *Biology and Philosophy* 16: 507–519, 2001.

[Simberloff, 2006] D. Simberloff. Invasional meltdown 6 years later: Important phenomenon, unfortunate metaphor, or both? *Ecology Letters* 9: 912–919, 2006.

[Sokal, 1996] A. Sokal. A physicist experiments with cultural studies. *Lingua Franca* May/June, 62–64, 1996.

[Soule, 1990] M. E. Soule. The onslaught of alien species, and other challenges in the coming decades. *Conservation Biology* 4: 233–239, 1990.

[Soulé and Lease, 1995] M. E. Soulé and G. Lease. *Reinventing Nature: Responses to Postmodern Deconstruction.* Washington: Island Press, 1995.

[Talmy, 2000] L. Talmy. *Toward a Cognitive Semantics. Volume 1. Concept Structuring Systems.* Cambridge: MIT Press, 2000.

[Taylor, 1988] P. J. Taylor. Technocratic optimism, H. T. Odum, and the partial transformation of ecological metaphor after World War II. *Journal of the History of Biology* 21: 213–244, 1988.

[Vermeij, 2005] G. J. Vermeij. Invasion as expectation: A historical fact of life. In D. F. Sax, J. J. Stachowicz and S. D. Gaines (eds.), *Species Invasions: Insights into Ecology, Evolution, and Biogeography*, pp. 315–336. Sunderland: Sinauer Associates, 2005.

[Von Holle and Simberloff, 2005] B. Von Holle and D. Simberloff. Ecological resistance to biological invasion overwhelmed by propagule pressure. *Ecology* 86: 3212–3218, 2005.

[Warren, 2007] C. R. Warren. Perspectives on the 'alien' versus 'native' species debate: A critique of concepts, language and practice. *Progress in Human Geography* 31: 427–446, 2007.

[Williamson and Fitter, 1996] M. H. Williamson and A. Fitter. The characters of successful invaders. *Biological Conservation* 78: 163–170, 1996.

[Woods and Moriarty, 2001] M. Woods and P. V. Moriarty. Strangers in a strange land: The problem of exotic species. *Environmental Values* 10: 163–191, 2001.

[Young, 1993] R. M. Young. Darwin's metaphor and the philosophy of science. *Science as Culture* 3: 375–403, 1993.

[Zinken *et al.*, 2008] J. Zinken, I. Hellsten, and B. Nerlich. Discourse metaphors. In R. M. Frank, R. Dirven, T. Ziemke and E. Bernárdez (eds.), *Body, Language and Mind. Volume 2: Sociocultural Situatedness*, pp. 363–385. New York: Mouton de Gruyter, 2008.

A CASE STUDY IN CONCEPT DETERMINATION: ECOLOGICAL DIVERSITY

James Justus

1 INTRODUCTION

Some biological communities are more complicated than others. For example, tropical communities usually contain more species [Pianka, 1966; Willig *et al.*, 2003], there is evidence their species interact more intensely [Janzen, 1970; Møller, 1998], these interactions are more variegated in form [Dyer and Coley, 2001], and they exhibit more trophic levels than high latitude communities [Oksanen *et al.*, 1981; Fretwell, 1987]. Ecologists often use the concept of diversity to represent differences in the "complicatedness" of communities: tropical communities are often said to be more ecologically diverse than tundra communities.

At a coarse level of description, the vague connotation accompanying the term 'diversity' adequately captures the imprecise judgments that some communities are more complicated than others. Disagreement arises, however, over how the concept should be operationalized. As early as 1969, Eberhardt [1969, p. 503] characterized the ecological literature on diversity as a "considerable confusion of concepts, definitions, models, and measures (or indices)." A few years later, Hurlbert [1971, p. 577] argued that, "the term 'species diversity' has been defined in such various and disparate ways that it now conveys no information other than 'something to do with community structure'." MacArthur [1972, p. 197] similarly suggested that the term 'diversity' should be excised from ecological vocabulary as doing more harm than good, and that ecologists had, "wasted a great deal of time in polemics about whether [Simpson's] or [Shannon's] or $\frac{N!}{N_1!N_2!...N_n!}$ or some other measure [of diversity] is 'best'."[1] As these remarks indicate, ecologists have proposed several mathematical measures that differ about what properties are given priority over others in assessing diversity and which differ in mathematical form. Disagreements about these issues raise the question of what properties of a community *should* be considered part of its diversity and, in turn, what adequacy conditions the concept should satisfy.

Section 2 describes and defends seven adequacy criteria for the concept of ecological diversity. It also argues two additional criteria found in the ecological

[1] See sections 3 and 4 for a discussion of these diversity measures.

Handbook of the Philosophy of Science. Volume 11: Philosophy of Ecology.
Volume editors: Kevin deLaplante, Bryson Brown and Kent A. Peacock. General editors: Dov M. Gabbay, Paul Thagard and John Woods.

literature are untenable. The primary focus is adequacy criteria for measures of diversity, such as Shannon's and Simpson's, that make no assumption about the underlying distribution of individual organisms among species in a community. For this reason, these indices are sometimes called *nonparametric* (*e.g.*, [Lande, 1996]) to distinguish them from indices derived from parameters of statistical models of species abundance, such as the log series [Fisher *et al.*, 1943] and log normal [Preston, 1948], or from biological models, such as the broken stick and overlapping niche model [MacArthur, 1957].[2] Unlike parametric indices, nonparametric diversity indices are applicable to any biological community with any species abundance distribution.[3] This analysis also assumes that communities have been exhaustively sampled, thereby avoiding complex issues about the adequacy of diversity indices given imperfect and incomplete sampling to focus on the problem of specifying the concept of ecological diversity when complete knowledge about the community's relevant properties is available.[4]

Like most ecological literature on the concept of diversity, the focus is on species richness and evenness as components of diversity, although issues about how other information (*e.g.*, taxonomic information) should affect assessments of diversity are occasionally touched upon. A myriad of indices combine species richness and evenness into a single measure of diversity, the two most popular being Simpson's and Shannon's. Sections 3 and 4 describe these indices and evaluate how they fare against the adequacy criteria defended in Section 2. Despite its greater popularity, Shannon's index performs worse than Simpson's. Section 5 concludes by assessing an influential criticism of the role of the diversity concept within ecology.

2 ADEQUACY CRITERIA FOR THE CONCEPT OF ECOLOGICAL DIVERSITY

Like most systems studied in science, biological communities can be represented with different degrees of specificity. With low specificity, a community can be represented simply in terms of the species it contains and how individual organisms of the community are distributed among these species.[5] This information is provided by the *proportional species abundance vector*, V_p, of a community:

$$V_p = \langle p_1, \ldots, p_i, \ldots, p_n \rangle; \tag{1}$$

[2]See [Preston, 1962a; 1962b; May, 1975; Rosenzweig, 1995] for reviews.

[3]In fact, Lande [1996, p. 5] suggests that being nonparametric, and thus applicable to all biological communities, is a defensible adequacy condition for a diversity index. Some ecologists have also criticized that there is no theoretical justification for statistical models of species abundance distribution, and only poor ones for most biological models [Krebs, 1989, Ch. 10].

[4]See [Horn, 1966; Pielou, 1975; 1977; Patil and Tallie, 1982a] for extensive discussions of these issues.

[5]The target of this analysis is quite modest. Greater representational specificity is achieved, for example, if interactions between species in a community are described with differential or difference equations in addition to how individuals are distributed among species (see [Justus, 2006]). For a much broader representational scope on diversity see Maclaurin and Sterelny's [2008] analysis of biodiversity.

in which n designates the number of species in the community, *i.e.*, its *species richness*; p_i designates the proportional abundance of the i-th species in the community; the p_i are ordered from most to least abundant (ties broken by random selection), *i.e.*, $p_1 \geq \ldots \geq p_i \geq \ldots \geq p_n$; and $\sum_{i=1}^{n} p_i = 1$. The only properties of species V_p represents are their proportional abundances. Functional, trophic, and taxonomic differences (besides the species level) are not represented. Proportional abundances of species in a community often change over time for a variety of reasons (*e.g.*, migration, interspecific interactions such as predation, competition, etc.) so V_p must be updated as communities change.

'Abundance' is an ambiguous term. Besides referring to the number of individual organisms of a species (a discrete quantity), it can also refer to their biomass (a continuous quantity). Accordingly, p_i can designate either: (i) the proportion of individuals of species i in a community given by $\frac{N_i}{N}$ where N_i is the number of individuals of species i and N is the total number of community individuals, such as the proportion of wolves in a community; or, (ii) species i's proportion of total community biomass, such as dry weight of a particular plant species in a forest community. p_i may differ significantly on these two interpretations, so ideally V_p should be calculated according to both interpretations for a given biological community. If it is unclear how to count individual organisms, as is the case for some clonal plant species or asexually reproducing marine species, the biomass interpretation of p_i is preferable.

Mathematically, components of V_p can take values of zero to represent species with zero abundance. Unlike species for which $p_i > 0$, adding or subtracting these terms from V_p need not change the other proportional abundances to ensure $\sum_{i=1}^{n} p_i = 1$. Biologically, however, species for which $p_i = 0$ cannot be part of the community represented by V_p. To be one of the species comprising a community, the community must contain at least one representative of that species. As a biological collection, to deny this stipulation for communities would require commitment to the idea that a community can be represented to contain a species not instantiated by any of its members. Depending on the interpretation of p_i (see above), $p_i \geq \frac{1}{N}$ or $p_i \geq \frac{b_i}{B}$ for all i is assumed where N designates the total number of individuals in the community, B designates the total community biomass, and b_i designates the minimum biomass of an individual of species i.[6] In modeling contexts with different goals, such as in studies of extinction and migration processes, it may be useful to allow zero V_p components to represent when species have gone locally extinct or have emigrated completely from a community. Once a species disappears from a community, however, it is no longer part of that community and does not contribute to its diversity.[7]

[6] For expositional convenience, only interpretation (i) from above will be discussed in the following unless specified otherwise.

[7] Alternatively, the stipulation against zero p_i could be replaced with the proviso that species richness is determined by the number of *nonzero* p_i in V_p and that only they determine community diversity.

Ecologists widely agree that two properties of a community should be part of its diversity: *species richness* and *evenness* [Pielou, 1966; 1975; 1977; Tramer, 1969; Patil and Taillie, 1982a; 1982b; Margurann, 1988; 2004].[8] Consider two simple communities, A and B, both composed of two species s_1 and s_2. A and B have the same species richness (two). If the proportions of individuals distributed among the two species are $p_1 = 0.02\%$ and $p_2 = 99.98\%$ for A and $p_1 = 50\%$ and $p_2 = 50\%$ for B, B is said to be more even than A.

The widespread beliefs that species richness and evenness are components of diversity reflect intuitive constraints on the concept that can be formulated as explicit adequacy conditions:

(A1) for a given evenness, diversity should increase as species richness increases (*i.e.*, as n from (1) above increases); and,

(A2) for a given species richness, diversity should increase as evenness increases.

Note that neither (A1) nor (A2) necessitate a particular mathematical form to the increase in diversity required.

The first condition codifies an incontestable feature of the diversity concept: the diversity of a collection increases as the number of different types of entities in the collection increases. Applied to a biological collection such as a community, (A1) therefore captures the intuitive idea that a community composed of one thousand species is more diverse than one composed of ten.

But there is a difficulty with (A1) as formulated. Unlike the clause "for a given species richness" in (A2), (A1) contains a qualification, "for a given evenness," for which V_p does not provide a quantitative characterization.[9] The problem is that increases in species richness (represented by new p_i) necessitate changes in the p_i comprising V_p since all the p_i must sum to one following any change in species richness. These changes do not *necessitate* a change in evenness, but the absence of a quantitative characterization makes it unclear how evenness can remain static as species richness changes.[10] To avoid this difficulty, (A1) is often reformulated as:

(A1′) of two maximally even communities, the more species rich community is more diverse [Pielou, 1975, p. 7].

A quantitative characterization of 'maximally even' is provided by a third adequacy condition (A3) below.

[8] McIntosh [1967] was probably the first to coin the term 'species richness' to refer to the number of species in a community. 'Evenness' and 'equitability' are used interchangeably in the ecological literature (*e.g.*, [Lloyd and Ghelardi, 1964; McIntosh, 1967; Tramer, 1969; Peet, 1974; 1975]).

[9] n from (1) provides a quantitative characterization of species richness.

[10] Similarly, although the simple examples discussed above and below provide an informal grasp of how communities can differ in evenness, absence of a quantitative characterization also makes the clause "as evenness increases" of (A2) unclear. The remainder of this section proposes adequacy conditions which help precisify the evenness concept.

Beyond its intuitive appeal, it is worth pausing over the reason (A2) should be accepted. Consider two communities, each composed of 100 species and 10,000 total individual organisms. A community in which there are 100 individuals of each species seems more diverse than one with 9,901 individuals of one species and one individual each of the other 99. The reason seems to be that besides a consideration of the number of types of entities in a collection, diversity also involves a consideration of how well they are represented. For this reason, diversity is often interpreted as the *apparent* or *effective* number of species present in a community (*e.g.*, [Hill, 1973; Peet, 1974]). For example, to an observer with imperfect faculties of perception, or an ecologist with insufficient field time, or employing sampling methods with unavoidable limitations, the first community with evenly distributed individuals will usually appear to contain more species than the second community, despite their identical richness.[11] In this way, (A2) captures the intuitive idea that community B is more diverse than community A from above.

Thus far, evenness has not been explicitly characterized and (A1) and (A2) place no constraints on the concept. For a given species richness, evenness is clearly maximized when individuals of the community are equally distributed among species, *i.e.*, when $p_i = \frac{1}{n}$ for all i. This constraint corresponds to another adequacy condition:

(A3) for a given species richness, diversity is maximal when individuals of the community are distributed equally among species (*i.e.*, when evenness is maximal).

Let $V_p^{\max}$ designate the maximally even proportional species abundance vector for a given richness.

Similarly, evenness is clearly minimized when community individuals are maximally unequally distributed. Specifically, diversity is minimal when all but $(n-1)$ of the individual organisms comprising the community are of one species and the rest are equally distributed (one each) among the other $(n-1)$ species, *i.e.*, when $p_1 = \frac{N-(n-1)}{N}$.[12] Formulated as an explicit adequacy condition:

(A4) for a given species richness, diversity is minimal when individuals of the community are distributed maximally unequally among species (*i.e.*, when evenness is minimal).

Let $V_p^{\min}$ designate the minimally even proportional species abundance vector for a given richness.

[11] A different line of thought also motivates (A2). A biological community is a set of organisms of different species. Sets are characterized by properties of their members. Members of an uneven community poorly represent some species, while each species of an even community is equally represented by its members. As a set, the characterization of a community with evenly distributed individuals therefore depends more significantly on a greater number of species-types than an uneven community.

[12] Recall that n designates the number of species and N designates the total number of individual organisms.

Adequacy conditions (A1)–(A4) are found throughout the ecological literature [Hill, 1973; Pielou, 1975; 1977; Magurran, 1988; 2004; Lande, 1996; Sarkar, 2007]. Building on (A2)–(A4), a further constraint on the evenness concept, and thus on diversity, can be formulated. Focusing on (A3),[13] if evenness is maximal for $V_p^{\max}$, evenness must decrease as V_p diverges from it. This decrease can be quantified in many ways, but one rationale for doing so restricts the range of possible methods of quantification. Recall that the only differences between species being considered are their proportional abundances; V_p does not represent taxonomic, trophic, functional, and other interspecific differences. Besides their proportional abundances, different species are therefore treated as equally important in assessing the diversity of a community. Thus, if evenness decreases because one species deviates from its maximally even proportional abundance ($\frac{1}{n}$), an equal deviation from the maximally even proportional abundance by another species should induce an identical decrease in evenness and thus in diversity. Formulated as an explicit adequacy condition:

(A5) for a given species richness, if V_p^i and V_p^j are proportional species abundance vectors that deviate from $V_p^{\max}$ because species i and species j, respectively, deviate equally from a $\frac{1}{n}$ proportional abundance, evenness decreases by the same amount in both cases.

Put informally, (A5) stipulates that assessment of diversity is blind to species identity. It thereby captures the frequently made assumption that evaluating community diversity requires treating species as equals in the absence of taxonomic, functional, or other data [Magurran, 2004, p. 11]. In such cases, only the extent a species' proportional abundance deviates, not what species it is, is relevant when assessing a community's diversity.[14] (A5) is neutral, however, about whether rare or abundant species are more important to the diversity of a community. It requires merely that *equal* changes in the abundances of two species from maximal evenness necessitate *equal* decreases in diversity.

Even with this further constraint, (A1)–(A5) are weak adequacy conditions in the sense that they do not determine a unique quantitative measure of diversity. In fact, most common quantitative indices satisfy them (see §§3–5). Distinct quantitative diversity indices result from different ways of integrating and quantifying species richness and evenness consistent with (A1)–(A5). Before discussing the two most common such indices in the next sections, it is therefore important to

[13]Similar reasoning applies for (A4).

[14]Strictly speaking, (A5) follows from the way V_p was constructed. Recall that the p_i are ordered from most to least abundant. This was intended to impose a nonarbitrary ordering on the components of V_p. It also entails, however, that V_p^i and V_p^j referred to in (A5) are identical because p_i and p_j would fall at the same place in the ordering for V_p^i and V_p^j, respectively. If the ordering constraint were not imposed and species were assigned indices in V_p prior to determination of their proportional abundances, (A5) would constitute an independent requirement on diversity. As is, (A5) is retained to make the requirement explicit. I owe Samir Okasha for this clarification.

consider whether any other defensible adequacy conditions would necessitate a particular quantitative index of diversity.

Notice that (A5) does not entail different *types* of deviations from $V_p^{\max}$ (or $V_p^{\min}$), such as those involving different numbers of species, must be accorded the same import for diversity. Of course, unequal deviations *of the same type* should necessitate different values of diversity. Consider, for example, the type of deviation in which one species i deviates from $\frac{1}{n}$. If p_i decreases from $\frac{1}{n}$ to $\frac{1}{2n}$, diversity should decrease less than if p_i decreases to $\frac{1}{3n}$ all else being equal because the decrease in evenness is greater in the latter case (see (A2)). If p_i decreases from $\frac{1}{n}$ to $\frac{1}{4n}$, however, (A5) entails nothing about whether diversity should decrease more or less than in a case in which p_i decreases from $\frac{1}{n}$ to $\frac{1}{2n}$ and some other p_k decreases from $\frac{1}{n}$ to $\frac{1}{2n}$.

What is needed is a method for evaluating evenness that would adjudicate between different types of deviations from $V_p^{\max}$ (or $V_p^{\min}$) for a given species richness. One natural method for doing so simply evaluates evenness in terms of the *distance* between V_p and $V_p^{\max}$. In general, a function $d : G \times G \to \mathbf{R}$ is a distance metric if it possesses three properties for all $x, y, z \in G$:

(P1) $d(x, y) \geq 0$, and $d(x, y) = 0$ if and only if $x = y$;

(P2) $d(x, y) = d(y, x)$ (symmetry); and,

(P3) $d(x, z) \leq d(x, y) + d(y, z)$ (triangle inequality) [Kaplansky, 1977].

An infinite number of different functions satisfy these conditions and could therefore be used to measure deviation of V_p from $V_p^{\max}$. For instance, an especially simplistic distance metric satisfying (P1)–(P3) is $d(x, y) = 0$ if $x = y$, and 1 otherwise. This is plainly inappropriate as a metric for measuring the distance between biological communities represented by V_p and $V_p^{\max}$. According to this metric, communities A and B from above are at the same distance from $V_p^{\max}$.

Since components of V_p and $V_p^{\max}$ take real values, an appropriate distance metric, and certainly the most common, is the Euclidean metric:

$$d(V_p, V_p^{\max}) = \sqrt{\sum_{i=1}^{n} \left(p_i - \frac{1}{n}\right)^2}. \quad (2)$$

[15]

Measured in this way, the idea is that evenness is inversely related to the Euclidean distance between its species abundance vector and $V_p^{\max}$.[16] For example, if the diversity of $V_p^{\max}$ and $V_p^{\min}$ for a given richness are set at 1 and 0, respectively (see (A3) and (A4) above), the diversity of a community represented by V_p would take values on [0, 1] determined by the Euclidean distance between V_p and $V_p^{\max}$. Species abundance vectors that deviate from $V_p^{\max}$ in different ways but at the

[15] Note that I am not claiming the Euclidean metric is *uniquely* defensible.

[16] Similarly, since evenness is inversely related to the Euclidean distance, evenness is directly related to the Euclidean distance from $V_p^{\min}$.

same distance from it would thereby have the same evenness; those at different distances from $V_p^{\max}$ would differ in evenness. In particular, for the two species abundance vectors discussed above, one in which p_i for one species decreases from $\frac{1}{n}$ to $\frac{1}{4n}$ and another in which p_i and p_k for two species decrease from $\frac{1}{n}$ to $\frac{1}{2n}$, the latter would be accorded greater evenness because its distance from $V_p^{\max}$ is smaller than the former. Codified in an explicit adequacy condition:

(A6) for a given species richness: (i) evenness decreases (increases) as the Euclidean distance from $V_p^{\max}$ ($V_p^{\min}$) increases; and, (ii) communities represented by species abundance vectors at the same Euclidean distance from $V_p^{\max}$ (or $V_p^{\min}$) have the same evenness.[17]

Similar to the way (A5) stipulates that diversity is blind to species identity, (A6) stipulates that diversity is blind to the type of deviation from $V_p^{\max}$ (or $V_p^{\min}$). (A6) does not necessitate a unique quantification of evenness because the decrease (or increase) required in clause (i) may take many mathematical forms (*e.g.*, concave vs. convex, linear vs. nonlinear, exponential vs. nonexponential, etc.) Note that (A6) holds only if (A5) does as well.

Together with (A2), (A6) imposes a significant constraint on the diversity concept. It requires treating changes in the proportional abundances of rare and common species as equally important to diversity. According to (A6), for instance, diversity must decrease the same amount with a decrease in p_i for an extremely rare species and with an identical decrease in a much more common species.[18] Thus, (A6) precludes diversity from being partially sensitive to the proportional abundances of rare or common species. In particular, it requires that species abundance vectors in which several species are very abundant and a few are very rare, and in which several species are very rare and a few are very abundant have the same diversity if their distances from $V_p^{\max}$ are identical. (A6) thereby captures the same idea underlying (A5): that diversity requires treating all species as equals. To illustrate, consider two species abundance vectors $\langle \frac{8}{15}, \frac{4}{15}, \frac{3}{15} \rangle$ and $\langle \frac{7}{15}, \frac{6}{15}, \frac{2}{15} \rangle$. The first contains one abundant species and two rarer species (the maximally even p_i for each species is $\frac{1}{3}$). The second contains two abundant species and one rare species. The distances between each vector and $V_p^{\max}$ are identical and hence both are accorded the same diversity by (A6).

For the same reason diversity should be blind to species identity (see (A5)) and blind to the type of deviation from $V_p^{\max}$ (or $V_p^{\min}$) (see (A6)), diversity should not be partial to particular distances between V_p and $V_p^{\max}$ (or $V_p^{\min}$) in the sense

[17] Since Euclidean distance is not the only defensible distance metric (see footnote above), (A6) need not be formulated with it. I focus on Euclidean distance because it is the most common metric for vectors with real components and is clearly defensible. Smith and Wilson [1996], for instance, describe several desirable properties of potential indices of evenness captured by (A6). However, the degree of general agreement between analyses of diversity indices based on formulations of (A6) and (A7) with different defensible distance metrics is currently unknown.

[18] This holds when all species have proportional abundances $< \frac{1}{n}$. For species with proportional abundances $> \frac{1}{n}$, evenness will increase as their proportional abundances decrease towards $\frac{1}{n}$.

that diversity should decrease *uniformly* (*i.e.*, linearly) as the distance between V_p and $V_p^{\max}$ increases. Formulated as an explicit adequacy condition:

(A7) for a given species richness, evenness decreases (increases) *linearly* as the Euclidean distance between V_p and $V_p^{\max}$ (or $V_p^{\min}$) increases.

(A7) requires equal intervals of distance (as measured by equation (2)) correspond to equal differences in diversity values, regardless of the specific distance V_p is from $V_p^{\max}$. Specifically, if $d(V_p^i, V_p^{\max}) = x$ and the difference in diversity value between V_p^i and $V_p^{\max}$ is y, then if $d(V_p^j, V_p^{\max}) = x$ also, the difference in diversity value between V_p^j and $V_p^{\max}$ is also y. Note that (A7) holds only if (A6) does as well.

(A6) and (A7) both follow from a principle sometimes mentioned in discussions of ecological diversity (*e.g.*, [Krebs, 1989; Magurran, 2004]). The principle is that diversity should not be partial among *individual organisms*, just as it should not be partial among *species* in a community. Specifically, in assessments of diversity in the absence of taxonomic, functional, and other types of information, individual organisms should contribute to diversity in proportion only to the proportional abundance of the species to which they belong. If different types of deviations from $V_p^{\max}$ are weighted differently than as dictated by equation (2), *i.e.*, (A6) is violated, some individuals will contribute more (or less) to diversity merely because they are a member of a species that has deviated from $\frac{1}{n}$ in a way favored (or disfavored) by the candidate diversity index. Similarly, if different distances from $V_p^{\max}$ are weighted differently in assessing diversity, *i.e.*, (A7) is violated, some individuals will contribute more (or less) to diversity merely because they are a member of a species with a proportional abundance at a distance from $\frac{1}{n}$ that is favored (or disfavored) by the candidate index. In either case, individual organisms would not be treated as equals in determining the diversity of a community composed of them.[19]

Before evaluating Shannon and Simpson's indices against (A1)–(A7) in the next section, this section concludes by considering another adequacy condition for diversity proposed by Lewontin [1972] and recently endorsed by Lande [1996].[20] It concerns the relationship between the diversity of individual communities and the diversity of sets of different communities. Specifically, if a super-collection of individuals is formed by pooling the individuals of several distinct smaller collections, the idea is that the diversity of the super-collection must be at least as great as the average diversity of the smaller collections. Applied to biological communities, this requires the diversity of the super-community $\bigcup_{i=1}^{z}\{C_i\}$ formed by pooling

[19] There *can* be reasons to treat individuals of different species differently. Individuals of rare species (and their proportional abundances) are usually weighed more significantly in assessing the diversity of communities in conservation biology, for instance. Rare species are typically more likely to go extinct. Indices that accord changes in their proportional abundances more import than changes in common species are favored in conservation contexts because changes in the former are more likely to influence species persistence than changes in the latter (see §§3–5).

[20] Lewontin [1972] may have been the first to formulate this as an adequacy condition for the concept of diversity.

the individuals of each community C_i to be greater than or equal to the weighted mean diversity of the C_i (z is an index of the communities). Stated formally for the case of equal weights, and using 'DIV' to represent diversity:

$$DIV\left(\bigcup_{i=1}^{z}\{C_i\}\right) \geq \frac{1}{n}\sum_{i=1}^{z} DIV(C_i). \tag{3}$$

Equality holds only if the C_i are *compositionally* identical, *i.e.*, ${V_p}^{C_i} = {V_p}^{C_j}$ for all $i \neq j$.

Lewontin did not provide a rationale for this constraint on ecological diversity, and there are reasons to reject it as an adequacy condition. Consider two simple communities C and D composed of four different species (two each) with *absolute* (not proportional) abundances $\langle 2, 2\rangle$ and $\langle 1000, 1000\rangle$. The absolute species abundance vector for the super community $C \cup D$ with species richness four is $\langle 1000, 1000, 2, 2\rangle$. Equation (3) requires the diversity of $C \cup D$ be greater than the average diversity of C and D, but it is unclear why this is defensible as an adequacy condition on the concept of diversity. $C \cup D$ contains more species than either C or D, and in this respect seems more diverse. But it is also highly uneven compared with C or D. The proportional species abundance vector for $C \cup D$ is approximately $\langle 0.499, 0.499, 0.001, 0.001\rangle$ which is a highly uneven distribution, unlike the highly even distribution of C and D, $\langle 0.5, 0.5\rangle$. Equation (3) therefore forces a strong rank order of species richness over evenness in assessments of diversity.

This *may* be a defensible property of a proposed diversity index.[21] In conservation biology, for example, there may be advantages to prioritizing species richness over evenness in assessments of the diversity of communities targeted for conservation. Equation (3) is not, however, a defensible constraint on *any* potential quantitative specification of diversity. Species richness and evenness are independent properties. Though this does not entail one is not more important than another in evaluations of a community's diversity, nothing about the pre-theoretic concept of ecological diversity seems to suggest otherwise. Pielou [1977, p. 292], for instance, explicitly rejected the constraint imposed by (3): "since diversity depends on two independent properties of a collection ... a collection with few species and high evenness could have the same diversity as another collection with many species and low evenness."[22]

[21]Note that this property is consistent with (A1)–(A7) from above, which do not compel any relationship between species richness and evenness in assessments of diversity.

[22]In passing, Lande [1996, p. 8] motivates this condition by pointing out that its denial, "implies the possibility of a negative diversity among communities." But why this is problematic is unclear. It does not, for instance, entail the diversity of any *individual* community is negative, which would clearly be problematic.

3 SIMPSON'S INDEX

The first index that included species richness and evenness as components of diversity found in the ecological literature was proposed by Simpson [1949]. Simpson claimed that the probability two individuals drawn at random (with replacement) from an indefinitely large collection are of the same group is $\sum_{i=1}^{n} p_i^2$, where n is the number of groups exhibited within the collection, and he called it a "measure of concentration." Applied to biological communities, $\sum_{i=1}^{n} p_i^2$ then measures the dominance (in terms of abundance) of species within the community (Pielou 1977) and is at its minimal value ($\frac{1}{n}$) for a given species richness n when individuals of the community are equally distributed among the n species, *i.e.*, when $V_p = V_p^{\max}$.

On Simpson's interpretation, the complement of the concentration measure:

$$(D) \qquad 1 - \sum_{i=1}^{n} p_i^2,$$ [23]

represents the probability two randomly selected individuals belong to different species, which is an intuitive measure of diversity.[24] D is at its maximal value for a given species richness $n (D = 1 - \frac{1}{n})$ when individuals are maximally equally distributed among species, *i.e.*, when $V_p = V_p^{\max}$, and at its minimal value when individuals are maximally unequally distributed among species, *i.e.*, when $V_p = V_p^{\min}$.[25]

Several ecologists have suggested other interpretations of D. Hurlbert [1971], for instance, claimed that D multiplied by $\frac{N}{N-1}$ represents the probability of interspecific encounter in the community, rather than just the probability two randomly selected individuals belong to different species. Patil and Taillie [1982a] made a similar claim and showed how quantities such as the waiting time for intra- and interspecific encounter are related to D on this stronger interpretation. Recently, Ricotta [2000, p. 246] has suggested the same interpretation. These interpretations are only sound, however, if an additional assumption is made about community structure. In response to Patil and Taillie's analysis, Sugihara [1982] correctly pointed out that D represents the probability of interspecific interaction only if

---

[23] The inverse of Simpson's concentration measure, $\frac{1}{\sum_{i=1}^{n} p_i^2}$, is also commonly used as an index of diversity [Williams, 1964; Levins, 1968; Hurlbert, 1971; MacArthur, 1972; Hill, 1973; May, 1975; Pielou, 1977; Magurran, 1988; 2004; Lande, 1996].

[24] The counterpart of D that does not make the idealization that the two individuals are drawn from an indefinitely large population is $1 - \sum_{i=1}^{n} \frac{N_i(N_i-1)}{N(N-1)}$ and will not be discussed here [Pielou, 1977; Magurran, 2004].

[25] This minimum is not the value of D, however, when all individuals of the community are of the same species, which is 0. $D = 0$ when all but one p_i is zero, but this set of proportional abundances violates the biological requirement on V_p that $p_i \geq \frac{1}{N}$ for all i stipulated above. See the discussion preceding (A4) in Section 2.

the frequencies of interspecific encounters are directly proportional to the relative abundances of the species interacting. The problem, Sugihara [1982, p. 565] emphasized, is that:

> None of these interpretations [of D], however, has yet proved to be very fruitful, as they suffer from such real-world concerns as spatial patchiness and clumping in species distributions, differential mobility, and problems associated with interpreting niche overlap between species from their spatial covariance. Approaching the study of species diversity through a priori models is a valid enterprise, but requires a clear intuition of how communities operate, which thus far seems to be lacking.

Hurlbert's modification of D *may* be a good *estimator* of the probability of interspecific encounter if species are spatially distributed relatively uniformly throughout the area occupied by the community. If distributed in this way, species are likely to interact in direct proportion to their abundances. But without knowing this, and without a thorough understanding of the factors Sugihara mentioned, the legitimacy of the stronger interpretation cannot be reliably verified.[26]

As has been noted throughout the ecological literature, D satisfies (A1)–(A4) [Hill, 1973; Pielou, 1975; 1977; Magurran, 1988; 2004; Lande, 1996]. D also clearly satisfies (A5). As the similar mathematical structure of D and (2) from above suggest, D also satisfies (A6). To see this, recall that for a given species richness (A6) requires that: (i) the diversity of the community represented by V_p be inversely related to $d(V_p, V_p^{\max})$; and, (ii) that if $d(V_p^i, V_p^{\max}) = d(V_p^j, V_p^{\max})$, the diversity of communities represented by V_p^i and V_p^j are equal. Diversity is directly related to $-d(V_p,V_p^{\max})$ given (i), so what is first needed is to show that D exhibits the same relationship with $-d(V_p, V_p^{\max})$. The following algebraic identities demonstrate the required relationship:

$$\begin{aligned}
d(V_p, V_p^{\max}) &= \sqrt{\sum_{i=1}^{n} \left(p_i - \tfrac{1}{n}\right)^2} \\
&= \sqrt{\sum_{i=1}^{n} \left(p_i^2 - \tfrac{2p_i}{n} + \tfrac{1}{n^2}\right)} \\
&= \sqrt{\sum_{i=1}^{n} (p_i^2) - \tfrac{2}{n}\sum_{i=1}^{n} (p_i) + \sum_{i=1}^{n} \left(\tfrac{1}{n^2}\right)} \\
&= \sqrt{\sum_{i=1}^{n} (p_i^2) - \tfrac{2}{n} + \tfrac{1}{n}} \\
&= \sqrt{\sum_{i=1}^{n} (p_i^2) - \tfrac{1}{n}}.
\end{aligned}$$

[26]Similarly, Simpson's interpretation of D implicitly assumes the probability of selecting two individuals from the same group is directly proportional to their relative abundances, which is also violated under a variety of plausible biological conditions. I owe Sahotra Sarkar for emphasizing this fact.

Substitution for $\sum_{i=1}^{n} p_i^2$ using the definition of D from above yields: $\sqrt{(1-D)-\frac{1}{n}}$, which equals $\sqrt{-D+\frac{n-1}{n}}$.[27] Since $\frac{n-1}{n}$ is constant and nonnegative for a given n, $d(V_p, V_p^{\max})$ decreases as D increases, establishing the required relationship (i). Similarly, if $d(V_p^i, V_p^{\max}) = d(V_p^j, V_p^{\max})$, then $\sqrt{-D_i+\frac{n-1}{n}} = \sqrt{-D_j+\frac{n-1}{n}}$ where D_i and D_j represent the complement of Simpson's index for the proportional abundances of V_p^i and V_p^j, respectively. From this it follows that $D_i = D_j$ (given that D must be positive), establishing (ii). (A6) is thereby satisfied.[28]

That $d(V_p, V_p^{\max}) = \sqrt{-D+\frac{n-1}{n}}$ also shows, however, that D does not satisfy (A7). (A7) requires diversity decrease linearly with $d(V_p, V_p^{\max})$, but D scales quadratically with $d(V_p, V_p^{\max})$. Thus, D violates (A7) because it is more sensitive to changes in proportional species abundance vectors at greater distances from $V_p^{\max}$. In particular, D is more sensitive to the proportional abundances of especially abundant or rare species. If diversity were specified with Simpson's index, (A7) would therefore not be satisfied and diversity would fail to be impartial among individual organisms comprising a community.

4 SHANNON'S INDEX

Probably the most popular index of community diversity is Shannon's index (H), so-called after Claude Shannon, who developed it in the context of what came to be called information theory.[29] The index was originally formulated to quantify the amount of information transmitted in a communication channel [Shannon, 1948; Shannon and Weaver, 1949] and, despite Margalef's [1958] claim to priority, Good [1953] was the first to use it as an index of ecological diversity. The index:

$$(H) \qquad -\sum_{i=1}^{n} p_i \ln p_i,$$ [30]

can be used to measure the information of a message composed of n types of symbols whose individual probability of occurrence is p_i, $i = 1, \ldots, n$. Within

---

[27] The term $\sqrt{-D+\frac{n-1}{n}}$ is well defined because $-D+\frac{n-1}{n}$ is nonnegative. Specifically, at its maximum $D = 1-\frac{1}{n}$, in which case $\sqrt{-D+\frac{n-1}{n}} = 0$.

[28] As a check on this result, it can be verified that the two species abundance vectors $\langle \frac{8}{15}, \frac{4}{15}, \frac{3}{15} \rangle$ and $\langle \frac{7}{15}, \frac{6}{15}, \frac{2}{15} \rangle$ from above, which have the same distance (0.3887) according to equation (2), also have the same value (0.6044) according to Simpson's index.

[29] Aczél and Daróczy [1975] suggest Norbert Wiener independently developed an index which is a special case of Shannon's more general index in 1948 [Wiener, 1948].

[30] Shannon's index assumes p_i is a proportional species abundance from an infinitely large community [Magurran, 2004]. This idealization, and the emendation needed to correct it for finite communities, will not be discussed here. The sampling estimator of H which does not make this assumption is Brillouin's index, $\frac{1}{N} \ln \left(\frac{N!}{N_1! N_2! \ldots N_n!} \right)$ [Brillouin, 1962].

ecology, however, p_i represents the familiar proportional abundance of species i and n represents the community's species richness.

H is at its maximal value ($\ln n$) for a given species richness n when individual organisms are equally distributed among species, thereby satisfying adequacy condition (A3) [Pielou, 1977]. Similar to D, that H satisfies (A1), (A2), and (A4) is well-known [Hill, 1973; Pielou, 1975; 1977; Magurran, 1988; 2004; Lande, 1996]. H obviously satisfies (A5). But, as a simple example demonstrates, Shannon's index is more sensitive than Simpson's index to the abundances of rare species [Peet, 1974] and therefore fails (A6).[31] Consider a four species community composed of one abundant, one rare, and two evenly distributed species such that $V_p = \langle \frac{12}{32}, \frac{8}{32}, \frac{8}{32}, \frac{4}{32} \rangle$. $D = 0.7188$ and $H = 1.3209$. If the rare species becomes rarer, so that V_p becomes $V_p^r = \langle \frac{13}{32}, \frac{9}{32}, \frac{9}{32}, \frac{1}{32} \rangle$, then evenness decreases and both D and H decrease to 0.6758 and 1.1878, respectively. If the abundant species becomes more abundant, so that V_p becomes $V_p^a = \langle \frac{15}{32}, \frac{7}{32}, \frac{7}{32}, \frac{3}{32} \rangle$, then evenness also decreases and both D and H decrease to 0.6758 and 1.2420, respectively. That H decreases more than D for V_p^r vs. V_p^a does not necessarily show it is more sensitive to rare or abundant species because the range of values D and H take between $V_p^{\min}$ and $V_p^{\max}$ differ. Differences in their values may be due merely to a scaling effect. But the way values of D and H change from V_p to $V_p^{\min}$ and to $V_p^{\max}$ reveal that H is more sensitive to abundances of rare species. Specifically, H decreases more between V_p and V_p^r (1.3209 - 1.1878 = 0.1331) than between V_p and V_p^a (1.3209 - 1.2420 = 0.0789), while D decreases by the same amount. H is therefore more sensitive to the abundances of rare species, unlike D, and contrary to (A6)(i). In addition, since $d(V_p^r, V_p^{\max}) = d(V_p^a, V_p^{\max})$, (A6)(ii) requires a diversity index assign the same value for V_p^r and V_p^a. Their H values, however, differ (1.1878 and 1.3209, respectively).[32] (A6) is a necessary condition of (A7), so H also fails to satisfy (A7).

How should H be *ecologically* interpreted? Pielou [1977] provided a particularly simple and probably the clearest interpretation. As an entropy measure, Pielou suggested H measures uncertainty, and that diversity and uncertainty are closely related concepts.[33] Specifically, as the diversity of a community increases,

---

[31] Compared with D, H is also less sensitive to proportional abundances of species (and hence evenness) and more sensitive to species richness [May, 1975; Magurran, 1988; 2004].

[32] Despite these and other differences (see below), Simpson's and Shannon's indices are both members of a family of entropy measures defined by: $H_q = \frac{\ln \sum_{i=1}^{n} p_i^{\alpha}}{1-\alpha}$, in which $\alpha > 0$ and $\alpha \neq 1$ [Rényi, 1961; Pielou, 1975]. Rényi [1961] showed, for example, that $\lim_{\alpha \to 1} (H_q) = -\sum_{i=1}^{n} p_i \ln p_i = H$. Pielou [1975] showed that if $\alpha = 2$, $H_q = -\ln \sum_{i=1}^{n} p_i^2$, which is equivalent to the inverse form of Simpson's diversity index after exponential transform.

[33] Pielou's view of the proper ecological interpretation of Shannon's index shifted markedly in the 1960s and 1970s. In 1966, Pielou suggested there was an "obvious analogy" between a biological community and a coded message, and that, "the actions of a biologist are formally identical with those of a man observing, one after another, the symbols of a message" [1966, p. 164]. (A very similar characterization had been suggested by Margalef [1958] almost a decade

the uncertainty about which species a randomly selected individual belongs to increases. It is difficult to deny this claim, but the crucial issue is whether H is the *uniquely* appropriate measure of uncertainty. After all, given the interpretation of D described earlier, it also seems to measure a similar, if not identical, kind of uncertainty about a biological community.

Pielou's argument that H is the uniquely appropriate index of ecological diversity relied on a mathematical fact about H proved in a non-ecological context by Shannon. Shannon [1948] showed that H is the only function (up to a multiplicative constant) which exhibits three properties he thought reasonable to require of the concept of information.[34] These properties include that an information function should be continuous in p_i, and that it should monotonically increase with n for maximally even p_i (see (A1′) from above). The most important was an additive property which was the basis of Pielou's argument for preferring H over other diversity indices and, even stronger, that it constitutes an adequacy condition for any index of diversity [Pielou, 1977, pp. 293–294]. The interpretation of this property as applied to biological communities requires some elaboration.

Just as V_p is based on a classification P of individuals of a biological community into n species, let V_q be a proportional abundance vector with m components based on a different classification Q. The second classification could be derived from further taxonomic information, information on habitat requirements or other properties of individual organisms, etc. As with the p_i, assume that the proportional abundances q_j of the second classification are such that $\sum_{j=1}^{m} q_j = 1$, and that each individual organism falls into only one class. In analogy with the information concept, if the two classifications are independent, Pielou [1977, p. 294] suggested that ecological diversity must satisfy:

$$DIV(PQ) = DIV(P) + DIV(Q); \tag{4}$$

in which $DIV(PQ)$ is the diversity of a biological community with individuals classified into both P and Q for a total of $m \times n$ classes, and $DIV(P)$ and $DIV(Q)$ are the diversity of the community with individuals classified by only P or Q, respectively.[35] Mathematically, it can be shown that H is the only continuous function of the p_i up to a multiplicative constant that satisfies (A1′), (A3), and for which equation (4) holds [Khinchin, 1957].

In general, (4) is a desirable property because it permits the additive decomposition of a function of two combined input arguments. Together with its intu-

before.) By 1975, however, Pielou's view of such analogies was decidedly negative: "it cannot be too strongly emphasized that fancied links between the information-theoretic concept of 'information' and the diversity of an ecological community are merely fancies and nothing more" [1975, p. 9].

[34]In a later review of information theory, Aczél and Daróczy [1975, p. 29] called these properties, "natural properties which are essential from the point of view of information theory." Shannon, however, only called them "reasonable" and emphasized that, "The real justification of these definitions, however, will reside in their implications," [Shannon and Weaver, 1949, p. 50].

[35]Equation (4) can be generalized to any finite number of independent classifications.

itive plausibility for the concept of information, this motivated Shannon [1948], Khinchin [1957], Rényi [1961], Aczél and Daróczy [1975], and others to stipulate it as an adequacy condition for any information measure. What is needed to show it is an appropriate adequacy condition for the concept of ecological diversity, however, is an account of why this is a *necessary* property of diversity. Pielou [1977] did not supply such a rationale. *If* diversity were specified as H, (4) logically follows given its logarithmic form, and it can be agreed that there are benefits of being able to additively decompose ecological diversity in this way [Pielou, 1977, pp. 303–307]. This is insufficient, however, to establish the stronger claim that (4) is a defensible *adequacy* condition, especially given that there are other non-additive methods with attractive features in which ecological properties besides proportional species abundances can be integrated into a measure of community diversity (*e.g.*, [Rao, 1982; Ricotta, 2002]).

5 THE ROLE OF THE DIVERSITY CONCEPT WITHIN ECOLOGY

Simpson's and Shannon's diversity indices emerged within ecology in the 1950s. By the late 1960s, a large number of diversity indices had been formulated, and numerous empirical studies of different ecological systems were being conducted to estimate diversity using these indices.[36] Table 1 lists some of these indices. One reaction to this diversity of diversity indices was the thought that anything goes, that the diversity concept was deeply and problematically unclear. The attention being devoted to indices of diversity sparked several such criticisms, perhaps the most incisive from Hurlbert [1971] who called diversity a "nonconcept." His influential critique of this research agenda targeted the fundamental vagueness of the underlying concept, which he thought ecologists had exacerbated by appropriating statistical measures of diversity developed in nonbiological contexts, such as information theory, with dubious ecological relevance. MacArthur [1972, p. 197] voiced the same criticism around the same time: "Applying a formula and calculating a 'species diversity' from a census does not reveal very much; only by relating this diversity to something else—something about the environment perhaps—does it become science." Other ecologists were similarly skeptical of the role of the diversity concept within ecology, and generally of the ecological utility of information theory (*e.g.*, [Hill, 1973]).

Rather than attempt to rehabilitate the concept by proposing adequacy conditions by which to evaluate relative merits and weaknesses of different diversity indices as attempted above, Hurlbert suggested the search for relationships between diversity and other community properties, such as stability, should be refocused on the relationship between those properties and indices that reflect biologically meaningful properties that might influence community dynamics.[37] His proposed index of the probability of interspecific encounter is one example (see §3). As a

[36]See [Pielou, 1975; 1977; Magurran, 1988; 2004; Sarkar, 2007; Drake, 2007] for reviews.

[37]See [Justus, 2008] for an analysis of the stability-diversity debate within ecology.

measure of ecological diversity, note that species richness alone fails this test since it is generally unlikely that extremely rare species (*e.g.*, s_1 in community A from §2) play an important role in community dynamics.[38] Since it does not consider evenness, furthermore, species richness fares very poorly as a specification of ecological diversity; it fails adequacy conditions (A2)–(A7). Species richness was and remains, however, the predominant surrogate for diversity in analyses of stability-diversity relationships (*e.g.*, [Tilman, 1996; 1999]). As such, it is unclear these studies offer significant insights into possible relationships between stability and diversity in biological communities.[39]

Since Hurlbert's critique, ecologists have proposed a multitude of new diversity indices to satisfy different proposed adequacy conditions besides those about species richness and evenness (see [Magurran, 2004; Ricotta, 2005; Sarkar, 2007] for reviews). Diversity indices should increase, for instance, as interspecific taxonomic and functional differences increase. Besides properties of species, spatial properties of their geographical distribution could also be included in an index of ecological diversity. Since species distributions are significantly influenced by regional geology and environmental gradients, however, including these properties would expand the scope of ecological diversity beyond just the biological properties of communities. Expanded in this way, the "diversity" of the physical environment in which a community resided would also contribute to the value of indices of *ecological* diversity.[40]

How compatible these additional adequacy conditions are with one another, or with the other conditions is not yet clear. Some conditions appear to be conceptually independent, but some formal diversity indices suggest that others are not. Rao's [1982] "quadratic entropy" diversity index, for instance, which generalizes the Simpson index [Ricotta and Avena, 2003], incorporates interspecific taxonomic and functional differences as well as evenness and species richness into a single quantitative measure. Unlike the Shannon and Simpson indices, however, quadratic entropy violates the adequacy condition (A3) [Ricotta, 2005]. (A4) is also violated. This is as it should be. If functional or taxonomic information is included in assessments of diversity, then high functional or taxonomic diversity may make a less even community more diverse overall than a more even one. In effect, functional or taxonomic diversity can trump evenness.

As new indices are devised, similar incompatibilities between other adequacy conditions may be revealed. Absent a general proof that plausible adequacy criteria are themselves incompatible, however, the formulation of a uniquely defensible

[38]Potentially important exceptions include keystone species and so-called "ecosystem engineers" (see [Paine, 1969; Jones *et al.*, 1994; Power *et al.*, 1996]).

[39]A recent study by Tilman *et al.* [2006] that uses the Shannon index to measure the diversity of a Minnesota grassland and finds the same positive correlation as between species richness and temporal stability is a rare exception.

[40]Properties of the spatial extent of species, in particular their geographical rarity, are clearly relevant to the concept of diversity utilized in the context of biodiversity conservation, and therefore must be integrated into any defensible measure of biodiversity [Sarkar, 2002; 2005; 2007].

diversity index satisfying all of them remains possible. Focusing on species richness and evenness, adequacy conditions (A1)–(A7) are compatible and thus a defensible diversity measure satisfying them exists. In fact, (A6) and (A7) suggest two obvious candidates: (i) $-\sqrt{\sum_{i=1}^{n}\left(p_i - \frac{1}{n}\right)^2}$; and, (ii) $\frac{1}{\sqrt{\sum_{i=1}^{n}\left(p_i-\frac{1}{n}\right)^2}}$, which both also satisfy (A1)–(A5). Despite the multitude of quantitative measures of ecological diversity in the literature, that a set of defensible adequacy conditions can be formulated shows the concept is not problematically obscure and that it may facilitate scientific insights into biological communities.

Glossary:

n = species richness;
p_i = the proportional abundance of species i;
N_i = the abundance of species i;
N = the abundance of all species in the community;
N_{max} = abundance of the most abundant species;
d_{ij} = distance (*e.g.*, taxonomic, functional, etc.) between species i and j;
r_i = rank of species i in V_p;
a = a constant ≥ 0.

Table 1. A list of some of the diversity indices in the ecological literature.

Diversity Index	Mathematical Operationalization	Origin
Simpson's Index (infinite community)	D: $1 - \sum_{i=1}^{n} p_i^2$ or $\frac{1}{\sum_{i=1}^{n} p_i^2}$	Simpson [1949]
Simpson's Index (finite community)	$1 - \sum_{i=1}^{n} \frac{N_i(N_i-1)}{N(N-1)}$ or $\frac{1}{\sum_{i=1}^{n} \frac{N_i(N_i-1)}{N(N-1)}}$	Simpson [1949]
Shannon's Index (infinite community)	H: $-\sum_{i=1}^{n} p_i \ln p_i$	Shannon [1948]
Brillouin's Index (Shannon's index for a finite community)	$\frac{1}{N} \ln \left(\frac{N!}{N_1!N_2!...N_n!} \right)$	Brillouin [1962]
Margalef's Index	$\frac{n-1}{\ln N}$	Margalef [1958]
Menhinick's Index	$\frac{n}{\sqrt{N}}$	Menhinick [1964]
McIntosh's Index	$\left(N - \sqrt{\sum_{i=1}^{n} N_i} \right) / \left(N - \sqrt{N} \right)$	McIntosh [1967]
Berger-Parker's Index	$1 - \frac{N_{\max}}{N}$ or $\frac{N}{N_{\max}}$	Berger and Parker [1970]
Hurlbert's "Interspecific Encounter" Index	$\left(\frac{N}{N-1} \right) \left(1 - \sum_{i=1}^{n} p_i^2 \right)$	Hurlbert [1971]
Fager's Index	$\frac{N(n+1)}{2} - \sum_{i=1}^{n} r_i N_i$	Fager [1972]
Hill's Family of Diversity Indices	$\left(\sum_{i=1}^{n} p_i^a \right)^{\frac{1}{(1-a)}}$	Hill [1973]
Keefe and Bergersen's Index	$1 - \frac{N}{(N-1)} \left(\sum_{i=1}^{n} p_i^2 - \frac{1}{N} \right)$	Keefe and Bergersen [1977]
Rao's "Quadratic Entropy" Index	$\mathbf{Q}$: $\sum_{i,j=1}^{n} d_{ij} p_i p_j$	Rao [1982]

ACKNOWLEDGEMENTS

Thanks to Russell Lande, Samir Okasha, Carl Salk, and Sahotra Sarkar for helpful comments. Members of the Florida State University biology department and Sydney-ANU philosophy of biology workshop also provided helpful feedback on a presentation based on this work.

BIBLIOGRAPHY

[Aczél and Daróczy, 1975] J. Aczél, and Z. Daróczy. *On Measures of Information and Their Characteristics.* New York: Academic Press, 1975.

[Berger and Parker, 1970] W. H. Berger and F. L. Parker. Diversity of Planktonic *Formaminifera* in Deep-sea Sediments. *Science* 168: 1345–1347, 1970.

[Brillouin, 1962] L. Brillouin. *Science and Information Theory*, 2nd Edition. New York: Academic Press, 1962.

[Drake, 2007] J. M. Drake. The Background of Biodiversity: Quantitative Measures of Species Diversity, 1943–1982. Unpublished Manuscript.

[Dyer and Coley, 2001] L. A. Dyer and P. D. Coley. Latitudinal Gradients in Tri-trophic Interactions." In Tscharntke, T. and Hawkins, B. A. (eds.), *Multitrophic Level Interactions*, pp. 67–88. New York, NY: Cambridge University Press, 2001.

[Eberhardt, 1969] L. L. Eberhardt. Some Aspects of Species Diversity Models. *Ecology* 50: 503–505, 1969.

[Fager, 1973] E. W. Fager. Diversity: A Sampling Study. *American Naturalist* 106: 293–310, 1973.

[Fisher *et al.*, 1943] R. A. Fisher; A. S. Corbet; and C. B. Williams. The Relationship Between the Number of Species and the Number of Individuals in a Random Sample of an Animal Population. *Journal of Animal Ecology* 12: 42–58, 1943.

[Good, 1953] I. J. Good. The Population Frequencies of Species and the Estimation of Population Parameters. *Biometrika* 40: 237–264, 1953.

[Hill, 1973] M. O. Hill. Diversity and Evenness: A Unifying Notation and Its Consequences. *Ecology* 54: 427–432, 1973.

[Hurlbert, 1971] S. Hurlbert. The Nonconcept of Species Diversity: A Critique and Alternative Parameters. *Ecology* 52: 577–586, 1971.

[Jones *et al.*, 1994] C. G. Jones; J. H. Lawton; M. Shachak. Organisms as Ecosystem Engineers. *Oikos* 69: 373–386, 1994.

[Justus, 2006] J. Justus. Loop analysis and Qualitative Modeling: Limitations and Merits. *Biology and Philosophy* 21: 647–666, 2006.

[Justus, 2008] J. Justus. Complexity, Diversity, and Stability. In Sarkar, Sahotra and Plutynski, Anya (eds.), *A Companion to the Philosophy of Biology*, pp. 321–350. Malden, MA: Blackwell, 2008.

[Kaplansky, 1977] I. Kaplansky. *Set Theory and Metric Spaces.* New York: Chelsea Publishing, 1977.

[Keefe and Bergersen, 1977] T. J. Keefe and E. P. Bergersen. A Simple Diversity Index Based on the Theory of Runs. *Water Research* 11: 689–691, 1977.

[Khinchin, 1957] A. I. Khinchin. *Mathematical Foundations of Information Theory.* New York, NY: Dover Publications, 1957.

[Lande, 1996] R. Lande. Statistics and Partitioning of Species Diversity, and Similarity among Multiple Communities. *Oikos* 76: 5–13, 1996.

[Lewontin, 1972] R. Lewontin. The Apportionment of Human Diversity. *Evolutionary Biology* 6: 381–398, 1972.

[Lloyd and Ghelardi, 1964] M. Lloyd and R. Ghelardi. A Table for Calculating the Equitability of Species Diversity. *Journal of Animal Ecology* 33: 217–225, 1964.

[MacArthur, 1957] R. H. MacArthur. On the Relative Abundance of Bird Species. *Proceedings of the National Academy of Sciences of the USA* 43: 293–295, 1957.

[MacArthur, 1972] R. H. MacArthur. *Geographical Ecology.* New York: Harper and Row, 1972.

[Maclaurin and Sterelny, 2008] J. Maclaurin and K. Sterelny. *What is Biodiversity?* Chicago: The University of Chicago Press, 2008.

[Magurran, 1988] A. E. Magurran. *Ecological Diversity and Its Measurement.* Princeton: Princeton University Press, 1988.

[Magurran, 2004] A. E. Magurran. *Measuring Biological Diversity.* Malden: Blackwell Publishing, 2004.

[Margalef, 1958] R. Margalef. Information Theory in Ecology. *General Systems* 3: 36–71, 1958.

[May, 1975] R. M. May. Patterns of Species Abundance and Diversity. In Cody, Michael L. and Diamond, Jared M. (eds.), *Ecology and Evolution of Communities*, pp. 81–120. Cambridge, MA: Harvard University Press, 1975.

[McIntosh, 1967] R. P. McIntosh. An Index of Diversity and the Relations of Certain Concepts to Diversity. *Ecology* 48: 392–404, 1967.

[Menhinick, 1964] E. P. Menhinick. A Comparison of some Species-Individuals Diversity Indices Applied to Samples of Field Insects. *Ecology* 45: 859–861, 1964.

[Møller, 1998] A. P. Møller. Evidence of Larger Impact of Parasites on Hosts in the Tropics: Investment in Immune Function Within and Outside the Tropics? *Oikos* 82: 265–270, 1998.

[Oksanen *et al.*, 1981] L. Oksanen; S. Fretwell; J. Arrunda; and P. Niemela. Exploitation Ecosystems in Gradients of Primary Productivity. *American Naturalist* 118: 240–261, 1981.

[Paine, 1969] R. T. Paine. A Note on Trophic Complexity and Species Diversity. *American Naturalist* 100: 91–93, 1969.

[Patil and Taillie, 1982a] G. P. Patil and C. Taillie. Diversity as a Concept and Its Measurement. *Journal of the American Statistical Association* 77: 548–561, 1982.

[Patil and Taillie, 1982b] G. P. Patil and C. Taillie. Diversity as a Concept and Its Measurement: Rejoinder. *Journal of the American Statistical Association* 77: 548–561, 1982.

[Peet, 1974] R. K. Peet. The Measurement of Species Diversity. *Annual Review of Ecology and Systematics* 5: 285–307, 1974.

[Peet, 1975] R. K. Peet. Relative Diversity Indices. *Ecology* 56: 496–498, 1975.

[Pianka, 1966] E. Pianka. Latitudinal Gradients in Species Diversity: A Review of Concepts. *American Naturalist* 100: 33–46, 1966.

[Pielou, 1966] E. C. Pielou. The Use of Information Theory in the Study of the Diversity of Biological Populations. *Proceedings of the Fifth Berkeley Symposium on Mathematical Statistics and Probability* 4: 163–177, 1966.

[Pielou, 1975] E. C. Pielou. *Ecological Diversity.* New York: John Wiley & Sons, 1975.

[Pielou, 1977] E. C. Pielou. *Mathematical Ecology*, 2nd Edition. John Wiley & Sons, New York, 1977.

[Preston, 1948] F. W. Preston. The Commonness and Rarity of Species. *Ecology* 29: 254–283, 1948.

[Preston, 1962a] F. W. Preston. The Canonical Distribution of Commonness and Rarity, Part I. *Ecology* 43: 185–215, 1962.

[Preston, 1962b] F. W. Preston. The Canonical Distribution of Commonness and Rarity, Part II. *Ecology* 43: 410–432, 1962.

[Rao, 1982] C. R. Rao. Diversity and Dissimilarity Coefficients—A Unified Approach. *Theoretical Population Biology* 21: 24–43, 1982.

[Rényi, 1961] A. Rényi. On Measures of Entropy and Information. *Proceedings of the 4th Berkeley Symposium on Mathematics, Statistics, and Probability* 1: 547–561, 1961.

[Ricotta, 2000] C. Ricotta. From Theoretical Ecology to Statistical Physics and Back: Self-Similar Landscape Metrics as a Synthesis of Ecological Diversity and Geometrical Complexity. *Ecological Modeling* 125: 245–253, 2000.

[Ricotta, 2002] C. Ricotta. Bridging the Gap between Ecological Diversity Indices and Measures of Biodiversity with Shannon's Entropy: Comment to Izsák and Papp. *Ecological Modeling* 152: 1–3, 2002.

[Ricotta, 2005] C. Ricotta. Through the Jungle of Biological Diversity. *Acta Biotheoretica* 53: 29–38, 2005.

[Ricotta and Avena, 2003] C. Ricotta and G. C. Avena. An Information-Theoretical Measure of Taxonomic Diversity. *Acta Biotheoretica* 51: 35–41, 2003.

[Sarkar, 2002] S. Sarkar. Defining "Biodiversity"; Assessing Biodiversity. *The Monist* 85: 131–155, 2002.

[Sarkar, 2005] S. Sarkar. *Biodiversity and Environmental Philosophy.* New York: Cambridge University Press, 2005.

[Sarkar, 2007] S. Sarkar. From Ecological Diversity to Biodiversity. In Hull, David and Ruse, Michael (eds.), *Cambridge Companion to the Philosophy of Biology*. Cambridge, MA: Cambridge University Press, 2007.

[Shannon, 1948] C. E. Shannon. A Mathematical Theory of Communication. *Bell Systems Technical Journal* 27: 379–423, 623–656, 1948.

[Shannon and Weaver, 1949] C. E. Shannon and W. Weaver. *The Mathematical Theory of Communication*. Urbana: University of Illinois Press, 1949.

[Simpson, 1949] E. H. Simpson. Measurement of Diversity. *Nature* 163: 688, 1949.

[Smith *et al.*, 1996] B. Smith and J. B. Wilson. A Consumer's Guide to Evenness Indices. *Oikos* 76: 70–82, 1996.

[Sugihara, 1982] G. Sugihara. Comment. *Journal of the American Statistical Association* 77: 564–565, 1982.

[Tilman, 1996] D. Tilman. Biodiversity: Population Versus Ecosystem Stability. *Ecology* 77: 350–363, 1996.

[Tilman, 1999] D. Tilman. The Ecological Consequences of Biodiversity: A Search for General Principles. *Ecology* 80: 1455–1474, 1999.

[Tilman *et al.*, 2006] D. Tilman; P. B. Reich; J. M. H. Knops. Biodiversity and Ecosystem Stability in a Decade-long Grassland Experiment. *Nature* 441: 629–632, 2006.

[Tramer, 1969] E. J. Tramer. Bird Species Diversity: Components of Shannon's Formula. *Ecology* 50: 927–929, 1969.

[Wiener, 1948] N. Wiener. *Cybernetics: Or the Control and Communication in the Animal and the Machine*. Cambridge: MIT Press, 1948.

[Willig *et al.*, 2003] M. R. Willig; D. M. Kaufmann; and R. D. Stevens. Latitudinal Gradients of Biodiversity: Pattern, Process, Scale and Synthesis. *Annual Review of Ecology and Systematics* 34: 273–309, 2003.

THE BIODIVERSITY–ECOSYSTEM FUNCTION DEBATE IN ECOLOGY

Kevin deLaplante and Valentin Picasso

1 INTRODUCTION

Population/community ecology and ecosystem ecology present very different perspectives on ecological phenomena. Over the course of the history of ecology there has been relatively little interaction between the two fields at a theoretical level, despite general acknowledgment that many ecosystem processes are both influenced by and constrain population- and community-level phenomena. However, recent years have seen a growing interest in theoretical models and experimental studies aimed at investigating the relationship between biological diversity and higher-level community and ecosystem properties, such as invasibility and productivity. This research on the relationship between biodiversity and ecosystem functioning has spawned a large and growing literature that holds great promise for productive engagement between community ecology and ecosystem ecology. Indeed, some have argued that the synthetic viewpoints developing out of this research represent a genuine "paradigm shift" in ecology [Naeem, 2002].

However, this research has also generated heated debate among ecologists over experimental methodology and interpretation of research results. The debate burst into the public sphere in 2000 when a group of critics of the biodiversity-ecosystem function experiments accused proponents of misrepresenting the scientific debate to the public for political purposes. One media source described it as a "full war among ecologists" [Kaiser, 2000]. Recent writings have been more conciliatory in tone, but the incident points to a broader socio-political context that has played an important role in both motivating and enabling research on biodiversity-ecosystem function relationships, a context that connects research in this field to debates in conservation science and environmental policy. A comprehensive overview of this debate needs to take account of this socio-political context.

Young ecologists beginning their research careers are often unaware of the intellectual history of their field, or the relevance of this history for understanding the scientific and socio-political environment within which their work is situated. The primary aim of this paper is to provide an historical and conceptual overview of the biodiversity-ecosystem function debate that will help to illuminate research that is currently being conducted in this field.

Handbook of the Philosophy of Science. Volume 11: Philosophy of Ecology.
Volume editors: Kevin deLaplante, Bryson Brown and Kent A. Peacock. General editors: Dov M. Gabbay, Paul Thagard and John Woods.

The biodiversity-ecosystem function literature employs concepts like "biodiversity", "ecosystem" and "function" that are themselves subjects of considerable debate in the foundational literature in ecology and in the philosophy of ecology and biology. It is a central thesis of this paper that a proper understanding of the biodiversity-ecosystem function debate requires an appreciation of this broader intellectual history. Consequently, one of the tasks of the paper is to critically assess the status of these concepts as they are used in ecology generally and in the biodiversity-ecosystem function literature in particular. In this respect the paper serves not only as an introduction to the biodiversity-ecosystem function debate, but also as an introduction to a number of central debates in the philosophy of ecology more broadly.

2 BACKGROUND: THE DIVERSITY-STABILITY DEBATE

The contemporary biodiversity-ecosystem function debate is best viewed against the background of the long-standing debate in ecology over the relationship between the diversity and stability of ecological systems.

Commentators on the history of the diversity-stability debate commonly distinguish three historical periods in the history of ecology, each characterized by a particular theoretical and empirical perspective on diversity-stability relationships, with the most recent third period identified with a shift toward what we now call "biodiversity-ecosystem function" relationships [Ives, 2005; McCann, 2005].

As we will see, the history of the diversity-stability debate has important lessons for contemporary research on biodiversity and ecosystem function.

2.1 The 1950s and 1960s

The view that diversity is positively correlated with stability was endorsed by a number of prominent ecologists in the 1950s and 1960s, including Eugene Odum [1953], Robert MacArthur [1955] and Charles Elton [1958].

Odum related the notions of diversity and stability to the flow of energy through the trophic links in an ecological network. A system with greater redundancy in energetic pathways will be more stable than one with lesser redundancy. For Odum, diversity is interpreted as diversity of network connections, and stability as stability of energetic throughput and organizational structure—the more stable system is the one that suffers the least change in energy flow with the removal of a random species. However, these ecosystem concepts have a rough correspondence to population and community concepts via the identification of network nodes with species populations and network connections with trophic links.

MacArthur [1955] followed Odum in understanding stability as a measure of "the amount of choice which the energy has in following the paths up through the food web". He sketched a series of food webs and described the ramifications of energy partitioning for stability using information theory. Formally, MacArthur's notion of stability is a measure of the response of a community to a perturbation that influences the density of at least one of the species. MacArthur gives a semi-formal argument that recapitulates Odum's conclusion—in general, more diverse communities will be more stable than less diverse communities.

Elton's [1958] arguments draw on a wider range of theoretical and empirical evidence, but he agrees that diversity and stability are positively correlated. Elton noted that both simple Lotka-Volterra models and simple laboratory microcosms suffered from instability, and argued that simpler food webs are more vulnerable to invaders. Elton's definitions of stability vacillate within his discussion, but they reflect his general interest in dynamic instabilities that drive destructive oscillations and population explosions in food webs.

To sum up, the broad consensus during this period was that stability of ecological systems is positively correlated with diversity, and indeed that diversity is a causal factor in generating stability.

2.2 The 1970s and 1980s

This consensus did not survive the next two decades. By the end of the 1980s the general consensus was that diversity is not, in general, positively correlated with stability. How did this shift in attitudes come about?

In the early 1970s mathematical ecologists began to systematically study diversity-stability relationships in model communities [Gardner and Ashby 1970; May 1973; Pimm 1980]. The conclusion of these studies undermined the conventional wisdom about diversity and stability.

The most influential work of this period was Robert May's seminal 1973 book *Stability and Complexity in Model Ecosystems*. May argued that diversity actually begets instability. More specifically, he showed that the chances of a randomly constructed Lotka-Volterra community being stable decreases with both the number of species in the community and the connectance among species, where connectance is measured by the probability that a pair of species interacts.

May's argument employed a very specific definition of stability: it is the probability that the population size of every species in the community would return to equilibrium if there were an arbitrarily small perturbation in the population size(s) of one of the species. It is important to note that this so-called "neighborhood stability" (or "Lyapunov stability") is an all-or-nothing property; for a given perturbation, either every population returns to equilibrium or it doesn't. May presents his results in terms of the probability that a community, randomly selected from a certain hypothetical population of communities, is neighborhood-stable.

Stuart Pimm [1980] came to a similar conclusion in his influential analysis of stability properties of food webs. However, Pimm's analysis employs a different definition of stability. He questioned the ecological relevance of May's "arbitrarily small perturbations" and chose to model instead the effects of a more significant perturbation, the permanent removal of one of the species in the community. This "species-deletion stability" is defined as follows: it is the probability that the removal of one species will not lead to any further local extinctions. Pimm's analysis showed that, indeed, communities with more species were less "species-deletion stable" than communities with fewer species.

These theoretical results were taken to have broad significance for ecology and lead to a general rejection of the diversity-stability hypothesis among ecologists. The significance of these results can be challenged, however.

Consider, for example, that a negative diversity-stability relationship *is an immediate statistical consequence* of the definitions of stability used by both May and Pimm. If every species population must return to equilibrium after a perturbation (May), or if every species population must survive the permanent deletion of one species (Pimm), then the criteria for stability *necessarily* becomes more and more strict as you add more species to the community. The conclusion is independent of any particular feature of ecological communities; indeed, it can be viewed as an artifact of probability theory.

This fact can be viewed as undermining the empirical significance of the conclusions; the stability definitions that are employed in the analysis turn what ought to be an empirical hypothesis into a probabilistic *tautology* in idealized systems that are unlikely to be realized in nature anyway [Mikkelson, 1999].

Moreover, it can be argued that these strict, population-level concepts of stability don't faithfully capture the original notions of stability expressed in the writings of Odum, MacArthur and Elton, which more often referred to functional properties of whole communities or ecosystems.

It is this intuition—that a proper test of the diversity-stability hypothesis should focus on functional properties of communities and ecosystems—that motivates more recent work on diversity-stability relations.

2.3 The 1990s

The 1990s saw a revival of the diversity-stability hypothesis in experimental studies that indicated a positive relationship between diversity and the stability of various functionally defined properties of communities and ecosystems. The leading figure in this revival was David Tilman [Tilman and Downing, 1994; Tilman *et al.*, 1996], though many researchers have since contributed to research in this field.

The general conclusion of these more recent studies is that increasing species diversity may well decrease the stability of individual plant populations, but it may simultaneously *increase* the stability of higher-level community and ecosystem properties. This is because the increased fluctuations in population size induced by increased diversity aren't in phase across all populations—while some populations are decreasing, others may be increasing. Within a more diverse community there is a greater chance that downward fluctuations will be balanced by upward swings elsewhere in the community, resulting in greater stability of community and ecosystem properties that are averaged over individual population sizes.

These studies typically employ one of two measures of stability: *resistance to invasion* by new species, or *temporal stability* of an ecosystem property like biomass or productivity. Here, "temporal stability" is the mean value of a variable divided by its standard deviation, both calculated over time; it is a measure of the degree of variability of a property over time. These concepts of stability self-consciously reflect the concerns with resistance to invasion and temporal variability that dominated pre-1970s thinking about diversity-stability relationships. Note that this shift in stability measures inspired a corresponding shift in terminology, from talking about the stability of *population sizes* to the stability of *ecosystem functions*.

Another feature of the recent literature on diversity-stability relations is a recognition that "diversity" itself has many possible measures other than species richness. There is considerable interest, for example, in studying relationships between the *functional diversity* of a community and the stability of ecosystem functions. Functional diversity represents the diversity of functional traits or groups. Examples of functional traits include properties like leaf size, seed size, dispersal mode, canopy structure, and capacity for symbiotic fixation of nitrogen. Examples of functional groups include trophic groups (*e.g.*, producers, consumers, decomposers), animal guilds (*e.g.*, granivores, sap suckers, leaf miners, pollinators) or plant groups (*e.g.*, legumes, cool season grasses, warm season grasses, woody forbs).

Consequently, recent work has moved toward a broader investigation of relationships between different measures of *biodiversity* and the stability properties of *ecosystem functions*. Thus do we arrive at the nomenclature of contemporary **biodiversity-ecosystem function** studies.

2.4 Diversity-Stability Relationships and Environmental Policy

We noted in the introduction that the biodiversity-ecosystem function debate burst into the public sphere in 1999 when a group of critics of the biodiversity-ecosystem function experiments accused proponents of misrepresenting the scientific debate to the public for political purposes [Kaiser, 2000]. We discuss the details of this event in section 4; here we wish to emphasize the general point that diversity-stability hypotheses do have implications for environmental policy, and this fact is relevant in evaluating how ecologists interpret and report research findings.

Consider, for example, (1) increasing concern over loss of biodiversity induced by environmental deterioration and loss of habitat, and (2) the growing perception that human impacts on the biosphere may significantly alter the behavior of ecosystems and threaten vital ecosystem services. Diversity-stability hypotheses are relevant to environmentalist and conservationist arguments in both areas of concern by linking issues in one area to issues in the other. If one believes that certain types or levels of biological diversity are necessary to maintain the stability of ecosystems and correlated ecosystem services, then one can easily develop an argument for placing a high instrumental value on biodiversity, and thereby motivate environmental policies that promote the conservation and restoration of biodiversity.

This observation highlights an important fact: *ecological research on diversity-stability relationships is conducted in a socio-political environment that favors certain outcomes over others.* People who endorse environmental protection policies often look to ecology for scientific support for their agendas. Indeed, ecologists themselves may be motivated for similar reasons to look for evidence that supports a positive diversity-stability relationship.

In the 1970s and 1980s the majority view among ecologists was broadly skeptical of diversity-stability hypotheses. It was easy to regard ecologists who continued to defend a positive relationship between diversity and stability in light of the evidence mounting against it as either stuck in an outmoded paradigm, engaged in wishful thinking, or overly beholden to environmentalist interests.

In the 1990s it once again became scientifically respectable to defend diversity-stability hypotheses, but many ecologists remained wary of the influence of environmental advocacy on the interpretation and presentation of scientific results. As will be shown in greater detail later, these concerns came to a head in 1999 when critics complained that an Ecological Society of American *Bulletin* presented a biased and politically motivated account of the biodiversity-ecosystem function research results.

2.5 Diversity-Stability Relationships and the Holism-Reductionism Debate in Ecology

The study of diversity-stability relationships also takes place in a context framed by the historical schism in ecology between holistic and reductionistic research traditions and worldviews. A belief in a positive diversity-stability relationship is commonly associated with some kind of commitment to holism, while skepticism is more commonly associated with reductionism. Thus, in addition to biases arising from environmental policy considerations as outlined above, we must also consider biases arising from philosophical predispositions toward holism or reductionism in ecology.

These claims require some elaboration. In ecology, holistic and reductionistic theses come in several varieties, but they can generally be divided into one of two categories depending on whether their focus is *ontological* or *epistemological.* Ontology pertains to the nature of reality, of what exists. Epistemology pertains to knowledge and the justification of beliefs about the world (in a scientific context, issues concerning scientific methodology fall into this category). For example, ecologists may differ on the ontological constitution of communities and ecosystems (*e.g.*, whether they have "emergent causal properties" at the community and ecosystem level), and they may differ on the best way to represent and analyze ecological systems in ecological theories (*e.g.*, whether community- and ecosystem-level phenomena can be exhaustively explained in terms of the behaviors of their component parts). The latter is an epistemological issue, the former an ontological issue.

There are at least two reasons why a belief in a positive diversity-stability relationship is commonly associated with holism:

1. There is an historical association between diversity-stability theses and traditional notions of the "balance of nature", the view that ecological systems are naturally driven toward an equilibrium state in which community composition persists and population sizes are (roughly) stable. In its original formulation with the Greeks, the balance of nature was explained in terms of teleological principles governing nature as a whole. In the Medieval period the common explanation was divine providence [Egerton, 1973]. In the modern period the favored explanations have referred either to density-dependent regulation or the stabilizing effects of network redundancy (as articulated, for example, in the arguments of Odum, MacArthur and Elton). Whether these modern explanations are properly described as "holistic" depends largely on how one defines the term, but the point is that the diversity-stability hypothesis has an historical association with worldviews that are widely regarded as holistic.

2. We noted that the diversity-stability hypothesis fell out of favor in the 1970s and 1980s in the wake of theoretical studies that seemed to undermine any positive relationship between diversity and stability. It is notable that this period also saw the rise to prominence of a new "non-equilibrium" paradigm in ecology that rejected the balance of nature hypothesis outright [Botkin, 1990]. This paradigm reconceptualized the default state of nature as one of constant flux and change, and its proponents were often motivated to label the paradigm as reductionistic to contrast it with the holism associated with equilibrium views of nature [Simberloff, 1980]. Proponents of the non-equilibrium paradigm were also inclined to associate the rejection of the diversity-stability hypothesis with the broader move toward reductionism during this time period.

These developments were, and continue to be, significant for research aimed at reviving the diversity-stability hypothesis. The fact is that within mainstream academic ecology—particularly plant ecology—there is a general bias toward reductionistic and away from holistic hypotheses and methods. The default view is to be skeptical of holistic hypotheses. Insofar as a positive diversity-stability relationship is associated with ecological holism one can expect it to face the same default skepticism.

As contemporary research on biodiversity-ecosystem function relationships continues to mature these default attitudes may slowly be changing, but among plant ecologists who continue to strongly identify with reductionism (*e.g.*, neo-Gleasonian views on plant dynamics) one is likely to encounter resistance to any diversity-stability hypothesis that is perceived as appealing to holistic mechanisms or properties to account for experimental results.

2.6 Lessons Learned

The purpose of this brief overview of the diversity-stability debate in ecology was to show how biodiversity-ecosystem function research may be viewed as both a consequence of and contribution to the long-standing debate in ecology over the relationship between the diversity and stability of ecological systems, and to outline a number of factors that played prominent roles in this debate. There are several lessons that can be learned from this overview for researchers in biodiversity and ecosystem functioning:

1. *Sensitivity to definitions.* The question of whether diversity begets stability is not well-posed until one stipulates a definition of the key terms. We have seen that certain definitions of diversity and stability in certain modeling contexts may yield a negative correlation, while other definitions in other modeling (and experimental) contexts may yield a positive correlation. Thus the relevant scientific question to ask is not "does diversity beget stability?", but rather "does diversity of type D beget stability of type S under conditions of type C?". As we will see, the same lesson applies to debates over biodiversity-ecosystem function relationships.

2. *Biases arising from ideological commitments relating to environmental policy.* Ecological research on diversity-stability relationships is conducted in a socio-political environment that favors certain outcomes over others. In particular, a positive diversity-stability relationship (*i.e.*, one showing a positive correlation between diversity and stability) will be a preferentially desired outcome for those looking for scientific support for biodiversity conservation policies. We should expect the same factors and biases to be in play in contemporary biodiversity-ecosystem function research.

3. *Biases arising from attitudes toward holism versus reductionism in science.* A positive diversity-stability relationship is historically more closely associated with holistic than reductionistic research programs in ecology. Consequently, skepticism about holistic interactions in ecological systems can translate into skepticism about positive diversity-stability relationships. Similarly, a commitment to the reality and ecological significance of holistic interactions in ecological systems can translate into a bias in favor of positive diversity-stability relationships.

3 BIODIVERSITY AND ECOSYSTEM FUNCTIONS: KEY CONCEPTS

As noted in section 2, one of the lessons learned from earlier studies of diversity-stability relationships is the importance of being clear about the definitions of key theoretical terms and their empirical measures. Biodiversity-ecosystem function research is particularly vulnerable to charges that their key concepts, "biodiversity" and "ecosystem function", are either too vague, multi-faceted or value-laden to properly serve the needs of empirical science. In this section we discuss the various meanings with which these terms are used in the ecological literature, identify some of the conceptual challenges facing the use of these terms in a scientific context, and clarify their usage in the biodiversity-ecosystem function literature.

3.1 Biodiversity

We begin with the concept of "biodiversity", central to conservation biology and the biodiversity-ecosystem function literature.

3.1.1 Biodiversity and Conservation

The concepts of biological and ecological diversity are as old as natural history, but the term "biodiversity" only appeared in the scientific lexicon in the late 1980s, coinciding with the emergence of conservation biology as an applied science aimed at preserving and conserving biological diversity in the face of a looming biodiversity "crisis" [Soulé, 1985].

Attempts to define "biodiversity" as an object of conservation have always been complicated by the fact that, in this context, the *objects* that comprise biodiversity are associated with conservation *values*, *i.e.*, those aspects of the natural environment that we value and wish to preserve for current and future generations (or for their own sake). In principle this can include any biological entity or process of interest. However, this move runs the risk of making biodiversity co-extensive with all of biology and consequently rendering biodiversity conservation impractical, since everything biological would become a goal of conservation

Definitions of biodiversity are also complicated by the fact that objects of biological and ecological interest don't fall under a single hierarchy of nature (see [Sarkar, 2005] for elaboration on the following). One can distinguish at least two distinct hierarchies: (i) a *taxonomic* hierarchy that includes genes and alleles, genotypes, subspecies, species, genera, families, orders, classes, phyla, and kingdoms; and (ii) a *spatial/compositional* hierarchy that includes biological molecules, cell organelles, cells, individuals, populations, meta-populations, communities and ecosystems (communities plus their physical environments), and extending ultimately to the entire biosphere. Biological entities of interest may not fall cleanly into any specific category in either hierarchy (consider fungi, or asexual species), and at every level of each hierarchy one finds significant variation.

Standard definitions of biodiversity address this problem by focusing on the diversity of entities at three levels of organization—alleles or genes, species, and ecosystems. The reasoning is that if you can preserve allelic diversity then you'll likely preserve most of the variation of interest below the level of the individual; if you preserve species diversity then you'll preserve all of the taxonomic entities above the species level; and if you preserve ecosystem diversity then you'll preserve most kinds of communities [Sarkar, 2005].

This traditional approach to defining biodiversity has been criticized for being overly focused on conserving biological *entities*—individuals, species, communities, etc. In addition to entities, conservation efforts are also (or should be) aimed at conservation of unique or valuable biological and ecological *phenomena* that don't fit into either the spatial or taxonomic hierarchies. A standard example is seasonal migration patterns, such as the migration of monarch butterflies in North America from the eastern and western regions of the US and Canada to Mexico and back. This migration pattern would disappear if overwintering sites were destroyed, though the species itself may persist. Conservation of unique biological phenomena isn't guaranteed by conservation of genetic, species and ecosystem diversity.

Conservation science and the associated literature on biological diversity has also been influenced by the rise to prominence of holistic conservation concepts like "biological integrity", "ecosystem integrity" and "ecosystem health". Here the focus is less on preserving individual species and more on preserving or restoring the biotic and abiotic conditions that allow different community and ecosystem types to persist. On this more holistic view, the targets of biological conservation also include ecosystem properties like network organization, characteristic rates of cycling and throughput of energy and materials, and dynamical properties related to adaptability and resilience.

These and other considerations have led many writers to suggest that the concept of biodiversity—in the context of conservation science and policy—is necessarily pluralistic and value-laden [Norton, 2000; Sarkar, 2005]. There is no single correct measure of biodiversity to be discovered but many, each representing different ways of valuing biotic and abiotic resources.

3.1.2 Biodiversity and Ecosystem Function Experiments

Many of the complicating factors noted above (relating to, for example, the association between biodiversity and conservation values) are fortunately not present in the context of the common forms of biodiversity-ecosystem function experiments. In this context we are concerned with determining empirical relationships between biodiversity and various measures of community or ecosystem stability and function. The experimental context requires that all biodiversity concepts be operationally measurable and controllable in such a way that empirically significant conclusions can be drawn. In practice this amounts to a severe restriction on the scope of possible biodiversity measures. Typical experiments focus on one taxonomic group (usually plants, but sometimes microorganisms) and then consider only the species level of biodiversity, leaving the genetic and ecosystem levels out of the discussion. At the species level, various measures of diversity may be used, such as the Shannon-Weiner index which takes into account two components, *richness* (the number of species in an area) and *evenness* (the relative abundance of different species in an area). (See Justus, this volume, for a detailed discussion of diversity measures in community ecology.)

Another class of biodiversity-ecosystem function studies focuses on relationships between *functional diversity* and ecosystem function. Functional diversity includes diversity of functional traits and groups. Functional *traits* are "the characteristics of an organism that are considered relevant to its response to the environment and/or its effects on ecosystem functioning" [Diaz and Cabido, 2001]. Examples include leaf size, seed size, dispersal mode and canopy structure. A functional *group* or *type* is a set of organisms sharing similar responses to the environment (*e.g.*, temperature, water availability, nutrients) or similar effects on ecosystem functioning (*e.g.*, productivity, nutrient cycling). Like species diversity, common measures of functional diversity include two components: i) *functional richness* (the number of different functional groups or the proportion of a multi-dimensional trait space covered by a particular suite of species) and ii) *functional composition* (presence or absence of certain functional groups or traits). Although functional diversity can apply to an indefinite number of traits, it is commonly measured by measuring the diversity of *functional groups*.

Though biodiversity-ecosystem function experiments involving functional diversity are becoming more common, it remains the case that for the majority of biodiversity-ecosystem function studies, *the proxy for biodiversity is nothing more than plant species richness*—the number of plant species in a plot. There are several practical reasons for this simplification: species are easy to identify; plant communities are easy to assemble, manipulate and maintain in pots and fields; and many interactions among plants are well documented in ecology. Also, policy makers tend to prefer single numerical measures over complex multidimensional indices to make decisions about conservation [Purvis and Hector, 2000].

Not surprisingly, this simplification imposes serious limitations on the inferences that can be drawn from biodiversity-ecosystem function studies. Claims about the significance of biodiversity *in general* for ecosystem functioning, or about the applicability of observed biodiversity-ecosystem function relationships for ecological systems *in general* (in both experimental and non-experimental contexts), will be extremely tentative at best. This is a potentially serious concern because, as noted in section 2.4, one of the motivations for the biodiversity-ecosystem function research program is the perception that this research has policy implications. Indeed, one of the criticisms of the controversial 1999 ESA Bulletin report was that the authors were too hasty in drawing *general conclusions for environmental policy* from the biodiversity-ecosystem function literature.

3.2 Ecosystem Function

For some ecologists the term "ecosystem function" is suspect because it carries with it associations of holism and teleology that are perceived to be outdated and unscientific. The term seems to presuppose the existence of ecosystems as integrated entities with emergent properties that can properly be said to fulfill "functions". However: (i) the general trajectory of plant ecology over the past thirty years has been away from strongly holistic conceptions of communities and ecosystems, and (ii) the concept of "function" in ecology is historically associated with Clementsian teleology and group-selection mechanisms of community and ecosystem development, both of which are now widely viewed by plant ecologists as empirically falsified and/or inconsistent with neo-Darwinian evolutionary theory[1] [Hagen, 1992; Glenn-Lewin *et al.*, 1992].

Defenders of the concept of "ecosystem function" should have something to say in response to objections such as these. In this section we take a closer look at these objections and clarify the meaning of the term "ecosystem function" as it is employed in the biodiversity-ecosystem function literature. We will see that, as with the case of "biodiversity", in the context of biodiversity-ecosystem function experiments the operational meaning of the term "ecosystem function" is usually rather tightly circumscribed, and consequently is less problematic than it might otherwise be. Nevertheless, ecologists need to become more aware of the conceptual issues surrounding the use of "function language" in science if they wish to avoid confusion and misreading of their work.

3.2.1 Modern Science and the Challenge to Natural Functions

Tools and other artifacts have obvious functions (a carpenter's hammer has the function of hammering nails, a coffee maker has the function of making coffee, etc.), but the function of these artifacts is grounded in the intelligent design of human beings—these objects are built and used for a conscious purpose. But do the objects studied by the natural sciences have functions? Do water molecules, chemical reactions, cells, frogs or lakes have functions? If an object is not the product of conscious intelligent design, can it have a function?

Greek and Medieval natural philosophers believed the answer was "yes": in fact, all natural systems have functions, and these functions are essential to any explanation of what they are and why they behave the way they do. Within Aristotle's philosophy of nature, every object has a "final cause" or "telos", which is the goal or purpose of the object, and every object strives to fulfill it's natural goal or purpose. This is what is meant by saying that Aristotle has a "teleological" worldview.

Indeed, Aristotle believed that natural systems possess a set of functions that reflects a hierarchical and teleological conception of the cosmos as a whole. The cosmos is an organic whole composed of many parts nested in various hierarchies. The functions of the parts are partly defined in relation to the role they play within the greater wholes that contain them. Thus, one function of plants is to grow and develop as plants do, but for Aristotle another function of plants is to provide food for animals, and this function is part of the explanation for why plants exist with the properties that they do.

[1]Among certain biologists and philosophers of biology, group selection has enjoyed a comeback in recent years under the label of "hierarchical" or "multi-level" selection theory [Wilson, 1983; Sober and Wilson, 1994]. However, it remains the case that most biologists and ecologists are *taught* that group selection is either incompatible with Darwinian evolutionary theory or that it occurs only rarely in natural systems, and it is this sociological fact that is relevant to the discussion here.

Greek and Medieval scholars working out of this teleological tradition agreed that dead, inert objects could not have natural functions of their own—any functions they have must be derived from some form of intelligent agency. For Plato and the Medieval theologians, this agency is derived from the creative work of an external designer (a "demiurge" for Plato, a theistic God for theologians). For Aristotle this agency is not external, but internal, immanent in the fundamental nature of objects. Thus, while not all objects are conscious in the way that higher animals and human beings are, all objects possess "mind-like" qualities in some sense [Lindberg, 1992]. Within this context, traditional ecological notions like the "balance of nature" were articulated in explicitly teleological language, appealing either to the immanent teleology of Aristotle or the external teleology of divine creation [Egerton, 1973].

However, the scientific revolution of the 16th and 17th centuries brought about a dramatic change in cosmological worldview. The "mechanical philosophy" developed by (among others) Bacon, Galileo, Kepler, Hobbes, Boyle, Gassendi, Descartes and Newton was grounded in the notion that the physical universe was entirely made up of small solid corpuscles in motion, and that these corpuscles are inert, devoid of any of the "psychic characteristics" that were common to the earlier frameworks. Within this framework, natural phenomena are explained as the result of mechanical interactions of inert particles. The immanent teleological principles of Aristotle were "squeezed out", and the origin of natural functions was consolidated in the external agency of God.

The more serious challenge to the concept of natural functions arose as scientific explanation became increasingly "naturalized" and explicit references to God were discouraged. Without reference to God or other forms of intelligent agency, how are we to understand natural functions?

3.2.2 Natural Functions, "Function Talk" and the Philosophy of Biology

The view that came to dominate the physical sciences was that appeal to natural functions could not be justified, and reference to them should be eliminated in scientific explanations. By the end of the 18th century the dominant research programs in physics and chemistry were mechanistic in orientation.

In the biological sciences the mechanical revolution had a less dramatic impact on the use of natural function concepts in scientific explanation. To most scientists there seemed no hope of explaining the striking adaptedness of organisms to their environments, or phenomena such as embryonic development, in purely mechanical terms. Darwinian evolutionary theory eventually offered a non-teleological explanation for biological adaptations, but in many areas of biology teleological explanations continued to flourish under the banners of vitalism, Lamarckism and orthogenesis.

It was not until the neo-Darwinian synthesis of the 1930s and 1940s and the discovery of the molecular basis of heredity that overt teleological explanations were eliminated from most areas of biology and the prevailing view in the physical sciences was finally endorsed: teleological explanations are illegitimate outside the context of human intentional explanation.

But of course function talk *didn't* disappear in the biological sciences. Biologists and ecologists continue to use expressions like "the function of", "the role of", "for the sake of", "serves as" and "for the purpose of" in discussing biological and ecological entities, processes and mechanisms. Function talk also persists in the social sciences and in medicine. This linguistic fact poses a puzzle: on the one hand, modern scientists officially disavow teleological explanations in science; on the other hand, they routinely use the language of functions in scientific description and explanation. Is this usage justified? And if so, how is it justified? This question has spawned a large philosophical literature on the relationship between function talk and teleology.

Early work by philosophers was uniformly hostile to teleology and attempted to show how function talk can be reinterpreted in non-teleological terms without loss of meaning [Hempel, 1959; Nagel, 1961]. This project had only limited success. The problem is that function talk—and especially reference to "natural functions"—seems to presuppose a degree of *normativity* that resists analysis in purely descriptive terms.

To give a standard example, we might say that the heart can perform a number of functions in virtue of its causal properties: it can produce rhythmic sounds, for instance; it can also be used to train medical students in physiology and dissection. But we also want to say that producing rhythmic sounds or assisting the training of medical students isn't the *proper* or *natural function* of the heart—the proper or natural function of the heart is to pump blood through the circulatory system of an organism. And when a heart fails to pump blood, then it's *malfunctioning*. The concepts of "natural function" and "malfunction" appear to be normative concepts in the sense that they refer not only to what hearts *in fact* do, but what they *should* do. This kind of normative function attribution is quite common in biology, but where and how does the normativity arise in the absence of immanent teleological properties (as in Aristotelian science) or intelligent design by an external agent like God?

More recent work on the philosophy of functions has attempted to *naturalize* the teleology that is evident in normative function ascriptions. The most discussed theory of normative functions is based on the observation that Darwin's theory of natural selection seems to justify a certain kind of teleology [Wright, 1973; Millikan, 1984]. We say that certain traits were "selected for". For what? For the effects of that trait that contributed to its persistence within a population over evolutionary time frames. Hearts haven't persisted in populations because they make rhythmic sounds; they persisted because they perform a particular adaptive function—pumping blood—that contributed to the survival of organisms; they were selected *for* this causal effect. Thus, the selection history of a trait allows us to distinguish between causal effects of a trait that are merely accidental and causal effects that contributed to survival because they performed an adaptive function. This conception of natural functions justifies a certain kind of normative teleological language without recourse to intelligent agencies or immanent teleological principles in nature.

However, not all philosophers are happy with theories of natural functions based on evolutionary history. If an organism didn't have any evolutionary history—if, say, it was an entirely new species created in a laboratory—but it still had a heart, wouldn't we still want to say that the heart has a function, and that function is to pump blood? Considerations such as these have motivated philosophers to develop alternative accounts of functions that are not based on evolutionary history (*e.g.*, Cummins [1975]; Boorse [2002])

For current purposes there is no need to survey the (vast) philosophical literature on functions any further (for an extended survey written for biologists see Wouters [2005]), suffice it to say that, while there is currently no consensus theory of functions among philosophers or biologists, there is widespread agreement that function talk is unlikely to be eliminated from biology, and that certain kinds of normative function attributions may be justified without presupposing Aristotelian or theological conceptions of nature.

3.2.3 Functions and Ecology: The Holism-Reductionism Split Once Again

Though biologists and ecologists have conducted their affairs largely in ignorance of the philosophical debate over functions, we should not conclude that philosophical attitudes toward functions and functional explanations have played no role in shaping the practices of scientists. These philosophical attitudes are revealed in general attitudes toward scientific methodology and holistic versus reductionistic research programs.

With respect to methodology, it is a generally accepted principle of modern scientific reasoning that a proper scientific explanation is either causal-mechanical in nature or grounded in general laws that describe uniform regularities; overt appeals to teleological principles in explaining the properties of natural systems are either discouraged or dismissed. This is the legacy that modern science has inherited from the scientific revolution of the 17th century.

In addition, the history of 20th century ecology is marked by a schism between holistic and reductionistic research programs that reveal differing views on the proper role of functions and function language in ecology. Put succinctly, holists are more willing than reductionists to attribute functions to higher-order ecological entities and processes.

Some of the reasons for these predilections should be obvious. In the non-human world, function talk is most naturally applied to well-organized systems with component parts that play distinctive roles in maintaining the structure and behavior of the system as a whole. Organisms are the quintessential example of such integrated systems and consequently function talk is most naturally applied to organisms. There is a long-standing tradition of holistic theorizing in ecology that is grounded in analogies between ecological systems and organisms. The most obvious historical example is the Clementsian concept of the plant community as a kind of "super-organism" that has an ontogeny and phylogeny directly analogous to that of individual organisms [Clements, 1916]. Organismal metaphors are also prominent in ecosystem ecology via the language of respiration, metabolism, growth, development and self-organization, and in the work of certain theorists who self-consciously defend non-trivial analogies between organismal and ecosystem development (*e.g.*, [Odum, 1969]). There are also holistic traditions of population and community ecology that emphasize the roles of individual species in contributing to the stability of higher level ecological properties, such as resistance to invasion [Elton, 1958; 1966]. It is within these holistic traditions of ecological theorizing where one is most likely to find the language of functions and functional roles applied to populations, communities and ecosystems.

By contrast, within more reductionistic approaches to ecology that are more strongly under the influence of either neo-Gleasonian individualist conceptions of plant communities and succession [Gleason, 1939; Egler, 1954], and/or the view that ecological principles must at least be consistent with, if not ultimately grounded in, neo-Darwinian evolutionary theory [Pianka 1999; Mayhew 2006], one is far less likely to find the language of functions applied to ecological entities above the levels of individuals and populations. And when it is used the tendency is to have the function language grounded in natural selection history.

There are at least two reasons for this. First, research within these traditions emphasizes the changing, stochastic, non-equilibrium aspects of ecological systems, and by and large rejects the holistic view of communities and ecosystems as coherent, organized entities with emergent causal properties. By rejecting the organismal metaphor they consequently reject function attributions that are predicated on strong analogies between ecological systems and organisms.

Second, attitudes toward function language in ecology have been influenced by the group selection debate that took place in the 1960s [Wynne-Edwards, 1962; Williams, 1966]. The critique of group selection was based on the affirmation that within orthodox evolutionary theory, natural selection acts primarily at the level of individual organisms (or, indeed, the level of individual genes), and rarely if ever at the level of groups. This debate raised awareness among ecologists of the broader implications of the theoretical perspective represented in population genetics and the neo-Darwinian synthesis, and was partly responsible for the rise of evolutionary ecology in the late 1960s and early 1970s. Evolutionary ecologists tend to associate the language of functions with organism-environment relationships relevant to selection and adaptation (*e.g.*, "functional traits"). But if natural selection only acts at the level of individuals within species populations, then the language of functions should only apply at this level (though we note again the point made in footnote 1, section 3.2). Consequently, evolutionary ecologists are inclined to be skeptical of function attributions at the community and ecosystem level.

To sum up, in ecology the language of functions is historically and conceptually tied to philosophical and theoretical debates between holists and reductionists that have played central roles in the intellectual history of the discipline. The biodiversity-ecosystem function literature is notable for its heavy use of function talk. It is an open question whether and to what extent differing philosophical attitudes toward functions (and their affiliation with holistic research traditions) influence the work of researchers within this field, but it would be naïve to assume that they play no role at all. There is no doubt, however, that some ecologists (generally, those not directly involved in biodiversity-ecosystem research) may view this research program with suspicion because of its affiliations with what they regard as a discredited ecological holism (*e.g.*, [Goldstein, 1999]).

3.2.4 Functions in the Biodiversity-Ecosystem Function Literature

We have seen that function attributions come with a certain amount of philosophical baggage associated with commitments to holism and the normativity of so-called "natural" or "proper" functions. But not all function talk in biology or ecology carries this baggage. In many cases the term "ecological function" is used synonymously with "ecological process", and merely refers to an ecologically relevant causal process. The biodiversity-ecosystem function literature uses the term "function" in a wide range of senses, some of which are innocuous and with no implications for the philosophical issues described earlier. But this is not always the case. In some cases the language of functions is used in ways that invoke the normative sense of function and that presuppose a certain kind of holism with respect to ecosystems.

Kurt Jax [2005] offers a helpful review of function language in ecology and specifically in the context of biodiversity-ecosystem function research. Jax distinguishes four major uses of the term "function" in ecology:

1. *to characterize processes and interactions between pairs of objects, and the causal relations that sustain them.* This sense of function refers to pair-wise interactions. Examples: a fox eats a mouse; a plant assimilates nutrients. In most cases the term "function" can be replaced by "process" or "interaction" without loss of meaning.

2. *to characterize processes and interactions between a collection of objects, and the causal relations that sustain them.* At this level we are viewing the objects as constituting or as situated within a larger system, and asking how the objects (now conceived as "parts") contribute to or relate to the larger system (now conceived as a "whole"). Examples: biomass production and phosphorus cycling within a lake; community population dynamics. These kinds of investigations are the stock-in-trade of a great deal of ecological research.

3. *to characterize the overall processes that sustain an ecological system as a whole, and the role of the component parts in these processes.* At this level the focus is on whole-system properties and processes. The parts of the system and their behaviors are reconceived as bearers of functions in relation to properties and processes of the whole. Examples: describing a plant

species as a "primary producer" or a bacterium as a "decomposer"; a species conceived in terms of its Eltonian "functional role" niche.

4. *to characterize those aspects of an ecological system that are useful or important to humans.* Examples: the concept of an "ecosystem service", such as providing oxygen or purifying water. Though this concept of function is most generally used in relation to human needs and interests, in principle it could be applied to other living beings.

Another important distinction that cross-cuts these categories is between functions conceived as "means" and as "ends". When conceiving of functions as "ends" we are simply focusing on the activity or performance of various objects within a temporal sequence or causal chain. When conceiving of functions as "means" we are asking about the role or contribution that an object makes for something else (*e.g.*, "what is the function of biodiversity to ecosystem functioning?"; "what function does species X play in the service of ecosystem property Y?"). Studies that focus on functions as ends are generally unproblematic since they involve nothing more than empirical investigation of a process (like productivity, or drought resistance). Studies that focus on functions as means are more problematic because they require that we consider the "aims", "goals" or "purposes" served by the function, and this brings into play the issues of teleology and normativity discussed earlier. We argued earlier that certain kinds of normative function attributions can be justified in biology, but raised questions about their applicability to ecosystem processes (we return to this issue below).

The question to be asked is this: How is the language of functions used in the biodiversity-ecosystem function literature? And are these uses problematic or unproblematic?

Jax distinguishes three kinds of research questions in the biodiversity-ecosystem function literature that employ different meanings of "function" [Jax, 2005, p. 644]:

1. How does biodiversity relate to *ecosystem processes* (= ecosystem function)?

2. How does biodiversity relate to the *functioning of ecosystems*?

3. How does biodiversity relate to *ecosystem services* (= ecosystem functions)?

The bulk of the experimental work on biodiversity-ecosystem function relations is focused on answering the first question, where the variables of interest (productivity, drought resistance, decomposition of litter, etc.) are treated as ends, not as means to some other end. This usage is largely unproblematic since it is does not invoke the normativity of functions conceived as means to some other end.

The second question employs a sense of "function" that can be problematic when the expression "functioning of ecosystems" (or "ecosystem functions") refers to the overall behavior or performance of an ecosystem, because this usage often presupposes a certain conception of ecosystems as entities in the world. Consider how the expression is used in the controversial ESA article on "Biodiversity and ecosystem functioning" (more on this in section 4):

> Ecosystem functioning reflects the collective life activities of plants, animals, and microbes and the effects these activities—feeding, growing, moving, excreting waste, etc.—have on the physical and chemical conditions of the environment. (Note that 'functioning' means 'showing activities' and does not imply that organisms perform purposeful roles in ecosystem-level processes.) A functioning ecosystem is one that exhibits biological and chemical activities characteristic of its type." [ESA, 1999, p. 3]

The authors try to head off worries about their use of function language but the last line betrays a normative interpretation of this language. A functioning ecosystem is one that "exhibits biological and chemical activities characteristic of its type". As Jax puts it,

> The aim of investigating "functioning" ecosystems here is clearly not to observe any activities of organisms in a particular area, but specific activities that sustain some "typical" ecosystem. Here "functioning" clearly receives a normative dimension in the sense that it refers to some pre-defined reference states of an ecosystem (those that "exhibit biological and chemical characteristics of its type"). The "functioning" of the ecosystem thus is a desirable state, and the organisms in fact are investigated as if they perform purposeful roles in its perpetuation. This is a legitimate aim of applied ecological research, but it goes beyond a pure description of processes that occur in some aspect of nature. [Jax, 2005, p. 644]

In short, this usage presumes that one can describe ecosystems as functioning or malfunctioning relative to some reference state that characterizes an idealized ecosystem "type". The problem here isn't so much the normativity of the function ascription as the conception of ecosystems and ecosystem individuation that is being presupposed. Very few ecologists believe that ecosystem "types" are part of the furniture of the world. By far the more common view (even among holists) is that the boundaries and variables that characterize an ecosystem *are chosen by observers*, they're not given in nature as such. Consequently, making statements about the functioning of ecosystems demands that observers delimit the ecosystem in question and specify the relevant reference states. The problem, as Jax sees it, is not that this is impossible, but that it is almost never done in a careful, explicit and motivated fashion. As a result, the concept of a "functioning ecosystem" is never operationally defined. This kind of usage lends support to critics who charge that expressions like "ecosystem function" are nothing more than trendy buzzwords that don't belong in the scientific lexicon of ecology.

Jax [2005] identifies a number of other examples in the biodiversity-ecosystem function literature where distinctions between ecosystem processes and ecosystem functions, and between normative and descriptive senses of function, are blurred, resulting in semantic confusions that hinder rather than help the empirical investigation of biodiversity-ecosystem function relationships. We agree that this research can benefit from a theoretical framework that encourages greater precision in the use of key concepts and that is more mindful of the historical and philosophical issues associated with the use of these concepts.

3.3 Summing Up

In the preceding sections we presented an overview of conceptual issues related to the use of the terms "biodiversity" and "ecosystem function". We saw that scientific investigations of biodiversity are challenged by a multiplicity of concepts and measures of biodiversity, and by associations of the concept with normative and political goals of conservation ethics and policy. And we saw that scientific investigations of ecosystem function are challenged by historical associations of "function talk" with teleological views of nature, discredited (or at least, marginalized) holistic views of the structure and organization of ecological systems, and by ambiguity in the usage of the term "ecosystem function". Consequently, we should not be surprised to find divided opinions on the status and interpretation of contemporary biodiversity-ecosystem function research.

4 THE BIODIVERSITY-ECOSYSTEM FUNCTION DEBATE

In this section we present an historical narrative leading up to the so-called "war among ecologists" that was reported in the journal *Nature* [Kaiser, 2000]. As we shall see, this more recent debate shares several features with earlier debates over diversity-stability relationships.

4.1 The Socio-Political Context

Concerns about biodiversity loss escalated in the 1980s and 1990s, along with a growing awareness that intact, functioning ecosystems perform a wide range of so-called "ecosystem services", among them the provisioning of food and clean water, crop pollination, pest and disease control, nutrient dispersal and cycling, and seed dispersal. It is not surprising that researchers would be interested to investigate whether loss of biodiversity might interfere with the ability of ecosystems to perform these vital functions, but the research program on biodiversity-ecosystem function relationships that emerged in the 1990s was driven not by scientific curiosity alone, but by an international group of scientific and policy organizations motivated by a range of policy concerns. These organizations included the following:

- *International Council for Science* (ICSU). An NGO founded in 1931, comprised of 112 national scientific bodies and 29 international scientific unions, to promote scientific activity applied for the benefit of humanity. ICSU's broad scientific expertise addresses major issues by creating interdisciplinary bodies and joint initiatives with other organizations.

- *United Nations Educational, Scientific and Cultural Organization* (UNESCO). Founded in 1945 with the goal of building peace though education, science, culture, and sustainable development.

- *Scientific Committee on Problems of the Environment* (SCOPE). An international scientific organization, comprised of 38 national science academies and 22 international scientific union. SCOPE develops scientific reviews of environmental issues in three cluster areas: "managing societal and natural resources", "ecosystem processes and biodiversity", and "health and environment."

- *International Geosphere Biosphere Program* (IGBP). One of ICSU's interdisciplinary boards charged with studying global change, started in 1987. One of its projects, Global Change in Terrestrial Ecosystems (GCTE), addressed how global change would affect terrestrial ecosystems and feedbacks to the climate system.

- *DIVERSITAS*. Joint initiative by SCOPE, UNESCO, ICSU, and other organizations, started in 1991. It provides an international multi-disciplinary framework for promoting integrative biodiversity science through synthesizing scientific knowledge, promoting new interdisciplinary research, and communicating policy implications.

- *National Science Foundation* (NSF). A US federal agency created in 1950 to promote the progress of science; to advance the national health, prosperity, and welfare; and to secure the national defense. The Directorate of Biological Sciences, Division of Environmental Biology, funded much of the American biodiversity-ecosystem function research of this period.

- *European Science Foundation* (ESF). Association of 75 member organizations (European national research councils) devoted to scientific research in 30 European countries. Established in 1974, it has coordinated a wide range of pan-European scientific initiatives. LINKECOL, a program to promote a synthesis between population, community, and ecosystem ecology, funded most European biodiversity-ecosystem function research between 1999 and 2004.

The reality is this: the biodiversity-ecosystem function research program that emerged in the mid-1990s was driven by an organized effort of the international scientific community, with the explicit goal of providing evidence for the utilitarian value of biodiversity for human society, in order to convince policy makers to take serious action towards conservation of biodiversity. This is the socio-political context in which this research was conducted, a context that from the very beginning was motivated by normative concerns about biodiversity loss and its impact on the planet. Loss of biodiversity alone was enough to motivate scientific and ethical concern, but if it could be established that biodiversity loss negatively impacted ecosystem functioning, then one had a powerful economic and self-interested argument that could be used to motivate broad conservation initiatives. We have seen such arguments before, in earlier debates over diversity–stability relationships, but not on the scale witnessed here.

4.2 The Experiments: ECOTRON, Cedar Creek, and BIODEPTH

The initial phase (mid-1990s) of the biodiversity-ecosystem function research program is dominated by three experiments: the ECOTRON (UK), Cedar Creek (USA), and BIODEPTH (Europe).

In these studies, diversity is manipulated by constructing multi-species assembled communities and the effects of these communities on ecosystem function subsequently determined. If the observed response of the multi-specific assemblages differs from the response predicted by simple summation of the single species responses, then it is concluded that diversity *per se* has had an effect on ecosystem functioning.

In the ECOTRON experiment, Naeem *et al.* [1994] assembled communities of plants, microorganisms, and animals (representing trophic levels of decomposers, producers, and consumers) with three different biodiversity levels (9, 15, and 31 species), in replicated controlled growth chambers, and observed an increase of plant productivity and community respiration in the more species-rich communities. They explained this positive association between biodiversity and ecosystem functioning by the mechanism of "niche complementarity". The idea is that at lower diversity, species are more likely to compete for a given resource, but as diversity increases, different species are forced to exploit the same environmental resource in different, non-competitive ways (*e.g.*, some animals feed off leaves at the tops of trees while others feed off the bottom; or some feed by day while others feed by night; etc.). This is expected to have an effect on overall system function. A simple example: a more diverse community of plants may have a canopy structure that intercepts more light at various heights, thereby capturing more energy that can be converted into biomass.

The second set of experiments was conducted at Cedar Creek, Minnesota, by Tilman and his colleagues. In one experiment they used different nitrogen fertilizer rates to alter the species composition and diversity of native grasslands, and observed an increase in stability with species richness (and fertilizer), which they measured as resistance and recovery after a major drought [Tilman and Downing, 1994]. In a second experiment they assembled communities of native grassland species with different species richness levels (1 to 24 species) drawing species at random from a list, and measured an increase in productivity and nutrient use with greater diversity [Tilman *et al.*, 1996]. A similar "niche complementarity" model was used to explain these results: diverse communities make more complete use of the resource space, increasing the resources available for ecosystem processes.

The third major experiment was the European BIODEPTH (Biodiversity and Ecological Processes in Terrestrial Herbaceous Ecosystems). Hector *et al.* [1999]

manipulated replicated artificially assembled grassland communities with varying species richness (1 to 32) at eight different sites (Silwood and Sheffield in UK, Sweden, Portugal, Ireland, Greece, Germany, and Switzerland) and observed a reduction in total plant productivity with decreasing diversity levels. They explained the results by niche complementarity and positive species interactions, as well as the selection effect (see 4.3 below).

These experiments and others were interpreted as providing evidence for a general and positive relationship between species richness and ecosystem productivity. Some were featured in prestigious scientific magazines and general news media, accompanied by calls to support biodiversity preservation.

However, some experiments looking at other ecosystem processes such as soil organic matter decomposition failed to provide evidence for a positive relationship between diversity and ecosystem functioning (*e.g.*, [Griffiths *et al.*, 2001]). Other studies highlighted the greater contribution of functional composition rather than species diversity to ecosystem processes [Hooper, 1997; Tilman *et al.*, 1997a]. Because species diversity and functional composition may not necessarily be correlated, the interpretation of the functional composition effects also became an issue of debate.

4.3 Critical Response (late 1990s)

By the late 1990s, two types of scientific criticism had arisen that challenged the results and interpretation of the previous experiments. First, there were observational studies that appeared to contradict the experimental results (*e.g.*, [Wardle *et al.*, 1997]). And second, there was growing recognition that the design of the experiments made the interpretation of results either ambiguous or impossible to extrapolate to natural ecosystems [Huston, 1997; Huston *et al.*, 2000]. We will consider these objections in turn.

First, most of the high productivity ecosystems in the world appear to have low species richness, an observation that runs counter to the general inference that ecologists wanted to draw from the biodiversity-productivity experiments [Huston and McBride, 2002]. If diversity was positively correlated with productivity, this association should be evident in natural ecosystems. But in community ecology it has long been recognized that productivity is generally a "hump-backed" function of diversity [Grime, 1973], *i.e.*, species numbers will be maximized in environments with *intermediate* productivity. The prevailing rationale for this result was that at low levels of environmental productivity (*e.g.*, poor soils), species diversity in natural ecosystems is low because few species can survive. Diversity increases as more resources become available for species to exploit, reaching a maximum at intermediate levels of productivity. Then diversity declines at higher levels of productivity because dominant species either out-compete others or are limited by growth in size of individual plants. But from this perspective, environmental conditions are the driver of diversity, and not the other way around. Only when the environment is controlled, say the critics, can the relatively small effects of species composition on productivity be distinguished [Huston and McBride, 2002].

Second, experiments with randomly assembled plant communities have several hidden effects that are confounded with the diversity effect. The most important of these is the *sampling effect* (now considered an example of the *selection effect*) [Huston, 1997; Tilman *et al.*, 1997b]. As a statistical necessity, the probability of including a highly productive species in a random pool increases as you add new species. Consequently, the increase in productivity may be due to the presence of a single highly productive species, rather than due to an increase in species diversity *per se*. To critics, the sampling effect is better viewed as an artifact of the way the experiments were conducted, not a biologically valid mechanism to explain an increase in productivity with species richness. Other design problems of these first experiments noted by critics included "quasi replication" (low diversity replicates are less represented, and there is more chance that the most productive individual species are not included) and variance reduction effects (high diversity replicates are more similar than low diversity replicates, confounding experimental error with the diversity effect). According to critics, these design problems rendered invalid any general conclusions about the relationship of diversity to ecosystem function based on this class of experiments. Interestingly, Tilman responded by granting that the sampling effect was indeed the simplest mechanism to explain the observed positive diversity-productivity relationship, but given that species extinction processes are poorly understood, and assuming that species loss is random, he asserted that this is a reasonable and legitimate scientific explanation of the effect [Tilman *et al.*, 1997b]. The interpretation of the role of the sampling effect in biodiversity experiments remained a contentious issue. As Grime [1999] put it, the debate "deepened".

4.4 "War among ecologists"

In 1999, a panel of ecologists reported in the Ecological Society of American *Bulletin*, a publication aimed at the general public and policy makers, that there was scientific evidence that loss of biodiversity impacted ecosystem functioning by reducing plant productivity, decreasing ecosystem resistance to environmental perturbations, and increasing the variability of soil nitrogen levels, water use, and pest cycles [Naeem *et al.*, 1999]. The report concluded that, because "both the magnitude and stability of ecosystem functioning are likely to be significantly altered by declines in local diversity," it recommends "the prudent strategy of preserving biodiversity in order to safeguard ecosystem processes vital to society."

A group of critics of the biodiversity-ecosystem function experiments subsequently wrote a letter to the ESA *Bulletin* heavily criticizing the report [Wardle *et al.*, 2000]. As Kaiser [2000] commented in *Science*,

> Huston and the other critics hit the roof. In a commentary published in the July 2000 ESA *Bulletin*, which goes to all 7,700 ESA members, they mince no words, charging that the pamphlet is "biased," "states opinions as facts," and sets "a dangerous precedent"—especially as it

> appears to represent the position of the entire society. It is "a propaganda document," they claimed, "and an advertisement for some authors' research." By promoting "unjustifiable actions" based on a "house of cards," they wrote, "scientific objectivity is being compromised." [p. 1283]

It is unusual for scientific disagreements to enter the public sphere in so dramatic a fashion. Certainly there were legitimate questions about the design and interpretation of the experiments that the authors of the original report failed to mention, but by itself these methodological facts don't account for the heat of the exchange. The full story has to recognize that basic ecological research has rarely been subject to such anticipation and scrutiny from professional associations, science and policy institutions, and the general media. In addition, the differing sides in this dispute were also professional rivals in a real sense, vying for hefty grant dollars and peer recognition (Consider: Tilman's Cedar Creek experiments have received over 10 million dollars in NSF grants over the past fifteen years). It is the environmental, socio-political and institutional context of the research that encouraged both the publication of the original ESA report and the critical response.

4.5 Conciliation and Synthesis

This story has a happy ending. In the wake of the flare-up over the ESA report, a conference was held in Paris in December 2000 in an attempt to bring everybody to the table and reach a consensus on the status and interpretation of the biodiversity–ecosystem function experiments. This "Synthesis Conference" was an effort to reconcile the different interpretations of the results and to arrive at a consensus framework for guiding new research and for framing the current understanding of the science for the general public. Participants described the conference as "a delight" [Naeem *et al.*, 2002]. "Perhaps it was the rich desserts and the French wine, but there were few signs of acrimony at the conference" [Hughes and Petchey, 2001].

The consensus framework was structured by pointing out the issues that were clear, and identifying questions that remained to be answered, so that the framework might serve as a guide for future research endeavors.

First, it was clear that a large number of species is required to maintain ecosystem functioning, but whether this is because more rich communities have some key species that differentially affect ecosystem function, or whether diversity effects arising from niche complementarity had an effect on ecosystem function, was unclear. This provided a goal for further studies, to separate and measure the effects of these two non-exclusive mechanisms, complementarity and selection effects. In addition, it was recognized that the biological relevance of the sampling effect turns in part on whether species extinctions are random, and research had to be conducted to address this question [Loreau *et al.*, 2001]. It was also agreed that a greater number of species may be needed to maintain stability in ecosystems (the "insurance hypothesis") and that further experiments were needed to test this hypothesis specifically controlling diversity and the environmental variation.

Another important question addressed in the conference was how to reconcile the observational and experimental data on diversity-productivity relationships. Recall that observational studies had repeatedly shown a hump-backed relationship, where productivity peaks at intermediate levels of diversity but declines at higher levels. By contrast, the biodiversity-ecosystem function experiments showed a positive relationship of increasing productivity with diversity. These results were reconciled by realizing that the observational studies were plotting diversity not against productivity in a fixed environment, but against productivity across a range of environmental gradients, such as soil fertility and disturbance regime. Consequently, decreasing productivity at higher diversity levels may be due (for example) to decreases in soil fertility in those environments, but if soil fertility was held constant, productivity may be observed to increase with diversity, as was observed in the controlled biodiversity-ecosystem function experiments. Thus, rather than being interpreted as contradictory results, the observational and experimental results are interpreted as revealing different mechanisms operating under different conditions. It was concluded that much further work needed to be done to investigate feedbacks between diversity, ecosystem functioning and environmental factors [Loreau *et al.*, 2001].

In addition, it was acknowledged that most of the experimental evidence came from grasslands ecosystems, where only plant diversity was manipulated. Therefore, before making generalizations to other ecosystems (*e.g.*, aquatic) and other trophic levels (*e.g.*, consumers, decomposers) further research was needed in these areas.

Finally, it was agreed that it is *functional traits* of species and their interactions that predominately affect ecosystem functioning. Consequently there was a call for more research on the relationship between species diversity and functional diversity, and in defining functional groups or types relevant for ecosystem functioning [Loreau *et al.*, 2001].

4.6 More Recent Work

The "synthesis conference" helped to frame a research agenda that has shaped more recent work on biodiversity-ecosystem function relationships. This work has helped to refine our understanding of the mechanisms relating diversity to ecosystem functioning, including the role of selection effects, such as interspecific competition that can cause one species to dominate a community (selection effects can be positive or negative depending whether the dominant species is positively or negatively associated with ecosystem functioning). The synthesis framework also helped initiate a second generation of biodiversity experiments, such as the Jena Project in Germany [Roscher *et al.*, 2007; Temperton *et al.*, 2007], and the forest biodiversity mega-project in Sabah, Malaysia [Scherer-Lorenzen *et al.*, 2005]. These experiments usually have some subset of the following characteristics: i) the treatments include as many monocultures as possible, in order to make comparisons of overyielding, complementarity, and selection, ii) the design is balanced to allow contrasts for plots with and without certain species or groups of species, iii) they are designed with the objective of testing specific mechanisms directly beyond the general overyielding in a specific function, iv) they extend for longer time periods, and larger spatial scales, v) experimental design includes replications and local environmental control, and vi) they consider biodiversity and ecosystem function effects across more trophic levels (producers, consumers, predators).

As a follow up to the synthesis conference, in 2005 a committee of scientists from the Ecological Society of America published a review in *Ecological Monographs* titled 'Effects of biodiversity on ecosystem functioning: a consensus of current knowledge' [Hooper *et al.*, 2005]. Like most papers in the literature this report starts by describing the threats that biodiversity loss and environmental degradation pose to society, and finishes by recommending to policy makers to set biodiversity as a priority for action. But the tone of the 2005 report is moderate and balanced, discussing uncertainties and contradictions present in the literature, avoiding generalizations and describing the many factors other than diversity that can influence ecosystem functioning.

The main points stressed in the report are: i) functional composition is more important than species richness in affecting ecosystem functioning; ii) abiotic controls (climate, resources, disturbance) interact with biodiversity to influence ecosystem properties, and the feedbacks between biotic and abiotic controls are central to understanding ecosystem functioning; and iii) diversity effects and the underlying mechanisms can differ among ecosystem properties and ecosystem types. The report notes that diversity may have no effect on some ecosystem processes (*e.g.*, when multiple species carry out similar functional roles or abiotic conditions primarily control the process) but as larger spatial and temporal scales are considered, greater diversity is needed to maximize functioning.

With less certainty, the authors assert that i) complementarity of resource use by certain combinations of species can increase productivity; ii) species richness decreases exotic species invasion under similar environmental conditions (though not across all environments); and iii) species diversity can stabilize ecosystem process in response to disturbances and variation in abiotic conditions. The authors note areas of uncertainty that need further research, including i) the relationships between taxonomic diversity, functional diversity, and community structure; ii) ecosystem response across multiple trophic levels to varying composition and diversity of consumer organisms; and iii) the need for long-term experiments to assess temporal stability and perturbations to assess response to and recovery from disturbances. Finally, meta-analyses of the more than 150 biodiversity experiments conducted in terrestrial and marine ecosystems conducted recently [Balvanera *et al.*, 2006; Cardinale *et al.*, 2006; Cardinale *et al.*, 2007; Stachowicz *et al.*, 2007] reported that on average the effect of biodiversity on ecosystem processes was positive, although effects varied with scale and hierarchical level (population, community, ecosystem). In most studies diverse communities performed better than the average of monocultures although in very few cases diverse communities were better than the best monoculture (*i.e.*, transgressive overyielding was infrequent). Other issues considered recently involve looking at the effect of measures of biodiversity other than richness, like evenness and diversity indices on ecosystem function [Wilsey *et al.*, 2005; Losure *et al.*, 2007; Kirwan *et al.*, 2007].

4.7 Discussion

The aim of section 4 was to provide an overview of research and debate over the relationship between biological diversity and ecosystem functioning. Here we pause to reflect on attributes of this debate that are illuminated by the discussion of the diversity-stability debate in section 2 and the discussion of biodiversity and ecosystem function concepts in section 3.

One of the lessons learned from the earlier diversity-stability debate was that apparently conflicting experimental and theoretical results may be in fact be compatible, because the arguments actually employ different concepts or measures of diversity or stability. We see this pattern in the biodiversity-ecosystem function debate as well. It shows up in several places, but in the review above we see it explicitly with respect to measures of productivity. In observational studies, ecosystem productivity is confounded with effects due to environmental variation, while in the biodiversity experiments environmental variation is controlled. These different measures of productivity resulted in different diversity-productivity curves, but the curves were really measuring different effects, and so were not genuinely incompatible.

We also saw that the earlier stability-diversity debate was subject to biases arising from ideological commitments relating to environmental policy and concern over biodiversity loss. The worry was that a desire to promote conservation policies would bias researchers to look for confirming evidence for positive diversity-stability relationships and downplay or ignore contrary evidence. This was precisely the charge made by the critics of the 1999 ESA report, that the authors of the report were driven by a desire to influence public policy in favor of conservation, and that this lead them to give a biased review of the biodiversity-ecosystem function literature and to make hasty generalizations about the implications of the research for conservation policy.

In sections 2 and 3 we also noted that philosophical attitudes toward holism and reductionism in ecology can predispose ecologists toward or away from positive diversity-stability relationships, because such relationships have an historical association with holistic views of ecological dynamics. And we noted that such views are likely to be aggravated by the language of ecosystem "functions", insofar as these are taken to imply that ecosystem behaviors are goal-directed in some sense, or that ecosystems have behaviors that may be judged against certain idealized ecosystem "types". It is difficult to judge the degree of influence that these sorts of philosophical biases have on biodiversity-ecosystem function research, since ecologists are unlikely to comment on such issues in their research activity. But there is anecdotal evidence that reductionistically-oriented, neo-Gleasonian plant ecologists are inclined to be more cautious about this research program and the general conclusions for environmental policy that many want to draw from it.

The public disagreement surrounding the 1999 ESA report was embarrassing for the institution and the participating ecologists, but as we described above, it resulted in a productive dialogue among scientists that helped to address misunderstandings and build a consensus framework for a research program that would work to resolve remaining uncertainties. Post-synthesis research has been much more conciliatory in tone and more cautious in its declarations, but also more productive in illuminating the various mechanisms at work, and in articulating a more unified vision of ecological science that spans the historical schism between population/community and ecosystem ecology.

5 CONCLUSION

In this paper we presented a survey of the debate over the relationship of biodiversity to ecosystem functioning. Our goal was to provide an overview that would help researchers and commentators to understand the various different sources of conflict that have played a role in structuring the debate. Some of these sources of conflict have roots in earlier debates in ecology over diversity-stability relationships, the relationship of ecology to environmental policy, and in the long-standing schism between reductionistic and holistic research traditions. Consequently, our review has focused on situating the biodiversity-ecosystem function debate within this broader intellectual history.

It is our conviction that members of any scientific field can benefit from instruction in the history and philosophy of their field. Such instruction can help researchers, teachers and students to better understand the conceptual issues they confront in their on-going research projects, and to appreciate the broader social and humanistic significance of their work. We hope that this overview of the historical and philosophical foundations of the biodiversity-ecosystem function debate will prove similarly helpful as a guide to the issues and controversies surrounding this exciting area of ecological research.

BIBLIOGRAPHY

[Balvanera *et al.*, 2006] P. Balvanera, A. B. Pfisterer, N. Buchmann, J.-S. He, T. Nakashizuka, D. Raffaelli, and B. Schmid. Quantifying the evidence for biodiversity effects on ecosystem functioning and services. *Ecology Letters* 9: 1146–1156, 2006.

[Boorse, 2002] C. Boorse. A rebuttal on functions. In A. Ariew, R. Cummins and M. Perlman (eds.) *Functions: New Essays in the Philosophy of Psychology and Biology*, pp. 63–112. New York: Oxford University Press, 2002.

[Botkin, 1990] D. Botkin. *Discordant Harmonies: A New Ecology for the Twenty-First Century*. New York: Oxford University Press, 1990.

[Cardinale *et al.*, 2006] B. Cardinale, D. Srivastava, J. Emmett Duffy, J. Wright, A. Downing, M. Sankaran, and C. Jouseau. Effects of biodiversity on the functioning of trophic groups and ecosystems. *Nature* 443: 989–992, 2006.

[Cardinale *et al.*, 2007] B. J. Cardinale, B, J. P. Wright, M. W. Cadotte, I. T. Carroll, A. Hector, D. S. Srivastava, M. Loreau, and J. J. Weis. Impacts of plant diversity on biomass production increase through time because of species complementarity. *Proceedings of the National Academy of Sciences* 104: 18123–18128, 2007.

[Clements, 1916] F. E. Clements. *Plant Succession: An Analysis of the Development of Vegetation.* Carnegie Institution of Washington, Washington DC, 1916.

[Cummins, 1975] R. Cummins. Functional analysis. *The Journal of Philosophy* 72(20): 741–764, 1975.

[Diaz and Cabido, 2001] S. Diaz and M. Cabido. Vive la difference: plant functional diversity matters to ecosystem processes. *Trends in Ecology and Evolution* 16: 646–655, 2001.

[Egerton, 1973] F. Egerton. Changing concepts of the balance of nature. *Quarterly Review of Biology* 48: 322–350, 1973.

[Egler, 1954] F. E. Egler. Vegetation science concepts: I: initial floristic composition, a factor in old-field vegetation development. *Vegetatio* 4: 412–417, 1954.

[Elton, 1958] C. S. Elton. *The Ecology of Invasions by Animals and Plants.* London: Chapman & Hall, 1958.

[Elton, 1966] C. S. Elton. *The Pattern of Animal Communities.* London: Methuen, 1966.

[Gardner and Ashby, 1970] M. R. Gardner and W. R. Ashby. Connectance of large dynamical (cybernetic) systems: critical value of stability. Nature 228, 784, 1970.

[Gleason, 1939] H. A. Gleason. The individualistic concept of the plant association. *American Midland Naturalist* 21: 92–110, 1939.

[Glenn-Lewin *et al.*, 1992] D. C. Glenn-Lewin, R. K. Peet and T. T. Veblen (eds.), *Plant Succession: Theory and Prediction.* New York: Springer-Verlag, 1992.

[Goldstein, 1999] P. Z. Goldstein. Functional ecosystems and biodiversity buzzwords. *Conservation Biology* 13: 247–255, 1999.

[Griffiths *et al.*, 2001] B. S. Griffiths, K. Ritz, R. E. Wheatley, H. L. Kuan, B. Boag, S. Christensen, F. Ekelund, S. Sorensen, S. Muller, and J. Bloem, J. An examination of the biodiversity-ecosystem function relationship in arable soil microbial communities. *Soil Biology & Biochemistry* 33: 1713–1722, 2001.

[Grime, 1973] J. P. Grime. Competitive exclusion in herbaceous vegetation. *Nature* 242(5396): 344–347, 1973.

[Grime, 1997] J. P. Grime. Biodiversity and ecosystem function: The debate deepens. *Science* 277(5330): 1260–1261, 1997.

[Hagen, 1992] J. B. Hagen. *An Entangled Bank: The Origins of Ecosystem Ecology.* New York: Rutgers University Press, 1992.

[Hector *et al.*, 1999] A. Hector, B. Schmid, C. Beierkuhnlein, M. C. Caldeira, M. Diemer, P. Dimitrakopoulos, J. Finn, H. Freitas, P. S. Giller, J. Good, R. Harris, P. Högberg, K. Huss-Danell, J. Joshi, A Jumpponen, C. Körner, P. W. Leadley, M. Loreau, A. Minns, C. P. H. Mulder, G. O'Donovan, S. J. Otway, J. S. Pereira, A. Prinz, D. J. Read, M. Scherer- Lorenzen, E. Schulze, A. S. Siamantziouras, E. M. Spehn, A. C. Terry, A. Y. Troumbis, F. I. Woodward, S. Yachi & J. H. Lawton. Plant diversity and productivity experiments in European grasslands. *Science* 286: 1123–1127, 1999.

[Hempel, 1959] C. G. Hempel. The logic of functional analysis. In L. Gross (ed.), *Symposium on Sociological Theory*, pp. 271–287. New York: Harper and Row, 1959.

[Hooper, 1997] D. Hooper. The effects of plant composition and diversity on ecosystem processes. *Science* 277(5330): 1302–1305, 1997.

[Hooper *et al.*, 2005] D. U. Hooper, F. S. Chapin, J. J. Ewel, A. Hector, P. Inchausti, S. Lavorel, J. H. Lawton, D. M. Lodge, M. Loreau, S. Naeem, B. Schmid, H. Setala, A. J. Symstad, J. Vandermeer, and D. A. Wardle. Effects of biodiversity on ecosystem functioning: a consensus of current knowledge. *Ecological Monographs* 75(1): 3–35, 2005.

[Hughes and Petchey, 2001] J. Hughes and O. L. Petchey. Merging perspectives on biodiversity and ecosystem functioning. *Trends in Ecology and Evolution* 16(5): 222–223, 2001.

[Huston, 1997] M. A. Huston, M. A. Hidden treatments in ecological experiments: re-evaluating the ecosystem function of biodiversity. *Oecologia* 110: 449–460, 1997.

[Huston *et al.*, 2000] M. A. Huston, L. W. Aarssen, M. P. Austin, B. S. Cade, J. D. Fridley, E. Garnier, J. P. Grime, J. Hodgson, W. K. Lauenroth, K. Thompson, J. H. Vandermeer, and D. A. Wardle. No consistent effect of plant diversity on productivity. *Science* 289(5483): 2000.

[Huston and McBride, 2002] M. A Huston, and A. C. McBride. Evaluating the relative strengths of biotic versus abiotic controls on ecosystem processes. In M. Loreau, S. Naeem, and P. Inchausti (eds.) *Biodiversity and Ecosystem Functioning: Synthesis and Perspectives*, pp. 47–60. Oxford: Oxford University Press, 2002.

[Ives, 2005] A. R. Ives. Community diversity and stability: changing perspectives and changing definitions. In K. Cuddington and B. Beisner (eds.), *Ecological Paradigms Lost: Routes of Theory Change*, pp. 159–182. San Diego: Elsevier, 2005.

[Jax, 2005] K. Jax. 'Function' and 'functioning' in ecology: what does it mean? *Oikos* 11: 641–648, 2005.

[Kaiser, 2000] J. Kaiser. Rift over species diversity divides ecologists. *Science* 289: 1282–1283, 2000.

[Kirwan *et al.*, 2007] L. Kirwan, A. Luscher, M. T. Sebastia, J. A. Finn, R. P. Collins, C. Porqueddu, A. Helgadottir, O. H. Baadshaug, C. Brophy, C. Coran, S. Dalmannsdottir, I. Delgado, A. Elgersma, M. Fothergill, B. E. Frankow-Lindberg, P. Golinski, P. Grieu, A. M. Gustavsson, M. Hoglind, O. Huguenin-Elie, C. Iliadis, M. Jorgensen, Z. Kadziuliene, T. Karyotis, T. Lunnan, M. Malengier, S. Maltoni, V. Meyer, D. Nyfeler, P. Nykanen-Kurki, J. Parente, H. J. Smit, U. Thumm, and J. Connolly. Evenness drives consistent diversity effects in intensive grassland systems across 28 European sites. *Journal of Ecology* 95: 530–539, 2007.

[Lindberg, 1992] D. C. Lindberg. *The Beginnings of Western Science: The European Scientific Tradition in Philosophical, Religious, and Institutional Context: 600 B.C. to A.D. 1450.* University of Chicago Press, Chicago, 1992.

[Loreau *et al.*, 2001] M. Loreau, S. Naeem, P. Inchausti, J. Bengtsson, J. P. Grime, A. Hector, D. U. Hooper, M. A. Huston, D. Raffaelli, B. Schmid, D. Tilman, and D. A. Wardle. Biodiversity and ecosystem functioning: current knowledge and future challenges. *Science* 294: 806–808, 2001.

[Losure *et al.*, 2007] D. Losure, B. Wilsey, and K. Moloney. Evenness-invasibility relationships differ between two extinction scenarios in tallgrass prairie. *Oikos* 116: 87–98, 2007.

[MacArthur, 1955] R. H. MacArthur. Fluctuations of animal populations, and a measure of community stability. *Ecology* 36: 533–536, 1955.

[May, 1973] R. M. May. *Stability and Complexity of Model Ecosystems.* Princeton, NJ: Princeton University Press, 1973.

[Mayhew, 2006] P. Mayhew. *Discovering Evolutionary Ecology: Bringing Together Ecology and Evolution.* New York: Oxford University Press, 2006.

[McCann, 2005] K. S. McCann. Perspectives on diversity, structure, and stability. In K. Cuddington and B. Beisner (eds.), *Ecological Paradigms Lost: Routes of Theory Change*, pp. 183–200. San Diego: Elsevier, 2005.

[Mikkelson, 1999] G. M. Mikkelson. Methods and metaphors in community ecology: the problem of defining stability. *Perspectives on Science* 5: 481–498, 1999.

[Millikan, 1984] R. G. Millikan. *Language, Thought, and Other Biological Categories.* Cambridge: MIT Press, 1984.

[Naeem, 2002] S. Naeem. Ecosystem consequences of biodiversity loss: the evolution of a paradigm. *Ecology* 83(6): 1537–1552, 2002.

[Naeem *et al.*, 1994] S. Naeem, L. J. Thompson, S. P. Lawler, J. H. Lawton, and R. M. Woodfin. Declining biodiversity can affect the functioning of ecosystems. *Nature* 368: 734–737, 1994.

[Naeem *et al.*, 1999] S. Naeem, F. S. Chapin, R. Costanza, P. R. Ehrlich, F. B. Golley, D. U. Hooper, J. W. Lawton, R. V. O'Neill, H. A. Mooney, O. E. Sala, A. J. Symstad, and D. Tilman. Biodiversity and ecosystem functioning: maintaining life support processes. *Issues in Ecology* 4: 2–11, 1999.

[Naeem *et al.*, 2002] S. Naeem, M. Loreau, and P. Inchausti. Biodiversity and ecosystem functioning: the emergence of a synthetic ecological framework. In S. Naeem, M. Loreau, and P. Inchausti (eds.), *Biodiversity and Ecosystem Functioning: Synthesis and Perspectives*, pp. 3–11. Oxford University Press, 2002.

[Nagel, 1961] E. Nagel. *The Structure of Science.* London: Routledge & Kegan Paul, 1961.

[Norton, 2000] B. G. Norton. Biodiversity and environmental values: in search of a universal earth ethic. *Biodiversity and Conservation* 9: 1029–1044, 2000.

[Odum, 1953] E. P. Odum. *Fundamentals of Ecology.* Philadelphia: Saunders, 1953.

[Odum, 1969] E. P. Odum. *The strategy of ecosystem development.* Science 164: 262–270, 1969.

[Pianka, 1999] E. Pianka. *Evolutionary Ecology*, 6th Edition. New York: Benjamin Cummings, 1999.

[Pimm, 1980] S. L. Pimm. Food web design and the effect of species deletion. *Oikos* 35: 139–149, 1980.

[Purvis and Hector, 2000] A. Purvis and A. Hector. Getting the measure of biodiversity. *Nature* 405: 212–219, 2000.

[Rorscher *et al.*, 2007] C. Roscher, C., J. Schumacher, W. W. Weisser, B. Schmid and E. Shulze. Detecting the role of individual species for overyielding in experimental grassland communities composed of potentially dominant species. *Oecologia* 154(3): 535–549, 2007.

[Sarkar, 2005] S. Sarkar. *Biodiversity and Environmental Philosophy: An Introduction.* New York: Cambridge University Press, 2005.

[Scherer-Lorenzen *et al.*, 2005] M. Scherer-Lorenzen, C. Potvin, J. Koricheva, B. Schmid, A. Hector, Z. Bornik, G. Reynolds and E. Schulze. The design of experimental tree plantations for functional biodiversity research. In M. Sherer-Lorenzen, C. Körner and E. Schulze (eds.), *Forest Diversity and Function*, Berlin: Springer-Verlag, 347–376, 2005.

[Simberloff, 1980] D. Simberloff. A succession of paradigms in ecology: essentialism to materialism and probabilism. *Synthese* 43: 3–39, 1980.

[Sober and Wilson, 1994] D. S. Wilson and E. Sober. A critical review of philosophical work on the units of selection problem. *Philosophy of Science* 61: 534–555, 1994.

[Soulé, 1985] M. Soulé. What is conservation biology? *Bioscience* 35: 737–734, 1985.

[Stachowicz *et al.*, 2007] J. J. Stachowicz, J. F. Bruno, and J. E. Duffy. Understanding the effects of marine biodiversity on communities and ecosystems. *Annual Review of Ecology, Evolution, and Systematics* 38: 739–766, 2007.

[Temperton *et al.*, 2007] V. Temperton, P. Mwangi, M. Sherer-Lorenzen, B. Schmid and N. Buchmann. Positive interactions between nitrogen-fixing legumes and four different neighbouring species in a biodiversity experiment. *Oecologia* 151(2): 190–205, 2007.

[Tilman and Downing, 1994] D. Tilman and J. A. Downing. Biodiversity and stability in grasslands. *Nature* 367(6461): 363–365, 1994.

[Tilman *et al.*, 1996] D. Tilman, D. Wedin, and J. Knops. Productivity and sustainability influenced by biodiversity in grassland ecosystems. *Nature* 379(6567): 718–720, 1996.

[Tilman *et al.*, 1997a] D. Tilman, J. Knops, D. Wedin, P. Reich, M. Ritchie and E. Siemann. The influence of functional diversity and composition on ecosystem processes. *Science* 277: 1300–1302, 1997.

[Tilman *et al.*, 1997b] D. Tilman, C. Lehman, and K. T. Thompson. Plant diversity and ecosystem productivity: Theoretical considerations. *Proceedings of the National Academy of Sciences* 94(5): 1857–1861, 1997.

[Wardle *et al.*, 1997] D. A. Wardle, O. Zackrisson, *et al.* The influence of island area on ecosystem properties. *Science* 277(5330): 1296–1299, 1997.

[Wardle *et al.*, 2000] D. A. Wardle, M. A. Huston, J. P. Grime, F. Berendse, E. Garnier, W. K. Lauenroth, H. Setala, and S. D. Wilson, Biodiversity and ecosystem function: an issue in Ecology. *Bulletin of the Ecological Society of America* 81(3): 235–239, 2000.

[Williams, 1966] G. C. Williams. *Adaptation and Natural Selection.* Princeton, NJ: Princeton University Press, 1966.

[Wilsey *et al.*, 2005] B. J. Wilsey, D. R. Chalcraft, C. M. Bowles, and M. R. Willig. Relationships among indices suggest that richness is an incomplete surrogate for grassland biodiversity. *Ecology* 86: 1178–1184, 2005.

[Wilson, 1983] D. S. Wilson. Group selection controversy: history and current status. *Annual Review of Ecology and Systematics* 14: 159–187, 1983.

[Wouters, 2005] A. Wouters. The function debate in philosophy. *Acta Biotheoretica* 53(2): 123–151, 2005.

[Wright, 1973] L. Wright. Functions. *Philosophical Review* 82: 139–168, 1973.

[Wynn-Edwards, 1962] V. C. Wynne-Edwards. *Animal Dispersion in Relation to Social Behaviour.* Edinburgh: Oliver & Boyd, 1962.

A DYNAMICAL APPROACH TO ECOSYSTEM IDENTITY

John Collier and Graeme Cumming

INTRODUCTION

Although various kinds of systems thinking have been present in ecology for many years (*e.g.*, [Tansley, 1935; Clements, 1936; Odum, 1983]), systems approaches in ecology have gained increasing prominence in recent times as a tool for the interdisciplinary exploration of complex human interactions with nature (*e.g.*, [Gunderson and Holling, 2002; Norberg and Cumming, 2008; Waltner-Toews *et al.*, 2008]). Complex systems that are capable of adaptation and learning, such as human societies, are of particular interest. For example, complex adaptive systems with high current relevance for human well-being include such diverse entities as the global climate system; rainforest ecosystems; the economic systems that underlie the banking and housing sectors of the economy; the social dynamics that lead to terrorism; and social-ecological systems that range from local harvesting networks through to global oceanic fisheries.

Despite their diversity, complex adaptive systems are considered to have a number of common properties. They are assembled from diverse components that interact with one another. Complexity evidences itself through system dynamics, which include non-linear relationships between key variables, the presence of local equilibria and thresholds, feedback loops, and the ability to self-organize, learn, and respond actively to environmental change.

Ecosystems are a particular kind of complex adaptive system. They are commonly understood to consist of organisms, an abiotic environment, and a set of interactions that occur between organisms and between organisms and their environment [Tansley, 1935]. Although we focus here on ecosystems, many of the same ideas are more generally relevant to other complex adaptive systems.

The concept of an ecosystem, as summarised above and by Tansley [1935], is deceptively simple. On closer inspection, various practical problems arise with applying this definition. Such problems include questions like (1) 'Who decides what constitutes an ecosystem, and how does their definition influence the outcome of an ecological or social-ecological analysis?' (2) 'What is inside the ecosystem and what is external to it? Where are its boundaries in space and time?' and (3) 'How do I know when the ecosystem that I have been observing is in fact a different ecosystem?'

Handbook of the Philosophy of Science. Volume 11: Philosophy of Ecology.
Volume editors: Kevin deLaplante, Bryson Brown and Kent A. Peacock. General editors: Dov M. Gabbay, Paul Thagard and John Woods.

In this chapter we consider these issues from a philosophical perspective, focusing on the need for ecosystem definitions to have a dynamical nature—that is, for our conceptual and empirical models of ecosystems to confront the processes of self-organization and adaptation that allow ecosystems to respond to change. After a brief discussion of the practical implications of ecosystem individuation, we describe the logic of dynamical system individuation and offer some specific observations of how this logic applies to ecosystems in particular. This section is followed by a description of various ecosystem meta-models and strategies for their use. Finally, we draw some general conclusions about how the apparatus we develop can be applied in practice.

1 THE PRACTICAL SIGNIFICANCE OF ECOSYSTEM INDIVIDUATION

Individuation refers to our ability to characterize an individual ecosystem. Before delving deeper into ecosystem individuation, however, we need to address the question, "Why bother?"

The determination of ecosystem identity has various practical consequences that bear on ecosystem sustainability and management, and ultimately (through ecosystem services and livelihoods) to human well-being [Millennium Assessment, 2005]. From a scientific perspective, clear definitions of ecosystems are necessary for comparisons in space and time. This need for comparability extends to both the science of ecology and its practical management applications. On the scientific side, generalities about pattern-process linkages in ecosystems can only be developed if potentially important similarities and differences between individual case studies are clear. For example, the tight relationship between rainfall and tree canopy cover in savanna ecosystems falls away in areas that experience over 700mm of rainfall per annum, creating a threshold beyond which interactions between soil type, herbivores, and fire can be expected to dominate ecosystem dynamics [Sankaran *et al.*, 2008].

On the practical side, spatial and temporal transferability of management models and approaches is contingent on system identity. If systems change in significant ways in space or time, there is no reason to expect that management approaches that have been successful in one place or time will be successful in other places or times. For example, a considerable amount of research on deforestation has been undertaken in the southern and eastern Amazon (*e.g.*, [Nepstad *et al.*, 2006]). As development marches along the TransAmazon highway towards the west, it is unclear whether existing models of deforestation and ecological impacts can simply be transferred from other study contexts, or whether there are aspects of the western Amazonian ecosystems (and social-ecological systems) that differ from those in other regions and could have significant impacts on outcomes. There is also the potential that the mechanisms underlying deforestation have changed in time, for instance through the development of new logging technologies or the implementation and enforcement of new laws, such that principles and data derived from research in the 1990s are no longer relevant to understanding deforestation in the

2010s. These kinds of question cannot be addressed without a clear definition of what constitutes the system and whether, or how, it has changed.

A further implication of ecosystem individuation concerns scale and hierarchies. Cumming *et al.* [2006] have argued that mismatches between the spatial, temporal and functional scales of ecological processes and management can lead to various management problems. Similarly, misunderstandings of the boundaries of an ecosystem or a social-ecological system can result in a narrow focus on optimising the management of a subset of a larger system, potentially leading to various pathologies in natural resource management [Holling and Meffe, 1996]. A dynamical account of ecosystem identity implies a dynamical account of its scale, which can potentially be matched with management processes. If there is a scale mismatch, then the scale-related nature of management problems will be easier to diagnose if a clear definition of system identity has been developed.

Finally, a dynamical account of ecosystem identity is helpful in understanding the nature and limits of ecosystem change. If we know the dynamical identity conditions, which are typically abstract organizational criteria, then we can more readily predict which sort of changes will be within the limits of the dynamical identity conditions and which will not. This allows a better understanding of how human interventions and natural changes like climate change will affect the stability and resilience of ecosystems, allowing better management and possible ameliorative actions, or in the worst case, better predictions of the impacts of human and natural factors. An example of the application of identity criteria to a real-world problem (the impacts of the TransAmazon highway on rainforest social-ecological systems) is presented in Cumming *et al.* [2005].

2 IDENTITY AND INDIVIDUATION OF DYNAMICAL SYSTEMS

Ecosystems are complex adaptive systems [Holland, 1995; Collier and Hooker, 1999] for which complete empirical descriptions are impossible [Rosen, 1991]. Although less systematic approaches exist, the incompleteness of empirical descriptions suggests that a systematic approach to ecosystem individuation is important for delineating analytical problems and strategies.

The difficulty of defining an ecosystem is complicated by the fact that any description of an ecosystem is from the perspective of an observer, and the focus of their description will be on the issues in which they are most interested [Weinberg, 1975; Kay, 2008]. In an era of postmodern science [Funtowicz and Ravetz, 1993], there are good theoretical and practical reasons for questioning whether there are general or specific kinds of models (either in terms of components or processes) that adequately cover all ecosystems, and for thinking that we need model all but the simplest ecosystems with a variety of kinds of models, or meta-models, simultaneously in order to obtain a reliable perspective on ecosystem dynamics and identity [Cumming and Collier, 2005]. Kay [2008] terms this approach "polyocular", in the sense that our understanding is more complete if we look through a number of different lenses.

Despite the need to entertain multiple perspectives and working models, there are a number of characteristics common to all satisfactory ecosystem meta-models that allow them to be coordinated to give a more complete picture of ecosystem identity in general and of the identity of particular ecosystems. In particular, system identity in general, and ecosystem identity in particular, is most usefully represented in dynamical terms [Collier and Hooker, 1999]. This is because both measurements and interactions with any system are dynamical processes, so a dynamical account of identity allows the account to be applied directly to empirical and practical interactions with the system. The need for a dynamic character is also reflected in the fact that some of the most powerful frameworks for the analysis of complex adaptive systems are process-oriented rather than merely descriptive or structure-oriented (*e.g.*, [Darwin, 1859; MacArthur and Wilson, 1967; Holland, 1995]).

The analysis of ecosystem individuation turns on three primary issues: identity, unity, and cohesion. We next discuss these concepts in detail.

1. Identity

   We start with the logical notion of identity [Collier, 2004a; 2004b], since the logical form is required of all satisfactory accounts of identity. It is straightforward, though there is some debate about condition (c), which we will address shortly.

   Identity, $A = B$:

   (a) Is a logical condition, same for all things.

   (b) Is an equivalence relation: symmetric, transitive, reflexive.

   (c) $A = B$ implies that B has every property that A has, and vice versa.

   This tells us virtually nothing, since it is a purely logical relation, but it does put some logical constraints on any concept of dynamical identity. Condition (a) rules out so-called relative identity, according to which things can be identical in different ways. This notion is awkward, and neither simplifies things nor adds clarity. Condition (b) just says that identity is an equivalence relation. Equivalence relations divide classes of entities into disjoint classes that all share the equivalence relation to each other. Identity is the strongest equivalence relation; its classes all have one member, and every member holds that relation to itself, so $(x)(x = x)$. Condition (c) is the one that distinguishes identity from all other equivalence relations. It says that $(x)(y)(P)$ ($x = y$ if and only if (Px if and only if Py)). Sufficiency, $(x)(y)(P)(x = y$ only if (Px if and only if Py)) is uncontroversial, and is often called Leibniz' Law. Leibniz in fact preferred the stronger version, since he thought there must be some sufficient reason for two objects to differ, and that this could only be in their properties. However his reasoning is controversial. But for dynamical identity if two objects do not differ in their dynamical properties there is no dynamical difference, so they cannot be distinguished dynamically. Barring nondynamical properties like *haecceity*, or bare otherness, there cannot be two particular distinct dynamical entities that do not differ dynamically. So even if the converse of Leibniz' Law is not true for everything that can be imagined (whatever the limits of that process are), that is irrelevant for dynamical purposes. If there are two objects with the same dynamical properties, then they cannot be distinguished by any interactions we might have with them.

2. Unity

   Dynamical objects are typically made of parts held together over space and time by dynamical processes. So the next move is to look at what makes parts of something parts of that thing. This is provided by the unity relation [Perry, 2002]:

   Unity is the relation among the parts of a thing A such that:

   (a) If a and b are parts of A, then aUb, and bUa (symmetric).

   (b) If a, b and c are parts of A, then aUb and bUc implies aUc (transitive).

   (c) If a is a part of A, then aUa (reflexive).

   (d) By (a), (b), and (c), U is an equivalence relation.

   (e) $U(A)$ is the closure of U, given any initial part.

   (f) By (a) to (d), $U(A)$ contains all and only the parts of A.

   It is an empirical question what satisfies $U(A)$ for a given A. Typically the type of unity relation will depend on the kind of thing A is.

3. Cohesion

   For dynamical objects, the parts and their relations must all be dynamical. In previous writing, Collier has called "dynamical unity" cohesion [Collier, 1986; 1988; 2003; 2004a; 2008; in press]:

   Cohesion $C(A)$ is the unity relation for dynamical objects, such that:

   (a) All parts aCb are dynamical

   (b) C is dynamical

   Cohesion both holds dynamical things together, and also individuates them from other dynamical things. For this reason it can be called it *the dividing glue* [Collier, 2004a]. Any dynamical account of individuation and diversity will be grounded in the formation and disruption of cohesion.

However, there is a lot more to cohesion than its formal definition. Details are spelled out at some length in Collier [2003], much of which derives from as yet unpublished work with C. A. Hooker. We will summarize the main points. First of all, a *dynamical system* is a set of interacting components that is characterized and individuated from other systems by its cohesion. It is therefore a natural object.

Its properties must be discovered, and its models must be tested. We need to have some basic idea of what the object is to begin with, and then we can use the properties of cohesion to sharpen our understanding. There are three ways that have been recognised in the literature on explanation (following [Salmon, 1984]) to explain natural unity. These are:

1. essential properties (natural kinds, archetypes)

2. stable properties (resistance to internal or external perturbations)

3. cohesion (causal relations that make physical wholes out of parts, or create *sine qua non* dependencies).

We can use the first two, or at least intuitions about the first two, to make a preliminary identification, and then use the cohesion concept and empirical investigation to home in on the appropriate properties. Then we can use cohesion to explain the individual essence of the system, and its stability. In a strict part-whole (nested) hierarchy, parts are integrated into wholes, and these wholes are further integrated into larger wholes, and so on. Cohesion increases as we go up to higher hierarchical levels. Things are somewhat more complicated if we have a non-nested hierarchy (such as a food chain, in which cohesion is provided by trophic relationships) or a hierarchy in which lower level members may belong to more than one cohesive higher level. Such systems are sometimes called *heterarchies*. For example, an individual actor in a natural resource management situation may belong simultaneously to a governmental agency, a political party, and a community action group. At higher hierarchical levels, these different memberships may serve to reinforce system cohesion in some circumstances and undermine it in others.

There are other pitfalls with the cohesion concept that must be minded. These can be divided into basic and derived properties (see [Collier, 2003] for more explicit detail). The basic properties derive from the nature of dynamical interactions and the concept of cohesion. B1: *The first basic property of cohesion is that it comes in degrees.* This is a direct consequence of its being grounded in forces and flows, which come in varying kinds, dimensions and strengths. Secondly, and following on from the first property together with the individuating role of cohesion, B2: *cohesion must involve a balance of the intensities of centrifugal and centripetal forces and flows*[1] *that favours the inward, or centripetal.* This balance is not absolute, but is probabilistic over the dimensions and boundaries of the cohesive entity. Just as there are intensities of forces and flows that must be balanced, there are, due to fluctuations, propensities of forces and flows that show some statistical distribution in space and time (or other relevant dynamical dimensions). B3: *Cohesion must involve a balance of propensities of centrifugal and centripetal forces and flows that favours the inward, or centripetal.* The asymmetry of this balance of tendencies implies a distinction between inner and outer, consistent with the

[1]We get the term *centripetal* from [Ulanowicz, 1997, pp. 47–50, 94]. Collier suggested the addition of the converse centrifugal flows and forces; it is implicit in [Ulanowicz, 1997].

role of cohesion in individuating something from its surroundings, but it also plays down rare events and emphasizes more common events (for specific application to ecosystems, see [Ulanowicz, 1997, pp. 47–50 and 94]).

The derived aspects of cohesion now follow from the basic properties as they apply to specific systems with many properties. From B1, only some properties are relevant to cohesion. Thus, A1: *In general, a dynamical system will display a mix of cohesive and non-cohesive properties.* Next, from B2 and B3, A2: *Cohesion is not just the presence of interaction.* Whence, A3: *A property is cohesive only where there is appropriate and sufficient restorative interaction to stabilize it.* From A1 and A4: *Cohesiveness is perturbation-context dependent with system properties varying in their cohesiveness as perturbation kinds and strengths are varied.*

Furthermore, A5: *The cohesive support of nominal system properties may extend across within-system, system-environment and within-environment interactions.* There is no reason to think that a cohesive system must be closed. Rather, A6: *cohesion characterizes all properties, including higher order process properties that are dynamically stabilized against relevant perturbations.* Living systems are primarily characterized in terms of their process organization. Their structures may change, and must change somewhat whenever their adaptability is manifested; the more organized their adaptability, the higher order the cohesive processes that characterize them. Properties A1–A6 are relevant to the discussion of the application of cohesion to ecosystems in the next section.

3 ECOSYSTEM INDIVIDUATION AND CHANGE

There are several definitions of ecosystems that take into consideration their parts and/or their flows. These are the beginnings of dynamical definitions, but are too limited in certain respects. Tansley [1935] defined the ecosystem as 'the fundamental concept appropriate to the biome considered together with all the effective inorganic factors of its environment'. In a more recent discussion of ecosystem definitions, Pickett and Cadenasso [2002] argue that 'the main components of the [ecosystem] concept are its abiotic and biotic features and the interactions between them'. They add that although the definition of ecosystems is scale independent, 'all instances of ecosystems have an explicit spatial extent'. So, Pickett and Cadenasso [2002] effectively argue that an ecosystem is defined by its materials, the relationships among them, and its location.

There are circumstances under which this definition is inappropriate or ambiguous. These problems are of particular importance when developing dynamic models of ecosystems. For example, as the global climate warms, we can expect to see a shifting of the spatial boundaries of ecosystems. If the boundaries of a deciduous forest gradually change until they lie 50 km to the north of its original location, does it remain the same ecosystem? Many ecologists would say that it does, but the 'explicit spatial extent' has changed. Or imagine a situation in which a large disturbance hits a particular sub-catchment and the entire flora and fauna of the area is destroyed. Recolonization from neighbouring areas occurs, and a community develops that has exactly the same species composition and ecological functions as the previous one. Is the new ecosystem the same, or different? Although it might be the same kind in all important respects (*e.g.*, structure, location, components, interactions, functions), we would argue that the new ecosystem is different as an individual because its cohesion has been disrupted. In practice, of course, exact reconstitution of an ecological community would be so unlikely as to be impossible; but the thought experiment nonetheless raises an important point.

These two examples illustrate a particular kind of idea that our current definition of ecosystems fails to capture; that of continuity through space and time as a central component of identity.[2] In evolutionary biology, a close parallel to ecosystems lies in species concepts. The old definition of immutable species having some essential property or set of properties that could be determined from a single type specimen was gradually transformed as systematists thought through the full implications of Darwin's ideas. Species change over time, making the identification of a species on the basis of a single individual problematic at best (B1 above). The key distinction that led to the formulation of the evolutionary and phylogenetic species concepts was that made by the biologist Michael Ghiselin [1966; 1974; 1987] and the philosopher David Hull [1976; 1978]: species are natural individuals, not natural kinds. They are not like gold or lead, which remain gold and lead and would do so even were it possible to transform one into the other. Species, like ecosystems, are mutable, dynamic things. However, unlike species, which are scattered as both individuals and separate populations, ecosystems are typically localized and spatio-temporally contiguous. The lesson from the Ghiselin-Hull approach to species is that mutable, dynamical entities need not have essential properties that are present in all of their parts, but their identity is a relational property. The problem is to find suitable dynamical relations that determine ecosystem identity by binding the system into one (A3 above). These are the sort of natural properties that we should look for, not localized properties that are found in every part of the ecosystem (A1 above).

Our aim in raising these issues is not to provide a new ecosystem definition, because the appropriate definition in each case is context-dependent (A4 above). The point that we wish to highlight is the lack of temporal competence in most current definitions. We need some guidelines that enable us to say whether or not the same ecosystem exists under a wider range of conditions and possible events than our current definition can cope with. We propose that a reasonable addition would be that ecosystem identity abides in the continued presence, in both space and time, of key components and key relationships, though these may be rather abstract compared to individual organism or even species and their local interactions (A6 above). This perspective on identity permits gradual (and not

[2]David Wiggins [1967] uses spatio-temporal contiguity as the defining characteristic of identity. This works well for many cases, but it requires many subtle qualifications to deal with things like spatially discontinuous nation states, and spatiotemporally overlapping natural objects that interact with each other only minimally, such as hybridisation zones between sister species.

necessarily linear) change from one kind of system to another, through a series of intermediate stages; but saltationary change will always result in a new system. Just as species can change gradually from one into another, however, ecosystems can also transform and split (and merge, unlike most species).

When an ecosystem definition is applied to a specific instance the temporal component of the system must be dealt with explicitly. If the preceding argument is sound, an adequate working specification of an ecosystem should encompass the following: (1) the ecosystem components, which may be defined in varying degrees of detail; (2) the relationships between ecosystem components; (3) the location and spatial scale at which the definition is applicable, and the importance (or lack of it) of spatial constancy; and (4) the temporal scale at which the definition is applicable, and the author's perspective on the question of identity through time. This final point is essential to the distinctions that we wish to make in the next section of the chapter. These four points are logically related and mutually constraining, so it is not enough to consider the fourth point alone. The relationships among the ecosystem components constrain the types of components that are suitable for maintaining identity. At the same time, the components determine the sort of relations that they can have with each other and still maintain a cohesive system. Unlike designed artefacts, an ecosystem is self-organized; it must emerge naturally from the interactions of its components and its environment, and its very possibility depends on both the nature and the existence of its components. Furthermore, the very notion of an ecosystem component itself depends on the mutual constraints of ecosystem relations and component nature. Although the atoms making up an ecosystem are constituents, they are not really components, since they can vary freely (and typically do) without changing the nature of the ecosystem (point A1). Being a component must be understood in terms of having a relevant role in overall functioning of the ecosystem, not just being there as a constituent of the system. Lastly, the scale and limits of the interactions will determine both the scale and limits of the ecosystem itself, both spatially and temporally, as well as determining the nature of the boundaries of the ecosystem, including how it is nested within larger ecosystems.

Given these points and their consequences, for the purposes of the next section we have adopted the view [Cumming and Collier, 2005] that ecosystems are determined by their main components (abiotic and biotic), the relationships of these components to one another, *and the maintenance of both spatial and temporal continuity* (ecosystems may move in space, and inevitably move in time, but saltation in either instance constitutes a loss of identity).

On this view, an ecosystem is a network of components connected by various relations. Given that the relations are dynamical, they constitute constraints and flows of various kinds, including inputs, outputs, feedbacks, and external constraints. The main problem of ecosystem identity, or unity, is to decide what is internal to the system and what is external. Collier [Collier, 1986; 1988; 2003; Collier and Hooker, 1999] has suggested in other contexts that the best way to decide dynamical unity is to compare the strength of internal relations among components with those of external relations. This not always possible, since the relations come in degrees (B1), and vary in kind. Furthermore, only some of the relations are relevant to system unity. Which these are is an empirical issue, and varies for each type of dynamical system. Another approach to assessing dynamical unity is through the three different lenses of asymmetries, networks, and information processing [Norberg and Cumming, 2008]; relations between components in the same system may be easier to clarify if one explores whether they share membership in a hierarchy, whether they are connected via some kind of network, and whether they contribute to information processing and/or systemic responses. Individuation of different types of ecosystems may require focusing on different kinds of relations; however, as we suggested in the last paragraph, all relationships that are included in the definition should have a role in the overall functioning of the system. The closure of such relations determines the dynamical unity of the system.

This closure is typically going to be immensely complex, and simplifications will be needed. Ulanowicz [1986] developed a network account that relies on the strength of flows of carbon, reasoning that carbon flows are a good stand-in for species interactions, though they don't directly capture behavioural interactions that may be important to ecosystem unity. Nonetheless, he was able to create workable models of trophic relations for complex estuary ecosystems using this model, as well as to come up with a measure of connectedness and ecosystem health based on a mutual information that could calculate the degree of connectedness (at least by way of carbon flows). It should be noted, though, that the sort of closure required for ecosystem identity is not complete; there can be, and will be, flows into and out of the ecosystem, at the very least sunlight and water, but also typically organism migration both in and out, and the flushing of wastes (point A5). Ecosystems are not like organisms, since they are not actively self-regulatory, but they are not mere collections of interacting things either. They depend for their continued existence on predictable interactions both within the system and without, and the latter may depend on predictable supporting processes within larger ecosystems.

The complexity of ecosystems, with their openness and nonlinear dynamical interactions, shows complexly organized behaviour (*sensu* [Collier and Hooker, 1999]). This in itself is not a problem for studies in many cases, in which we can segment and focus on specific issues, but it becomes an issue if we are interested in whole ecosystem function. Even where specific issues like predator-prey relations are studied, it is well known that they can show highly unpredictable behaviour (*e.g.*, [Barkai and McQuaid, 1988]). It is well known now that complex dynamical system are emergent from their components and their local relations. Specifically, they cannot be circumscribed by single closed models. Rosen [1991] explains this in detail, in full logical form, though he identifies such systems with living systems, which is probably not correct, since complexly organized systems are found in physics (*e.g.*, [Bénard cells; Chandrasekhar, 1961]), and ecosystems cannot be said to be alive in anything like the sense in which organisms are alive. His argument that complexly organized systems cannot be reduced to the local interactions of their components or to input-output relations is sound, however. It is a direct consequence that no single model can mathematically capture all the possible behaviour of such a system. This means that more open models will be required, and typically more than one (and even then we can't get a fully circumscribed combination of models). The reasoning follows from the nature of complex dynamical systems, especially self-organizing ones, and issues in logic stemming from Gödel and Turing. It becomes somewhat of a pragmatic issue which models to use. For that reason it is useful to have a set of kinds of models available to use and guide empirical work.

4 ECOSYSTEM META-MODELS

The reason why there are so few truly general ecosystem models is undoubtedly the irreducible complexity of ecosystems.[3] At the heart of cohesive models of ecosystems are a few extremely complex issues. Ecosystems are dynamic entities that span multiple spatial and temporal scales; the distinction between endogenous and exogenous dynamics is not always clear; and because of their many components, the outcome of manipulations on the system may differ depending on relatively small differences in starting conditions.

Despite these complexities, however, ecology has made some progress towards developing a more general framework for understanding ecosystems. The many specific models of ecosystems together with accumulating empirical evidence have begun to produce a few more general models that incorporate and summarize the findings of many specific models. Such models are a step back from the immediate process of prediction; they are simple, often tantalizing statements that hint at an underlying order to the workings of the world. Their value comes from the way in which they somehow capture the essential ingredients of many interrelated models in symbolic form. Consequently, we term them 'meta-models'.[4]

Meta-models are not hypotheses in the commonly-used sense. They are not necessarily rigorous quantitative statements, although they must be supported by rigorous quantitative studies. Indeed, they are more a kind of specific metaphor; a way of thinking about things that serves as a powerful tool for the generation of specific hypotheses in specific cases. In this respect they are more like Kuhnian paradigms, or Lakatosian research programmes. Their value is measured more in terms of their impacts and their usefulness than their immediate scientific testabil-

[3]However, general models that deal solely with complexity issues and their consequences can be very general. Robert Ulanowicz [1986; 1997] has used such models to explain very general features of ecosystems that can be applied powerfully to draw conclusions about the growth and development of ecosystems and the stability of specific ecosystems such as Chesapeake Bay and the Baltic Sea. These models, however, require a wealth of specific information about flows throughout the ecosystem, and cannot be constructed directly from individual trophic relations and resource and waste flows.

[4]There is a more detailed discussion of the meta-models in [Cumming and Collier, 2005], along with helpful animated diagrams and a comparative table in the .pdf version. The discussion here follows that discussion.

ity. However, although they have that certain vagueness that is bred of generality, meta-models must be clearly and unambiguously defined. They are not models of specific systems; but at the same time they are not as broad as the 'world views' or paradigms outlined by Holling and Gunderson [2002]. Again, although they are less explicit than the 'ecosystem models' discussed by Pickett and Cadenasso [2002], they are considerably less vague than their 'ecosystem metaphors'. Recognition of the strengths and weaknesses of our own meta-models, and consideration of alternative meta-models, should serve a useful purpose in refining concepts and highlighting the key distinctions between them.

Holling's *adaptive cycle* [Holling, 1986; 1987; 2001; Holling and Gunderson, 2002] is one of the few well-defined, well-supported interpretations of ecosystem dynamics. The behaviour of systems of a certain kind has been shown to closely match the adaptive cycle. Because it seems to fit many ecological and social systems, and few or no counter-examples have been described, the adaptive cycle has been criticized for being too broad. Few critics have appreciated that the adaptive cycle is really a meta-model; a broader class of model that encapsulates the key dynamics of numerous other models.[5] We have argued firstly that there are other meta-models of ecosystem function; and secondly, that these meta-models should not be expected to pick out the same aspects of system dynamics as the adaptive cycle, because they are models of a fundamentally different kind. Evaluation of the adaptive cycle has yet to move beyond systems or models that have essentially the same dynamics as the models from which the meta-model was constructed; from this comes the illusion that the adaptive cycle explains everything. By defining rigorously the properties that are expected of systems that match different kinds of meta-model, we can move a step closer to understanding what the central ingredients of particular system behaviours are and develop an improved appreciation of their commonalities and differences. Furthermore, the various meta-models give us a set of tools to use when it is unlikely that one meta-model, even one as successful as the adaptive cycle, will be complete.

The adaptive cycle is defined by phases that follow one another sequentially. These can be summarized as resource accumulation; resource release followed immediately by system reorganization and reconfiguration; and re-entry into an accumulation trajectory. It is a meta-model of a continuous dynamic process, in which complex interactions between system components result in a long, slow build-up that contains the seeds of its own subsequent collapse. Other essential ingredients of the adaptive cycle meta-model include a focus on the role of endogenous dynamics; a view of systems as continuous entities in both space and time; and an emphasis on periodic reorganization, through endogenous or exogeneous drivers.

Although the adaptive cycle offers a persuasive approach to characterizing and understanding ecosystem dynamics, it is only one of a set of possible meta-models that might explain or clarify different aspects of ecosystem function. We propose that further attempts to develop, refine and examine alternative meta-models will

---

[5] See Ulanowicz [1997] for an explanation of Holling's adaptive cycles that is compatible with the context of this chapter.

help us to make further progress in ecosystem ecology. To find exceptions, we must look for systems that are discontinuous; that exhibit few or no relevant internal dynamics or are continuously overwhelmed by external forces; and that have little or no self-organizational ability or 'adaptive capacity'. In the next section we consider some candidates for alternative meta-models that may explain different kinds of ecological phenomena. Some of the most interesting alternative meta-models for complex systems may be those that mirror many of the dynamics of the adaptive cycle but can be distinguished from it in one or more crucial ways.

Alternative meta-models will be relevant wherever a system is in clear violation of one of the central features of the adaptive cycle. We use a strict definition of the adaptive cycle, believing that it is only through making the details of each meta-model clear and explicit that we will be able to progress towards a consistent framework. Continuous modification of the adaptive cycle to encapsulate all possible ecosystem dynamics is neither useful nor desirable.

1. Random walk
   The most obvious alternative meta-model is encapsulated in unpredictability. Under this model, ecosystems wander randomly through a multivariate space. Their dynamics and components would undergo continuous, stochastic changes at irregular intervals of time. There is no cycling, and no particular regularity in system properties. This model is primarily a null hypothesis that exists to be disproven, and has been disproven in many cases. Nonetheless, it is worth stating explicitly because it is a null model against which other models must be contrasted; alternative meta-models must encapsulate some form of order or repetition. A topical example of a largely stochastic ecological process is that of the location and timing of species invasions [May, 1976]. These can act as profound constraints on adaptive cycles, changing the dynamics beyond recognition.

2. Replacement
   The adaptive cycle is not an appropriate meta-model for systems that lose their continuous identity in either space or time. Such systems may follow after one another, be similar to one another, and occur in the same location as one another; but they are not true examples of a single system that undergoes a periodic cycle of growth and reorganization. Replacement may occur with a predictable or semi-predictable frequency, and may be weakly reinforced by internal dynamics. These characteristics make it distinct from a purely stochastic meta-model. Nonetheless, cohesion criteria require that the old and new systems are not the same ecosystem.

   An example of a biological system that fits a replacement meta-model better than it fits an adaptive cycle meta-model is that of a lotic (flowing water) ecosystem. The quantity of water flowing in a stream is largely an exogeneous property of the system. Following a severe flood, sediments are rearranged and many organisms are swept away. The community that remains or is reconstituted after the disturbance is a combination of legacies ('ecological memory') from the previous community, plus new colonizers. There may be profound changes in the components from which the system is constructed and their relationships to one another. According to the continuity criterion, what remains is a different system. There is no fundamental dynamic of reorganisation, no return to the previous trajectory, and no obvious accumulation of 'capital' (in the sense that forests accumulate wood or companies accumulate money) between disturbance events. The system is dynamic, but the adaptive cycle does not offer an adequate summary of it. Obviously, at smaller scales, alternative kinds of system dynamic (including the adaptive cycle) may be possible.

Systems in which substantial legacies are left after disturbances fall into a grey area between replacement and reorganization. The ends of the continuum (disturbances leave no legacy, or disturbances leave a legacy of the entire system) are easy to classify as instances of replacement or the adaptive cycle respectively. At locations mid-way between these two extremes, there is no simple answer. The solution will depend on the proportion of the subsequent biotic community that is endogenous, the extent to which the abiotic environment was altered by the disturbance, and the degree to which biotic interactions in the new ecosystem have changed. Of course the cohesion criterion of ecosystem identity implies that there will be intermediate cases, just as there are intermediate cases between species.

3. Succession

The adaptive cycle uses the older meta-model of succession as its fundamental dynamic. Holling's important insight was to recognize the process of reorganization that occurs between successional events as an integral part of ecosystems, and to make it explicit; a natural extension of successional theory. Any system that does not undergo both succession and a subsequent reorganization phase of some kind does not fit the adaptive cycle meta-model.

As a thought experiment, imagine that through careful management, a system could be kept in the 'r to k' phase of the adaptive cycle indefinitely. Next, imagine that the manager could gradually remove his or her influence by developing the self-organizational capacity of the system. And finally, imagine that the manager could completely withdraw and leave the system perpetually stuck in the r to k phase. To argue that this situation is only possible by maintenance of adaptive cycles at a smaller scale is to miss the point. The point is that such a system, if it existed, would fit the successional meta-model better than it does the adaptive cycle. Decades of work have shown that few or no real-world systems fall into this category [Holling and Meffe, 1996]; but without these rigorous tests of real-world dynamics, we would not be able to dismiss the successional meta-model so readily. This sort of model implies a much higher degree of regulation than is typically found in natural ecologies, and is more typical of that found in organisms. The existence of self-regulating systems suggests that the succession model is not impossible.

4. Dynamic limitation
   Another potential meta-model is encapsulated in the idea that ecosystems are constrained by external drivers. This can be visualized as a case in which the ecological system dynamics leading to growth and expansion, for example, are constantly pushing against external limits. As the system boundaries change along any of the multiple axes that pertain (such as in space, substrate or temperature), components of the ecosystem either go extinct or expand to exploit the full plausible state space. In this meta-model there is no accumulation or reorganization, and cycling is not a necessary condition; limitation comprises a set of forward and backward movements as if between two dance partners, with an occasional 'explosion' or release when constraints are removed.

   The process of dynamic limitation is also distinct from the replacement model. The internal dynamics of the system will depend heavily on ecological processes, and there is no reason why the endogenous or finer-grained exogenous dynamics should not follow the adaptive cycle meta-model, but the dynamic limitation model is applied at a broader scale than this. Dynamic limitation is primarily a boundary condition, not a system-wide driver. Changes in limitation do not produce an entirely new system; there is no obvious replacement event, except possibly through some kind of accumulation of small changes. In this meta-model, exogenous drivers 'tinker' with some of the pieces of the system, and endogenous variation occurs at such a fine scale that it is largely irrelevant.

5. System Evolution
   The theory of evolution provides us with another example of a meta-model, and has been criticized in a similar fashion ('it's not falsifiable') to the adaptive cycle. Holling and Gunderson [2002] incorporate 'nature resilient' within a world view of 'nature evolving', suggesting perhaps that they see the adaptive cycle as one member of a subset of evolutionary meta-models.

   In the strict sense, it is not obvious that ecosystems can be said to evolve. Darwinian evolution implies a mechanism by which variations are generated and selection removes individuals that are poorly suited to current conditions. Although there may be ecological parallels to anagenesis, cladogenesis at an ecosystem level would be difficult to demonstrate. Applying the assumptions of a rigid evolutionary meta-model of adaptation to entire ecosystems leads inevitably to the murky arena of group selection. Since many ecosystems are unique, and there is little opportunity for one ecosystem to displace another (anthropogenic impacts aside), it seems that the evolutionary meta-model is not relevant in this context. Rather than dilute the clear insights of Darwin's theory by applying it outside its original context, it seems wiser to capture change in ecosystems using other conceptual frameworks.

However, selection with cladogenesis is not the only way to get directed change. Ulanowicz [1997] argues that ecosystems have a tendency to increase *ascendency*, which he defines as the product of the total system throughput (analogous to the economic GDP) and the average mutual information of the trophic network. The limiting factor is the overhead, or manoeuvring room resulting from endogenous and exogenous factors. One limit is too much diversity, which leads to collapse of the system, but if this can be managed, gradual increases in ascendency are possible, leading to a version of the succession model.

We have focused here on the adaptive cycle and a set of meta-models of ecosystems that offer alternatives to the same kind of dynamic description. It is important to note that a wide variety of other kinds of dynamic meta-model (many of which are quite different from the adaptive cycle in their framing and intent) have been published for complex adaptive systems. For example, Kay and Boyle [2008] present a model that uses thermodynamic principles and ideas about dissipation and exergy to set the stage for self-organization in social-ecological systems; and Holland [1995] focused on agency and adaptation as central processes in the development of complexity from simpler building blocks. These different views lead to different kinds of insight into system individuation, and together with other examples, demonstrate how the consideration of multiple meta-models can be useful for understanding ecosystem processes.

5 CONCLUSIONS

The issue of ecosystem individuation is of both theoretical and practical importance. Ecosystems are dynamical systems, so a dynamical account of ecosystem is more appropriate than a static definition. Dynamical definitions are also more useful if we want to study ecosystem change and the possible limits of that change. A dynamical account is especially useful for ecosystem management and intervention, since, aside from the issue of matching management scale with ecosystem scale, these are dynamical interactions themselves, and their dynamics must be incorporated into the existing ecosystem dynamics. Because ecosystems are typically complexly organized, and thus not subject to one grand model, it is useful to develop a number of working models that can be applied in specific cases as appropriate. In many cases more than one model or meta-model will apply, and different models can be used to constrain each other, especially in cases where ecosystems skirt the borders of specific meta-models.

ACKNOWLEDGEMENTS

John Collier would like to acknowledge the support of the South African National Research Foundation through its Incentive Funding for Rated Researchers program.

BIBLIOGRAPHY

[Barkai and McQuaid, 1988] A. Barkai and C. McQuaid. Predator-prey inversion in a marine benthic ecosystem. *Science* 242: 62–64, 1988.

[Chandrasekhar, 1961] S. Chandrasekhar. *Hydrodynamics and Hydromagnetic Stability.* Oxford, UK: Oxford University Press, 1961.

[Clements, 1936] F. E. Clements. Nature and structure of the climax. *Journal of Ecology* 24: 252–284, 1936.

[Collier, 1986] J. Collier. Entropy in evolution. *Biology and Philosophy* 1: 5–24, 1986.

[Collier, 1988] J. Collier. Supervenience and reduction in biological hierarchies. In Matthen, M. and Linsky, B. (eds.), *Philosophy and Biology: Canadian Journal of Philosophy Supplementary Volume* 14: 209–234, 1988.

[Collier, 2003] J. Collier. Hierarchical dynamical information systems, with a focus on biology. *Entropy* 5: 100–124, 2003.

[Collier, 2004a] J. Collier. Self-organization, individuation and identity. *Revue Internationale de Philosophie* 59, 151–172, 2004.

[Collier, 2004b] J. Collier. Autonomy, anticipation and novel adaptations. In Irina Dubranova (ed.), *Science of Self-Organization and Self-Organization of Science*, pp. 64–89. Kyiv: Abris, 2004.

[Collier, 2008] J. Collier. Simulating autonomous anticipation: the importance of Dubois' conjecture. *BioSystems* 91: 346–354, 2008.

[Collier, in press] J. Collier. A dynamical account of individuation and diversity. In Paul Cilliers (ed.), *Complexity, Difference and Identity.* New York: Springer, in press.

[Collier and Hooker, 1999] J. Collier and C. A. Hooker. Complexly organised dynamical systems. *Open Systems and Information Dynamics* 6: 241–302, 1999.

[Cumming and Collier, 2005] G. S. Cumming and J. Collier. Change and identity in complex systems. *Ecology and Society* 10(1): 29, 2005. `http://www.ecologyandsociety.org/vol10/iss1/art29/`

[Cumming *et al.*, 2006] G. S. Cumming, D. H. M. Cumming, and C. L. Redman. Scale mismatches in social-ecological systems: causes, consequences, and solutions . *Ecology and Society* 11(1): 14, 2006. `http://www.ecologyandsociety.org/vol11/iss1/art14/`

[Cumming *et al.*, 2005] G. S. Cumming, G. Barnes, S. Perz, M. Schmink, K. Sieving, J. Southworth, M. Binford, R. D. Holt, C. Stickler, and T. Van Holt. An exploratory framework for the empirical measurement of resilience. *Ecosystems* 8: 975–987, 2005.

[Darwin, 1859] C. Darwin. *On the Origin of Species by Means of Natural Selection, or the Preservation of Favoured Races in the Struggle for Life.* London: John Murray, 1859.

[Funtowicz and Ravetz, 1993] S. O. Funtowicz and J. R. Ravetz. Science for the post-normal age. *Futures* 25: 739–755, 1993.

[Ghiselin, 1966] M. T. Ghiselin. On psychologism in the logic of taxonomic controversies. *Systematic Zoology* 15: 207–215, 1966.

[Ghiselin, 1974] M. T. Ghiselin. A radical solution to the species problem. *Systematic Zoology* 23: 536–544, 1974.

[Ghiselin, 1987] M. T. Ghiselin. Species concepts, individuality, and objectivity. *Biology and Philosophy* 2: 127–144, 1987.

[Gunderson and Holling, 2001] L. H. Gunderson and C. S. Holling. *Panarchy: Understanding Transformations in Human and Natural Systems.* Washington DC: Island Press, 2001.

[Holland, 1995] J. H. Holland. *How Adaptation Builds Complexity.* New York: Perseus Books, 1995.

[Holling, 1986] C. S. Holling. The resilience of terrestrial ecosystems: local surprise and global change. In W. C. Clark and R. E. Munn (eds.), *Sustainable development of the biosphere.* Cambridge, UK: Cambridge University Press, 1986.

[Holling, 1987] C. S. Holling. Simplifying the complex: the paradigms of ecological function and structure. *European Journal of Operations Research* 30: 139–146, 1987.

[Holling and Meffe, 1996] C. S. Holling and G. K. Meffe. Command and Control and the Pathology of Natural Resource Management. *Conservation Biology* 10: 328–337, 1996.

[Holling, 2001] C. S. Holling. Understanding the complexity of economic, social and ecological systems. *Ecosystems* 4: 390–405, 2001.

[Holling and Gunderson, 2002] C. S. Holling and L. H. Gunderson. Resilience and adaptive cycles. In Gunderson, L. H., and Holling, C. S. *Panarchy: Understanding Transformations in Human and Natural Systems.* Washington DC: Island Press, 2002.

[Hull, 1976] D. L. Hull. Are species really individuals? *Systematic Zoology* 25: 174–191, 1976.

[Hull, 1978] D. L. Hull. A matter of individuality. *Philosophy of Science* 45: 335–360, 1978.

[Kay, 2008] J. J. Kay. Framing the situation: developing a system description. In Waltner-Toews, D., Kay, J. J., and Lister, N-M. 2008. *The Ecosystem Approach: Complexity, Uncertainty, and Managing for Sustainability*, pp. 16–34. New York: Columbia University Press, 2008.

[Kay and Boyle, 2008] J. J. Kay and M. Boyle. Self-organizing, holarchic, open systems (SOHOs). In Waltner-Toews, D, Kay, J. J., and Lister, N-M. 2008. *The Ecosystem Approach: Complexity, Uncertainty, and Managing for Sustainability.* New York: Columbia University Press, 2008.

[MacArthur and Wilson, 1967] R. H. MacArthur and E. O. Wilson. *The Theory of Island Biogeography.* Princeton, NJ: Princeton University Press, 1967.

[May, 1976] R. May. Simple mathematical models with very complicated dynamics. *Nature* 261: 459–467, 1976.

[Millennium Assessment, 2005] Millennium Assessment. *Ecosystems and Human Wellbeing: Biodiversity Synthesis.* New York: Island Press, 2005.

[Nepstad *et al.*, 2006] D. Nepstad, S. Schwartzman, B. Bamberger, M. Santilli, P. Schlesinger, P. Lefebvre, A. Alencar, D. Ray, E. Prinz, A. Rolla. Inhibition of Amazon deforestation and fire by parks and indigenous reserves. *Conservation Biology* 20: 65–73, 2006.

[Norberg and Cumming, 2008] J. Norberg and G. S. Cumming. *Complexity Theory for a Sustainable Future.* New York: Columbia Press, 2008.

[Odum, 1983] H. T. Odum. *Systems Ecology: An Introduction*, Wiley-Interscience, 1983.

[Perry, 2002] J. Perry. *Identity, Personal Identity, and the Self.* Indianapolis: Hackett, 2002.

[Pickett and Cadenasso, 2002] S. T. A. Pickett and M. L. Cadenasso. The ecosystem as a multidimensional concept: meaning, model, and metaphor. *Ecosystems* 5: 1–10, 2002.

[Rosen, 1991] R. Rosen. *Life Itself.* New York: Columbia University Press, 1991.

[Sankaran *et al.*, 2008] M. Sankaran, J. Ratnam, and N. Hanan. Woody cover in African savannas: the role of resources, fire and herbivory. *Global Ecology and Biogeography* 17: 236–45, 2008.

[Salmon, 1984] W. C. Salmon. *Scientific Explanation and the Causal Structure of the World.* Princeton: Princeton University Press, 1984.

[Tansley, 1935] A. G. Tansley. The use and abuse of vegetational concepts and terms. *Ecology* 16: 284–307, 1935.

[Ulanowicz, 1986] R. E. Ulanowicz. *Growth and Development: Ecosystems Phenomenology.* New York: Springer, 1986.

[Ulanowicz, 1997] R. E. Ulanowicz. *Ecology, the Ascendant Perspective.* New York: Columbia University Press, 1997.

[Waltner-Toews *et al.*, 2008] D. Waltner-Toews, J. J. Kay, and N.-M. Lister. The ecosystem approach: complexity, uncertainty, and managing for sustainability. New York: Columbia University Press, 2008.

[Weinberg, 1975] G. M. Weinberg. *An Introduction to Systems Thinking.* New York: Wiley, 1975.

[Wiggins, 1967] D. Wiggins. *Identity and Spatio-Temporal Continuity.* Oxford, UK: Oxford University Press, 1967.

SYMBIOSIS IN ECOLOGY AND EVOLUTION

Kent A. Peacock

1 SYMBIOSIS—THE NEGLECTED LINK BETWEEN ECOLOGY AND EVOLUTION

In their pioneering text on symbiosis, Ahmadjian and Paracer state,

> There is a growing awareness of the fundamental importance of symbiosis as a unifying theme in biology, an awareness that organisms function only in relation to other organisms. [Paracer and Ahmadjian, 2000, p. 13]

Despite this widening appreciation of both the scientific and philosophical interest of symbiosis, it is still not unusual to find thick compendia on the philosophy of biology in which the very term "symbiosis" is not mentioned at all [Hull and Ruse, 1998] or is mentioned only briefly by a few deviant authors [Sarkar and Plutynski, 2008]. The marginalization of symbiosis in mainstream evolutionary thinking and ecology is not due, however, merely to a general suspicion of holism on the part of reductionistically-inclined biologists and philosophers of biology, for it still remains importantly unclear precisely what symbiosis is and how it works. In particular, it has been difficult to see the sense in which symbiotic associations can be favoured by natural selection. Many evolutionary biologists remain under the spell of some version of Garrett Hardin's "tragedy of the commons" argument [Hardin, 1968], according to which cooperative behaviour is selectively self-defeating. Closely related to this is the unit of selection problem; even James Lovelock, co-founder (with Lynn Margulis) of the controversial Gaia hypothesis (which amounts to the proposal of a planetary-scale symbiosis) has stated that he accepts the criticism of Ford Doolittle and Richard Dawkins that "global self-regulation could never have evolved, as the organism was the unit of selection, not the biosphere" [Lovelock, 2003, p. 769]. As we shall see, Lovelock has conceded to his critics far too much, although it is beyond the scope of this paper to fully explicate or defend the Gaia hypothesis. Rather, my aim is to outline directions in which a comprehensive theory of symbiosis could be constructed and suggest its application to several problems within evolutionary theory, biology, and ecology, including punctuated equilibrium, group selection, and the origin of cancer. The aim will be to support and strengthen the claim made by Ahmadjian and Paracer, for symbiosis, as I hope to show, serves as a link between ecology and evolutionary biology. I will argue

Handbook of the Philosophy of Science. Volume 11: Philosophy of Ecology.
Volume editors: Kevin deLaplante, Bryson Brown and Kent A. Peacock. General editors: Dov M. Gabbay, Paul Thagard and John Woods.

that the concept of symbiosis must at last be taken as seriously in evolutionary theory as it is in ecology—and that it is not always taken as seriously in ecology (especially human ecology) as it could and should have been. I will conclude by arguing that if the notion of sustainability is to mean anything more than a vague aspiration it needs to be thought of as the attainment of a globally mutualistic symbiosis between the human species and the planetary system.

2 HISTORY OF THE CONCEPT

Historical review is not the major purpose of this paper, but sketching some of the main turning points in the growth of the concept of symbiosis will help to clarify the conceptual problems the study of symbiosis still faces today. In this section I rely heavily on Sapp's indispensable *Evolution by Association: A History of Symbiosis* [Sapp, 1994]; see also his [Sapp, 2004].

An awareness of the interdependency of life must be ancient. As a convenient historical reference point, however, we will mark the beginning of the modern scientific investigations of symbiosis with the work of Simon Schwendener, who in 1868 proposed his "dual hypothesis" that lichen are an intimate association of fungi and algae [Sapp, 1994, pp. 4–5]. His radical suggestion was received with general shock and disapproval; it is now, of course, a commonplace of botany. The term "symbiosis" is usually credited to Anton de Bary, although it seems to have first been coined by Albert Bernhard Frank (as "symbiotismus") in 1877 [Sapp, 1994, pp. 6–7], a year before de Bary (who later credited Frank) used it publicly. De Bary defined symbiosis as "the living together of unlike named organisms" [Sapp, 1994, p. 7]. (Later in this paper I shall have occasion both to sharpen the sense in which symbionts "live together," and advocate the broadening of the scope of the concept to include associated organisms of all degrees of genetic likeness or unlikeness.)

Around this time several investigators, including de Bary, realized that often (although not invariably) symbionts can become unable to live on their own; their interdependency with their symbiotic partners can become so complete that their *combination* functions very nearly as a new species of life. De Bary was also among the first to argue that symbiosis is a driving factor in evolution [Sapp, 1994, pp. 9–10].

The concept of *mutualism* or *mutual aid* was introduced to biology by Pierre-Joseph van Beneden in 1873 [Sapp, 1994, p. 7]. Van Beneden drew many of his examples from the animal kingdom. To some extent the literature on mutualism has, even fairly recently, developed independently of the literature on symbiosis. However, de Bary and van Beneden early recognized that there is a gradation from parasitism to mutualism throughout nature, and de Bary realized that both extremes of the scale can be thought of as varieties of symbiosis.

Several biologists, notably Petr Kropotkin [Kropotkin, 1989] studied the phenomenon of "mutual aid" or mutualism. Kropotkin debated Thomas Huxley, who had described nature as a "gladiator's show" [Huxley, 1989]; Kropotkin insisted that cooperation (mutual aid) was as important a factor in survival as competition, especially in harsh or constrained environments. Kropotkin's reasoning was inspired in part by his fieldwork in Russia and Siberia. In Sapp's words:

> In an immense underpopulated country, for the most part a harsh land, competition was more likely to find organism pitted against environment than organism against organism. Malthusian principles seemed to be simply irrelevant [Sapp, 1994, p. 22].

Kropotkin and others had therefore exposed the problem of defining the difference between those ecological contexts in which cooperation gives a greater selective advantage, and those in which competition is the best survival strategy.

By the late nineteenth century the biological literature was "peppered" [Sapp, 1994, p. 34] with suggestions that the numerous small bodies within cells (such as plastids and mitochondria) might be endosymbionts—and this at a point in the history of biology where cell theory itself was barely established. In 1893, for instance, Shosaburo Watase described intracellular symbionts as "physiological complements" of one another "in the struggle for existence" [Sapp, 1994, p. 77]. Extensive work in support of the hypothesis of symbiogenesis, the idea that symbiotic unions can lead to new forms of life, was carried out by the Russian botanists K. S. Merezhkovskii and A. S. Famintsyn in the early years of the 20th century. (Merezhkovskii himself coined the term "symbiogenesis.") Despite this widespread interest in the idea that the nucleated cell is a symbiotic association, by the early 20th century nucleocentrism—the doctrine that all heredity in the cell is concentrated in the nucleus—became dominant in most of cell biology. This probably occurred because, in the absence of any means of detailed study of cellular organelles at the molecular level, nucleocentrism seemed like the simplest and most conservative hypothesis. (In the best light microscopes of 1900 the mitochondrion was an indistinct splodge.) Hand in hand with nucleocentrism were the notions (by now quaint) that bacteria are primarily or entirely parasites and that healthy tissue should be entirely "aseptic."

It has been suggested by Anne Fausto-Sterling that Russian biologists were more ready to accept the importance of symbiosis because Russian thinkers had more socialistic or communal political sympathies than Western scientists [Fausto-Sterling, 1993]. However, as Sapp explains [Sapp, 1994], the picture of the symbiotic tradition as something exclusively carried on by Russian thinkers is an oversimplification. The French scientists Yves Delage and Paul Portier kept the symbiotic torch alive, and the German Hermann Reinheimer wrote extensively on symbiogenesis from a (probably misguided) Lamackian perspective. Before 1920 Portier developed a quite modern picture of symbiosis, and insisted in the face of ridicule that mitochondria are symbiotic bacteria, a point that even Merezhkovskii had been unwilling to concede.

In the 1920s the American biologist Ivan Wallin developed his own comprehensive theory of what he called "symbionticism." Wallin misunderstood some of Portier's ideas but independently arrived at many of the same conclusions. He argued that symbiosis played a central role in evolution, even in the evolution of the nuclear genome. Most point mutations are deleterious, and it was hard to understand how the course of evolution could lead to the acquisition of genes that conferred a selective advantage. It was also unclear how mutation alone could explain the increase in size of the genome in more complex organisms. Wallin (and also William Bateson a few years earlier) proposed the idea that the nuclei of cells could incorporate genes from endosymbionts; the genome of a complex organism could therefore have been built up piece by piece from those of simpler organisms. This was very advanced thinking for their time. It is now known that bacterial and viral genes can be read into the genome of the host cell, but the question of the importance of symbiosis in the construction of complex genomes remains open.

Wallin proposed that evolution was driven by his symbionticism, which he defined as a "taxis" toward association. A taxis is usually understood as a type of behavioral response, and while many organisms do indeed tend to aggregate under various conditions, it seems to be too specialized an explanation for the tendency toward symbiosis, which arguably occurs even at the molecular level where there can be no question of behavior as such. The idea of evolution being driven by a poorly-defined taxis may have contributed to the rejection of Wallin's thinking. The problem remained (and to some extent still remains) to explain how it is that natural selection can account for the increasingly unavoidable fact that symbiotic association is adaptively favoured in a multitude of ecological contexts. I will return to this point below.

Wallin's ideas were ridiculed or ignored until they were revived by Lynn Margulis in the 1960s [Margulis, 1993] and called by her serial endosymbiosis theory (SET). At last, the ideas of SET and the importance of symbiosis generally gained acceptance; Fausto-Sterling suggests, perhaps facetiously, that this could be due to the fact that the "flower children of the 1960s are the working scientists of the 1990s" [Fausto-Sterling, 1993]. However, the transition of SET from heresy to a well-confirmed theory had much more to do with the availability of experimental techniques that allow the theory to be tested; for instance, with the electron microscope it is immediately evident that mitochondria are structurally similar to bacteria, and it has become possible to study the tRNA present in organelles such as mitochondria and note their similarities with bacterial tRNA [Gray, 1992]. The acceptance of SET also had much to do with the dedicated work and intellectual courage of Lynn Margulis. Modern cell biology affords spectacular confirmation of the early speculations of Watase, Poirier, Wallin and others that the eucaryotic (nucleated) cell is a highly obligate symbiotic colony of procaryotes (bacteria). In the meantime patient work by investigators too numerous to list here continues to fill in the details of the extent and importance of symbiotic interactions in the plant, animal, and microbial world [Douglas, 1994; Paracer and Ahmadjian, 2000].

3 WHAT, PRECISELY, *IS* SYMBIOSIS?

No modern investigator has done more than Lynn Margulis in making clear the importance of symbiosis in biology [Margulis, 1993]. And yet, even her definition of symbiosis exposes common misunderstandings of the term:

> ... symbiosis is simply the living together in physical contact of organisms of different species ... literally touching each other ... [Margulis, 1998, p. 2]

This is neither a precise enough nor a general enough conception of symbiosis. First, the mere fact of "living together" is not what counts for symbiosis. What makes a relationship symbiotic is that the organisms involved include each other in their life cycles—that is, their reproductive, metabolic, or trophic cycles. For a relationship to count as symbiotic it is not enough that it be merely an occasional or accidental encounter or juxtaposition. Rather, it is something that tends to happen in a regular or even periodic way, and is therefore something that could have been reinforced by natural selection (in ways I will explore below). Second, the notion that symbionts must be in direct physical contact, which I will call the *contact* interpretation of symbiosis, is both imprecise and far too restrictive, even though many symbionts (including many belonging to the symbioses first studied, such as the lichen) do indeed live in very intimate contact. It is imprecise because the notion of "literally touching" is poorly-defined and highly scale-dependent; protists could be living within the gut of a termite, for instance, and yet they could be swimming freely of each other and rarely directly touching the host's tissues at the molecular level. More important, what counts for symbiosis is that there be *causal interaction* between symbionts, and this is something that can be mediated at distances in space and time in complicated and often quite indirect ways. Let us call this the *causal link* interpretation of symbiosis. It makes perfect sense to say that birds of prey, for instance, are in a symbiotic relationship with the burrowing mammals they feed on, or that whales are in a symbiotic relationship with schools of krill. This is because the life cycles of such predators can be affected by and linked with the life cycles of their prey even if the prey are in direct physical contact with the predators only when one literally eats the other. The insistence that symbionts must be in close physical contact with one another makes it easier to miss the pervasiveness of symbiotic relations throughout biology at *all* scales.

A likely objection to the causal link interpretation of symbiosis could be that it trivializes the notion of symbiosis since essentially all organisms on the earth are linked causally with each other in some fashion, directly or indirectly. The objector is correct that on the causal link view virtually all life on Earth is symbiotically entangled to some degree. However, some causal links are stronger than others, or work on shorter scales in distance or time; thus, even if all biota on the Earth constitute one grand symbiotic system when viewed on a large enough scale, many subsystems are partially independent to varying degrees and can be studied with varyingly useful degrees of accuracy in partial isolation. Thus, for instance, the life cycles of African elephants probably have little short- or medium-term impact on the life cycles of (say) Antarctic penguins, even though both penguins and elephants may indirectly affect each other *via* planetary-scale factors such as climate. Human activities in particular, for better or worse, cannot help but affect essentially all life on earth. The objector must also be reminded that life is inherently complex, and it does not manage its business in order that it can be conveniently classified and described by human biologists.

Expanding the Terminology Suitable terminology can help to bring a concept into focus (just as excessive terminology can obscure it). Biologists currently recognize a two-fold classification of symbiotic relations: *endosymbiosis*, in which some of the partners in a symbiosis live inside another, and *ectosymbiosis*, in which one or more partners live on the surface of others. (Margulis refers to endosymbiosis as a "topological condition" [Margulis, 2004, p. 172].) Let us add to this *exosymbiosis*, in which some members of a symbiotic association are distant in time or space from others. Symbionts may cycle between all three modes at various stages of their life cycles. Whether one organism is inside the physical envelope of the other is scale-invariant, but the distinction between ectosymbiosis and exosymbiosis is to some degree a matter of scale; for instance, bacterial symbionts swimming freely within a large protist are exosymbiotic with respect to each other, and even humans can be considered to be exosymbiotic with respect to the plants and animals with which they are interdependent.

The generalization of the notion of symbiosis to include exosymbiosis is in the spirit of early remarks by de Bary, who

> recognized that the term symbiosis might equally apply to looser associations such as that between pollinating insects and flowers and those between animals that search for food or shelter and the animals and plants that supply it [Sapp, 1994, p. 9].

The central idea of symbiosis is that organisms live together in the sense that they include each other in their life cycles, and this can arise in any case in which organisms can directly or indirectly have causal effects on each other, regardless of their physical distance apart.

I will also take advantage here of the useful term *symbiome* which Sapp has proposed to denote any kind of symbiotic association, whether loosely facultative or tightly obligate [Sapp, 2004].

3.1 Methodological Challenges

Sapp [Sapp, 2004] lists several reasons why symbiosis has been too often marginalized in modern biology, especially evolutionary biology. Some of these are sociological and I will not directly address them here, save to note Sapp's concern that academic specialization has probably hindered the acceptance of symbiosis because the study of that subject is unavoidably cross-disciplinary.

There are also aspects of symbiosis that make it inherently difficult to investigate scientifically. One disincentive to the investigation of symbiosis is simply the fact that symbiotic interdependencies can be of enormous complexity. Also, the symbionts in many symbiomes, especially at the microbial level, cannot be grown or cultured independently of their partners. (Mitochondria are an important example.) We know now that this is due to the fact that there is selective pressure for the elimination of genetic redundancy, but this makes it difficult to establish that partners in a highly obligate symbiosis were once independent organisms, even if (like mitochondria) they still contain *some* of their original DNA. As Nancy Moran observes,

> [t]he organisms that are easiest to grow and study in the laboratory... are weedy species adapted to show fast growth in temporary niches. But most microorganisms in natural communities are likely to have obligate dependencies on other species... explaining why 99% of microorganisms are difficult or impossible to culture. Similarly, most symbionts of plants and animals cannot be readily cultured independently of hosts, precluding most conventional microbiological analyses [Moran, 2006, p. R866].

Cell and molecular biologists, who have had quite enough work to do as it is, have tended to focus on those systems that are easiest to probe, an illustration of Medawar's observation that science is naturally opportunistic and indeed owes much of its success to this fact [Medawar, 1982]. Symbiosis challenges scientific reductionism not only through the difficulty of isolating the partners in an obligate symbiosis, but more generally because of the web of dynamic feedbacks that typify complex symbiotic associations. Science has followed the advice of Descartes (especially in *The Discourse on Method*), who advised the inquirer to understand a whole by identifying all of its parts and grasping fully the relations between them. Scientists accordingly prefer to work mainly on those entities and factors that can be isolated and tested by manipulating independent parameters. There is no question that these analytical methods are enormously effective *where they can be carried out.* However, in the study of symbiosis (and other areas of biology) they may be reaching their limits, since not all biological systems can be separated into distinct parts, and there really are *no such things* as genuinely independent parameters in some of the most important types of interdependent systems in biology and ecology. (Of course some parameters are *approximately* independent in many useful contexts.) In the study of symbiosis one therefore encounters a challenge similar to a methodological problem (still not completely solved) encountered in quantum mechanics, which is the impossibility (deplored by Einstein) of fully isolating certain kinds of systems for study [Born and Einstein, 1971, pp. 170–171]. This does not mean that such systems do not exist or that they should not be studied; it is just that they should not be studied with unrealistic expectations of completeness. It is essential to avoid the tendency to regard things that cannot be isolated and manipulated in canonically acceptable ways as not legitimate objects of scientific inquiry. Taking symbiosis seriously may lead us not only to a broader conception of evolution but of science itself.

3.2 The Scale of Symbiosis

Symbiosis is sometimes taken loosely to suggest a cooperative or mutually beneficial relationship. This is not necessarily the case; parasites are in a symbiotic relationship with their usual hosts, though they may not do the hosts much good, or at least much immediate good, at all. It is helpful and not entirely misleading to array the various kinds or degrees of symbiotic cohesion on a scale, running from extreme pathogenic parasitism at one end to symbiogenesis (the formation of new species by symbiotic merger) at the other [Peacock, 1999a]; de Bary seems to have been the first to explicitly make this suggestion [Sapp, 1994, p. 7].

In pathogenic parasitism an emergent or mutant parasite overwhelms the defences of its host, destroying both the host and sometimes itself in the process. Unpleasant examples such as necrotizing fasciitis and metastatic cancer come to mind, but the sort of runaway population crisis first indicated by Malthus [Malthus, 1798] is also an important example of pathogenicity. (Malthus' mistake was to suppose that because life should be an "ordeal of virtue," that this was the only sort of population dynamic that was morally acceptable for humans.) In chronic or symbiotic parasitism the parasite harms its host but the harm is tolerated either because the parasite to some degree restrains its attack upon the host, or because the harm can be absorbed or compensated for in some way by the host species.

Parasitism shades into commensalism, which in effect is a low-grade, tolerable parasitism in which the commensal has a more-or-less neutral effect on its host. Commensalism is enormously pervasive in nature. Amusing examples of commensals are the two *Demodex* species, the human forehead mites [Wilson, 1992]. In fact, *Demodex* teeters on the brink of pathogenicity [Harwood, 1979], which illustrates the fact that many symbiotes may seem to be neutral commensals only because we do not understand the subtle details of their interactions with their hosts. DNA testing and other molecular techniques now make it possible to individuate the species of bacteria present in a shovelful of topsoil or the crook of a person's elbow, and it has been shown that humans carry an enormous number of bacterial commensals, the surprising variety of which is only recently beginning to be appreciated [Sapp, 2004; Grice *et al.*, 2008]. It is by no means clear that these armies of commensals do not play a role in the normal functioning of their hosts.

Commensalism shades into mutualism, in which a symbiotic association is of mutual benefit to its members. Below I discuss the difficult question of what constitutes "benefit." Mutualistic associations can be obligate (physiologically obligatory) versus facultative (optional). It will often be a lot easier to tell whether a relationship is symbiotic than whether it is specifically mutualistic, since the former can often be identified from overt phenomenology, while demonstrating mutualism may be more indirect. Mutualism between organisms with complex neurologies (such as humans) and other organisms at a similar or larger scale tends strongly to be facultative and thus to an important extent dependent upon learned behavior, a fact wherein lies our great peril today.

At the extreme mutualistic end of the scale is *symbiogenesis*, the process in which two or more distinct species form a mutualistic association that is so well-amplified by natural selection that it defines a new species. Symbiogenesis amounts literally to the formation of anastomoses, a merger of branches, on the tree of life. It is extraordinary that the phenomenon of symbiogenesis has received so little comment or notice from philosophers of biology.

The most ambitious notion of symbiosis is the Gaia hypothesis of James Lovelock and Lynn Margulis [Lovelock and Margulis, 1974; Lovelock, 1988], according to which the entire biosphere (or "earth system") can be regarded as a single coherent, self-regulating biological system. Lovelock himself rarely if ever uses the term "symbiosis," and tends to describe Gaia in almost engineering terms as a biologically-mediated control system. Margulis, however, refers to Gaia as "symbiosis as seen from space" [Margulis, 1998], and emphasizes the parallels between what occurs on the cellular and the planetary scale.

Symbiotic shifts up and down the scale can occur within the life cycles of a single organism; an organism can be a predator or parasite in one ecological setting, a mutualist in another. (Predation can be thought of as a kind of parasitism in which the host is consumed all at once.) Especially philosophically interesting are the symbiotic shifts studied by Margulis and other cell biologists in which microorganisms move from opportunistic parasite to endosymbiote. This rather common phenomenon is apparently the basis of serial endosymbiosis, since plastids and mitochondria can now be traced with some confidence to precursor bacteria that in the first instance invaded certain other cells as parasites [Gray, 1992; Margulis, 1993; Margulis, 2004; Sapp, 2004].

It is also possible to think of ecosystems as mutualistic symbiomes. This approach goes at least as far back as A. G. Tansley [Tansley, 1935] and Eugene Odum [Odum, 1971]. This viewpoint, although very influential, is not universally accepted, essentially for the same reasons that the pervasiveness of symbiosis itself is still not generally accepted. For review, see [Peacock, 2008].

The conditions under which the symbiotic transition from parasite to mutualist can occur are not well enough understood, although there is reason to think that outright parasitism tends to be favoured or at least tolerated in an ecology large enough to absorb its deleterious effects, while the shift to mutualism seems more likely in restricted or harsh environments where there would be obvious advantages to cooperation [Kropotkin, 1989]. A striking instance of this pattern occurs in the work of Jeon and Jeon [Jeon and Jeon, 1976; Smith, 1979; Margulis and Sagan, 1995], in which parasitical bacteria accidentally introduced to a culture of *Amoeba* became, after many cell generations, obligate organelles of the protists. It would be worthwhile to conduct a parallel experiment designed to see whether the same translation from parasite to mutualist would occur in a less constrained environment, such as, perhaps, a much larger container where there would presumably be less adaptive advantage in cooperation for the bacteria. One of the more important problems in biology is to better understand the conditions under which mutualism is favoured and when it is not, and the transition regions between mutualism and other symbiotic phases.

The difficult question of what constitutes benefit or harm is at the heart of understanding symbiosis. Some mutualistic associations are little more than opportunistic mutual parasitism, almost like the relations between rival street gangs, but it is a mistake to suppose that this is as far as mutualism goes. The sense in which one symbiote may benefit another has a lot to do with reproductive success. D. C. Smith remarks,

> If such colonization [of one organism by another] is to the selective disadvantage of the host, it is called parasitism. If it is of advantage, it is often called mutualism... [Smith, 1979, p. 115].

What complicates the matter is that, as Smith goes on to say, "evolutionary processes can lead to such a degree of morphological modification and integration of symbiont into the cellular habitat [provided by its host] that it becomes no longer easily recognizable as a foreign intrusion" [p. 116]. Such cases of symbiotic fusion may well be to the selective advantage of the *combined* system, but it is less clear that they are to the advantage of either colonizer or host individually, except that in a successful symbiotic fusion *some portion* of the symbiont's genome is likely to survive for quite a long time. In the formation of such tight symbiotic associations, we see a shift in what counts as the unit of selection. In many (though not all) symbiomes, the symbionts literally give up the capacity to reproduce independently, and it is no longer meaningful to speak of the association serving their individual reproductive interests. It is not even clear that it is meaningful to think of the association as serving the needs of the "selfish genes" carried by the individual symbionts, since the formation of obligate associations often lead to the loss of redundant genes. What is frequently (though not invariably) "seen" by natural selection is the symbiotic unit as a whole, not the genes or the (often vestigial) component organisms out of which it was constructed.

There is a rough but instructive parallel between symbiogenesis and certain features of entangled states in quantum mechanics. It is demonstrable that quantum mechanically entangled particles cannot be described as sets (Boolean combinations) of simpler independent entities with fully definite physical properties [Bub, 1997]. Similarly, while it is often helpful to think of the organization of the various forms of life including symbiomes as nested hierarchies, it is, as Eldredge points out, "incorrect to call them nested sets" [Eldredge, 1985, p. 141]. Rather, as Eldredge explains, "higher-level units are themselves individuals, although not *ipso facto*, as the ontological status of each putative individual needs to be independently established" [Eldredge, 1985, p. 141]. I would reiterate that what specifically establishes a given symbiotic association as an individual is the dynamic interactions within it.

From a more abstract physical viewpoint the scale of symbiosis can also be defined in terms of thermodynamic synergy [Peacock, 1999a]. A symbiome (especially a tightly coupled mutualism) can be regarded as a sort of battery or energy circuit, capturing and recirculated external flows of energy provided by sources such as sunlight, maintaining low internal entropy through active constructive processes, and actively exporting lots of entropy so as to satisfy the Second Law of Thermodynamics [Schneider and Kay, 1994]. In a mutualism, symbionts feed free (usable) energy to each other and thereby maintain each other's structure and functioning; a parasite, by contrast, draws down the free energy of its host, physically degrading the host's structure and function. Thus the notions of harm and benefit, and thereby the distinction between mutualism and parasitism, is definable in thermodynamic terms; however, the thermodynamic aspect of symbiosis, and its relation to the evolutionary aspects of symbiosis, merits much further study. Steven A. Frank has made a promising contribution to this study by investigating the dynamic conditions that favour the transition to cooperative from individual evolution. Frank argues that crossing the threshold to cooperation is difficult, but "cooperative evolution proceeds rapidly once a symbiosis overcomes the threshold" [Frank, 1995, p. 403].

The members or components of a symbiome can exchange information as well as nutrients, and this could be an important part of how the symbiome is maintained. This aspect of symbiosis also depends upon the ability to interchange materials or free energy since "all information is physical" [Landauer, 1991].

Natural selection is one of a class of recursive or feedback processes which lead to the formation of stable or quasi-stable dissipative structures (such as species and symbiotic complexes). Such processes are widespread in nature because they are very efficient ways to generate entropy [Schneider and Kay, 1994; Schneider and Sagan, 2005]. Understanding the pervasiveness of symbiosis is thus an extension of the thermodynamic approach to understanding life itself pioneered by Schrödinger [Schrödinger, 1944]. On this statistical-mechanical interpretation, symbiosis, like life itself, is probabilistically favoured given the availability of a generous external flow of free energy and a broad range of sufficiently benign physical conditions. Indeed, it has been argued that the very origin of life can be understood as a symbiotic process [King, 1977].

4 SYMBIOSIS AND EVOLUTION

In this section I will explore some of the interactions between the concept of symbiosis and neo-Darwinism, the modern received view of how evolution works. There is a widely quoted remark by Dobzhansky that "nothing in biology makes sense except in the light of evolution" [Dobzhansky, 1964]. A major theme of this paper is that there are aspects of evolution (especially in its relation to ecology) that make sense only in the light of symbiosis.

4.1 Evolution as an Ecological Phenomenon

Evolution can be presented to the beginner in terms of a simplistic model in which organisms adapt (*via* natural selection) to relatively fixed and stable ecological conditions. This might be called the "Post Office" theory of ecology, because it imagines that species fit neatly into ecological niches the way letters fit into pre-made post office boxes. The reality is, of course, much more complex. Evolution is an ecological phenomenon with a molecular basis. How it works cannot be adequately grasped without seeing that organisms not only adapt to their environments but alter their environments [Jones *et al.*, 1994; Odling-Smee *et al.*, 2003]: as Simpson put it,

> There is not simply a given environment to which organisms adapt. Their own activities change the environment and are part of the environment [Simpson, 1953, p. 182].

This further implies that organisms must in turn adapt to the changes they themselves have caused in those environments. Obviously, some events having their origin outside the biotic sphere, such as changes in solar output, bolide impacts, and massive volcanism, can have a drastic effect on the fortunes of earthly life, and there could be no life on this planet if it had not had the good luck to be about the right size, with abundant supplies of water and suitable minerals, and be orbiting a comfortably stable Main Sequence star at about the right distance. However, many environmental conditions at local, regional, and global levels are partially or wholly biological byproducts, including the atmosphere, soil, and many structures in the crust of the Earth itself. Organisms on the Earth are therefore themselves important causes of the selective pressures they ultimately face. In the case of humans this is further complicated by the fact that human preferences and choices, whether coherent and principled, or expedient and short-sighted, determine how we impact our environment and thus how it impacts us and therefore, ultimately, how we must also evolve. Subtle aspects of human culture (even such factors as architecture, literature, or music) could be amplified by feedbacks between culture, environment, and evolution in ways that determine the very sorts of organisms we ourselves become [Peacock, 1999b]. Winston Churchill once observed, "We shape our buildings and afterwards our buildings shape us" [Churchill, 1943]. Even more broadly, we must now say, we shape our ecologies and they shape us, ultimately even at the genetic level.

4.2 The Roles of Symbiosis in Evolution

Evolution occurs when heritable (and thus ultimately genetic) variations are amplified or damped by environmental (natural) selection. Thus, how novelty becomes established genetically, once it appears, is to be explained by natural selection, and the theory of symbiosis has little to add to that fact *per se*. However, an awareness of symbiosis adds to our understanding of how natural selection can operate in several ways:

1. Symbiosis plays an obvious role in the generation of functional novelty, and it may be an essential part of the explanation both of rapid bursts in evolution, and the very existence of certain types of organisms.

2. More important, the fact of symbiosis broadens the spectrum of the types of selective pressures that matter for survival. It is still insufficiently appreciated that certain kinds of cooperative and constructive functionality can be reinforced by selection.

3. Despite the fact that Wallin and others introduced the idea nearly a century ago, not nearly enough attention is paid to the possibility that symbiosis plays a major role in the genesis of both functional and genetic novelty. There are a number of well-studied genetic mechanisms, such as point mutation, recombination, and genetic drift, which are known to generate evolutionary novelty [Brown, 2007]. However, it is still not clear that these can fully explain the sudden appearance of novel functionality or the general increase in the size and complexity of the genome as one moves up the evolutionary tree.

4. Symbiosis forces us to broaden our notions of what is heritable. Some symbiotic associations are themselves heritable since the genomes of the symbionts are passed on (usually maternally) to the offspring; it is not only nuclear genes which are inherited [Sapp, 2004]. As well, symbiotic functionality and behavior can be selected for, quite likely even in many organisms which are only facultatively symbiotic (although this suggestion requires further study).

I explore aspects of these points in more detail below.

4.3 Symbiosis, Punctuated Equilibrium, and the Mousetrap Problem

Gould and Eldredge [1972] have noted the phenomenon of punctuated equilibrium; that is, the fact that evolution does not always occur at a smooth rate, as might be expected from a naïve understanding of Darwinism. Rather, the fossil record seems to suggest that species may be relatively stable for long periods of time and then undergo rather quick shifts with the (geologically) sudden appearances of new species. Darwin himself [Chapter XV, *Origin of Species*] put this down to gaps in the fossil record, and in more recent years some of those gaps have been filled by the painstaking work of paleontologists. It is now possible to see that while much evolution occurs in a succession of small steps, precisely as Darwin insisted, the rate of evolution is indeed variable. There can be bursts of rapid speciation, often but not necessarily following extinctions (which are also often sudden in the fossil record). It is beyond the scope of this paper to fully examine the large and occasionally contentious literature on punctuated equilibrium. However, it is clear by now that the history of life on Earth is defined by *both* gradual change and catastrophe [Hsü, 1986]. Environmental conditions can be stable for very long periods of time and then shift rapidly, sometimes because of catastrophes of external origin (such as impacts) and sometimes because of still incompletely-understood nonlinearities within the Earth's biotic system. It seems reasonable to infer that the sudden appearance of a new species could itself be a sharply non-linear response to certain kinds of changes in environmental conditions. Sometimes, therefore, the rapid appearance of a new species is likely to be best explained by a rapid change in habitat. However, the formation of symbiotic associations does provide one obvious mechanism for rapid evolutionary change, especially during conditions of environmental stress, and it has certainly played a central role in at least some occasions when new forms of life have rather suddenly appeared on Earth.

The importance of symbiosis in generating evolutionary novelty is recognized by Angela Douglas, who argues that symbiosis "is a route by which organisms gain access to novel metabolic capabilities, such as photosynthesis, nitrogen fixation, and cellulose degradation" [Douglas, 1994, p. v]. This viewpoint can be broadened: symbiosis is a route to novel *survival* possibilities, which would include, of course, metabolic capabilities but need not be limited to them. Novel symbiotic associations could also allow organisms ways of responding to rapid changes in habitat and climate. Symbiosis is arguably a source of novelty comparable in importance (though working in importantly different ways) to mutation and other well-studied mechanisms of direct genetic change.

The way in which cooperation can generate novel functionality can be illustrated with homely examples. A circular saw plus a hammer gives a carpenter the ability to frame a wall, which neither the saw nor the hammer alone can do *at all*. The point, almost too obvious to mention except that it is not clearly enough kept sight of, is that cooperation can produce full-blown novel functionality *instantaneously*. If this new functionality confers a survival advantage on the cooperating organisms so long as they continue to cooperate in the relevant way, and if any aspect of the cooperative behavior or functionality is heritable, then it could be quite quickly reinforced by natural selection.

This fact helps to resolve what intelligent design apologist Michael Behe [Behe, 1996] has called the *Mousetrap Problem*, which is to explain the evolution of functionality that does not seem to be capable of having evolved by numerous small variations from earlier components. A number of small parts only becomes a working mousetrap when those parts are assembled in a certain way. Behe dismisses the notion that symbiosis could play a role in solving the mousetrap problem, since he says that the functionality of the symbiotic parts has to have already been present to begin with. But this is obviously false in general; there are innumerable examples in which the recombination of given parts and functions produces *entirely* novel function. It is too much to say that symbiosis is the only explanation for the sudden appearance of novel functionality in evolution, but it has to be one of the major mechanisms by which this occurs.

At the microbial level new associations would begin with a genetic variation that (essentially by chance) happens to conduce to adaptive cooperation. However, this is unlikely to be the usual explanation for new cooperative associations at the complex metazoan level, where new symbiotic associations would often begin with a behavioral change; in humans and likely some other species with especially rich neurosystems even forethought can play a role. One does not normally think of novel behavior as being heritable; however, if the associative behavior confers a survival advantage in the precise sense that it conduces to the survival *of the association* then it, or the neural adaptability that makes it possible, could quite quickly be codified or reinforced by natural selection at the genetic level.

One of the lessons of evolution is how quickly natural selection can occur if a variation confers a survival advantage. This is apparent especially at the micro level: bacteria, for instance, can acquire resistance to toxins so quickly that biologists have (perhaps with tongue in cheek) toyed with the notion of "directed mutation." And yet it is clear that this seemingly near-clairvoyant ability of bacteria to anticipate which variations would be favourable is essentially due to the rapidity of the amplification of mutations in response to natural selection. This is partially a reflection of how quickly bacteria can reproduce, but there is also evidence that the rate of favourable bacterial mutation can increase when bacteria are stressed. Indeed, there is evidence that there are "mutator alleles" which "hitchhike" with the genes they may benefit [Moxon and Thaler, 1997], indicating that the very process of mutation itself may depend partially upon mutualistic functionality at the genetic level. (See also [Beardsley, 1997].)

Nothing I have said here is meant to deny that evolution can and does occur by the usually-cited process: that is, small heritable genetic variations produced by a variety of "blind" mechanisms being amplified in a population by natural selection (often with remarkable rapidity in micro-organisms) if those variations are in some way favourable to survival. The exceedingly interesting and important question remains to elucidate the relative importance of these two evolutionary processes.

4.4 Natural Selection and the Symbiome

Some of the things I am going to say in the following section are going to sound like a defence of the Gaia hypothesis against the sort of selectionist critique that (as noted above) seems to have given pause even to Lovelock himself. However, it is not the purpose of this paper to fully explicate the Gaia hypothesis of James Lovelock and Lynn Margulis [Lovelock, 1988; Lovelock and Margulis, 1974; Margulis, 1998]. (Thinking of Gaia as a mutualistic symbiome rather than as a single living organism may make the concept more palatable for some.) The major aim of this section is to explicate how the evolution of symbiotic associations of organisms (whether Gaia or something on a less grand scale) could be favoured by natural selection. Some of what I say here takes advantage of an analysis by Timothy Lenton [1998].

I will take as my nominal target a token of Dawkins' influential critique of the Gaia hypothesis:

> I don't think Lovelock was clear—in his first book, at least—on the

> kind of natural-selection process that was supposed to put together the adaptive unit, which in his case was the whole world. If you're going to talk about a unit at any level in the hierarchy of life as being adaptive, then there has to be some sort of selection going on among self-replicating information. And we have to ask, What is the equivalent of DNA? What are the units of code? What are the units of copy-me code which are being replicated? ... I don't think for a moment that it occurred to Lovelock to ask himself that question. And so I'm skeptical of the rhetoric of the Gaia hypothesis, when it comes down to particular applications of it, like explaining the amount of methane there is in the atmosphere, or saying there will be some gas produced by bacteria which is good for the world at large and so the bacteria go to the trouble of producing it, for the good of the world. That can't happen in a Darwinian world, as long as we think that natural selection is going on at the level of individual bacterial genes. Because those individual bacteria who don't put themselves to the trouble of manufacturing this gas for the good of the world will do better. Of course, if the individual bacteria who manufacture the gas are really doing themselves better by doing so, and the gas is just an incidental consequence, obviously I have no problem with that, but in that case you don't need a Gaia hypothesis to explain it. You explain it at the level of what's good for the individual bacteria and their genes. [Dawkins, 1995]

In fairness to Dawkins, these remarks were apparently made *ex tempore* at a conference. However, they illustrate a lack of clarity about symbiosis that is endemic to the thinking of evolutionary biologists.

The first thing to clear out of the way is to remind ourselves that we need to take care to avoid teleological language which is applicable only to conscious organisms such as humans who can plan ahead on the basis of imaginative representations of goals. Dawkins, who should know better, gets sloppy this way when he suggests that his hypothetical bacteria might produce a gas "for the good of the world". No bacteria produce gases or anything else for the sake of anything, *even themselves*, while humans do all sorts of things for the sake of goals and purposes. (It would probably be better as well if biologists were to avoid the term "altruism" for the self-sacrificial behavior that sometimes occurs in mutualistic functioning, since that word is most accurately applied to certain human motivations.)

In a mutualistic system a species of bacteria may well have the *function* of producing a certain gas that facilitates the operation of the system as a whole; functional language is perfectly appropriate for coordinated living systems from protozoans to ecosystems [Allen, 2004]. But the fact that a system has evolved in such a way that some of its components have recognizable functions in the economy of the whole does not mean that they have purposes in the sense that things done intentionally by humans have a purpose, nor that they have their function for the sake of the whole. (This was expressed clearly by Simpson; see [Simpson,

1953, p. 181].) To say that (for instance) the cells in my kidneys cooperate in a certain way is to say that they happen to function in concert in a certain way, not that they cooperate in the sense that humans can (on selected occasions) *choose* to cooperate. My kidneys have the function of eliminating excess water and certain toxins from my body, but they do these things because these activities are supported by a complex network of feedback loops; they do not do them for my sake *or even for their own*. This is an important part of the answer to Paley [Paley, 1802] and other champions of "intelligent design": the fact that parts of a complex system have recognizable functions does not by itself imply that they were products of intentionality.

A much tougher question is to say what constitutes a replicator. Dawkins thinks that it does not make sense to say that Gaia has a genome. But of course Gaia has a genome; the genome of Gaia and any other sort of symbiotic complex is comprised of the combined DNA and RNA of all of the myriad organisms of which it is composed. A *distributed genome* is very common at the eukaryotic cellular level. By now there is no controversy about the fact that there is cytoplasmic DNA, namely the DNA belonging to organelles of endosymbiotic origin such as the mitochondria and plastids. The genome of virtually all metazoan cell lines consists not only of nuclear genes but of the genetic heritage of an often bewilderingly complex suite of endosymbiotes. The genome of an organism does not have to be concentrated in *one spot* within the organism, and it rarely is.

A good illustration of this fact is the protozoan (or more properly protist) *Mixotricha paradoxa*, an extraordinarily beautiful organism often cited by Margulis (*e.g.*, in [Margulis, 1998; Margulis and Sagan, 2001]) as an exemplar of symbiogenesis. *M. paradoxa* lives in the gut of certain termites, and apparently serves its hosts by digesting cellulose and lignin. But it is a symbiote built out of symbiotes: as well as its own nucleus, each *M. paradoxa* contains several hundred thousand individuals of at least four other species of bacteria [Margulis and Sagan, 2001]. (Curiously, the one type of symbiotic organelle it does not contain is the mitochondrion, probably because the termite gut is anoxic.) Each individual *M. paradoxa* is a populous community, cooperating as a mutualistic whole. So what, in such a case, is the unit of selection?

Dawkins is right that any chunk of genetic code that in effect says "make more of me" can be a replicator and will succeed in being replicated if it says this in just the right way to resonate with the demands of its environment. However, networks of cooperative behaviors can and often are sufficiently successful that they are amplified by natural selection into a coherent, reproducing whole. This can occur not only in the cases of endosymbiosis studied by Margulis; complex associations of metazoa can form such symbiotic networks as well, some of which may be more tightly coupled (that is, causally interactive) than others. To further complicate the story, it is increasingly evident that complex metazoa such as mammals are host to a rich array of microbial symbiotes, so much so that microbiologists are beginning to describe multicellular organisms as *metagenomic* [Grice *et al.*, 2008; Ley *et al.*, 2008]. If a symbiotic network is sufficiently coherent and coordinated that it reproduces as a whole, then *its entire genetic code* is a replicator. So the question of the unit of selection, the question of what is "seen" by natural selection, is not simple; it is not just the gene (whatever that is) unless by "gene" one means simply any replicator. A sufficiently well coordinated symbiotic association can itself become a unit of selection.

Most of Dawkins' objections to Gaia apply to *Mixotricha paradoxa* as well, and if he were right, there ought to be no such thing. In fact, the way that *M. paradoxa* reproduces can give us some insight into the sense in which a hypothetical planetary-scale symbiotic unit could evolve. In symbiotic protists like *M. paradoxa* the orchestration of reproduction is complex and not yet well-understood. However, there is no reason to suppose that all the component symbionts of such organisms reproduce in perfect concert, even though the host cell is capable of division as a unit. Endobacterial symbionts within a larger complex could well run through many reproductive cycles of their own during one reproductive cycle of the larger complex. Their survival would depend upon adapting to the constraints within the larger organism, just as all organisms on Earth have to adapt to the often-inorganic but sometimes organic constraints of the larger world. (An important example of such a constraint is climate, which might best be described as an organically-mediated inorganic constraint. Clearly when one is speaking of an environmental parameter such as temperature, which is partially controlled by solar input and partially controlled by carbon dioxide concentration, the dividing line between the organic and the inorganic is often fuzzy.) To a single bacterium within *M. paradoxa*, one cell generation of its host is an entire cosmological cycle which defines a world to which the bacterium must adapt like any other organism in nature. Such symbionts within an organism such as *M. paradoxa* would often be subject to natural selection that would tend to favour their ability to contribute to the economy of the whole organism. Complex symbiotic associations like *M. paradoxa* therefore also can evolve piecemeal in response to *internal* constraints as well as all at once in the usually understood fashion, in which the composite organism evolves as a whole in response to *external* constraints. One can therefore distinguish between *external* evolution (which is well-studied) and *internal* evolution—evolution of the *components* of a complex symbiotic association in response to survival challenges and opportunities acting internally to the association.

A key difference between Gaia (as hypothesized by Lovelock and Margulis) and the kinds of organisms to which the usual model of natural selection applies is, therefore, that Gaia does not reproduce as a *unit* as do its component organisms, including *M. paradoxa*. Rather, Gaia evolves because evolution occurs *within* it, just as it does within *M. paradoxa*. Gaia reproduces gradually, part by part, in a process of growth, regeneration, adaptation, and decay, almost like an organic version of Neurath's boat of knowledge which is rebuilt piece by piece as it floats along. Gaia as a whole adapts to its external environment over millions of years in a piecemeal, not-perfectly-coordinated way as its component organisms adapt to the constraints of the external environment and the internal constraints imposed on them by the other organisms in the system. In a remarkably English manner, Gaia muddles through and remains tough and resilient despite its jury-rigged nature. Although the details must be very complex and may never be fully elucidated, there is no reason why we cannot suppose that Gaia (viewed as something like a planetary-scale *M. paradoxa*) cannot be supposed to evolve in the piecemeal way that a complex symbiotic association like *M. paradoxa* can evolve, even though it neither has a nucleus which partially coordinates its activities, nor reproduces as a *unit* the way a protist can.

Now, Dawkins suggests that we imagine that some mutant bacteria happen to start producing a certain gas that is beneficial to the symbiotic complex as a whole. He makes a very odd claim: "those individual bacteria who don't put themselves to the trouble of manufacturing this gas for the good of the world will do better." (This is more or less Garrett Hardin's tragedy of the commons at the cellular level.) But it should be clear that this is not *necessarily* the case; an organism manufacturing some component that increases the overall suitability of the environment *for that organism* could very well increase the reproductive success of that organism even if the manufacturing process has costs and risks associated with it. There is no guarantee that this would happen in all cases, but there is no a priori reason that it would not, either.

Some parasitical "free-riders" can be tolerated so long as the functionality of the system is maintained; indeed, some parasitism may benefit the system in indirect ways if it maintains variability. But if *all* organisms in an ecosystem are parasitical in the sense that they do not put themselves to the trouble of contributing something to the system, they certainly will not do better since the whole system will ultimately degrade.

Perhaps the notion of a cost-benefit analysis would be helpful here. Any conceivable activity by an organism has a cost. This need not be only in terms of energy and materials; adaptation to any particular environment also exposes an organism to the hazards typical of that environment, such as the predators peculiar to it. There are also opportunity costs: if an organism becomes adapted to the Arctic cold, for instance, then it may have given up survival options suitable to warmer weather. It is elementary that cooperative behaviour carries costs and risks precisely as Hardin indicated; for instance, if the organism shares some of its resources with others it will have less for itself, and it opens itself up to the risk that it may be out-reproduced or otherwise out-competed by others of its species or other species who are less inclined to share the goods. However, an action can be advantageous even if it has a cost, so long as its benefits outweigh its cost, while *failure* to cooperate may have costs as well, which could include (as in Hardin's tragic scenario) subversion of the very environmental conditions that made life possible for that organism in the first place. Again, at the risk of repetition, the existence of a co-operative symbiotic modality does not imply intentionality (as with co-operation between humans) but rather coherence of functionality.

As Lenton observes,

> Organisms possess environment-altering traits because the benefit that these traits confer (to the fitness of the organisms) outweighs the cost

in energy [emphasis added] to the individual [Lenton, 1998, p. 440].

This remark suggests a clarification of the sense in which benefit flows back to a symbiont. The most general sense of "benefit" to an organism is the availability of free energy; this can translate into reproductive opportunities or simply an increased survival probability for the individual (since more free energy allows for a wider repertoire of survival strategies and modalities). We see here again an instance in which thermodynamics can illuminate the workings of evolution.

If Hardin's scenario were the normal pattern—that is, if life *typically* subverts the conditions for its existence—how could there be life on Earth at all? Earthly life has proved remarkably resilient for over 3.5 billion years, despite celestial impacts, episodes of massive volcanism (and the occasional runaway greenhouse catastrophes possibly consequent upon them [Ward, 2007]), and steadily increasing solar output. This could only be possible if the persistence of complex life is somehow probabilistically favoured within the broad range of physical conditions that have been available on Earth for about the past four billion years, and that is only possible if life (despite the constant recurrence of endemic parasitism at all scales from the viruses to human society) has had (so far at least) a net tendency to co-operate in order to maintain the conditions necessary for its continuance. This is especially clear if we understand parasitism from the biophysical (thermodynamic) point of view as something that results in the physical degradation of the host; if life on Earth *in net* degraded its habitats then it would have destroyed itself long ago. Furthermore, if life in net were balanced on the knife-edge of commensalism, it is hard to understand how such a precarious state could have persisted for so long. A planetary-scale, rough-and-ready mutualism seems to be the only possibility, and this observation could be thought of as a minimal Gaia hypothesis.

Suppose that the cost of a new trait is that it requires self-sacrificial behavior for some members of the species. If a strain of mutant organisms simply commits suicide *en masse* then its evolutionary story is over. However, if the self-sacrificial behavior greatly facilitates the reproduction of the survivors, even if there are rather few of them, then it will tend to be amplified by natural selection. The importance of mechanisms of this sort has been emphasized by Bonner who has described, for example, the self-sacrificial behavior of slime mold amoeba (in vast numbers) in the formation of a slime mold fruiting body [Bonner, 1998]. There is nothing unusual about this sort of thing; it occurs throughout nature from the bacterial level on upward. Again, the fact that cooperative behaviour has costs and risks does not imply that it puts its possessor at a selective disadvantage, so long as there is a sufficient reward for the behaviour as well.

Apparently-altruistic behavior need not be explained merely as "kin selection"; an organism need not be in a mutualism merely with its cousins. It could be in mutualism with any other conceivable form of life at all, so long as the net effect is to provide a modality of survival for the organism. Here, by the way is the basis for so-called group selection, which is nothing more than selection in favour of mutualistic symbiosis. This is a point that even the most sympathetic and well-informed apologists for group selection do not bring out as clearly as they could [Sober and Wilson, 1998]. (Whether *species selection* can be understood in symbiotic terms is a different and difficult question since it is not clear that a species can always be thought of as a symbiome; I will not address this question further here. [Stanley, 1979].)

Dawkins' distressingly sloppy argument is above all a crashing *non sequitur*—for from the fact that cooperation must inevitably have costs and incur risks it does not follow that it cannot have benefits as well, and indeed net benefits. What really matters is the *timing* of those benefits: the feedback from the environment has to return to the organism soon enough to make a difference to its reproductive success or probability of survival. Therein lies the real tragedy of Hardin's medieval commons: a social pathology that prevented sufficient rewards for cooperative behaviour from flowing back to the beleaguered peasants soon enough for those rewards to make a difference to their well-being.

Natural selection can be understood as a process involving feedbacks. If a trait increases reproductive success that process can be described as the amplification of the trait by positive feedback from the environment. On the other hand, if a trait triggers a chain of events that decreases the probability of its own recurrence then it will be damped out by that negative feedback from the environment. In order for the effect of the altered trait on the environment to make any difference to reproductive success, it has to feed back to the organism in time to affect its reproduction; it does not have to feed back within just one reproductive cycle, but the feedback cannot take forever or be so attenuated that it makes no difference to the reproductive or survival probability of the organism. (As in so many endeavors, timing is almost everything.) Such feedbacks can certainly reward cooperative as well as competitive behaviour. And once again, by "cooperative" behavior we do not mean activity that is motivated by warm feelings of fellowship, but coherence of functionality.

4.5 Symbiosis and Fitness

A full treatment of the complex and important topic of fitness is beyond the scope of this paper. The term "fitness" is ambiguous and has been used in many ways. Elliott Sober usefully distinguishes between fitness as *viability* (the tendency of an individual organism to survive) and as *fertility* (the fecundity of an organism), and he explores ways in which one could treat overall fitness as a product or some other mathematical function of measures of viability and fertility [Sober, 2001]. The problem with this approach is that it tends to focus on the fitness of individual organisms or species. The study of symbiosis shows that this emphasis is too narrow, since as noted organisms can combine symbiotically to form new organisms in which the genomes of the component symbionts interact coherently or even partially disappear. It might be more appropriate to define fitness in the broadest sense as a measure of the tendency of life itself to survive and flourish. I will not attempt that large task here. Rather, here I want to think of fitness as whatever combination of traits, qualities, propensities, or properties it may be that enables an organism to adapt to the feedback from its environment; from this viewpoint, therefore, talk of fitness is not so much a description of adaptive success as an attempt to explain it. I will outline reasons to think that in this sense there are *three* "faces" of fitness—one of which has not received the attention it merits.

Darwin, Huxley and Spencer took fitness (in the sense of the term as an explanation for success in the "struggle for existence") as primarily the ability to compete with other organisms for a larger slice of the ecological pie. Kropotkin and other biologists interested in mutualism and symbiosis insisted that the ability to cooperate, to share the ecological pie in a way that optimizes survival for all concerned, is at least as important for survival as the ability to compete in many ecological contexts (especially where resources of space, materials, and energy may be limited). Both viewpoints tacitly assume that organisms have no option but to survive within an ecology possessing only a fixed budget of resources. In 1922 A. J. Lotka pointed out that natural selection can favour the ability of organisms to enlarge the ecological pies upon which they depend:

> But the species possessing superior energy-capturing and directing devices may accomplish something more than merely to divert to its own advantage energy for which others are competing with it. If sources are presented, capable of supplying available energy in excess of that actually being tapped by the entire system of living organisms, then an opportunity is furnished for suitably constituted organisms to enlarge the total energy flux through the system. Whenever such organisms arise, natural selection will operate to preserve and increase them. The result, in this case, is not a mere diversion of the energy flux through the system of organic nature along a new path, but an increase of the total flux through that system [Lotka, 1922, p. 147].

Lotka's claim is obvious in the case of autotrophic organisms, especially the all-important photosynthesizers. No life is possible without a generous external supply of free energy, whether it is supplied by the sun, nuclear reactions within the Earth, or some other source of energy outside the biosphere. From the abstract thermodynamic point of view, the autotrophs act like valves; they divert some of the external flows of energy into the ecosystems in which they participate. The crucial point is that in general they divert more energy into the system than they need for their own metabolisms. Their way of being mutualistic with other life on earth is that they capture more free energy than they need themselves and distribute the excess in the form of carbohydrates and oxygen. The photosynthesizers can build up the amount of free energy circulating in the earth system, thereby multiplying the survival possibilities for themselves and other forms of life. Let us call the ability of organisms to enlarge the carrying capacity of their supporting systems *constructive* or *Lotkan* fitness. This is one of the most important ways in which organisms can be mutualistic: by the ability to capture, store, and recirculate more energy than they themselves need, organisms that exhibit Lotkan fitness benefit themselves and their offspring by building up the physical supporting capacity of the systems that they *and other organisms* depend upon. An organism that can do this could end up with its gene frequency in the ecosystem at a given time unchanged but its longer-term probability of survival enhanced, simply because it has increased the carrying capacity of the system as a whole. A number of authors since Lotka have noted the existence of this constructive sense of fitness [Wicken, 1987; Depew and Weber, 1995], but its importance remains underappreciated even though the diverse panoply of life on Earth today could not exist without it.

It is not generally appreciated that heterotrophs, including humans, can also build up the capacity of their supporting ecosystems. They cannot directly convert energy from inorganic sources into useable form through biochemical mechanisms within their own bodies, as can the autotrophs, but through a variety of constructive activities they can greatly increase the niches available for autotrophic life and thereby indirectly cause photosynthesis and other energy-capturing processes to occur [Peacock, 1999a]. I will return to this important point at the end.

5 SYMBIOSIS AND CANCER

5.1 Cancer as a Breakdown of Mutualism

The symbiotic way of thinking may offer a door to a deeper understanding of the evolutionary basis of cancer. However, in order to open this door we need to consider one more expansion of the concept of symbiosis. Symbiosis came to the attention of biologists as the association of different species of organisms, often species that are not even closely related taxonomically or genetically (as with many of the charismatic examples of symbiosis such as the lichens). This restriction of the term symbiosis to relations between identifiably-different species is too narrow. First, the distinction between species is not always sharp, especially at the cellular or micro-organismal level. Second, there are associations between cells of the same or very nearly the same genome that could be reasonably thought of on other grounds as symbiotic (the important example of the slime molds will be discussed below), and it could therefore be useful to think of the highly orchestrated cooperative relation between the cells of a metazoan body as a mutualistic symbiosis of clones of a zygote. If this is correct, then cancer is a breakdown of mutualism which arises when a cell undergoes a transformation into something analogous to a free-living, predatory amoebic state.

It seems natural, especially from the viewpoint of the human cancer patient, to interpret cancer as nothing more than some sort of failure of normal cellular function, like a car engine breaking down on the highway due to wear and tear or a manufacturing flaw. There is no question that to some extent cancer happens simply because it *can* happen. However, the transformation of normal mutualistic metazoan cells into parasitic cancer cells is something that occurs throughout virtually the whole range of metazoan life. If something is this widespread it is reasonable to seek an adaptive explanation for the mechanism behind it; it is unlikely that such a mechanism would be merely an oft-recurring accident or breakdown of normal functioning. Even if the occurrence of cancer is not presently adaptive (except insofar as it provides a check on population growth) then perhaps the transformation of benign cells to malignant cells that underlies cancer once was. The transformation to malignancy is mediated by specific genes, the oncogenes, which are mutated or differently-activated versions of genes (the proto-oncogenes) which have normal functions in the cell [Weinberg, 1998]. This also suggests that the transformation to malignancy is something that is genetically programmed into the cell and is not merely an aberration. This section will sketch a "just so" story that could provide an evolutionary explanation for cancer, and suggest ways in which this story could lead to testable consequences.

Very early in the evolutionary history of metazoan life the benefits of multicellularity would have been mixed. While multicellularity offers all of the advantages that come with specialization and increased mobility, it has certain risks as well. Metazoans can be consumed all at once by a predator, and there are numerous hazards including starvation, radiation, chemical toxins or infection that can cause all members of the metazoan association to die at once. It is plausible to suppose that the cells that composed early metazoans evolved a molecular switch or series of switches that allowed them to toggle between multicellular and unicellular modes of existence. Such a switch could only be triggered on a cell-by-cell basis by local biochemical signals. There exists a group of well-studied organisms that have such switches, the cellular slime molds [Bonner, 1998]. These organisms can alternate between differentiated, multicellular fruiting bodies and dedifferentiated unicellular amoeba in response to environmental conditions. In the single-celled phase, *Dictyostelium* species prey largely on bacteria. Under certain conditions (including when prey gets scarce) a chemical signal or *acrasin* is emitted which causes the amoeba to congregate and differentiate into a multi-celled fruiting body. If the acrasin is absent the cells can become amoebic again. It is probably too much to hope that there is a single acrasin-like compound that mediates cellular aggregation in humans, and which could be administered to flip cancer cells back to the metazoan state, but the possibility may be worth investigating. However, it is quite reasonable to suppose that cancer could be fundamentally the consequence of the triggering (by a variety of mechanisms) of an ancient molecular switch that causes mutualistic metazoan cells to revert to a single-celled, parasitical state.

What would be likely to cause the switch to flip in highly evolved complex organisms which are primarily metazoan? Some types of cancer could occur because viruses exploit the switch to their own reproductive advantage. However, the switch may often be flipped due to the biochemical signals of chronic stress. The role of chemical toxins, radiation, and viruses in cancer activation has been fruitfully studied [Weinberg, 1998], but insufficient attention seems to have been paid to the fact that cancer can often be provoked merely by chronic mechanical irritation [Roe, 1966]. A bit of chemically inert glass, sponge, or asbestos fibre implanted in tissue can, over many cell generations, cause a tumour to form. This could be due to the reactivation of an ancient molecular mechanism that allowed a primitive, loosely aggregated metazoan to break up into individual amoeba when conditions were too stressful for the compound structure to exist. The mechanism would be triggered by molecular signatures of chronic stress or possibly inflammation, signs that the multi-celled organism was threatened *as a whole* to the extent that individual cells might have a higher probability of leaving progeny were they to dedifferentiate and scavenge freely again. It is conceivable that this picture could have implications for therapy, if ways could be found to deactivate this hypothetical molecular switch or flip it back to the metazoan state.

To return to the earlier discussion of the evolutionary basis of mutualism, organisms like slime mold cells or any metazoan cells do not "decide" to aggregate "in order to" increase their offspring's chances. Superficially what happens looks like altruistic behaviour since the majority of the cells that congregate thereby forego their chance of reproducing. It is also simplistic, however (although of course somewhat closer to the truth), to suppose that cellular aggregation happens because "selfish genes" have a greater chance of propagating themselves through time if the *Dictyostelium* cells they inhabit occasionally participate in a fruiting body. It is still more accurate to say that it is the *process* (the alternation between differentiation and dedifferentiation according to environmental conditions) that is favoured and reinforced by natural selection. The process replicates itself because it happens to work for replication rather well, and it is at least as true to say that the process takes advantage as it may of the individual genes of the organisms which participate in it, as it is to say that those individual genes take advantage of the process.

From a broader perspective cancer can be understood as an example of the incoherence that can occur between adaptivity at different scales within an organism. The problem of understanding cancer is therefore a facet of the larger problem of understanding the conditions when mutualistic associations are favoured, and when they are not.

5.2 Are Anti-Cancer Viruses Human Symbiotes?

The report by Shafren *et al.* that injection of coxsackievirus causes remission of melanoma tumours highlights the fact that suppression of cancer by viruses is widespread [Shafren *et al.*, 2004; Russell and Peng, 2007]. When a biological phenomenon is this common in nature it is worthwhile not only to investigate its clinical applications (which in this case are highly promising), but (as mentioned in the previous section) to ask whether it has an adaptive explanation. It is quite possible that coxsackieviruses, adenoviruses, REO viruses, and other viruses which have been found to be antagonistic to cancer have evolved into a symbiotic relationship with humans and possibly many other metazoans. Presumably the symbiotic tradeoff would be that in return for cancer suppression, the hosts provide the viruses with a longer-lasting and mobile habitat and thereby facilitate viral replication. If a high-dosage direct application of virus is sufficient to send melanomas and other cancers into remission, it may be that the low-level, diffuse viral infections that are endemic to the human population serve to suppress many tumour cells before they have had a chance to develop beyond the microscopic level. It might be possible to check this hypothesis by seeing whether populations of humans or animals in which various common viruses were missing had higher incidences of cancer. And if this hypothesis is correct, it would lead one to suppose (whimsically, but the point has in fact a quite serious basis) that perhaps the last thing we would want to do is cure the common cold.

6 SYMBIOSIS AND HUMAN ECOLOGY

6.1 Moving Up the Symbiotic Scale

Ecology—and particularly human ecology, which centres on the question of how humans do and could continue to survive on this planet—possesses a peculiar urgency not attached to many other scientific and philosophical subjects. With our still-exponentially burgeoning population, the rapidly dwindling supplies of petroleum, fresh water, topsoil, timber, fish, and other fruits of the "found" ecology upon which we depend, and the accelerating impact of human exploitation on climate and the whole fabric of planetary life, we as a species are approaching a crisis point in our evolutionary history. It is a mistake, however, to blame this entirely on modern industrialization; where we are now is the product of the way that humans have mostly interacted with their supporting environments, and often each other, since modern *H. sapiens* burst upon the evolutionary scene sometime during the last glaciation. Historian William McNeill offers a not very flattering assessment of the human condition:

> It is not absurd to class the ecological role of humankind in its relationship to other life forms as ... an acute epidemic disease, whose occasional lapses into less virulent forms of behavior have never yet sufficed to permit any really stable, chronic relationship to establish itself [McNeill, 1976, p. 23].

It is crucial to realize that from the viewpoint presented in this paper, McNeill's characterization of humans as "macroparasites" is painfully accurate and not merely metaphorical. There is little question that if the pathogenic phase of human evolution continues on its present pace, then the end result (as with any unmitigated pathogenic attack) can only be the severe curtailment of the prospects of the pathogen or host—or both.

Humanity can also be viewed (perhaps ironically, but also more hopefully) as an evolutionary experiment: could a species with the neurological capacity to possess technology, language, and culture have a future? As McNeill explains, it is our technological ingenuity and language (which allows the accumulation of knowledge) that have enabled our largely successful parasitism to date; it could only be our language and ingenuity that will allow a movement to another phase on the symbiotic scale. That such a movement is possible is not entirely out of the question, for the general picture of symbiotic dynamics that has been revealed by many investigators from Frank and de Bary onwards shows that it is rather common (though by no means guaranteed) that emergent parasites can and often do reach states of mutualistic *rapprochement* with their hosts. At the micro level this occurs by a variety of biochemical feedbacks; at the human level, a symbiotic modality must be *culturally constructed* and *learned*. Certainly any mode of human-Gaian interaction that could be genuinely sustainable (tending to support rather than undermine itself) would have to be some sort of mutualistic symbiosis [Peacock, 1995; Peacock, 1999a]. The fact that such transitions from parasite to mutualist are *generally* possible and often favoured, and the capacity of the human organism to learn when it really has to, give some cause for cautious optimism about the prospects for humanity, despite our present increasingly-worrisome predicament.

In "The Land Ethic," one of the foundational documents in modern environmental ethics, Aldo Leopold argued that the key to the establishment of any effective human-land (or human-Gaian) symbiotic modality is an ethic:

> An ethic, ecologically, is a limitation on freedom of action in the struggle for existence. An ethic, philosophically, is a differentiation of social from anti-social conduct. These are two definitions of one thing [which] has its origin in the tendency of interdependent individuals or groups to evolve modes of co-operation. The ecologist calls these symbioses [Leopold, 1996, p. 212].

Leopold insisted that a land (or environmental) ethic is "an evolutionary possibility and an ecological necessity" [212]. On Leopold's view, the practice of an ethic, broadly speaking, is just the human way of being symbiotic [Leopold, 1996; Peacock, 1999b]. A similar view is found in the writings of Eugene Odum:

> ...if understanding of ecological systems and moral responsibility among mankind can keep pace with man's power to effect changes, the present-day concept of 'unlimited exploitation of resources' will give way to 'unlimited ingenuity in perpetuating a cyclic abundance of resources' [Odum, 1971, p. 36].

Grant Whatmough describes what he calls an *artifactual ecology*:

> No parasitic species has ever, nor can ever, prosper expansively [except for short periods of time!]. Our species, like the ancient stromatolitic algae so long before us, must either accomplish a symbiotic adaptation, or perish... Thus the only *serious* question is whether we can actually manage any such adaptation within the context of those genetic features that distinguish our species. Amongst those is our uniquely receptive neurology—our 'open' and experientially structured synaptic system—that has given us our 'minds,' our 'souls,' our consciousness, and ingenuity. And already... that has—on two small islands with dense populations and limited resources (England and Japan)—created for a time a 'horticulturally' modified ecology that proved itself

> well able to provide a prosperous abundance of food, clothing, and civilized shelter for substantial populations, *by way of an intensified ecology*... [an] increase in the density and luxuriance of the whole spectrum of local flora and fauna, as an entailed consequence of the techniques by which those populations then produced their necessary supplies. Those were primarily artifactual ecologies (however accidental)... utterly dependent on the essential contribution of their human element... It can only be by some such means that our species can possibly transform our present parasitic dependence on the found ecology to some kind of symbiotic alternative [Whatmough, 1996, pp. 418–419].

In such an ecology humans would be doing just what photosynthetic algae are doing: benefitting the larger ecology by precisely the means with which they benefit themselves. Ecological fitness for humans is not merely cooperative but Lotkan.

6.2 The Methodological Challenge

I will not conclude this paper with an exhortation to environmental responsibility—such rhetoric can be found in abundance elsewhere—but rather I will attempt to define and highlight the *methodological* problem that follows from our current ecological predicament. Whether or not we can devise more effective modalities for interacting with the planetary system is not a question of purely theoretical interest, to be studied at a leisurely and cautious pace over the coming generations. Today's ecologist is something like an emergency room physician who has to act immediately to save a patient's life, but who does not have the luxury of a fool-proof and complete diagnosis of the patient's condition and possibilities for treatment. The ancient Hippocratic injunction is above all else to do no harm; but the emergency room doctor knows that taking no action may itself guarantee a very negative outcome for the patient. Some remediation has to be risked. Similarly, with respect to human-Gaian interactions, some remediation has to be risked. The long-range goal is a culturally-mediated, mutualistic artifactual ecology in which there is no contradiction between the goals of caring for and protecting the viability of the earth system, and the goal of nourishing and housing the human species. But how we get there is far from obvious.

There can be no such thing as working out a grand, fool-proof plan beforehand. Even if such a thing could be done (which it could not) we don't have the time. I would like to propose a *recursive* approach to environmental remediation, with the overall goal in mind of achieving a mutualistic state such as that envisioned by Odum, Whatmough, and Leopold. The methodology of remediation would be a step-wise on-going process, in which the first steps are actions that (however modest) are highly likely to produce good results and relatively unlikely to backfire (though that could never be fully guaranteed). Successive steps are guided by the response of the system, and this part of the recursive process is absolutely crucial: ecological history tends to show that those few past societies that were able to construct relatively sustainable modalities were those willing to learn from their mistakes [Diamond, 2005]. The zeroeth term is doing nothing (although even this has consequences, of course). The first-order terms could include things like applying known techniques of soil restoration where they are most likely to be effective, massive reforestation world-wide with an emphasis on restoring diversity, and greatly increasing the effectiveness of recycling techniques (such as composting, using agricultural waste for soil restoration and fuel, and recycling of materials). Other first-order steps should include rigorous preservation of those areas of forest and other high-value biomes that are not yet totally despoiled by human intervention, but this will be politically difficult, at least until improvements in agriculture (flowing from soil restoration and reforestation) make it less necessary to mine the remaining wild places of the world for sheer sustenance. The seas are a special case: the best zeroeth order method of remediation in many cases will simply be *hands off!*—at least until we have a far better understanding than we do now of how the deep seas could be positively helped. Again, this will be very difficult politically. There must also be a diversity of creative research, first into methods that are modest extensions of known technology, but also into more advanced possibilities like fusion that do have a reasonable prospect of success in the nearer term, and "blue sky" proposals as well. It is essential, also, to learn as much as possible from the wealth of indigenous and grass-roots technologies available around the world.

With nearly seven billion people on this planet, the only way of avoiding a massive die-off of the human species or a climate catastrophe or both is to work out a mutualism between humans and the Earth system that takes full advantage of human ingenuity in all its facets, and in which, as Odum and many other authors have insisted, a sense of ethical responsibility plays a central role—not just because that would be the "right" thing to do (if that expression means anything at all), but because that's the only way that humans *could* be mutualistic.

ACKNOWLEDGEMENTS

The work reported in this paper was supported by the University of Lethbridge, the University of Western Ontario, and the Social Sciences and Humanities Research Council of Canada. I am grateful to the following people for helpful discussions, comments, criticism, and advice: Frédéric Bouchard, Bryson Brown, Sol Candel, Dawn Collins, Richard Delisle, Gail Greer, Dan Johnson, Kevin deLaplante, James Kaye, Martin Ogle, Cody Perrin, Jane Spurr, Matt Voroney, Grant A. Whatmough, and John Woods. None of these good people are responsible for any errors that remain.

BIBLIOGRAPHY

[Allen, 2004] Colin Allen. Teleological notions in biology. In Richard N. Zalta (ed.), *Stanford Encyclopedia of Philosophy*, 2004.

[Beardsley, 1997] Tim Beardsley. Evolution evolving. *Scientific American*, (September): 15–18, 1997.

[Behe, 1996] Michael J. Behe. *Darwin's Black Box: The Biochemical Challenge to Evolution*. The Free Press, New York, 1996.

[Bonner, 1998] John Tyler Bonner. The origins of multicellularity. *Integrative Biology* 1(1): 27–36, 1998.

[Born and Einstein, 1971] Max Born and Albert Einstein. *The Born-Einstein Letters*. Walker, New York, 1971. Trans. I. Born.

[Brown, 2007] Bryson Brown. *Evolution: A Historical Perspective*. Greenwood Guides to Great Ideas in Science. Greenwood Press, Westport, CT, 2007.

[Bub, 1997] Jeffrey Bub. *Interpreting the Quantum World*. Cambridge University Press, Cambridge, 1997.

[Churchill, 1943] Winston S. Churchill. The House of Commons (1943). In C. P. Curtis and F. Greenslet (eds.), *The Practical Cogitator, or, The Thinker's Anthology*, pp. 423–426. Dell Laurel, New York, 3rd Edition, 1975.

[Dawkins, 1995] Richard Dawkins. Comment on 'Gaia is a Tough Bitch' by Lynn Margulis. In John Brockman (ed.), *The Third Culture: Beyond the Scientific Revolution*, 1995.

[Depew and Weber, 1995] D. Depew and B. Weber. *Darwinism Evolving*. MIT Press, Cambridge, MA, 1995.

[Diamond, 2005] Jared M. Diamond. *Collapse: How Societies Choose to Fail or Succeed*. Viking, New York, 2005.

[Dobzhansky, 1964] Theodosius Dobzhansky. Biology, molecular and organismic. *American Zoologist* 4(4): 443–452, 1964.

[Douglas, 1994] Angela E. Douglas. *Symbiotic Interactions*. Oxford University Press, Oxford, 1994.

[Eldredge and Gould, 1972] Niles Eldredge and S. J. Gould. Punctuated equilibria: an alternative to phyletic gradualism. In T. J. M Schopf (ed.), *Models in Paleobiology*, pp. 82–115. Freeman Cooper, San Francisco, 1972.

[Eldredge, 1985] Niles Eldredge. *Unfinished Synthesis: Biological Heirarchies and Modern Evolutionary Thought*. Oxford University Press, New York, 1985.

[Fausto-Sterling, 1993] Anne Fausto-Sterling. Is nature really red in tooth and claw? *Discover* 14(4): 24–27, 1993.

[Frank, 1995] Steven A. Frank. The origin of synergistic symbiosis. *Journal of Theoretical Biology* 176: 403–410, 1995.

[Gray, 1992] Michael W. Gray. The endosymbiont hypothesis revisited. *International Review of Cytology* 141: 233–357, 1992.

[Grice *et al.*, 2008] E. A. Grice, H. H. Kong, G. Renaud, A. C. Young, G. G. Bouffard, R. W. Blakesley, T. G. Wolfsberg, M. L. Turner, and J. A. Segre. A diversity profile of the human skin microbiota. *Genome Research* 19(5): 1043–1050, 2008.

[Hardin, 1968] Garrett Hardin. The tragedy of the commons. *Science* 162(3859): 1243–1248, 1968.

[Harwood, 1979] Robert Frederick Harwood. *Entomology in human and animal health*. Macmillan, New York, 1979.

[Hsü, 1986] Kenneth J. Hsü. *The Great Dying*. Harcourt Brace Jovanovich, San Diego, 1986.

[Hull and Ruse, 1998] David L. Hull and Michael Ruse (eds). *The Philosophy of Biology*. Oxford University Press, New York and Oxford, 1998.

[Huxley, 1989] Thomas H. Huxley. The struggle for existence in human society. In George Woodcock (ed.), *Mutual Aid: A Factor of Evolution*, pp. 329–341. Black Rose Books, Montréal, 1989. First publication in *The Nineteenth Century*, February, 1888.

[Jeon and Jeon, 1976] K. W. Jeon and M. S. Jeon. Endosymbiosis in amoebae: Recently established endosymbionts have become required cytoplasmic components. *Journal of Cell Physiology* 89: 337–344, 1976.

[Jones *et al.*, 1994] C. G. Jones, J. H. Lawton and M. Shachak. Organisms as ecosystems engineers. *Oikos* 69: 373–386.

[King, 1977] G. A. M. King. Symbiosis and the origin of life. *Origins of Life* 8: 39–53, 1977.
[Kropotkin, 1989] Peter Kropotkin. *Mutual Aid: A Factor of Evolution.* Black Rose Books, Montréal, 1989. First published London, 1902.
[Landauer, 1991] Rolf Landauer. Information is physical. *Physics Today* 44(5): 23–29, 1991.
[Lenton, 1998] T. M. Lenton. Gaia and natural selection. *Nature* 394(30 July): 439–447, 1998.
[Leopold, 1996] Aldo Leopold. The land ethic. In Kent A. Peacock (ed.), *Living With the Earth: An Introduction to Environmental Philosophy*, pp. 211–226. Harcourt Brace Canada, Toronto, 1996. First published in 1949 in *A Sand County Almanac, and Sketches Here and There*, New York, Oxford University Press.
[Ley *et al.*, 2008] R. E. Ley, M. Hamady, C. Lozupone, P. J. Turnbaugh, R. R. Ramey, J. S. Bircher, M. L. Schlegel, T. A. Tucker, M. D. Schrenzel, R. Knight, and J. I. Gordon. Evolution of mammals and their gut microbes. *Science* 320(5883): 1647–1651, 2008.
[Lotka, 1922] A. J. Lotka. Contribution to the energetics of evolution. *Proceedings of the National Academy of Sciences USA* 8: 147–151, 1922.
[Lovelock and Margulis, 1974] James E. Lovelock and Lynn Margulis. Atmospheric homeostasis by and for the biosphere: The Gaia Hypothesis. *Tellus* 26(1–2): 2–9, 1974.
[Lovelock, 1988] James E. Lovelock. *The Ages of Gaia.* W. W. Norton, New York, 1988.
[Lovelock, 2003] James E. Lovelock. The living earth. *Nature* 426(18/25 December): 769–770, 2003.
[Malthus, 1798] Thomas R. Malthus. *An Essay on the Principle of Population.* London, 1798.
[Margulis and Sagan, 1995] Lynn Margulis and Dorion Sagan. *What Is Life?* Simon and Schuster, New York, 1995.
[Margulis and Sagan, 2001] Lynn Margulis and Dorian Sagan. The beast with five genomes. *Natural History Magazine*, (June 2001), 2001.
[Margulis, 1993] Lynn Margulis. *Symbiosis in Cell Evolution.* W.H. Freeman, New York, Second Edition, 1993.
[Margulis, 1998] Lynn Margulis. *Symbiotic Planet: A New Look at Evolution.* Basic Books, New York, 1998.
[Margulis, 2004] Lynn Margulis. Serial endosymbiotic theory (SET) and composite individuality: Transition from bacterial to eukaryotic genomes. *Microbiology Today* 31: 172–174, 2004.
[McNeill, 1976] William H. McNeill. *Plagues and Peoples.* Doubleday, New York, 1976.
[Medawar, 1982] Peter B. Medawar. *Pluto's Republic.* Oxford University Press, Oxford, 1982.
[Moran, 2006] Nancy A. Moran. Symbiosis. *Current Biology* 16(20):R866–R871, 2006.
[Moxon and Thaler, 1997] E. R. Moxon and D. S. Thaler. The tinkerer's evolving tool-box. *Nature*, 387(12 June): 659–662, 1997.
[Odling-Smee *et al.*, 2003] F. John Odling-Smee, Kevin N. Laland, and Marcus W. Feldman. *Niche Construction: The Neglected Process in Evolution.* Princeton University Press, Princeton and Oxford, 2003.
[Odum, 1971] Eugene P. Odum. *Fundamentals of Ecology.* Saunders, Orlando, FL, Third Edition, 1971.
[Paley, 1802] William Paley. Selection from *Natural Theology.* In Tim Crane and Katalin Farkas (eds.), *Metaphysics: A Guide and Anthology*, pp. 34–40. Oxford University Press, Oxford & New York, 2004. First publication, London, 1802.
[Paracer and Ahmadjian, 2000] Surindar Paracer and Vernon Ahmadjian. *Symbiosis: An Introduction to Biological Associations. Second Edition.* Oxford University Press, Oxford and New York, 2000.
[Peacock, 1995] Kent A. Peacock. Sustainability as symbiosis: Why we can't be the forehead mites of gaia. *Alternatives: Perspectives on Society, Technology, and the Environment* 21(4): 16–22, 1995.
[Peacock, 1999a] Kent A. Peacock. Staying out of the lifeboat. *Ecosystem Health* 5(2): 91–103, 1999.
[Peacock, 1999b] Kent A. Peacock. Symbiosis and the ecological role of philosophy. *Dialogue* 38(4): 699–717, 1999.
[Peacock, 2008] Kent A. Peacock. Ecosystems. In Sahotra Sarkar and Anya Plutynski (eds.), *A Companion to the Philosophy of Biology*, pp. 351–367. Blackwell, Oxford, 2008.
[Roe, 1966] Francis C. J. Roe. Cancer inducing agents. *Science Journal* 2(5, May): 38–42, 1966.
[Russell and Peng, 2007] S. J. Russell and K. Peng. Viruses as anticancer drugs. *Trends in Pharmacological Sciences* 28(7): 326–333, 2007.

[Sapp, 1994] Jan Sapp. *Evolution by Association: A History of Symbiosis.* Oxford University Press, New York and Oxford, 1994.

[Sapp, 2004] Jan Sapp. The dynamics of symbiosis: an historical overview. *Canadian Journal of Botany* 82: 1046–1056, 2004.

[Sarkar and Plutynski, 2008] Sahotra Sarkar and Anya Plutynski (eds.), *A Companion to the Philosophy of Biology.* Blackwell, Oxford, 2008.

[Schneider and Kay, 1994] E. D. Schneider and J. J. Kay. Life as a manifestation of the Second Law of Thermodynamics. *Mathematical and Computational Modelling* 19: 25–48, 1994.

[Schneider and Sagan, 2005] E. D. Schneider and Dorion Sagan. *Into the Cool: Energy flow, thermodynamics, and life.* University of Chicago Press, Chicago & London, 2005.

[Schrödinger, 1944] Erwin Schrödinger. *What Is Life? The Physical Aspects of the Living Cell.* Cambridge University Press, Cambridge, 1944.

[Shafren *et al.*, 2004] D. R. Shafren, G. Au, T. Nguyen, N. G. Newcombe, E. S. Haley, L. Beagley, E. S. Johansson, P. Hersey, and R. D. Barry. Systemic therapy of malignant human melanoma tumors by a common cold-producing enterovirus, Coxsackievirus A21. *Clinical Cancer Research* 10: 53–60, 2004.

[Simpson, 1953] George Gaylord Simpson. *The Major Features of Evolution.* Columbia University Press, New York, 1953.

[Smith, 1979] D. C. Smith. From extracellular to intracellular: The establishment of a symbiosis. *Proceedings of the Royal Society of London. Series B, Biological Sciences* 204(1155): 115–130, 1979.

[Sober and Wilson, 1998] E. Sober and D. Wilson. *Unto Others: The Evolution and Psychology of Unselfish Behavior.* Harvard University Press, Cambridge, MA, 1998.

[Sober, 2001] Elliott Sober. The two faces of fitness. In R. Singh, D. Paul, C. Krimbas, and J. Beatty (eds.), *Thinking about Evolution: Historical, Philosophical, and Political Perspectives*, pp. 309–321. Cambridge University Press, Cambridge, MA, 2001.

[Stanley, 1979] Steven M. Stanley. *Macroevolution: Pattern and Process.* W.H. Freeman, San Francisco, 1979.

[Tansley, 1935] A. G. Tansley. The use and abuse of vegetational concepts and terms. *Ecology* 16(3): 284–307, 1935.

[Ward, 2007] Peter D. Ward. *Under a Green Sky: Global Warming, the Mass Extinctions of the Past, and What They Can Tell Us About our Future.* HarperCollins, New York, 2007.

[Weinberg, 1998] Robert A. Weinberg. *One Renegade Cell: How Cancer Begins.* Basic Books, New York, 1998.

[Whatmough, 1996] Grant A. Whatmough. The artifactual ecology: An ecological necessity. In Kent A. Peacock (ed.), *Living With the Earth: An Introduction to Environmental Philosophy*, pp. 417–420. Harcourt Brace Canada, Toronto, 1996.

[Wicken, 1987] Jeffrey S. Wicken. *Evolution, Thermodynamics, and Information: Extending the Darwinian Paradigm.* Oxford University Press, Oxford, 1987.

[Wilson, 1992] Edward O. Wilson. *The Diversity of Life.* W.W. Norton, New York, 1992.

ECOLOGY AS HISTORICAL SCIENCE

Bryson Brown

1 HISTORY IN SCIENCE AND HISTORICAL SCIENCES

There are two long-term issues about the nature of science that I want to address in this essay. The first concerns the overall *shape* of our scientific understanding of the world—an issue that was once central to philosophy of science, but has been largely eclipsed during the twentieth century. The second concerns the different subject matters of the sciences we distinguish today, and the implications of these for the forms of scientific understanding we seek.

The shape of science as a whole was once a central preoccupation of philosophers of science, who developed systematic views about the subject matters of the different sciences and their relations, about their relative standing in terms of authority and fundamentality and their special methodological characters. Taxonomies of human knowledge about the natural world traditionally distinguished *natural history* from *natural philosophy*, with natural history conceived as taxonomic and descriptive while natural philosophy dealt with causal relations, and aimed to produce not just descriptions but *explanations* of its phenomena.

However, during the eighteenth and nineteenth centuries, causal questions began to arise in fields that had been part of natural history. In biology, taxonomic work by Linnaeus, John Ray and others led to a plethora of newly recognized species, giving rise to puzzles about the distribution of different forms of life around the world, about the nature of species (arising especially from the difficulty of distinguishing well-marked subspecies from closely resembling but distinct species), about relations between different species (Linnaeus suggested some species had arisen by hybridization of other species), about extinction (arising from a growing recognition that numerous fossil species seemed to lack living representatives), and about the origin of species (as growing knowledge of the fossil record suggested the familiar species of today did not exist during earlier periods in the earth's history) (see [Young, 76f; Rudwick, 2007, 349f.]). In geology, the description of formations and their spatial relations (driven in large part by the practical concerns of miners) led scientists to an increasingly historical vision of their origins (see [Rudwick, 2007 181f.]), consciously developed in parallel with antiquarian history, in which fossils and other traces of the earlier earth served in place of ancient monuments and buildings to illuminate a distant, unrecorded past.

Since then, the sciences of natural history have become both explicitly causal and truly historical. As a result, the kind of explanation that we find in these sciences is notably different from the ideal of explanation inherited from the western

Handbook of the Philosophy of Science. Volume 11: Philosophy of Ecology.
Volume editors: Kevin deLaplante, Bryson Brown and Kent A. Peacock. General editors: Dov M. Gabbay, Paul Thagard and John Woods.

philosophical tradition. Ancient ideas about epistemology focused on what is fixed and unchanging as the proper objects of knowledge; the ever-changing world of individuals and their particular stories, inconstant and imperfect as they are, were regarded, at best, as second-class subjects of knowledge. Thus geometry, mathematics, biological and other species (conceived as fixed kinds to which various concrete and imperfect individuals belong), and physics (as a universal science of motion and change) provided true material for knowledge. This kind of knowledge is fully understood because it is grounded in the unchanging, necessary nature of things.

There is a special sort of explanation that seems a credible goal for such sciences, and unachievable for a truly historical science. On the assumption that fundamental principles are self-explanatory,[1] such explanations are *closed*, that is, they appeal to nothing that is not itself explained. But historical explanations are always open, appealing to conditions and circumstances and sequences of events as boundary conditions that themselves remain unexplained. Such explanations have traditionally been regarded as *incomplete* (hence various regresses of explanation, central to some forms of cosmological argument). But if we regard an explanandum as truly contingent, we cannot expect it to be explained except by appeal to other contingencies.

In what follows we will examine the distinction between the *natural sciences*, generally regarded as including physics, chemistry, and some aspects of biology and even ecology when directed towards present life and considered aside from the historical/evolutionary origins of living things and long-term changes in ecosystems, and the *historical sciences*, including cosmology, geology, earth systems science, evolutionary biology and ecology. This division, which is familiar in outline but perhaps not in detail, along with the motives for drawing a line separating these sciences will be examined carefully here. I will argue that there are indeed close parallels connecting biology and especially ecology to the historical sciences, some with important methodological implications, although the most important parallels are not really about history at all.

As the eighteenth century French savant the Comte de Buffon understood it, ecology represented an interesting middle stage in the emergence of historical science. Buffon envisaged a natural course of development for the earth and for life on earth—so his vision is clearly causal. And this natural course of development provides a narrative for a history of life on earth. According to Buffon, life formed as soon as the temperature of the cooling earth was low enough to allow it, arising first at the poles. As the earth continued to cool, the first forms of life migrated towards the equator while new forms, adapted to colder conditions, arose at the poles. In general, climatic and soil conditions determine all the rest: "Ainsi la terre

[1]See Aristotle, *An Post.* A.2: "We suppose ourselves to possess unqualified scientific knowledge of a thing, as opposed to knowing it in the accidental way in which the sophist knows, when we think that we know the cause on which the fact depends, as the cause of that fact and of no other, and, further, that the fact could not be other than it is." The intellect ('nous') is involved in the special grasp we have of principles, which cannot be demonstrated; this grasp assures us that the principles are among the things that 'could not be other' than they are.

fait les plantes, la terre et les plantes font les animaux, la terre, les plantes et les animaux font l'homme" (Thus the land makes the plants, the land and the plants make the animals, the land, the plants and the animals make man) [Buffon, 1756, p. 58]. Later biogeographical research made it clear that climate and soil were not enough to determine flora and fauna. But Buffon's notion of a tight link between physical conditions and life forms persisted in Lyell's extreme uniformitarianism, when he proposed that should the climate revert to Mesozoic conditions, it might bring back the prehistoric beasts of those times: "The huge iguanodon might reappear in the woods, and the ichthyosaur in the sea, while the pterodactyle might flit again through umbrageous groves of tree-ferns" [Lyell, 1830, p. 123].

But Buffon's narrative is not fully historical in the sense set out above: on Buffon's account, there is only one possible course for the history of life, only a single possible ecology for any specific geographical region (soil and climate). This is closely related to an attractive explanatory ambition, viz. to arrive at an account of things that (at least in principle) *rules out* alternatives, not conditionally, but absolutely; such an explanation is only possible if the present state of things is regarded as somehow inevitable, rather than contingent. Admittedly, there *is* one dramatic contingency at the very outset of Buffon's theory of the earth, as the material of the planets is dragged out of the sun by the gravitational force of a passing comet. But given that one cosmic accident (itself, perhaps, bound to occur *somewhere* in the vastness of our universe), the subsequent course of events is firmly fixed—something entirely foreign to the richly contingent narratives of the historical sciences.

The goal of developing an over-view of the entirety of human knowledge, and the special place and contribution that each science has in it, faded in importance for twentieth century philosophers of science. Nevertheless, some questions about relations between different sciences have remained important. First, there are logical and metaphysical questions about the relations between different sciences, revolving closely around the ideas of *reduction* and *supervenience.* Second, there are questions about evidence and methodology, with implications for the special methods and the relative authority of different sciences (an issue that becomes particularly important when conflicts arise, but also contributes to debates about, and misunderstandings of, the scientific world view). Finally, there are questions about the nature of the *explanations* that the different sciences provide for the phenomena they address.

Immense and wide-ranging effort is required to produce, evaluate and extend theories like evolution by natural selection or plate tectonics. Great effort is needed even to establish important empirical regularities, such as William Smith's insight into fossils and their regular appearance in certain geological formations, in turn related by superposition and thus by time. Similarly, great intellectual insight was required to connect the layered structure of rock formations to the temporal ordering of the processes that deposited them (and the principles of stratigraphy that follow from that connection between spatial structure and temporal sequence) [Albritton, p. 34, Rudwick, pp. 203–214]. The historical sciences have been able to transform our understanding of the natural world so dramatically because their practitioners have built a detailed and richly connected body of knowledge about the natural world and the processes by which many of its features have developed over time.

But the contribution the historical sciences have made to our understanding of the world is notably different from that of the more theory-centered sciences. Rather than focus on theory, which provides an account of basic concepts and their inferential relations[2] (an account that is intuitively freighted with a kind of necessity, encouraging metaphysical notions about natural laws and essences), the historical sciences aim first at *applications*, via the construction of narratives, contingent from the start, which coherently encompass and account for many patterns and features of life and the earth. The coherence of this narrative is not a matter of having demonstrated that it follows from first principles, and it is often largely insensitive to the details of those principles; it arises instead from the fact that the processes invoked in the narrative are grounded in and cohere with a rich and ever-growing variety of observations. Though these observations and the narrative we embed them in are contingent in themselves, they often fit together so intricately that it would be difficult (at best), and impossible (practically speaking) to invent an equally rich and coherent fiction.

In this essay I argue that we should think of ecology as an historical science, despite the fact that ecological models do not, in general, appeal to long periods of time as part of the story they tell about the populations, communities or ecosystems they represent. Ecology shares important features with evolutionary biology, geology and other historical sciences—features that illuminate the epistemic contact between ecological models and the phenomena we apply them to, the limitations of ecological models as predictive tools and the kinds of explanation we can expect from ecological models. By these criteria, ecology fits with the historical sciences—more generally, it emerges that these epistemological and explanatory characteristics that it shares with the historical sciences provide a more interesting dividing line within science than the element of historicity itself (in its contemporary sense), which turns out to be less central in our taxonomy. Finally, our conclusions have implications for what we should expect of ecology, and even for how ecological research ought to be done.

2 STATUS AND AUTHORITY AMONG THE SCIENCES

In terms of status, the historical sciences, including geology, paleontology, physical anthropology, taxonomy and ecology, have often had to take a back seat to the natural sciences and especially physics. To choose a particularly egregious example of a physics-centered view of science, Ernest Rutherford once famously remarked, "[i]n science there is only physics; all the rest is stamp collecting." This is obviously

[2]Scientific theories structured in this way provide a logico-mathematical framework within which we can state observations and make inferences from them.

a case of (perhaps deliberately exaggerated) physics chauvinism. It is true that the historical sciences lack the formal unity and elegance of mathematical physics—but they make up for their looser, more eclectic conceptual structure in the breadth of their scope, the rich variety of concepts, principles and processes they invoke, and the beautiful and subtle inferences they are able to make. Further, the historical vision of the earth and of life on earth that has emerged from geology and biology since the eighteenth century constitutes as important a change in our world view (and especially our understanding of our own place in the world) as the Copernican revolution.

For a long time many philosophers of science sided with Rutherford—to the point that Karl Popper once claimed that evolutionary biology was a pseudo-science.[3] There are obvious reasons, both internal and external, for this preference. First, where science is conducted through the use of a clear set of rich and unified theoretical principles, logically-minded philosophers find a happy hunting ground. Second, there is a sense that many philosophers share, that physics gives us insight into fundamental ontological questions about the make-up of the natural world. This makes the claims and evidence of physics particularly interesting from a metaphysical point of view. Third, the great success of physics in illuminating, revealing and producing a wide range of striking phenomena makes physics extremely interesting from an epistemic point of view as well. Finally, from an external point of view, the great prestige enjoyed by physics (the 'queen of the sciences') since Newton—especially in the English-speaking world—made it a natural focus for philosophical studies of science.

More recently this imbalance in favour of physics has been set at least partly right. Since the mid-twentieth century philosophy of science has become a recognized specialization in philosophy, often pursued by scholars with both scientific and philosophical training. At the same time, the attention of many philosophers of science has turned towards a wider view of the sciences, including detailed and careful studies of the history of science, and work on a wide range of specific sciences including chemistry, biology and geology. The historical sciences are now recognized as clearly worthy of philosophical study in their own right. As a result, we are now in a position to appreciate more fully what distinguishes historical sciences from the more theoretical sciences.

Some have even tried to reverse Rutherford's invidious ranking, arguing that while physics may be able to claim authority over the most basic principles governing nature as a whole, biology (and ecology) are more *comprehensive* because they deal with a much richer variety of processes that require the *fullest* collection of natural principles to be understood. This debate, obviously enough, points towards the long and complex literature on reduction, supervenience, emergence

[3]It's particularly interesting that it is Popper, with his rigorous insistence on falsification as the touchstone of science, who took this position. I argue below that there are some important methodological differences between the historical sciences and the natural sciences, that some of these differences might be mistaken for flaws in the methodology of the historical sciences, but that the testability of claims in the historical sciences is still robust.

and related concepts; here we will treat these topics, if only in passing, from an *inferentialist* point of view, that is, by focusing on the inferences that scientists make as they work with theories, models and observations.

Concern about the comparative authority of different sciences is especially acute when tensions arise between them. One striking example is the late 19th century debate over the age of the earth. Thermodynamic calculations by William Thompson (later Lord Kelvin), assuming a solid earth with an initial temperature at the melting point of its main constituents and a gravitational theory of the sun's energy, suggested that the earth was between about 40 and 400 million years old, and that the sun could not radiate energy at its present rate for more than roughly 100 million years. Despite arguments by Perry showing that a molten interior with convection currents could allow a much greater age for the earth, Kelvin insisted on his own model (arguing that transverse earthquake waves demonstrated the earth's solidity), and later tightened the limits on the earth's age to between 10 and 40 million years, based on new data for the heat capacity and melting temperature of various kinds of rock.[4]

The tension between these arguments of Kelvin's (and Kelvin's stature as a leader in physics) and the views of geologists (even some like Croll, who had been content to live within the limits of Kelvin's earlier results) became quite sharp in the last years of the nineteenth century. These geologists were convinced by their own evidence, most dramatically in the sheer thickness of past sedimentary formations, accumulated slowly as erosion wore down previous rocks and accumulated new beds of sediment, of a much longer history for the earth. Still, there was something *elastic* about the rough measures of time the geological evidence provided. In response to Kelvin's sophisticated calculations, geologists could offer only the crudest of hour-glass equations:

$$\text{minimum time elapsed} = \frac{\text{minimum accumulation (of sedimentary rock, erosion or other)}}{\text{maximum average rate of accumulation.}}$$

Certainly judgments about the minimum total accumulation and the maximum average rate of accumulation were variable—still, that this relation between accumulations and time holds is incontestable. This connection between the geological evidence and a minimum age for the earth is extremely robust (even young-earth creationists' 'flood geology' only raises—to absurd levels—the maximum average rate of accumulation). By contrast the relation between Kelvin's calculations and the age of the earth (and the sun) depended critically on the details of Kelvin's models. Perry's model vastly extended the age of the earth by invoking convection currents to speed up the transfer of deep heat from the earth's core to the surface, maintaining a higher temperature gradient at the surface. More radically, T. C. Chamberlin suggested that atoms might be 'seats of enormous energies' [Chamberlin, p. 18], able to replenish the energy radiated by the sun—this suggestion, subsequently borne out, breaks the connection between Kelvin's evidence and the age of the sun altogether.

[4] See [Burchfield, 1975, especially chapter II].

The evidence that geologists relied on was (and is) robustly connected to the age of the earth, while the evidence Kelvin appealed to depended on his particular (solid) model of the earth and on the assumption that there was no source of energy that could replace (a significant part of) the heat radiated into space by the earth and the sun. However, the inferences connecting Kelvin's assumptions and his models to the conclusions he drew from them provided much tighter constraints on the ages of the earth and the sun than the geological evidence could give, and the theoretical sophistication of his calculations combined with the general prestige of physics added still more weight to Kelvin's argument. A simple illustration of the authority Kelvin wielded is that in his 1903 essay, 'Was the earth made for man,' Mark Twain took it for granted that Kelvin's (early) figure of approximately 100 million years was the best science could offer.

This contrast suggests that our evaluations of the status and authority of different sciences depend on a multi-dimensional comparison that is by no means easy to reduce to a one-dimensional measure, even with respect to a single question. The natural sciences, exemplified by Kelvin's calculations, provide well-tested, mathematically powerful models for a wide range of phenomena. But their application in particular cases depends crucially on whether the models applied really fit the case, and whether basic theoretical assumptions that have been successful to date can be reliably applied in contexts where basic parameter values are extreme and/or where so-far undetectable levels of violations of the assumptions would be sufficient to invalidate the model. The application of natural science models in such cases might be described as *brittle*, because it can be shattered by new evidence demanding distinct models with very different implications and by new phenomena that occur only at low frequencies or under extreme conditions.

The historical sciences are generally less vulnerable to shifts in the detailed models of various natural processes. This is partly because there are so many coherence checks that can be applied to test and confirm their conclusions, and partly because often the fine details of processes, such as the mechanics and chemistry of surfaces, weathering, frost cycles and so on, or of burial, decay, permineralization, etc., don't threaten to transform the broader observable effects of erosion or fossilization. That a river valley was excavated after a volcanic eruption is sufficiently demonstrated by noting that the valley cuts through a flow of lava from the eruption, regardless of the details of how the water flowing through the river managed to cut through the rock or whether the valley was cut by steady, gradual river flow or by one or a few massive floods; that a small, three-toed animal with some characteristics now found only among horses lived during a certain period is demonstrated by the fossil remains and the formation they were found in, regardless of the details of how the remains were preserved until the present or the precise chemical processes involved in cementing the rock it was found in. The coherence of these inferences with familiar and straightforward observations about how rivers flow and alter the landscape,[5] about how objects with the shape (and

[5]See [Twain, 1883, especially chapters VI-XIII], in which Twain describes learning to 'read the river' as a pilot.

other characteristics) of bone or shell or wood come to be and what can happen to them after an organism dies, convinced scientists that these basic inferences of geomorphology and paleontology were correct long before refined accounts of the details of these processes became available. Further, the bare possibility that these inferences might be mistaken is extremely remote: no credible alternative model will revise these conclusions, even if it substantially alters our understanding of the detailed physical and chemical processes involved. Only a thoroughly radical large-scale transformation of our understanding of the world could lead to the surrender of these basic principles of geology and paleontology.[6]

This reliance on coherence checks to ground our inferences directs those inferences towards the past, tracing backwards towards Reichenbachian 'forks': in general past events leave multiple traces of different kinds, which we can compare against each other now to test historical hypotheses. Predictive power in this context is typically limited to a kind of *retrospective* prediction—that is, traces of a process (say, the iridium-rich layer at the KT boundary construed as a trace of the impact of an asteroid or comet) allow us to predict other traces as well (such as the possibility of finding an impact crater, evidence of a massive tsunami along fossil coastlines if the point of impact was in a sea or ocean, shocked quartz crystals in the boundary layer, and evidence of widespread fires in the boundary layer). As more of these other traces were found, the impact hypothesis became practically certain. Moreover, the resulting establishment of various phenomena as reliable indicators of certain past events makes further inferences stronger: systematic study of such traces both refines our understanding of how they are produced and the special, detailed features they display, providing still more secure ways of making the case for (or against) similar events having occurred in other cases.

The prediction of a future impact is much harder; of course what has happened once may well happen again, but the evidence we would use to predict a particular impact (as opposed to merely evaluating the likelihood of such an impact occurring within some interval of time based on the historical record of impacts) has to draw on the theories and inferences of celestial mechanics to detect an asteroid or comet on an orbit that will intersect the earth's. The observation of such a body really can provide a reliable prediction, but only because we are able to exclude as highly unlikely any dramatic alteration of its expected orbit within the time frame of the prediction: the principle gravitational influences of the sun and planets are well-understood, and the probability of some other body coming close enough to substantially change its orbit is extremely low given the prevailing conditions in our solar system. Further, there is enough uncertainty about the details of these orbits that, for any timescale greater than some tens of years, the prediction of an event as precisely constrained as a collision becomes effectively impossible—so,

[6] As an illustration of just how far such an hypothesis has to go, consider *Darwinia*, by Robert Charles Wilson. (Spoiler alert!) In this imaginative story, Darwinian evolution is undermined by the sudden replacement of Europe with a new continent inhabited by forms of life utterly unlike anything else on earth. In the end, this is explained by the fact that the world is really a kind of cosmic computer program.

while we can retroactively establish that a collision occurred millions of years in the past, we have no means by which we could hope to predict a collision as little as a thousand years hence.

For the purposes of confirmation, one advantage of predictive inferences is that there is little chance that the prediction is actually *ad hoc*.[7] But a retrodictive success could just be a disguised bit of ad hocery. Nevertheless, it's not difficult to find cases where this suggestion is extremely fanciful: for example, consider fossils showing traits intermediate, in various respects, between modern humans and the great apes. The probable existence of some such fossils follows from our evolution from a common ancestor with the apes, a hypothesis which in turn has been massively confirmed by the wealth of hominin fossils discovered since Darwin first claimed that we share a recent common ancestor with the great apes. It would be silly (at best) to suggest that Darwin actually had access to such fossils and deliberately shaped his theory of evolution to ensure that it predicted such fossils are likely to be found—silly both because the fossils were unknown at the time and because there is no room or need for such adjustment of Darwin's theory.

Aside from the often dismissible risk of such *ad hoc* manoeuvres, the epistemology of the historical sciences is on a very firm footing; in fact, they are arguably better supported by their evidence than the theories of the natural sciences, since, as we've already seen, the narratives that we arrive at in the historical sciences tend to survive changes in detail that drastically alter the theoretical principles of our physics and chemistry.[8] Still, it's worth pausing here to respond to an objection that is often heard, though rarely in academic circles. The objection concerns the special role of laboratory results in the natural sciences; in particular, some creationists have argued that the absence of replicated laboratory tests undermines the evidence for the historical narratives of both evolution and geology.[9]

My response is two-pronged. The first prong points out that this is just wrong. Many laboratory tests of both evolutionary and geological ideas have been conducted. Long-term experiments with bacteria have demonstrated evolution by natural selection over thousands of generations, including the development of new metabolic capabilities.[10] Laboratory experiments have explored the properties of many kinds of rock, including details of mineral composition, structure, melting temperatures etc., as well as the processes involved in sedimentation, earthquake dynamics and many other central issues in geology. Laboratory work on multiple forms of radiological dating has confirmed the ordering of formations and forms of life that emerged from stratigraphic work beginning in the 18th century.

[7]Someone could 'gin up' a prediction of some sort of dramatic event deliberately, on the outside chance that it might come true. The familiar 'psychic's' strategy of making multiple predictions and then publicizing only the successful ones comes to mind here.

[8]This independence is symmetrical, of course: changes in accepted historical narratives can occur without requiring changes in basic physics or chemistry. But it is more striking in the other direction, because the historical narratives are ultimately grounded on processes that are described in terms of these sciences.

[9]See the list of creationist claims at `talkorigins.org`, for references.

[10]For a recent and dramatic example see [Blount *et al.*, 2008].

But this response is unconvincing for most of those who raise the objection. It's tempting, and partly right, to diagnose this response as a purely defensive refusal to understand the evidence for evolution and geology. But that isn't all there is to it. There is a real difference here between the natural sciences and the historical sciences, and, though the difference doesn't undermine the evidence for the historical sciences at all, one can see why some would be tempted to think that it does. The temptation arises from the fact that many of the central principles of the natural sciences are directly tested in laboratory: we can, after all, precisely measure many kinds of basic physical interactions and their results there. Moreover, we have also successfully applied the results to predict the behaviour of many systems, both in the lab and in nature.

There certainly are important lab results in the historical sciences—lab work on genetics and biochemistry has been central to the development and refinement of evolutionary biology since the nineteenth century, and lab work has been similarly central to many geological questions as well. However, what people tend to think of as the *main principles* of these historical sciences are broad, long-term historical claims that aren't open to direct testing, in the lab or outside of it. What underlies these challenges to historical science (though it is generally not made explicit) is the notion that the distant past is a proper subject for skeptical worries, while what happens in laboratories is not. Consequently, while the laboratory tests of various processes and principles are taken to establish those processes and principles as reliable aspects of how the natural world operates, their application to unravel the distant past is regarded as dubious at best.

This concern combines with the relatively weak predictive powers of the historical sciences: while we can predict that living things will go on changing over time, that various geological processes, including the movements of tectonic plates, slippage and occasional earthquakes on active faults and various forms of erosion and sedimentation will continue, detailed predictions of specific events (the emergence of new adaptations, or the timing and exact locations of earthquakes, eruptions, etc.) are extremely difficult to make, and appeal to longer term processes and predictions about them (such as dramatic shifts in geography over millions of years) are treated with the same skepticism as claims about the deep past, despite their elegant fit with so much current evidence. The upshot is that laboratory science, celestial mechanics, and immediately testable claims are seen as far better supported by the evidence than any science whose principle claims concern the course of events in deep time could be.

The second prong of my response addresses this challenge directly, asking what justifies this special skepticism about distant periods of time and the long integration that accumulates small changes of the kinds observed over short periods into the dramatic changes that make up the history of our planet. As Charles Lyell argued in his *Principles of Geology*, 'drafts on the bank of time' are far less troublesome, when it comes to understanding their significance and implications, than the invocation of processes (whether natural or not) that can't be observed in detail now because they no longer occur.

Any narrative about the past (or anticipation of the future) is justified, if at all, by its coherence with the various traces we can find now together with our understanding of the processes linking those traces to the various events they can tell us about. As we've already noted here, the historical sciences have produced many remarkably rich, detailed and coherent narratives describing the development of the universe, the solar system, the earth and life on earth. The fact that other narratives can be imagined (including fanciful 'false-past' narratives) is no more evidence for skepticism about the past than the fact that other courses of events (in and outside of laboratories) can be imagined is evidence for skepticism about the present workings of the natural world. The many observational tests these narratives have passed, as the present traces of the processes they invoke were tracked down and documented, make them convincing parts of the natural history of our world.

Ecology shares many of these characteristics of historical sciences. It has been subject to criticism from partisans of the 'hard' sciences. Its subject matter involves rich and complexly connected processes, and predictions in ecology are well-known to be difficult at best. Finally, the retroactive construction of explanatory narratives plays a central role in ecological investigation. So, in both its epistemology and its methodology, ecology groups naturally with the historical sciences.

3 AIMS OF EXPLANATION

Another important contrast between the historical and natural sciences is the focus of the historical sciences on particular applications as opposed to general principles. Of course both principles and applications are part of every science. But in the natural sciences particular applications are typically concerned with how to account for some phenomena using specific theoretical principles—it is generally presumed that the features to be modeled can be expressed within the language of that theory, and that the principles of the theory should provide all the necessary constraints to make the model 'work' (if it emerges that they don't, the theory is in trouble). Further, we expect that the phenomena will be *the same* in any similar case: a successful account of the phenomena will be, in that sense, entirely general.

By contrast, applications in the historical sciences are not theoretically *pure*, often involving rich interactions between processes that are described in terms of different collections of basic principles. Further, they are not treated as *closed*; we expect the historical processes of geological and biological change to be interrupted, altered, even transformed by outside influences such as the eruption of a volcano many kilometres away, a planet-wide climatological shift that gives rise to an ice age, or the sudden impact of an extra-terrestrial body. The task is to unravel a particular sequence of events, not to identify a type of process that will be regularly repeated in every similar case. Differences over preferred models or conceptual outlooks often have more to do with preferences concerning starting points and which factors are treated as intrinsic features of the models and which as exogenous influences altering the course of events. See, for example, C. Eliot's discussion of Clement and Gleason's views of ecological succession, in this volume.

Of course the processes modeled by the natural sciences are also, as concrete individual processes, subject to such external influences. But the aims of the natural sciences don't include a systematic account of such external influences and their roles in the development of particular systems—when our interests do turn to the particular, we will certainly seek out the particular circumstances that explain what happens in an unusual case, but in the natural sciences our interests are not typically focused on the particular. We notice and try to explain unusual cases precisely because we see them as exceptions to the rule, and consequently important tests of it: the usefulness of the basic principles is first illustrated in relatively pure cases, but ultimately every case has to be reconciled with the principles.

Consider the erosion of a particular geological formation. In general, many kinds of processes will be involved, from small-scale mechanical and chemical goings-on to large-scale meteorological and climatological phenomena. The results will have effects on water conditions in the watershed, the soils in the region, and on plants and animals; in turn, plants and soils will alter erosional processes. More significantly, the processes will be *local* and *contingent*: the results will depend on the detailed history of that particular formation and the broader context (climatological, geological and biological) in which its erosion took place. Small features, such as the location and orientation of cracks in the bedrock, can have substantial influence on the direction of water flows and the subsequent development of a drainage system. Not only do the details matter here, but also the course of events outside the region, which often intrude on and alter the processes under study. Any explanation of the erosion of this formation will draw on many contingencies, both in the detailed interaction and feedback processes influencing events, and in the impacts of external events. As a result, detailed and reliable predictions are difficult, if not impossible.

This is partly a matter of complexity. Complex feedback interactions can occur in any science, but in the historical sciences as well as in biology and especially ecology, they are inevitable: even very simple population models can generate chaotic behaviour.[11] The upshot is that, even if we begin with very similar circumstances, the results we obtain from our models, and very probably the outcomes in the natural world, can vary widely. But it is also a matter of the focus on the particular: rather than begin with a universal course of events that will be characteristic of such situations in general though it may, in particular cases, be interrupted or altered, we aim to identify the particular sequence of events that has produced the erosional effects observable in this particular case. Finally, it is also due to

[11]For example, consider the simple logistic equation, $P_{n+1} = r(P_{\max} - P_n)$, where $P_{\max}$ represents the maximum population that can be sustained, r the rate of growth and P_n an initial population. If we normalize the population measure by setting $P_{\max}$ to 1, the result is chaotic for r greater than about 3.57.

the general openness of such systems, and the infrequency of useful explanatory narratives driven by a single isolated process that we can characterize by appeal to theoretically pure principles.

4 THE ROLE OF PREDICTION

Many events and processes studied by the natural sciences allow for powerful models that produce highly constraining, reliable predictions; in fact, the predictive success of such models is often taken as a paradigmatic example of (and a principal type of evidence for) good work in those sciences. This is closely related to the role of general principles used to produce the predictions: in the natural sciences, these principles are believed, at least in many cases, to specify all the relevant quantities and how they influence the system: they aim to be closed, in this sense. Even if the systems being described are not strictly closed and may, in some cases, be disrupted by external interventions, such external disruptions don't undermine the model, although predictive failure in the absence of external disruption does.

This sort of closure leaves aside the challenge of determining the right values to assign to the relevant quantities, as well as the often very difficult problem of calculating or inferring the consequences of such a set of conditions for a system. It also sets aside the fact that the systems we are describing are vulnerable to external influences that we may not be able to anticipate even when we take those external influences to be subject to and describable in terms of the same fundamental principles and quantities.

Nevertheless, the natural sciences do manage to assign values and perform reliable, detailed calculations predicting the behaviour of some important real systems.[12] This success in isolating[13] the course of certain kinds of processes underwrites some of the more metaphysical elements in scientific thought—what we see here is the 'natural' development of such systems in the absence of external interference (though even on a billiard table, a standard illustration of basic mechanical processes, the subtle effects of gravity together with rapid amplification of deviations in the motions of the colliding balls ensure that after more than a few collisions the state of the table is dramatically different from what it would be without the minuscule gravitational influence of the moon). Still, while a closed, predictive account of events on the table is not entirely possible, the external influences that affect it can be expressed, in principle, in terms provided by our theory

[12]See [Cartwright, 1999], *The Dappled World*, for some limits and interesting comments on this issue—the result is often that what we model is the behaviour of very special systems which are developed by experimentalists precisely to isolate/demonstrate certain basic processes and features. See also [Sellars, 1963] "Scientific Realism or Irenic Instrumentalism" for remarks on the role of metaphors as involving second-order similarities in science.

[13]Nature must be 'put to the question', Bacon infamously suggested, in order to achieve this kind of isolation—I see this remark more charitably than some, as drawing a contrast between Bacon and those who, like Descartes (cf. *Principles*) held that science should begin with familiar phenomena in natural contexts, which, Descartes claimed (erroneously, on the evidence) would display the simplest combinations of 'natures'.

of mechanics, and they are small enough not to matter for simple cases involving just a few collisions. This encourages the hope that in principle, a full calculation of all such influences would allow a perfect modeling of the system, a hope beautifully expressed by Galileo when he declared that imperfect results could be obtained from a correct mathematical model only because of the imprecision of our measurements and our own failure to account for all influences on the system:

> Just as the accountant who wants his calculations to deal with sugar, silk and wool must subtract the boxes, bales and other packings, so the mathematical physicist, when he wants to recognize in the concrete the effects which he has proved in the abstract, must deduct the material hindrances; and if he is able to do that, I assure you that matters are in no less agreement than for arithmetical computations.... The sources of error, then, lie not in abstractness or concreteness, not in geometry or physics, but in a calculator who does not know how to make a true accounting. (Galileo, *Dialogue of Two World-Systems*, cited in [Drake, 1970, pp. 68–69].)

Certainly our efforts to apply ever-higher levels of precision in measurement and accounting for small influences on mechanical systems have been well-rewarded, from lens grinding to using pendulums to measure the gravitational attraction of the earth to Hamilton's chronometer and on to today's efforts to detect gravitational waves or the Higgs boson.

5 A CASE IN POINT

Consider the famous debate over the relation between the extinct Dodo and the vanishing Tambalacoque tree, *Sideroxylon grandiflorum* (formerly *Calvaria major*). In a very influential paper, Stanley A. Temple [1977] proposed that the unusually heavy seeds of the Tambalacoque could not germinate unless they had been abraded by passing through a Dodo's gizzard. Temple's argument drew on the apparent absence of young Tambalacoque trees in Mauritius' forests as well as an experiment in which Temple fed fresh Tambalacoque fruit to turkeys (a somewhat smaller bird than the Dodo): three of ten seeds that were either regurgitated or passed whole through the turkeys' digestive tracts did germinate (though seven were crushed). But in a vigorous critique of Temple's work, Mark Witmer and Anthony Cheke [1991] drew on a richer range of evidence to argue that Temple's hypothesis of an obligate mutualism could not be right. Though it's clear that germination is rare in the field, Tambalacoque seeds have been reported to germinate without such treatment and unpublished trials showed no difference in germination rates between abraded and unabraded seeds; though Tambalacoque trees younger than the extinction of the Dodo are rare, they are not unknown; the hard endocarps of Tambalacoque seeds have a natural line of weakness along which the endocarp can split, allowing the seed to germinate (a characteristic Tambalacoque seeds share with *Canarium paniculatum*). Witmer and Cheke's investigations suggest that Tambalacoque seeds are very susceptible to fungal infections, and may (like those of many other tropical trees) need to be cleaned of pulp before the fruit begins to rot, in order to germinate. But other lost or reduced fauna of Mauritius, including an extinct parrot and two species of tortoise, may have eaten and dispersed the seeds. Further losses to introduced species at the vulnerable seedling and sapling stages, along with habitat degradation and competition from newly introduced tree species may also have contributed to the Tambalacoque's decline.

The evidence for this richer account of one ecological change on Maritius closely parallels the kinds of evidence that ground narratives in the historical sciences. The reasoning is clearly retrospective: the decline of Tambalacoque trees is well known; the process(es) that have led to that decline are what is in question. Certain kinds of processes—seeds' failure to germinate through disease or the absence of some helpful factor previously present, seedlings' and/or saplings' failure to survive, are known to be potential factors in such a decline; various tests and signs indicating the importance (or lack of importance) of these factors are explored. An account is supported by our evidence when it coherently connects the results of such studies into an account of the trees' decline; the more fully an account fits details in our evidence, integrating what we can discover about changes in population structure over time and how various processes can affect germination and survival of young trees over the last 300 years, the more satisfactory our explanation of this ecological change. One natural way to construe the reasoning involved is in terms of *eliminative induction*:[14] we accept a particular explanatory narrative when (and only when) the initial constraints on credible types of explanation and the accumulated evidence rule out other narratives; in general, such acceptance leaves open only the possibility that the problem was mis-posed from the outset.

Clearly enough, the result in this case will be a retrospective explanation for what has happened; there is little reason to expect that a prediction of the decline (or of other, similar declines) would be practically possible: much of the evidence used in testing and confirming our explanation wouldn't be available in advance. Moreover, ecologists would have little reason to pursue the evidence that might be available prior to the trees' decline. Many different kinds of ecological changes occur when an island is invaded by so many foreign species and subjected to new forms of agricultural exploitation. Attempting to anticipate them in advance would require extremely high levels of initial information, including detailed and complete models of complex ecological interactions and a rich variety of precise data to apply those models predictively. Finally, the bearing of that evidence would be hard to sort out in advance given the complexity of the web of interactions affecting the survival of various kinds of trees in such circumstances.

---

[14]See [Norton, 1993; 2003].

6 REDUCTION AND SUPERVENIENCE

The metaphysical relations between high-level sciences dealing with complex objects and processes and more 'fundamental' sciences have long been an important topic for philosophers of science. The importance of unraveling the relations between the different sciences in order to clarify the sense in which they may be said to collectively represent our best effort at describing the world and at identifying the best methods for producing such descriptions, make this topic worth addressing briefly here. But rather than focus attention on the metaphysical questions, my chief concern will be with the practical constraints that make an account of the world in terms of physics 'all the way up' impossible, and what prospects there are (in absence of this ideally completed unity) for giving substantive expression to the idea of the unity of science.

Ontologically, it often seems intuitively appealing (given the course that our scientific inquiries have taken) to regard the objects of theories applying on larger scales as composed of (and so at least *ontologically* reducible to) the objects of our microtheories. But giving in to the temptations of this metaphysical intuition is light work compared to the hard slogging involved in translating assertions expressed in terms of macrotheory vocabulary into the vocabulary of an ontologically-preferred microtheory. A full theoretical reduction (*cf.* [Bonevac, 1982]), in which the *inferences* made within the reduced theory are captured as special instances of inferences within the reducing theory demands more still. So part of the challenge here is to sort out the different ways in which we might seek to 'reduce' one theory to another.

Wilfrid Sellars distinguishes these types of reduction in "Philosophy and the Scientific Image of Man" ([Sellars 1956], reprinted in [Sellars 1963]), when he comments on the unity of the scientific image: "There is relatively little difficulty in telescoping *some* of the 'partial' images into one image...we can unify the biochemical and the physical images; for to do so requires only an appreciation of the sense in which the objects of biochemical discourse can be equated with complex patterns of the objects of theoretical physics. To make this equation, of course, is not to equate the sciences, for as sciences they have different procedures and connect their theoretical entities via different instruments to intersubjectively accessible features of the manifest world." In this passage two different kinds of reduction are contemplated—equation of the ontology of two sciences, and equation "of the sciences". Sellars elaborates, "[f]or to make this identification is simply to say that the *two* theoretical structures, each with its own connection to the perceptible world, could be replaced by *one* theoretical framework connected *at two levels of complexity* via different instruments and procedures to the world as perceived" [1963, p. 21]. The equation of the two sciences occurs at the level of *vocabulary*, as a function of the 'telescoping' relation: given this replacement, the *reports* we make and conclusions we draw as biochemists come to employ a vocabulary that is based on the vocabulary of theoretical physics.

Sellars further distinguishes this unification of the entities and vocabularies from unification of the *theoretical principles* of the two sciences:

> ...while to say that biochemical substances are complexes of physical particles is in an important sense to imply that the laws obeyed by biochemical substances are 'special cases' of the laws obeyed by physical particles, there is a real danger that the sense in which this is so may be misunderstood. Obviously a specific pattern of physical particles cannot obey different laws in biochemistry than it does in physics. It may, however, be the case that the behaviour of very complex patterns of physical particles is related in no simple way to the behaviour of less complex patterns...There is, consequently, an ambiguity in the statement: The laws of biochemistry are 'special cases' of the laws of physics. It may mean: (a) biochemistry needs no variables which cannot be defined in terms of the variables of atomic physics; (b) the laws relative to certain complex patterns of sub-atomic particles, the counterparts of biochemical compounds, are related in a simple way to laws pertaining to less complex patterns. [p. 21]

Inferential unification is very different from the telescoping unification that arises just from applying the same *language* (*i.e.*, the vocabulary) to report observations and inferences. An inferential reduction would require the basic *inferences* of particle physics to generate the inferences of biochemistry as well; not only the entities and vocabulary, but the science (and language) of biochemistry would then be fully unified with (*i.e.*, reduced to) particle physics.[15]

Distinguishing these different aspects of reduction is particularly helpful because it focuses attention on two separate elements in our use of scientific theories: first, the application of a theory to the world, both in observation, when we respond to situations in the world with assertions in the language of the theory, and in practice, when we use assertions expressed in the language of the theory to guide practical activity, and second, the theoretical inferences that make some assertions in the language follow from others, which provide opportunities to test the theory's ability to coherently represent some situations in the world.

Together, these aspects of reduction engage with the three main elements of Sellars' inferential view of language: norms governing language-entry, language-

[15]The distinction between vocabulary and language drawn here draws on Sellars' ideas about *material inference*; when vocabulary but not language has been 'telescoped' to unite two sciences, the same vocabulary is governed by two systems of material inference rules, one capturing the inferences of each science. The system of the reducing science will be tightly tied to the basic vocabulary, while the other system independently adds inferences applying to the complex systems of basic objects which are described by the reduced science. But with full unification of the language, we will have only one system of material inferences; the inferences of the reduced system will then be understood in terms of certain (generally complex) inferences in the reducing science. It's worth noting as a caveat here that this story does not yet deal with the subtle interactions that arise when not only reduction but also *correction* comes in: the inferences of the reduced science may well be highly reliable even though they are not precisely correct, from the point of view of an acceptable reduction—and they may depend on circumstances that had not yet been identified in the reduced science as required for its success. Consider as an example here the relation between classical and statistical thermodynamics.

language and language-exit 'moves'. Being able to express distinct theories (considered as systems of inference) in a common vocabulary is certainly an advance towards a unified scientific view of things, but the further step of inferential reduction is by far the hardest. Once it is achieved, the reduced science's inferences now appear as a consequence of the reducing science's principles, rather than merely being expressible in the language of the reducing science. Only this much stronger sort of reduction could satisfy those who follow in Rutherford's rhetorical footsteps, holding that only physics makes real contributions to the principles in whose terms we understand our world.

Sellars' account points towards the important practical limits of reductive efforts. We strongly suspect that there will often be no 'simple relations' between laws governing basic entities and laws governing various complex systems of basic entities. This suspicion is particularly well-founded when it comes to organisms and ecologies. Even at the large-scale level, when we try to model interactions between predators, food supplies, population density and disease to capture how a population changes over time, the resulting models are extremely sensitive to the details of these factors. Absent some radical breakthrough in our inferential capacities, it's obvious that there is no serious prospect of an inferential reduction of any of the other sciences, including ecology, to physics.

This observation underscores the independence of explanations in the historical sciences from changes in the fundamental principles of the natural sciences. Many of the inferential links that unify the narratives of the historical sciences have been formed independently of these fundamental principles, and they connect the observations of the historical sciences in ways we cannot replicate using only these fundamental principles: our understanding of the different processes involved in producing the phenomena of the historical sciences, and of their signs and symptoms, developed in a very empirical way from studies of these phenomena. However, this is far from saying that physics does not constrain the processes that the historical sciences describe and explain—it is a very modest sort of emergence that we are discussing. Further, at the micro-level, physics and chemistry do illuminate processes like erosion, fossilization and glaciation, while at the macro-level principles like conservation of mass and energy and the laws of thermodynamics often provide important constraints on our models. For illustrations from ecology, consider the importance of isotope-based measurements in efforts to track ecosystem productivity in the past, and models that trace flows of energy and materials. These refinements, drawing on the principles and applications of the natural sciences, have greatly extended our ability to measure these processes and learn from the traces and patterns they produce.

Here we find a practical kind of unity in the sciences. While the narratives of the historical sciences are well-grounded in their own evidence and the understanding of a wide range of various kinds of complex processes that has emerged from that evidence, they have also been substantially refined and extended, and (not coincidentally) more stringently tested by the application of principles and observational methods drawn from the natural sciences. The confirmation of the established geological column in the light of a multiplicity of radiological dating techniques illustrates both the advances that can be made with the help of observational techniques grounded in the natural sciences and the reliability of well-grounded results in the historical sciences; by way of contrast, the history of continental drift and plate tectonics illustrates a much more complex interaction, in which geophysics at first motivated wide resistance to Wegener's ideas, but later physical measurements (including magnetic surveys around mid-ocean ridges) confirmed the reality of plate motions, establishing a mechanism for 'drift' that escaped the traditional objections. Of course plate motions have since been richly integrated into geological narratives, including much of the evidence Wegener first identified as well as immense amounts of data from subsequent studies of the histories of continents and ocean basins.

7 MODELS AND EVIDENCE

The need for each science to articulate and apply its own inferential structure brings us to a discussion of models. The importance of models as intermediaries between theories and the phenomena we apply them to has been a hot topic in recent philosophy of science. A number of authors, including Nancy Cartwright [1983; 1999], Margaret Morrison [1999] and Naomi Oreskes [2003] have argued that science requires some such intermediary—the logical notion of a theory, *i.e.*, a set of sentences closed under the consequence relation, can't carry the load in practice, both because it does nothing to indicate how the theory is to be applied to the world,[16] and because, even assuming that we know what actual phenomena we want to apply the theory to and how to connect observations of those phenomena to assertions in the language of the theory, providing a full description of the phenomena and determining the implications of that description according to the theory are, in general, beyond us. Simplified descriptions, approximations and selective inferences are inevitable elements in the actual account of the world that our theories inform. *Models* are supposed to embody these descriptions, approximations and inferences.

The notion of a *model* here is intuitively straightforward, though there is room for many subtleties. First, models in this sense are not models in the sense of semantic theory, because they generally involve approximations and simplifications that, strictly speaking, are incompatible with the truth of the theory. Instead, they are attempts to capture or express, usually approximately, some of the *implications* of a theory (or theories) for a particular system or type of system. Models in this sense include familiar models of molecules that use sticks and springs joining different coloured wooden balls to represent some aspects of molecular structure, sophisticated computer programs that attempt to capture long-term change in the earth's climate, attempts to calculate the running speed of a *T. rex* based on a model describing the mass, bone structure and muscle strength of the fearsome

[16]See [Brown, 2004].

predator, and Kepler's elliptical model of the orbit of Mars.

Many questions come up here. Just how seriously should we take these models? Are they just a pragmatic element in our scientific practice, serving to link abstract theory to concrete applications, or do they play a substantive role in the content of our scientific accounts of the phenomena we apply them to? How much variation is there in the purposes we use them for? Do different models play different roles in our representation of the world? What sort of evidence do we need to justify adopting a model for the different uses that we put models to? But our concern here is with the special features and challenges of models in ecology, and their relations to features and challenges of models in the historical sciences.

Ecological models are generally divided into three basic types: population models, community models and ecosystem models. Population models focus on capturing changes in the numbers of individual species over time; they vary in the number of factors that they include, whether the model is deterministic (suited only to large populations where chance fluctuations are small enough to be ignored) or stochastic, and whether the model includes any representation of the different properties of individual members of the population (including, for example, a range of values for fecundity, ages of individuals, and links between these and other factors including probability of death within some time period). Community models treat populations of more than one species, including interactions between them (for example, predator-prey relations). Finally, ecosystem models extend (very ambitiously) to the flows of energy and material that link communities to the surrounding, non-living environment; recently, computational models using geographical information systems that provide a representation of the spatial distribution of conditions in the environment have emerged as an important new class of ecological models [Sarkar, 2005].

Even at the level of population models, substantial difficulties arise. Very small differences in input (or *boundary*) conditions can have large effects on the model's results, as can small differences between models.

Four basic challenges that these models face are the challenge of data, the challenge of model complexity, the challenge of natural complexity and the challenge of openness. For the first, we can't expect to have precise and accurate data on an actual population's numbers, the resources the population depends on, the threats its members face or the distribution of each of these in a particular region. Consequently, assigning values to the parameters of our models involves substantial uncertainty. As to the second, the mathematical analysis of ecological models is extremely difficult. They are often exquisitely sensitive both to details of boundary conditions and to the precise structure of the models. Consequently, the uncertainties arising from the first challenge are, at least in general, important to our ability to rely on the models' predictions. Third, our models don't generally include parameters and interactions rich and detailed enough to provide a true picture of all the elements involved in the development of a population, community or, still less, an ecosystem over time. Fourth and finally, the systems we apply these models to are subject to perturbation by external causes, *i.e.*, causes not included in the model.

These challenges are mutually reinforcing: small differences of structure or input can have substantial impacts on the conclusions we would draw from a model, and increasing the detailed structure of the model to provide an intuitively more realistic account of the actual phenomena only adds to the mathematical complexity of the model and to the difficulty of assigning values to an increased number of parameters based on good empirical evidence while uncertainties about interference due to causes not included in a model weaken our ability to evaluate models, and the complexities entailed by attempts to include them limit how far we can go in modeling even known causal factors while threatening (on the other hand) to allow so much unconstrained flexibility as to render 'agreement with the data' an all-too-easy hoop to jump. The result, when these challenges are summed up, is a high level of uncertainty regarding the relation between the development of parameters over time in ecological models and the actual course of real populations, communities and ecosystems.

When we consider these challenges, pessimism about ecological models seems unavoidable. But this pessimism is only justified to the extent that these challenges make *success* for ecological models unlikely—and we can't settle that issue until we've sorted out just what we want these models to do. Any account of *success* for a model will have to begin with what we actually expect it to do. At the empirical and methodological levels, sometimes we expect models to predict future observations, but sometimes we aim at subtler ends. These might include questions about the models themselves, such as identifying constraints on the states the modeled system can achieve or demonstrating the need for further elements in a modified model if the model is to produce reasonable results, and questions about the theory the model draws on for its inferential structure. These kinds of information can serve as premises in important scientific arguments even without substantial predictive success.

One standard view of the logical status of models [Kyburg, 1983; Oreskes, 2003] treats them as contributing certain *conditional premises* to our account of some phenomena; an argument drawing on such a model is broadly a *modus ponendo ponens* inference, with the model telling us that if certain boundary conditions hold, then certain results will follow. This fits nicely with an inferentialist view of models: the key point is that a model allows us to infer from certain premises to certain conclusions. On this account, models are *inferential machinery.* But the uses we put these inferences to vary widely, and we normally draw a pragmatic distinction between inferences that are considered reliable and inferences that, while endorsed by the model, are not regarded as reliable.

8 THE STANDING OF ECOLOGICAL MODELS

Ecological models have come in for some pretty vigorous criticism. For reasons of space we'll focus here on one critique, due to Naomi Oreskes [2003]. Oreskes' main concerns about our attempts to model complex natural systems are straightforward: she argues that uncertainties in such models are inevitable, as they cannot capture the full complexity of the systems they model. Further to this, she argues that efforts to enrich our models and make them more realistic make testing them harder, as the resulting complexity allows increasing room to adjust the model and input parameters to 'fit' our observations, casting doubt on the value of successful predictions as confirmation of the model. Oreskes concludes that it is a mistake to expect models to provide deterministic predictions of outcomes in the natural systems they are meant to represent, although they can illuminate by serving as 'what if' scenarios, to illustrate possible best and worst-cases, and suggest possible outcomes of different sorts of interventions.

This conclusion is well-taken, though I think it's important to add that it reflects a tension between reasonable scientific aims and the practical, public-policy aims which predominate in the examples Oreskes treats. As we've noted, models are not always used to produce predictions, and not all predictions that models apparently give rise to are regarded as *significant* (*i.e.*, sufficiently reliable to guide practical reasoning).

One alternative use of models begins with a deliberately crude model to launch a process of refinement and correction leading to a modified model that we do regard as predictive in at least some respects (see the discussions of pendulum models in [Morgan and Morrison, 1999]). The inferential process in such cases begins with the crude initial model, but then reflects on the model's limitations, contrasted with a more detailed theoretical understanding of the actual processes involved, and introduces step-by-step modifications meant to improve on the initial model. The result may be a model that really is taken to be predictively reliable or even an approximately accurate representation of the real system (or just guidance for building a better clock). However, the inferential process that leads to that final model depends on the crude initial model as well as the subsequent critique and refinement. Further, even the final model may not be used predictively or regarded as a realistic representation—it may still be aimed at identifying constraints on the system modeled, or at serving as a test bed for still more refinement. It may also be used to explore aspects of the theory the model is based on, as in the paradoxes of mechanics that result from Laraudagoitia's Zeno-style puzzles (see [Earman and Norton, 1998]).

A simple but ecologically interesting example is the exponential growth model of a population with unlimited resources. Such models are predictively useless for established wild populations, though they can be predictively successful over a limited number of generations in specific circumstances, such as the introduction of yeast into a fresh barrel of wort. But for Darwin (and later Wallace) such models led to an obvious Malthusian conclusion: in the long run, most organisms cannot survive and reproduce successfully. Here the predictive failure of the model provides a key premise in a convincing argument for an important conclusion: populations of organisms often undergo substantial selective pressure. This important conclusion can be reached without a predictively successful model of any natural population; it requires only the failure of simple models of populations that include no selective pressure, in the light of the exponential reproductive potential shared by all organisms.

Our focus here, however, is on using models to capture certain implications of a theory for some actual phenomena. The idea is that the model should capture, at least approximately, some of the inferences that a good theory would license. This is separate from whether we believe the theory, and from whether we want the model to 'correspond' in some sense to the natural system responsible for the phenomena. It does suppose that there is a theory in the background here, which is not always the case except in a trivial sense of 'theory'. More generally, we may have only the model along with some rough ideas about important features of the system to be modeled—but even then we can then use the model to produce inferences about a particular system or collection of systems, and reflectively evaluate the model in the light of the resulting inferences and how they work out.

As an example of this, consider a purely phenomenological 'model', meant only to capture dynamic relations between certain observable parameters. In principle such models can be very successful. For example, a model of the stock market would be a brilliant (and extremely profitable) success if it were merely predictively successful, regardless of whether it captured any of the 'real' economic dynamics underlying the changes in stock prices it had predicted. (See Vonnegut, *The Sirens of Titan* for an amusing but silly example; more serious examples include technical models of the stock market based on observed cyclical patterns of market changes.)

Nevertheless, in many cases we aim to produce models that do represent, if imperfectly, the systems we apply them to, and sometimes we actually think we've succeeded—further, we can have reasons to suppose that we have succeeded at this aim even while successful prediction remains elusive. After all, as Yogi Berra famously noted, "[i]t's tough to make predictions, especially about the future." This is, in effect, the flip side of the concerns about complexity and sensitive dependence raised above, since they imply that (at least with respect to some features of the phenomena) a model of the phenomena can be extremely realistic and still fail to make successful predictions.

Kyburg [1983] argues that predictive success need not be essential to a model's success (*i.e.*, to fulfilling our intentions for the model): we may aim to explore the implications of certain constraints on a system even though we recognize those constraints may not in fact obtain, and that there are other constraints ensuring that the system will not, in fact, develop as 'predicted' by the model. From Kyburg's perspective the Club of Rome model, widely disparaged as an example of a failed model, is of considerably more interest than that characterization suggests: given that there are natural limits of the kind that the model proposed on certain resources, whether the actual limits assumed for the purposes of the model are accurate, the *kinds* of constraints that the model predicted on the sustainability of economic growth (though not their timing) remain significant. Further, despite the model's failure to consider political and social factors that would certainly become significant as resource shortages begin to affect the economy, the constraints it does include are worth exploring on their own.

In another striking example, climatological modeling does not focus on a single *simulation*, *i.e.*, a global circulation model (GCM) together with a set of initial conditions, but an *initial condition* ensemble, which involves a single GCM together with a wide range of initial conditions. Beyond this, climatologists also employ *multi-model* ensembles combining a number of initial condition ensembles based on different GCMs, which turn out to give a better match to climatological observations than initial condition ensembles do. A *climate* for such a model or model ensemble is defined as the features of a simulation that are constant across the ensemble—typically, averages of certain quantities and other statistical measures, along with relations between certain variables. Model *weather*, on the other hand, is those features of a simulation that differ across members of the ensemble. Given the chaotic behaviour of real weather and the obvious limits on data for setting initial conditions and the models as representations of a much more complex real system, model weather has next to no predictive value (weather prediction efforts are guided by more detailed local models). But model climates do have substantial predictive skill, as retrodictive testing shows.

Moreover, when physical details can be successfully added, producing a more constrained model that has improved skill on such measures, we may have reason to hold that the resulting sequence of models demonstrates the characteristics explored by Morgan and Morrison in the case of the pendulum: we are refining the model in the light of a physical understanding of the processes that are actually occurring, and thereby improving its reliability and applicability. This progression may never produce an entirely realistic model, but it can improve and extend the inferences we can make with the model's help.

It's also important to point out that the inferences we regard as supported by a model do not depend just on the 'if-thens' we can extract by using the model. When a real system is known to be predictively intractable in some respects, we may regard a particular model—for example, a fluid mechanics-based model of a bill's trajectory across a public square—as a realistic depiction of the *kinds* of causes at work in a phenomenon despite the failure of such a model to predict certain observations, such as the path of the bill.[17] There can be different kinds of predictions at stake here—fluid mechanics predicts the very unpredictability of

[17]See [Cartwright, 1999, 27f. Of course this makes me and those who agree with me here 'fundamentalists' in Cartwright's sense [Cartwright, 1999, pp. 24–28]. But is such fundamentalism as unreasonable as Cartwright maintains? My view is that, when a theory provides detailed predictive successes in contexts where, by its own lights, such successes are to be expected, and the natural parameters of some other cases where it *predicts* predictive failure still fall within the range where, when prediction is reasonably expected according to the theory, the theory has been shown to be successful, then it's reasonable, absent further contrary evidence, to hold that the theory offers an acceptable account of what's going on in the predictively intractable case. This point is closely related to another fascinating issue, *viz.* the distinction between the inductive skepticism characteristic of science and Humean skepticism. It's clear that even Cartwright rejects Humean skepticism, accepting as she does that many models provide very reliable predictions of the behaviour of certain kinds of systems. By Humean standards, such predictions are just as questionable as the 'fundamentalist' notion that classical fluid mechanics is a sound basis for understanding in outline, though not predicting in detail, the motion of a bill blowing across a square.

such trajectories, a prediction that, so far, is borne out. If classical mechanics made similar predictions of unpredictability about planetary orbits in our solar system, the mere success of positional planetary astronomy would be a convincing counterexample to classical mechanics. Similarly, any model of turbulence that managed to reliably predict such a bill's trajectory, even a successful phenomenological model, would count as powerful evidence against classical fluid mechanics.

Of course when a theory like fluid mechanics takes such a risk and the risk (so far) pans out, the result is (at least) weakly confirming for the theory. On the other hand, current models of turbulent flow also have shared features that can be tested against real turbulent flows.[18] Once again, claims of predictive failure need to distinguish predictions that have failed from those (perhaps subtler or more general) that have succeeded; they must also distinguish those 'predictions' (*i.e.*, claims inferred from a particular model) that are *robust*, *i.e.*, likely to hold if the model is indeed meeting our expectations, from those that are *frail*, *i.e.*, unlikely to hold in the system being modeled even if the model is as accurate and realistic as we can reasonably expect it to be.

A related point arises in a rarely considered argument for modest realism about our cognitive commitments to science. The argument begins with the combination of *confidence* that scientists often express regarding applying a hitherto successful theory or model under the kinds of conditions it has succeeded in and their *reluctance* to put faith in its success under other kinds of conditions. The empirical parameters involved in distinguishing these types of conditions, such as spatial and temporal scales, velocities, temperatures, intensity of gravitational fields, etc., are believed to be (and have indeed turned out to be) predictive of when our theories will and won't fail at various tasks. This both involves and, as I see it, justifies a modest realism about the *significance* of the quantities measured by these parameters (and the types of circumstances distinguished) to how various processes proceed. While scientists are often skeptical about the *truth* of currently successful scientific theories, which would entail their reliability on any scale at all, they are often confident about their applications within the range of parameter values where they have been successful, as well as about the criteria by which we distinguish those established applications from relevantly new applications. Obviously enough, a Humean skeptic would not share this confidence.

An example from ecology is worth examining here, to see how models can be used in non-predictive ways. In this case, the subject is the ecology of cane toads (*Bufo marinus*). In [Lampo and Leo, 1998], a model of the cane toad population is used to help determine what explains the extremely high population densities of cane toads in Australia, where they are an invasive species, compared to their population densities in native habitats. The model is a simple time-based matrix model, distinguishing juvenile from adult (reproductive) stages and including parameters for fecundity, for successful transition from the juvenile to adult stages

[18] Consider the evidence for Kolmogorov's energy spectrum function (see [Frisch, 1995]) and the evidence against the scale-independence of turbulence in the inertial regime that undermines Kolmogorov's account [Mathieu and Scott, 2000].

and for year-to-year survival of adults. This model is extremely schematic—it does not provide a continuous account of the population's size from day to day, distinguish any specific causes of death, include models of resources that cane toads depend on, or their interactions with each other or with other species. But it embodies some obvious constraints all the same: clearly, the reproductive capacity of a species that breeds seasonally turns, in part, on the adult (reproductively active) population at the beginning of the breeding season, and in part on the fecundity of those adults; equally clearly, the size of that population depends on recruitment of new adults and the survival of animals that are already adult. The authors remark, in their summary, "[t]he model presented here is by no means a predictive, but rather an analytical tool" [Lampo and Leo, 1998, p. 395]. Field data provide evidence constraining fecundity and both recruitment and survival. Analysis of the model shows that, at high population densities, equilibrium densities were much more sensitive to adult survival rates than to variations in the other parameters; field data also show that adult survival is indeed much higher in Australia than in South America. The authors conclude, "...there is a general consistency between predicted and observed patterns. Thus we believe our study brings important insights on factors driving the enormous reproductive success of Australian toads and on strategies to control their rate of increase" [Lampo and Leo, 1998, p. 395]. Here we see a clear example of a non-predictive but still cognitive use for ecological models, as well as a significant role for a background, practical aim. Further proposals regarding the usefulness of false models in biology have been made by Wimsatt [1987, p. 28], who suggests that false models "can (1) lead to the detection and estimation of other relevant variables, (2) help to answer questions about more realistic models, (3) lead us to consider other models as ways of asking new questions about models we already have, and (4) (in evolutionary or other historical contexts) determine the efficacy of forces that may not be present in the system under investigation but that may have had a role in producing the form that it has."

9 EPISTEMIC REMARKS

A simple hypothetico-deductive picture of the epistemic situation doesn't fit either the historical sciences or ecology. In these sciences we rarely have a well-characterized (let alone formally specified) model whose applicability to some phenomena is in doubt until a healthy range of predictions have been successful. Much more common is a situation in which we know that a number of processes play a role in the phenomena we wish to understand. We then try to build a useful model by representing those processes in more or less detailed ways. Such models are typically tested by comparing patterns of behaviour in the resulting model with various patterns in our observations.

Although detailed predictions of outcomes are rarely expected, a broader grasp of patterns that make sense in the light of our modeling efforts can still be attained. The upshot, when these efforts are successful, is a *retrospective* understanding of the patterns and some features of individual events and cases. The growth of these sciences over time provides an increasingly rich range of significant observable patterns along with increasingly refined understanding of the various processes that are responsible for them.

Of course as Hume argued, there is no general logical license for inferences from the truth of a conclusion to the truth of some set of premises it follows from; to choose a trivial example, when the conclusion is a theorem of the language, it follows from every premise set, but those premises certainly aren't confirmed by the truth of the theorem.[19]

Probabilistic accounts of how observations support hypotheses that we've inferred them from have considerable intuitive appeal, but Bayesian methods provide no help with the initial probabilities such accounts depend on; other theories of probability have made heavy going in their attempts to provide a basis for initial probability assignments. In the absence of a general account of how initial probabilities are arrived at, it is all too easy to explain the intuitive appeal of conditionalization as a simple reflection of our intuitions about what evidence (successful predictions in particular) tells for or against a particular hypothesis, rather than as providing a *justification* for those intuitions. Further, the status of consistency constraints only becomes more difficult on a probabilistic approach: the challenge of maintaining consistency in the set of sentences we endorse becomes far more demanding when we're faced with a demand for *coherence*, the probabilistic generalization of consistency. The calculational burden of maintaining coherence in a large set of commitments is very heavy indeed; this renders downgrading the status of coherence to an ideal rarely met except in very specific contexts even more inevitable than the parallel downgrading of consistency.

Equally well established is the point that it's rare for a hypothesis all on its own to entail something that we can test independently. We need to draw on other premises in addition to the hypothesis to arrive at conclusions we can compare against observation (or some other independent and credible source of information). This gives rise to the familiar Duhem-Quine problem. Making a convincing case against skepticism in this context is extremely challenging, though (as noted above) this problem is not any harder for the historical sciences than it is for the natural sciences. In fact, because the importance of certain basic processes invoked in the historical sciences (if not the particular forms they take in a given model) is taken to be settled, skepticism about hypotheses in the historical sciences can be extremely unattractive even when detailed predictions remain impossible.

Because the historical sciences typically appeal to rich narratives involving multiple, interacting processes to provide an explanation for some phenomena and each process involved in such a narrative is often well understood on independent grounds, many *features* of the resulting models are not hypotheses up for test, but components that, in some form or other, *must* belong to any credible model of the phenomena. What tends to be in doubt are questions including how well

[19]See [Norton, 2003] for an account of *local* induction, rejecting the idea that there is any general formal structure that distinguishes all and only good inductive arguments.

our model captures the relevant features, how to represent their interaction, and whether they constitute a *complete* collection of the basic processes a realistic (or useful) model needs to include.

However, even lacking detailed predictive success with regard to the outcomes of particular cases, retrospective refinement and the exploration of a range of different models can lead us to convincing explanations of some observed patterns. Specific predictions are often not the aim, no more than specific predictions of the course of evolution in populations of organisms are the aim of evolutionary biologists.

Success at constructing such explanations emerges from retrospective testing and refinement of models in the light of ongoing observations. Further input from observation (*e.g.*, evidence of new mutations in a population of bacteria and the processes that lead them to be selected for or against)[20] is often needed to link the models provided by our theory to the outcomes in actual populations. But *patterns* of change in many different populations, and the historical patterns that emerge from those changes over time (such as the patterns of resemblance and difference found in the structurally parallel hierarchical trees of organisms that emerge from taxonomy, embryology, biogeography and paleontology) can be explained with the help of such observations. Scale factors and the *Reynolds number* in fluid mechanics are another illustration—though we cannot, as Cartwright emphasizes, predict the trajectory of a bill being blown across a square by turbulent winds, we can predict when flows will become turbulent (at Reynolds numbers greater than 4,000) and how the scales of vortices and eddies relate to the distribution of kinetic energy amongst them (cf. Kolmogorov's statistical theory of turbulence). Why is success at predicting such general and statistical features of turbulent flow to be discounted?

Of course, things would be different if alternative models offered detailed predictions of the bill's path. But does anyone think this is a likely accomplishment? Even regarding it as credible requires strong skepticism about classical fluid mechanics despite its successes in so many applications, including its account of the general features of turbulent flow and the circumstances in which it occurs. Further, there are concrete practical applications associated with contemporary work on turbulence applying computational fluid dynamics to refine Kolmogorov's account [Moin and Kim, 1997].

Obviously, whether a model is successful depends on what we expect of it. Less obviously, we should not assume that models always aim at the same sorts of ends, even when they belong to the same science and paradigmatically successful models in that science achieve certain ends.

Even when actual observations don't fit such a model (as the famous Club of Rome model of resource depletion failed to fit actual economic developments) the model can be very revealing: that the processes involved in this model correctly captured part of what was going on is not in dispute. Its predictive failure shows that other processes (including the conservation and discovery of new sources of essential resources along with development of alternative resources) are also

[20] See [Blount, *et al.*, 2008].

taking place. Given the constraints we believe are in place (natural limits on resource availability) we may use the obvious failure in a *modus tollendo tollens* instead—and follow that step by modifying the models. As the late Henry Kyburg noted, the 'if-then' inferences embodied in the model are important constraints on the system, despite the falsehood of the antecedent: "The weakness of the data base, the obvious inadequacy of the world model as a mechanism for categorical predictions, the fact that the model takes account neither of political feasibility nor moral acceptability—none of these things keep the model from being informative, none of them suggest we need not take the model seriously, none of them provide an excuse for not getting on with the next step, which is that of trying to discover what alternatives are open to us, what courses of action can in fact be implemented, and by whom, and what ethical, political, and social constraints it is possible to impose on those alternatives without eliminating them all" [Kyburg, 1983, p. 11]. Kyburg concluded, "I suggest that in either case, evidence is evidence and we should attend to it. I suggest that the results of programming world models in computers are relevant to our decisions, even though they are not—and were never intended to be—categorical forecasts of the future".

10 CONCLUSION: ECOLOGY AS A HISTORICAL SCIENCE

To this point we've identified a number of differences between the historical and natural sciences and considered some examples drawn from ecology. On each of these points of difference I think ecology falls more naturally on the side of the historical sciences.

First, ecological explanations generally share in the contingency of historical explanations, turning on a wealth of details that are clearly contingent. Consider as an example here cases of invasive species, where facts including a suitable climate, food sources, lack of predators, diseases and other constraints normally faced by species in their home territory serve to explain the dramatic spread of some newly introduced species. *Bufo maritimus*, the cane toad, reaches densities as much as 100 times those typical of its native habitats in some areas of Queensland Australia; the lack of predators able to cope with its poisonous glands plays a substantial role in the high adult survival that (chiefly) explains these high population densities. Such explanations are, sadly, all too often retrospective—a lack of caution and scientific input is not the only reason why invasive species have been deliberately introduced in so many cases, with such unhappy results: it is far from easy to anticipate such disasters, and perhaps the best lesson we can learn from them is simple caution.

Second, ecological explanations share the narrative structure of explanations typical in the historical sciences. Different processes and contingent features of the circumstances come together to provide an ecological account of (for example) the successful migratory habits of water birds, or the demise of the Dodo. These accounts depend on detailed work, which continues to turn up interesting connections that illuminate the natural interactions—the natural history—that extend and refine our explanations.

Third, our understanding of basic ecological processes is not grounded in theoretically pure systems of principles. Instead, our initial grasp of these processes is largely the result of independent, straightforward observations: that organisms tend to reproduce at rates that lead them to outstrip the available resources; that they have certain needs (metabolic, social, climatological); that they often have predators and suffer from various diseases; that each predator and disease poses different levels of risks under different circumstances; that both resources and threats are distributed in space and time in complex ways, and so on. The role of models is to find useful ways to capture such facts and to gather them together in a single inferential tool. Further, the results of such efforts are often useful in ways other than generating predictions based on the model and some observationally-grounded 'input' parameters.

Fourth, the evaluation of ecological models and hypotheses turns less on fundamental principles and more on the set of processes, modeled in different ways, which are combined to explain features of the phenomena. Data limitations are answered, in large part, by retrospective evaluation, which can provide many separate back-tracking inferential paths that coherently support a narrative explanation.

Fifth, ecological models have (at best) very limited predictive power—they are too complex, the phenomena they model are still more complex, our ability to gather and adequately represent the relevant data are too limited, and the systems they involve are open to many different kinds of processes that appear (in our models) as purely exogenous forcings. Nevertheless, we can be justifiably confident that they do capture important features of ecological processes, and that certain particular explanations arrived at in ecological studies are well-founded, due to the combination of confidence about many of the basic processes involved and the rich cross-checking that retrospective investigations can provide.

Sixth, Sellars' distinction between ontological and inferential reductions makes clear how the explanations offered by historical sciences including ecology can be substantively independent of the detailed principles that underlie various ecological and historical processes. The inferences we make are, and must be, shaped by the demands of particular domains[21] even if we maintain, as ontological reductionists, that the items we are describing are ultimately based on the ontology of physics.

Seventh and last, the use of ecological models displays all the features typical of models in the historical sciences as well, including openness, problems of both model and natural complexity, and substantial limitations on the data we can gather, when compared with the richness and detail of the natural phenomena we are describing. Finding grounds for taking such models seriously turns not on predicting detailed outcomes, but on finding cases in which the models are able to capture various features of the phenomena in ways that are independently credible, because support for the different processes and their interactions are drawn from separate lines of evidence. The result is, as Kyburg emphasized, a matter of *constraints* on the phenomena, rather than detailed predictions—the

[21]See [Shapere, 1974].

resulting inferences can tell us not what will happen but instead such things as how some of the factors and processes involved interact, what would happen if other processes and factors did not intervene, what sorts of processes may be involved in one case or another and (in many cases) how we can draw on various kinds of evidence to reconstruct what happened in particular cases.

BIBLIOGRAPHY

[Albritton, 1980] C. Albritton. *The Abyss of Time* San Francisco: Freeman Cooper & Co., 1980.

[Aristotle, 350 BCE] Aristotle, *Posterior Analytics*, Translated by G. R. G. Mure. `http://classics.mit.edu/Aristotle/posterior.1.i.html`.

[Blount *et al.*, 2008] Z. D. Blount, C. Z. Borland, and R. E. Lenski. Historical contingency and the evolution of a key innovation in an experimental population of *Escherichia coli*. *Proceedings of the National Academy of Sciences USA* 105: 7899–7906, 2008.

[Bonevac, 1982] D. Bonevac. *Reduction in the Abstract Sciences*. Indianapolis, IN: Hackett, 1982.

[Brown, 2004] B. Brown. The Pragmatics of Empirical Adequacy. *Australasian Journal of Philosophy* 82: 242–263, 2004.

[Buffon, 1756] Buffon, *Histoire Naturelle, Générale et Particulière, avec la description du cabinet du roi*. Tome Sixième. `www.buffon.cnrs.fr`.

[Burchfield, 1975] J. D. Burchfield. *Lord Kelvin and the Age of the Earth*. London: MacMillan, 1975.

[Cartwright, 1999] N. Cartwright. *The Dappled World: A Study of the Boundaries of Science*. Cambridge: Cambridge University Press, 1999.

[Chamberlin, 1899] T. C. Chamberlin. Lord Kelvin's Address on the age of the earth as an abode fitted for life (Part Two). *Science* 10: 11–18, 1899.

[Drake, 1970] S. Drake. *Galileo Studies*. Ann Arbor: University of Michigan Press, 1970.

[Earman and Norton, 1998] J. Earman, and J. D. Norton. Comments on Laraudogoitia's "Classical Particle Dynamics, Indeterminism and a Supertask". *British Journal for the Philosophy of Science* 49: 123–133, 1998.

[Frisch, 1995] U. Frisch. *Turbulence: The Legacy of A. N. Kolmogorov*. Cambridge: Cambridge University Press, 1995.

[Hallam, 1983] A. Hallam. *Great Geological Controversies*. Oxford: Oxford University Press, 1983.

[Kyburg, 1983] H. Kyburg. Prophecies and Pretensions. In H. Kyburg, *Epistemology and Inference*, pp. 3–17. Minneapolis: University of Minnesota Press, 1983.

[Lampo and Leo, 1998] M. Lampo and G. A. de Leo. The invasion ecology of the toad, *Bufo marinus*: From South America to Australia. *Ecological Applications* 8(2): 388–396, 1998.

[Lyell, 1830] C. Lyell. *Principles of Geology*, Volume 1. London: John Murray, 1830.

[Mathieu and Scott, 2000] J. Mathieu and J. Scott. *An Introduction to Turbulent Flow*. Cambridge: Cambridge University Press, 2000.

[Moin and Kim, 1997] P. Moin and J. Kim. Tackling Turbulence with Supercomputers. *Scientific American* 276(1): 62–68, 1997.

[Morgan and Morrison, 1999] M. S. Morgan and M. Morrison. *Models as Mediators*. Cambridge: Cambridge University Press, 1999.

[Norton, 1993] J. D. Norton. The determination of theory by evidence: The case for quantum discontinuity, 1900–1915. *Synthese* 97(1): 1–31, 1993.

[Norton, 2003] J. D. Norton. A Material Theory of Induction. *Philosophy of Science* 70: 647–70, 2003.

[Oreskes, 2003] N. Oreskes. The Role of Quantitative Models in Science. In Charles Draper William Canham, Charles D. Canham, Jonathan Cole and William K. Lauenroth. *Models in Ecosystem Science*, Ch. 2, pp. 13–32. Princeton: Princeton University Press, 2003.

[Rudwidk, 2007] M. J. S. Rudwick. *Bursting the Limits of Time: The Reconstruction of Geohistory in the Age of Revolution*. Chicago: University of Chicago Press, 2007.

[Sarkar, 2005] S. Sarkar. "Ecology" Stanford Encyclopedia of Philosophy, `http://plato.stanford.edu/entries/ecology/`.

[Sellars, 1963] W. Sellars. Philosophy and the Scientific Image of Man in W. Sellars, *Science, Perception and Reality*, pp. 1–40. London: Routledge and Kegan Paul, 1963.

[Shapere, 1974] D. Shapere. Scientific theories and their domains. In Suppe, F., *The Structure of Scientific Theories*, pp. 518–565. Chicago: University of Illinois Press, 1974.

[Twain, 1883] M. Twain. *Life on the Mississippi.* Boston: James R. Osgood and Company, 1883.

[Wimsatt, 1987] W. C. Wimsatt. False Models as Means to Truer Theories. In Nitecki, Matthew H. and Hofman, Antoni (eds.), *Neutral Models in Biology*, pp. 23–55. Oxford: Oxford University Press, 1987.

Part 2

Philosophical Issues in Applied Ecology and Conservation Science

ENVIRONMENTAL ETHICS AND DECISION THEORY: FELLOW TRAVELLERS OR BITTER ENEMIES?

Mark Colyvan and Katie Steele

1 INTRODUCTION

On the face of it, ethics and decision theory give quite different advice about what is the best course of action in a given situation: one says to do what's right while the other says to maximise expected utility.[1] We could say that the perceived conflict is about doing what is "right, and for the right reasons" versus pursuing a strategy that is merely pragmatic/expedient/economically efficient.[2] In this paper, we examine this alleged conflict in the realm of environmental decision-making. There is a great deal of disagreement in the community when it comes to environmental issues and at least some of this disagreement appears to be a result of disagreement about the role of ethics in decision making. Looking carefully at a couple of controversial cases will help shed light on the nature of the roles of ethics and decision theory in environmental decision making, and help us to better understand the relationship between the two.

The two examples of environmental decision-making we will focus on are environmental triage and carbon trading. Environmental triage is so-named because it mirrors the kind of triage strategy that is familiar in medical contexts, where waiting times and even treatment is determined by seriousness of the illness and expectations of recovery. There is no sense, for example, in wasting valuable medical resources on a patient who is likely to die, irrespective of the treatment. In triage, in the conservation setting, the idea is that in the face of potential species extinction, say, when resources are limited, we should allocate resources so as to minimise the number of extinctions. That is, we may need to "give up" on some species because either those species do not have a high enough chance of recovery,

[1] See [Jeffrey, 1990; Resnik, 1987] for introductions to decision theory. See [Pojman and Pojman, 2008] for many of the classic readings in environmental ethics.

[2] The latter are also often thought to be rational. The conflict might thus be seen as an apparent conflict between norms: between what is ethically right and what is rational. Alternatively, it could be seen as the recasting of a familiar debate in ethics about whether right action is about the actions themselves (broadly deontological views) or about the outcomes of actions (broadly consequentialist views).

Handbook of the Philosophy of Science. Volume 11: Philosophy of Ecology.
Volume editors: Kevin deLaplante, Bryson Brown and Kent A. Peacock. General editors: Dov M. Gabbay, Paul Thagard and John Woods.

or the price for their recovery is too high. More precisely, we want to minimise the expected number of extinctions, and this may involve allowing some species to go extinct in order to save others.[3]

The other example we will discuss is carbon trading and/or offsetting. This is a way of controlling emissions of carbon dioxide. Companies/economic agents are allowed a certain quantity of carbon dioxide emissions; those companies that emit more than their quota are penalised—they must buy carbon credits from others who have a surplus, or else offset their extra carbon dioxide emissions via carbon sequestration projects. Companies that emit less than their quota are rewarded, because they may sell their credits to other companies. The idea is that, once we establish what the overall carbon dioxide emissions target should be, the most efficient way to achieve the target is to let the market determine who reduces their emissions and by how much. It is assumed that individual economic players will choose to engage in emissions–reductions programs to the extent that it is economically advantageous for them to do so.[4]

From these brief descriptions of these schemes, it may seem that environmental triage and carbon trading are entirely different environmental strategies and will raise entirely different issues. Certainly, the specifics of these policy instruments will be rather different, and different problems will arise in their implementation. But what they have in common is that they both enjoy some support, and yet also some fundamental opposition within the conservation community. More important, the reasons that both evoke strong negative reactions amongst some conservationists seem to be much the same. Or at least, we will argue that this is the case. Both triage and carbon trading amount to strategies for efficient, cost-effective environmental conservation. On the face of it, they seem to have a firm decision-theoretical basis, and yet might be thought to ride roughshod over some environmental ethical issues.

In Section 2 we outline the case in favour of both triage and carbon trading. In Section 3 we present and dismiss some commonly-heard, but nevertheless poor, arguments against these strategies. The subsequent sections of the paper are an attempt to construct a cogent argument against triage and carbon trading.

---

[3]Ecological triage was first proposed in relation to species preservation in [Walker, 1992]. This approach is further developed and defended in, for example, [Possingham, 2001; Field *et al.*, 2004; Wilson *et al.*, 2006; Marris, 2007; Colyvan, 2007; Colyvan *et al.*, forthcoming b]. Also see [Richards *et al.*, 1999] for an application of similar operations-research methods in a real conservation management application.

[4]We focus on carbon trading, but there are similar disincentive schemes for other sorts of environmental pollutants (see, for instance, [Kneese and Schultze, 1975]). Note also that there is an assortment of policy options for regulating carbon dioxide emissions. Carbon trading (with or without the option of gaining extra credit via carbon sequestration) is very prominent amongst these (see [Capoor and Ambrosi, 2007]). Discussions of carbon trading proposals can be found in, for instance, [Ackerman and Stewart, 1988; Grubb, 1990; Hahn and Hester, 1989; Pearce *et al.*, 1989, pp. 165–166]). A different approach is for governments to impose a tax on carbon dioxide emissions that would allow agents to emit as much of the pollutant as they can afford to pay for. See [Epstein and Gupta, 1990; Weimer, 1990] for details on such "green tax" proposals. Alternatively, governments might simply stipulate the pollution rights and duties of economic actors, with no trading or buy-out options for excessive polluting.

We conclude that there *are* good arguments to be advanced against triage and carbon trading. At least, there are good arguments against particular versions or implementations of these strategies in some situations. Whether triage and carbon trading are justifiable will depend on the details of the case at hand. This should not be seen as a conflict between decision theory and ethics but, rather, as an internal dispute about the appropriate decision-theoretic representation of the decision situations confronting environmental managers and policy makers.

2 THE CASE FOR TRIAGE AND CARBON TRADING

Both triage and carbon trading invoke a kind of instrumental rationality that seems beyond reproach. Take triage first. Here we have fixed resources and predetermined conservation goals—typically conserving as many endangered species as possible. All that triage amounts to is the optimal allocation of the resources in the pursuit of the goal in question. Why would we choose to spend our resources in any other way? Now consider carbon trading. Here there is a choice between achieving a particular and predetermined environmental target—restricting carbon emissions to below a certain target—by one or another means. The central insight of the carbon-trading strategy is to allow market forces to determine the most efficient means of achieving the target in question. This means that we do not incur greater costs than required. Why would we not go for this option? In each case there is a constraint—whether this is a fixed set of resources and/or a fixed target outcome—and we are advised to make the best decision that satisfies the constraint.

Of course, for the case of triage, it may be difficult to determine what is the best way to spend the limited resources in question. To begin with, there are various, often competing, conservation goals [Margules and Pressey, 2000; Possingham, 2001]; managers must decide whether the appropriate focus is the persistence of selected species, or else the representation of some other biological entity like terrestrial habitat types or reef types in a marine ecosystem, or else some combination of biodiversity indicators. Secondly, we are dealing here with complex ecological phenomena, and any probabilities that enter into the decision problem will be largely based on subjective expert judgment. One would expect that it would be very difficult for an ecologist to determine how likely it is that, say, some critically endangered species will recover, given some chosen management strategy (perhaps captive breeding, perhaps larger reserve systems, perhaps something else). The point is just that we must estimate, as well as we can, the probabilities that are relevant to our decision problems. To just choose an action (like trying to save all endangered species, starting from the most critically endangered), without trying to estimate the relevant probabilities of survival, amounts to an implicit assumption about the probabilities that may be way off the mark. It is, in effect, accepting whatever probabilities are required to make this the best course of action. So we cannot escape probability judgments in our conservation planning. Better to consciously determine what the relevant probabilities are than to ignore them and inadvertently accept implausible probability assignments. Environmental triage, then, just amounts to the principle of maximising expected utility. To give an example (and one that will recur in this paper): if utility is taken to be directly proportional to the number of persisting species, triage dictates that we choose the management option that has the greatest expected number of persisting species, where this calculation rests on our best-informed probabilities regarding the survival of the species of interest under the various management options.

As mentioned, carbon trading is a little different because the constraint here is the conservation goal; given a pre-specified emissions target, we want to meet that target in the most efficient way possible. In a sense, carbon trading is, from the start, a more substantial suggestion than triage. It does not just counsel us to choose the strategy that is most efficient for reducing emissions by a given amount, it also embraces the stronger claim that given any target for emissions, the most efficient way to achieve that target is to harness the efficiency of the market. We will take this point for granted in this paper—that it is, indeed, most efficient to use market instruments to reach an emissions target.[5] Of course, it will be difficult to settle on a target for carbon-dioxide emissions. This involves thinking about how important the climate issue is, relative to other human concerns—a very significant and difficult question, to say the least—and to determine what levels of carbon emissions correspond to various climate change scenarios. But to try to avoid these prickly issues and carry on with the status quo, or some other measure for reducing greenhouse emissions, is just to implicitly accept some arbitrary target. If we want to take action, as a society, on air pollution and climate change, then we need to articulate goals. And the argument for carbon trading is that once these goals have been articulated, we want to achieve them in the most cost effective and efficient way possible.

It is important to note that the cases outlined in this section for triage and carbon trading are in terms of the basic premises behind these schemes, rather than the specifics of their implementation. Of course, in practice, there will be many different ways of implementing either of these policies, and some of these will be better than others, depending on things like the quality of data collection and monitoring, and, for carbon trading, the legal framework for handling compliance.[6] So far we have been abstracting away from these issues, and have been focussing on the basic rationales for triage and carbon trading. Although we have depicted this basic rationale as beyond dispute, many do oppose triage and carbon trading at the most basic level, regardless of the particulars. One of our aims here is to try to shed light on why this is so. We begin in the next section by presenting what we regard as weak arguments against triage and carbon trading. There is some room for cogent criticism of triage and carbon trading, but such criticism turns on

[5]In any case, while there may be some reason to question this economic assumption (recall the alternatives to carbon trading mentioned in a previous footnote), this does not seem to be the source of the opposition to carbon trading that we have in mind, and which we will get to in the next section.

[6]See [Bekessy *et al.*, forthcoming] for discussion of some of the pitfalls of various implementation strategies for bio-trading.

at least some of the details about how the schemes are implemented. Some may be unwilling to engage in debate about triage or carbon trading if the most basic constraints involved—conservation resources available and emissions targets—are not satisfactory. We discuss when such a position would be defensible in Section 4.

3 SOME ARGUMENTS AGAINST TRIAGE AND CARBON TRADING

Some conservationists/environmentally-concerned citizens express a strong negative reaction towards triage and carbon trading. And this is before any of the particulars of the schemes have been tabled. The basic idea seems to be that it is wrong to think strategically when it comes to matters of such importance as the environment: when we are dealing with matters of extinction and persistence of species/ecosystems, some seem to think there can be no negotiating. Presumably, these opponents would not endorse giving up on good decisions when the stakes are high, and instead act in an aimless, *ad hoc* way. The claim must be that there are principled reasons why decision-theoretic reasoning breaks down in these serious, life-and-death-type cases. Perhaps biodiversity and environmental well-being are thought to be the kinds of goods that cannot be valued in the usual way; they are set apart from other human interests, and cannot be traded with or substituted by any other sorts of goods. In particular, they cannot be traded for material wealth. Or so the argument might go.

Proponents of this sort of argument might appeal to particular environmental ethical positions to support their views. To give an obvious example, they might identify as "deep ecologists" who claim that nature/biodiversity has value in and of itself, independent of any value that we humans might attribute to it (see [Naess, 1973]; for a critical survey of deep ecology, see [Sylvan, 1985]). This kind of value would, indeed, be difficult to account for in human-centred decisions. By its very nature it is a value that is not for humans to apportion and trade with other values. There are also more "shallow" environmental ethical positions that nonetheless recognise the natural environment as having a value that goes well beyond humanity's short-sighted material needs. Goodin [1992], for instance, describes a "green theory of value" that ultimately celebrates the otherness of natural processes for allowing humans to feel part of something larger than themselves. On this account, the natural environment stands apart from anything human-made by virtue of its very naturalness, and is thus, to some extent, irreplaceable.

Whatever the theoretical underpinnings, there are a couple of ways one might formalise the value of biodiversity/the natural environment so that this kind of good is set apart from other human interests. The first—an appeal to infinite value—cripples decision-making right from the start. We illustrate how this would go for the triage case (which in fact only involves environmental goods). The second—an appeal to incommensurate value—can lead to stalemates. Incommensurability is more relevant to the carbon-trading debate so we use this as our example. We argue that there are problems with invoking either of these two kinds of value. Indeed, to the extent that the infinite-value or incommensurability formalism represents any particular position in environmental ethics, such a position is shown to be problematic.

At least some opposition to triage seems to go as follows: all threatened species are extremely important and we should not give up on any; if there is some possibility that we can recover a species from near extinction, then we should try to do so, starting with the most needy/threatened case. This may well be the right strategy were there no limitations on resources. Perhaps some opponents of triage simply do not appreciate that, even in an ideal world in which everyone places considerable value on biodiversity, there will still be limits to the resources that can be committed to conservation. The bottom line is that there are always resource constraints and once this is appreciated, triage is the only rational way to proceed.

But now consider how infinite utilities might bear on this. Suppose that each species is so important (whether to humans, or intrinsically) that there is infinite value in it being extant. If every species has infinite value, then there would be no good reason to simply abandon the "hopeless" cases, because an action that had even the slightest chance of leading to the survival of the most threatened/needy would have infinite expected value. In such case, we could not rationally prioritise some courses of action over others—at least not by the means we have been discussing so far. Indeed, it might be argued that we must appeal to other ethical considerations in order to decide a course of action, and that these further considerations favour treating the most needy species first.

Assigning every species infinite value might amount to a principled reason for objecting to the kind of expected utility calculations that underpin triage, but this move introduces a host of problems, and is simply untenable. To begin with, we have no way of distinguishing between conservation outcomes. One recovered species has the same value as one hundred recovered species. And worse still, any action that has some chance, however small, of saving one species is as good as any other: hunting black rhinos is no better or worse than captive breeding or allocating reserves for the rhinos. With the introduction of infinite values, conservation decision-making is no longer able to discriminate between various conservation strategies and goals. Moreover, it is not clear what the moral rules are that might come to the rescue and tell us how to proceed. After all, why save the most endangered first? Why not the least endangered? The situation gets even worse. Not only does the introduction of infinite values cripple conservation decision-making, it also cripples decision making elsewhere: so long as there is some non-zero probability that a positive conservation outcome will eventuate, the action in question will have infinite expected utility.[7]

Perhaps the attitude that some have towards carbon permits and carbon off-

[7]See [Hájek, 2003; Sorensen, 1994] for more on the problems associated with infinite values, and [Colyvan, *et al.*, to appear; Goodin, 1996] for more on problems with assigning infinite values to environmental outcomes. Justus *et al.* [2009] discusses problems associated with entertaining intrinsic values in conservation management decisions.

setting is also best explained by appeal to the infinite value of an unchanged environment, or the infinite disutility of carbon dioxide emissions, such that no amount of cash or offsetting (even in the form of carbon sequestration projects) can make up for the initial damage. If so, this stance will have the same problems as just described.[8] In a similar vein, but without the problems posed by the appeal to infinite value, it might be argued that no specific monetary value (or range of monetary values), and even no specific amount of carbon sequestration, balances a given amount of carbon dioxide emissions, whatever the existing concentration of carbon dioxide in the atmosphere and level of social welfare. The idea would be that the two sorts of goods are completely incommensurable. Like the infinite-value case, incommensurability might be seen as explaining why it is impossible to make the sorts of decisions required for carbon offsetting.

Invoking incommensurability, however, does not amount to a good argument against carbon offsetting. For a start, invoking incommensurability is dangerous. It effectively makes certain kinds of decisions inconclusive. If apples and oranges are genuinely incommensurable then there is simply no common currency to trade between the two. An orange is neither more valuable, less valuable, nor the same value as an apple. One cannot compare the two and so decisions involving apples and oranges in the outcomes of different actions will be inconclusive. Although it is sometimes suggested that environmental value is incommensurable with other values (perhaps because the former is understood as an intrinsic value, or else because environmental goods cannot be replaced or substituted), this position needs qualification if it is to be taken seriously. If environmental values were completely incommensurable with other values, it is not clear how we could motivate even the most modest conservation efforts. Nature would be neither more valuable, less valuable nor the same value as a parking lot. Anyone who shares the view that at least some portions of nature are more valuable than some parking lots, denies that the two are entirely incommensurable. Indeed, such incommensurability is utterly implausible and runs counter to the whole business of conservation. If the natural environment is to be preserved it must be recognised that it is valuable and that we are willing to allocate resources (*e.g.*, money) to its preservation. This cannot be done if natural resources are thought to have incommensurable value, for such values cannot be compared with any others.[9]

[8] In any case, it is likely that the exchange rate between carbon dioxide emissions and derived social goods will vary, depending on existing levels of both carbon dioxide pollution and social welfare.

[9] We should perhaps distinguish two kinds of incommensurability here. The first is where the value of one item is measured on a scale orthogonal to the scale for the value of the other. In this case, not only will there be no way of comparing the values of the two items, there will be no way of comparing the value of any item *of the first kind* with any item *of the second kind*. This kind of incommensurability is like trying to compare temperature with length. This is what we are calling *complete incommensurability*. The other, *partial incommensurability*, is where the values of the two items are on the same scale but each may lack a precise value. If the values of items are represented by (perhaps overlapping) intervals on the same scale, the values will not be totally ordered. That is, some items will be neither of equal value, of greater value, nor of lesser value than some items. With partial incommensurability, some comparisons can be made

Back to incommensurability and carbon trading. First we need to be careful not to confuse incommensurability with epistemic difficulties associated with determining the right substitution between emissions and public money/carbon sequestration. Despite our ignorance of what the right substitution between emissions and sequestration is, for instance, we can still settle on something, depending on how vigilant or risk-averse we want to be about carbon dioxide pollution.[10] In any case, it is plausible that the problems are not entirely epistemic; there is likely to be some degree of incommensurability between existing environmental well-being and restoration projects (*e.g.*, sequestration) or other social goods. The point is just that these values are not *entirely* incommensurable, because that would make any decision that involved conflicts between them inconclusive. At any given time/state of the world, there may be a number of exchange rates between carbon dioxide emissions and carbon sequestration/social goods that cannot, in principle, be decided between. So long as any such incommensurability is only limited, however, we will still be able to make conclusive decisions in a large number of cases. Indeed, some have proposed comprehensive theories of rational choice for conditions of partial incommensurability or indeterminacy (see, in particular, [Levi, 1986]). Moreover, when it comes to legislation that requires precise exchange rates/permit prices, we can simply settle on something, as in the case of epistemic uncertainty, depending on how risk-averse we want to be about carbon dioxide pollution.[11]

While some may concede that partial incommensurability/uncertainty with respect to the relative standing of environmental and social values should not obstruct rational decision-making, they may nonetheless resist the idea of paying to pollute. Goodin [1994] offers a defence of this view that involves comparing carbon trading with the much-criticised practice of "selling indulgences" within the medieval church. But as Goodin himself points out, it all depends on how carbon trading/offsetting schemes are perceived. The problematic interpretation is to regard a carbon permit as a payment to society that completely absolves any harm done to the environment and/or society, such that one may act with a clean conscience. This is a dangerous way of looking at things because, in practice, it is likely that the payment for carbon dioxide pollution will not, at least in the early stages of such a scheme, be as demanding as it should, and will only go some way towards compensating for environmental damage. But even if the payments were very stringent, there would still be cause for moral regret if one pursued a particular course of action when, all other things being equal, there were other more environmentally benign options available. It might be argued that the market takes care of this problem—provided all externalities are accounted for, markets achieve the most efficient or optimal outcomes. But even if this is true in the

---

but there will always be some decisions that will be inconclusive.

[10]There are various methods available for representing different kinds of uncertainty in environmental and other decision problems, and not all of these methods are probabilistic [Regan *et al.*, 2002; Burgman, 2005].

[11]See [Steele, 2006] for a discussion of the Precautionary Principle and the issue of uncertainty in environmental decision-making.

"ideal" situation, where fully rational agents pursuing self-interest alone act under conditions of perfect competition, it is a long stretch to claim that it is generally true in practice.[12]

As Goodin acknowledges, there is a less morally loaded way to perceive carbon trading/offsetting schemes and it is this interpretation we have been emphasising. The idea is that carbon trading/offsetting is just a good economic instrument for achieving a pre-determined carbon emissions target. The choice of target is not something that is determined posthoc by the market, but is rather a political decision that ideally represents the values and goals of the community at large. Individual agents who abide by the regulatory system can regard themselves as acting fairly and in the interests of the community; whether or not they are morally "clean" when it comes to the environment is a much more complex issue.

It should be apparent from our discussion thus far that there is ample scope for community values to enter into any triage or carbon trading proposal. Those who are anxious to incorporate environmental and other non-monetary social goods into the decision making need not resort to assigning infinite value to these goods, or to overstating the case for incommensurability. We need not throw out our best decision-making tools just because they are, in some instances, badly used.[13] In the case of triage, there is a significant value judgment in deciding how much of the community's shared resources should be directly devoted to protecting biodiversity. More fine-tuned value judgments then enter into the choice of measures for biodiversity and thus the kind of utility that we seek to maximise.[14] Such judgements turn on questions in environmental ethics. (In practice, however, biodiversity estimates will be somewhat crude given the constraints of data collection.) Value judgments, whether explicit or implicit, figure no less in carbon trading proposals. As mentioned, carbon trading requires the articulation of community goals for emissions reductions. Beyond this significant value issue, there are a host of other choices to be made regarding fairness and equality. For instance, the community needs to decide how carbon-emission permits should be distributed in the first instance, and also whether there should be periodic redistribution of permits (such that the right to pollute can only ever be leased temporarily).[15] Indeed, rather than being anathema to value considerations, decision modelling, in the

[12]See [Hausman and McPherson, 1996, pp. 43–44] for a discussion of this perception of the market. Goodin [1994] resists the idea that optimal emissions levels can be determined by the market once a suitable per unit price is set, on the grounds that there will always be too much (in principle) uncertainty about what is the right price for pollution.

[13]Of course there are many technical difficulties encountered in assessing the values and probabilities in question, especially when it is appreciated that it is the expected value of society as a whole that we seek to maximise.

[14]See [Maclaurin and Sterelny, 2008; Sarkar, 2002] on the merits of different theoretical definitions of "biodiversity". Regan *et al.* [2007] documents the various components of biodiversity or environmental well-being deemed important by a group of ecologists and other stakeholders.

[15]The issues become even more complex when we consider how much the wealthy, carbon-hungry countries as a whole, rather than individual companies, should compensate developing countries. Grubb [1990] discusses some of the justice issues that arise in the distribution of emissions permits.

form of social welfare functions, has proven invaluable in addressing these kinds of distributive justice issues (see, for example, [Sen, 1979; 1997; 1999; Hausman and McPherson, 1996]).

Finally, it might be argued that the whole decision-theoretic approach is politically dangerous in environmental contexts, in that it involves value judgements and (estimates of) probabilities. Each of these, the argument continues, is difficult to determine and open to revision. So, it would seem that an opponent of some environmental endeavour can derail proceedings, rather easily, by casting doubt on the value and probability assignments in question. A climate-change sceptic, for instance, might stall action on the reduction of greenhouse-gas emissions by emphasising the extent of the uncertainty in all parts of the relevant science, with the aim of casting doubt on the probability assignments used in the decision to reduce green-house gasses. According to this line of thought, the decision theory approach might well be right, in some sense, but it is easily subverted and is thus not an appropriate tool for conservation management.

The first thing to say in response is that scepticism cuts both ways: sometimes environmentally-unpalatable decisions can be undermined by questioning the science involved. For instance, an environmentalist might cast doubt on the thoroughness or impartiality of an environmental impact statement that cleared the way for industrial use of a piece of natural environment. Decision theory does not stack things against the environment; it can equally well be used to stack things in *favour* of the environment. The second point in response is that, just because decision theory is open to manipulation in these ways, does not mean it should be abandoned. After all, if we are talking about unsupported scepticism, then the science will help settle matters (as, indeed, it largely has in the climate change debate). And just because there is doubt about the values and (perhaps subjective) probabilities, does not mean that anything goes. If there is uncertainty, it should be acknowledged and treated accordingly. Even in cases where there is genuine, unresolvable uncertainty (such as model uncertainty—uncertainty about the details of the models used to derive the predictions and probabilities), sensitivity analysis will help to show how robust or volatile the decisions in question are.[16] A method that is explicit about uncertainty and provides the means to deal with it strikes us as less open to political manipulation than alternatives where uncertainty is ignored or not treated in an appropriate fashion.

4 A DECISION-THEORETIC CASE AGAINST TRIAGE AND CARBON TRADING

In this section we outline a more substantial argument against triage and carbon trading/offsetting. It is not an argument against these schemes outright. Our discussion so far should have made it clear that in our view a blanket dismissal

[16]Sensitivity analysis is a method for testing whether plausible changes to the scientific model will lead to different decisions. See [Regan *et al.*, 2002; Burgman, 2005] for more on the treatment of the various kinds of uncertainty and meta-uncertainty in ecology and conservation settings.

of these schemes is untenable—when stripped to their core, triage and carbon trading/offsetting are simply instances of a very uncontroversial kind of practical rationality. But one still might have concerns about a particular triage or a particular carbon-trading proposal. There are several related reasons for unhappiness about such schemes and they all revolve around the optimality of the long-term payoffs of such schemes.

Take triage first. Recall, that here the strategy is to treat the available resources as fixed, and then optimise the expected recoveries of species (to continue with our example conservation goal). Note that, in effect, such a decision is treated as a one-off decision. But, presumably, there will be another allocation of resources next year (or whenever). According to the standard triage strategy, the optimisation is performed every time there is a new allocation of resources. Each decision is treated in isolation, and yet they are a part of a series of decisions, the timing of which may well be highly predictable, depending on administrative processes. Local optimisation at each stage of a sequential decision process does not always result in the overall optimal outcome. One way to see this is to note that conservation budgets are typically not fixed from year to year. Surely, in ensuring optimal long-term conservation outcomes, one of the agenda items should be the securing of adequate resources for the conservation efforts required. Blindly accepting an inadequate budget, treating it as fixed, and then optimising outcomes based on this, may be the best you can do in any given year, but may well jeopardise future conservation efforts. It might, for instance, be in the best interests of conservation to refuse an inadequate budget and hold out for more. It all depends on how other parties are predicted to respond to pressure from conservationists. The problem thus becomes game theoretic rather than decision theoretic.[17] To put the point in a slightly different way, the triage strategy is based on an optimisation model that presupposes that the budget is fixed and that the decision is one-off. In the face of iterated conservation decisions and variable budgets, the triage strategy at the very least needs refining. It seems that the standard triage strategy concedes too much to funding agencies in accepting whatever is allocated and making do with that.[18] In short it is not always optimal in the long term to make the best of a bad lot; sometimes it is better to reject the bad lot or refuse to cooperate until things are improved.[19]

There is also the issue of the reallocation of resources. Triage assumes that reallocation is possible and cost free. Suppose, for example, that triage recommends withholding resources initially intended for the preservation of a particular

[17]Game theory is the branch of rational choice theory that deals with bargaining situations. See [Osborne, 2004; Resnik, 1987] for an introduction to game theory. Skyrms [2004] explores how iterated games can shed light on the development of social contracts.

[18]Of course a great deal of effort does go into negotiations over resource allocation, but this is quite separate from the optimisation performed in triage. The point being made here is that these two aspects of conservation management should not be disconnected.

[19]The analogy with workers strikes seems apt here. What is optimal in the long term for workers is sometimes to refuse to work for unfair wages, despite needing the money in the short term. There are fairly standard game-theoretic, bargaining treatments of such cases.

species and instead recommends redirecting those resources elsewhere. But often the original resources are provided by a particular funding agency, in a particular country, and for a particular purpose. It may not be possible to reallocate those resources to another purpose in another country. And even when such reallocations are possible, they may result in significant costs. This and other such idealisation of the triage model might well give us reason to reject that model in favour of a more sophisticated one, where resources are not fixed, and there are non-trivial reallocation costs. But relaxing such assumptions does not amount to a rejection of the decision-theoretic approach, for, as Hugh Possingham [2007] points out, all these issues are amenable to decision-theoretic (or in some cases game-theoretic) treatment. Indeed, it is hard to see any other way to approach issues involving tradeoffs.

Now reconsider carbon trading. Here, one might have concerns about the emissions target in a particular carbon-trading scheme. After all, there is no mechanism for the market to lower the target; the market merely optimises meeting the target. There is room for disagreement about the target that has been set, and it seems perfectly reasonable to push for the lowering of targets over subsequent years. Depending on the social and political environment at the time, it may well be strategic for the conservationist to show strong opposition to the basic proposal, and not participate in any further discussions of the schemes until the issue of adequate targets are dealt with in a satisfactory manner. Again, this can be seen as a case of attempting to achieve a better global result (that is, a better conservation outcome in the long term).[20]

Some have also argued that in the long run, trading schemes for carbon and other environmental pollutants may have a negative effect on basic attitudes towards the environment. The claim is that monetary rewards for good action, can, under particular conditions, undermine people's motivation to perform the action for its own sake (see [Kelman, 1981; Frey, 1986; 1993]).[21] In the case of carbon trading, the idea is that certain kinds of incentives for emissions reductions may spoil the potential for firms to develop a more mature sense of corporate responsibility that would lead them to reduce their emissions voluntarily. The main fear is that a weak sense of responsibility as regards carbon dioxide emissions would "spill over" into other domains where there is not the possibility of instituting payments for environmental damage. In other words, the concern is that the perceived worth of conservation efforts in all areas, not just with regard to air pollution, would lesson with time. So even though some varieties of carbon-trading scheme may produce better conservation results in the short term, they may not, on balance, be optimal in the long term, if general attitudes towards the environment become more lax and this leads to significant environmental degradation that would not

[20]There are other details of particular triage and carbon-trading schemes that might also give rise to opposition. The methods of policing compliance in carbon trading, for example, might at first blush seem like a mere detail, but an opponent might reject the whole scheme until such details have been provided and shown to be appropriate.

[21]Goodin [1994] also makes this point to support his argument concerning "selling environmental indulgences".

have otherwise occurred.

5 CONCLUDING REMARKS

What the objections in the previous section have in common is that they focus attention on the apparent short-sighted focus of triage and carbon trading—at least as these strategies are standardly presented. What is required is more long-range or sustainable thinking with regard to conservation strategies in the broader political setting. Moreover, such long-range thinking may also recommend considerable efforts to change people's attitudes towards the environment. This, in turn, might involve: spending resources on high profile species that are not always best suited (ecologically) for saving, or encouraging companies to undershoot carbon-emission targets, not so they can profit by selling the offsets, but in order to develop more robust and global environmental sensibilities. The value of education and a genuine concern for the environment are what seem to be missing from (or are at worst undermined by) the triage and carbon-trading strategies. We are thus led back to the apparent conflict between decision theory and ethics.

This apparent conflict, though, is *merely* apparent. All of these issues—the role of education, the potential benefits of attempting to save a high-profile species, the value of genuine green companies, or the potential gains from holding out for more conservation resources or more stringent pollution targets—can, and should, be incorporated into the decision-theoretic approach. These issues just amount to additional options or future choices and accompanying social interactions that must be incorporated when we are considering possible conservation strategies and their long-range consequences. It is likely that many disputes that look to be about conflicting core values will turn out to depend upon scientific issues—how to appropriately model the consequences of the various management options, and what are the best predictions about future social behaviour under the different scenarios.[22]

It might seem that we are nonetheless sidelining ethical considerations by forcing them into the decision-theory framework, but this is simply a misrepresentation of our project. We discussed earlier how ethics may play a role in determining the appropriate utility functions to use in particular management decisions, or to help settle what the goals of conservation efforts should be—maximising biodiversity, preserving our favourite species, or something else. What environmental ethics cannot do, however, is determine the right course of action on its own.[23] For the latter involves trade offs, uncertainty and optimisations, and these all require decision theory.

[22]See [Baron, 2006] for a similar deflationary account of bioethics.

[23]For a start, ethical theories typically do not give advice about what to do in the face of uncertainty [Colyvan *et al.*, forthcoming a].

ACKNOWLEDGEMENTS

The work on this paper was supported by project funding from the Australian Government's Commonwealth Environment Research Facilities Research Hub: Applied Environmental Decision Analysis and by the Australian Centre of Excellence for Risk Analysis and by an Australian Research Council Discovery Grant (grant number DP0879681).

BIBLIOGRAPHY

[Ackerman and Stewart, 1988] B. A. Ackerman and R. B. Stewart. Reforming Environmental Law: The Democratic Case for Market Incentives. *Colombia Journal of Environmental Law* 13: 171–199, 1988.

[Baron, 2006] J. Baron. *Against Bioethics*. Cambridge MA: MIT Press, 2006.

[Bekessy *et al.*, forthcoming] S. A. Bekessy, B. A. Wintle, M. Colyvan, B. Langford, D. B. Lindenmayer, and H. P. Possingham. The Biodiversity Bank Cannot be a Lending Bank. *Ecological Economics*, forthcoming.

[Burgman, 2005] M. A. Burgman. *Risks and Decisions for Conservation and Environmental Management*. Cambridge: Cambridge University Press, 2005.

[Capoor and Ambrosi, 2007] K. Capoor and P. Ambrosi. State and Trends of the Carbon Market 2007. Retrieved December, 2007, from `http://carbonfinance.org/docs/Carbon_Trends_2007-_FINAL_-_May_2.pdf`.

[Colyvan, 2007] M. Colyvan. Environmental Philosophy: Beyond Environmental Ethics. *Arts* 29: 95–104, 2007.

[Colyvan *et al.*, forthcoming a] M. Colyvan, D. Cox, and K. Steele. Modelling the Moral Dimension of Decisions. *Noûs*, forthcoming.

[Colyvan *et al.*, to appear] M. Colyvan, J. Justus, and H. M. Regan. The Natural Environment is Valuable but Not Infinitely Valuable, to appear.

[Colyvan *et al.*, forthcoming b] M. Colyvan, S. Linquist, W. Grey, P. E. Griffiths, J. Odenbaugh, and H. P. Possingham. Philosophical Issues in Ecology: Recent Trends and Future Directions. *Ecology and Society*, forthcoming.

[Epstein and Gupta, 1990] J. M. Epstein and R. Gupta. *Controlling the Greenhouse Effect: Five Global Regimes Compared*. Washington DC: Brookings Institution, 1990.

[Field *et al.*, 2004] S. A. Field, A. J. Tyre, N. Jonzen, J. R. Rhodes, and H. P. Possingham. Minimizing the Cost of Environmental Management Decisions by Optimizing Statistical Thresholds. *Ecological Letters* 7: 669–675, 2004.

[Frey, 1986] B. S. Frey. Economists Favour the Price System—Who Else Does? *Kyklos* 39: 537–563, 1986.

[Fey, 1993] B. S. Frey. Motivation as a Limit to Pricing. *Journal of Economic Psychology* 14: 635–664, 1993.

[Goodin, 1992] R. E. Goodin. *Green Political Theory*. Cambridge: Polity Press, 1992.

[Goodin, 1994] R. E. Goodin. Selling Environmental Indulgences. *Kyklos* 47: 573–596, 1994.

[Goodin, 1996] R. E. Goodin. Equity and the Environment, Environmental Economics Seminar Series, Department of the Environment, Sport and Territories, `http://www.environment.gov.au/about/publications/economics/equity/discuss2.html`.

[Grubb, 1990] M. Grubb. The Greenhouse Effect: Negotiating Targets. *International Affairs* 66(1): 67–89, 1990.

[Hahn and Hester, 1989] R. W. Hahn and G. L. Hester. Marketable Permits: Lessons for Theory & Practice. *Ecological Law Quarterly* 16: 361–406, 1989.

[Hájek, 2003] A. Hájek. Waging War on Pascal's Wager. *Philosophical Review* 112(1): 27–56, 2003.

[Hausman and McPherson, 1996] D. M. Hausman and M. S. McPherson. *Economic Analysis and Moral Philosophy*. Cambridge & New York: Cambridge University Press, 1996.

[Jeffrey, 1990] R. C. Jeffrey. *The Logic of Decision*, 2nd Edition. Chicago: University of Chicago Press, 1990.

[Justus *et al.*, 2009] J. Justus, M. Colyvan, H. M. Regan, and L. A. Maguire. Buying Into Conservation: Intrinsic Versus Instrumental Value. *Trends in Ecology and Evolution* 24(4): 187–191, 2009.

[Kelman, 1981] S. Kelman. *What Price Incentives? Economists and the Environment.* Boston: Auburn House, 1981.

[Kneese and Schultze, 1975] A. V. Kneese and C. L. Schultze. *Pollution, Prices and Public Policy* Washington DC: Brookings Institution, 1975.

[Levi, 1986] I. Levi. *Hard Choices: Decision Making Under Unresolved Conflict* Cambridge & New York: Cambridge University Press, 1986.

[Maclaurin and Sterelny, 2008] J. Maclaurin and K. Sterelny. *What is Biodiversity?* Chicago: University of Chicago Press, 2008.

[Margules and Pressey, 2000] C. Margules and R. Pressey. Systematic Conservation Planning, *Nature* 405: 243–253, 2000.

[Marris, 2007] E. Marris. What to Let Go. *Nature* 450: 152–155, 2007.

[Naess, 1973] A. Naess. The Shallow and the Deep, Long-Range Ecology Movement. *Inquiry* 16: 95–100, 1973.

[Osborne, 2004] M. J. Osborne. *An Introduction to Game Theory.* Oxford: Oxford University Press, 2004.

[Pearce *et al.*, 1989] D. Pearce, A. Markandya, and E. B. Barbie. *A Blueprint for a Green Economy: A Report to the UK Department of Environment.* London: Earthscan, 1989.

[Pojman and Pojman, 2008] L. P. Pojman and P. Pojman. *Environmental Ethics: Readings in Theory and Application.* Belmont, CA: Wadsworth, 2008.

[Possingham, 2001] H. P. Possingham. The Business of Biodiversity: Applying Decision Theory Principles to Nature Conservation. *Tela* 9: 1–37, 2001.

[Possingham, 2007] H. P. Possingham. Triage—Some Rational Counter Arguments. *AEDA News* 14 (30 Nov 2007): 1–2, 2007. `http://www.aeda.edu.au//docs/Newsletters/AEDA_News_14.pdf`.

[Regan *et al.*, 2002] H. M. Regan, M. Colyvan, and M. A. Burgman (2002) 'A Taxonomy and Treatment of Uncertainty for Ecology and Conservation Biology'. *Ecological Applications* 12[2): 618–628.

[Regan *et al.*, 2007] H. M. Regan, F. W. Davis, S. J. Andelman, A. Widyanata, and M. Freese. Comprehensive Criteria for Biodiversity Evaluation in Conservation Planning. *Biodiversity Conservation* 16: 2715–2728, 2007.

[Resnik, 1987] M. D. Resnik. *Choices: An Introduction to Decision Theory.* Minneapolis: University of Minnesota Press, 1987.

[Richards *et al.*, 1999] S. A. Richards, H. P. Possingham and J. Tizard. Optimal Fire Management for Maintaining Community Diversity. *Ecological Applications* 9(3): 880–892, 1999.

[Sarkar, 2002] S. Sarkar. Defining "Biodiversity"; Assessing Biodiversity. *The Monist* 85(1): 131–155, 2002.

[Sen, 1979] A. K. Sen. *Collective Choice and Social Welfare.* Amsterdam & New York: North-Holland, 1979.

[Sen, 1997] A. K. Sen. *On Economic Inequality.* Oxford: Clarendon Press, 1997.

[Sen, 1999] A. K. Sen. *Development as Freedom.* New York: Anchor Books, 1999.

[Skyrms, 2004] B. Skyrms. *The Stag Hunt and the Evolution of the Social Contract.* Cambridge: Cambridge University Press, 2004.

[Spremsem, 1994] R. A. Sorensen. Infinite Decision Theory. In J. Jordan (ed.), *Gambling on God: Essays on Pascal's Wager*, pp. 139–159. Savage, MD: Rowman & Littlefield, 1994.

[Steele, 2006] K. Steele. The Precautionary Principle: A New Approach to Public Decision-making? *Law, Probability and Risk* 5(1): 19–31, 2006.

[Sylvan, 1985] R. Sylvan. A Critique of Deep Ecology. *Radical Philosophy* 40; 41: 2–12; 10–22, 1985.

[Walker, 1992] B. H. Walker. Biodiversity and Ecological Redundancy. *Conservation Biology* 6(1): 18–23, 1992.

[Weimer, 1990] D. L. Weimer. An Earmarked Fossil Fuels Tax to Save the Rain Forests. *Journal of Policy Analysis and Management* 9: 254–259, 1990.

[Wilson *et al.*, 2006] K. A. Wilson, M. McBride, M. Bode, and H. P. Possingham. Prioritising Global Conservation Efforts. *Nature* 440: 337–340, 2006.

POSTMODERN ECOLOGICAL RESTORATION: CHOOSING APPROPRIATE TEMPORAL AND SPATIAL SCALES

J. Baird Callicott

INTRODUCTION: CLASSIC ECOLOGICAL RESTORATION

This chapter is not about the art of ecological restoration. Nor is it about restoration ecology, the science that informs the art. Rather, this chapter is about the *philosophy* of ecological restoration. The philosophy of ecological restoration is examined in a fairly long historical perspective, ranging from the first quarter of the twentieth century to the present. It concerns, primarily, the aim of ecological restoration, as it was originally conceived in the 1930s, and the ecological worldview in which that conception of ecological restoration was embedded and which, at that time, was assumed to be true. That's what I mean by *classic* ecological restoration, a legacy inherited by contemporary practitioners of the art, upon which critical philosophical reflection might be illuminating. In this chapter, more particularly, I argue that the classic target of ecological restoration—the classic "reference system"—has become problematic after a profound paradigm shift in ecology was consolidated during the mid-1970s. That then raises the question: So, what past ecological state or condition should be the target of restoration efforts? I consider and critically assess several alternatives, prominent among them the suggestion that ecological restoration in the Western Hemisphere should aim to re-establish biotic communities that existed in the hemisphere before the arrival of *Homo sapiens* as a keystone species at the Pleistocene-Holocene boundary some thirteen thousand years ago.

Philosophers often employ a device called the "thought experiment." Philosophers' thought experiments often range from the fabulous to the absurd. What, for example, would it be like if two psyches swapped bodies?—such stuff as that. Few boots-on-the-ground ecological restorationists have proposed restoring extant surrogates—such as camels, cheetahs, and elephants—of the extinct Pleistocene megafauna of the Western Hemisphere to the Western Hemisphere. Such a restoration project would appear at a minimum quixotic to work-a-day restorationists, if not altogether preposterous. But this time this philosopher does not have to make such a proposal up as a "thought experiment." Restoring surrogates of the extinct Pleistocene megafauna to the Americas has been seriously proposed by

Handbook of the Philosophy of Science. Volume 11: Philosophy of Ecology.
Volume editors: Kevin deLaplante, Bryson Brown and Kent A. Peacock. General editors: Dov M. Gabbay, Paul Thagard and John Woods.

credentialed professional scientists and, moreover, taken seriously by such unimpeachable sources of scientific authority as *Nature* and *Science News* and considered newsworthy by the *New York Times*. Apart from matters of affordability and both ecological and political feasibility, I try to explain here why—philosophically why; not ecologically why, not financially why, not politically why—real-world restorationists are likely to regard Pleistocene parks with horror.

The norm or target for ecological restoration seems straightforward and obvious. A given site has been manhandled by the saw, plow, cow or by some other instrument(s) of anthropogenic transformation. It has now been abandoned or retired and, by good fortune or foresight, it has become a locus for ecological restoration. To what ecological condition should it be restored? Its "original" condition, of course. And how do we know what its "original" condition was? The condition in which it was found at "settlement." This is, after all, exactly what one of the first and arguably the most famous of ecological-restoration projects was all about—the University of Wisconsin Arboretum and Wild Life Refuge. And none other than Aldo Leopold was the mastermind who conceived its purpose. Leopold [1999a] gave a brief speech at the dedication ceremony of the UW arboretum in which he outlined the project and provided a rationale for it. Leopold's statement on that occasion is the first clear articulation of the concept of ecological restoration. Curt Meine [1988, p. 328] sets the scene and quotes the key passage in Leopold's speech:

> On the morning of June 17, 1934, civic leaders and university officials gathered in a barn on the south edge of Madison and officially dedicated the University of Wisconsin Arboretum and Wild Life Refuge. The university had acquired five hundred acres of typical post-settlement Wisconsin farmland: pasturelands, grazed woodlots, plowed prairie, marshes, and fens. Indian burial mounds dotted the perimeter of Lake Wingra, on whose southern shores the lands lay.... Leopold was one of several speakers that morning. In his talk he described what he and other faculty overseers envisioned for the arboretum. It was not going to be just a collection of trees, like other arboreta, but "something new and different"—a collection of landscapes, a recreation of the land as it once existed. It would be replanted not simply with individual species, but with entire plant communities: prairies, hardwood forest, coniferous forest, marsh. "Our idea, in a nutshell, [Leopold said] is to reconstruct, primarily for the use of the University, a sample of original Wisconsin—a sample of what Dane County looked like when our ancestors arrived here in the 1840s."

At the moment he first defined ecological restoration, Leopold was under the sway of two then prevailing myths: (1) the colonial myth of wilderness; and (2) the scientific myth of Clementsian equilibrium ecology.

THE WILDERNESS MYTH AND THE EQUILIBRIUM-ECOLOGY MYTH

In January of the next year, Leopold would join Robert Marshall, Benton McKay, Harvey Broome, Bernard Frank, Harold Anderson, Ernest Oberholtzer, and Robert Sterling Yard to found the Wilderness Society [Meine, 1988]. In an article published in 1930 in *The Scientific Monthly*, Marshall [1998, p. 86] beautifully articulates the colonial wilderness myth:

> When Columbus effected his immortal debarkation, he touched upon a wilderness which embraced virtually a hemisphere. The philosophy that progress is proportional to the amount of alteration imposed upon nature never seemed to have occurred to the Indians. Even such tribes as the Incas, Aztecs, and Pueblos made few changes in the environment in which they were born. The land and all that it bore they treated with consideration, not attempting to improve it, they never degraded it. Consequently, over billions of acres the aboriginal wanderers still spun out their peripatetic careers, the wild animals still browsed in unmolested meadows, and the forests still grew and mouldered and grew again precisely as they had done for undeterminable centuries.

According to the wilderness myth, the entire Western Hemisphere was in a natural condition free from significant human influence when "discovered" by Columbus. What about the American Indians? Well, yes, they were here already, but there were too few of them and they were either too technologically backward or too environmentally ethical to have a serious impact on the primeval, original ecological conditions persisting in the hemisphere.

In the absence of significant human disturbance, those conditions would remain the same. Sure, trees and other organisms go through life cycles and die, but they are replaced by the same species, generation after generation. And sure, occasionally cataclysmic natural disturbances befall a whole biotic community, but after a series of successional stages, the climax community would reestablish itself. Therefore, overall, the Western Hemisphere remained unchanged for "undeterminable centuries." This is the ecological equilibrium myth in a nutshell.

Frederic Clements was arguably the most influential ecologist of the first half of the twentieth century [Worster, 1994]. He represented nature in the following way: Each region of the world, which he called a "biome," had a natural plant "formation," which he called the "climax," because it was determined by the climate, which he supposed to be stable [Clements, 1916]. Climate consists of two principal gradients, moisture and temperature. In North America, for example, the moisture gradient runs from the Sierra rain shadow eastward to the Atlantic: in the dry Southwest, a formation dominated by saguaro cactus is the climax; a little farther east the climax is short-grass steppe; still farther east, it's long-grass prairie; from the Mississippi valley on eastward, it's forests. Similarly the temperature gradient determines forest types from southern oak-hickory hardwoods to northern spruce-fir softwoods. Elevation complicates this picture. Going upslope is like going north, and in North America, like going east: the microclimate is cooler and wetter at higher elevations. Thus forests grow in the high lands of the American Southwest overlooking the lower-elevation deserts. Later, ecologists in the Clementsian tradition would include edaphic as well as climatic conditions as determinate of the climax plant formation [Tobey, 1981].

In any case, from time to time climax formations experience catastrophic external disturbances—volcanic eruption, wild fire, flood, wind storm. There follows a series of plant formations until the climax is reestablished. Clements [1916] called this process "succession." Moreover, he viewed this process as a kind of organismic development, an ontogeny. It was the climax "sere" that he believed to be a highly integrated superorganism. Ecology is the study of its anatomy, physiology, and metabolism.

> The developmental study of vegetation necessarily rests upon the assumption that the unit or climax formation is an organic entity. As an organism the formation arises, grows, matures, and dies.... Furthermore, each climax formation is able to reproduce itself, repeating with essential fidelity the stages of its development. The life history of a formation is a complex but definite process, comparable in its chief features with the life history of an individual plant. [Clements, 1916, p. 2]

Clements's study area was the prairie just at the time it was being settled by European-American agriculturists [Tobey, 1981]. Clements minimized the ecological significance of the indigenous peoples of the Americas, providing the scientific authority for Marshall doing so in service of wilderness preservation. In regard to "Indian tribes," to Clements [1936, p. 253], "it seems improbable that the total population within the grassland ever exceeded half a million ... while the influence of fires set by the Indians was even less significant" than "effects from overgrazing and trampling" by bison. And "[a]s to forests, those of the Northwest were still primeval and in the east they were yet to be changed over wide areas by lumbering and burning on a wide scale" [1936, p. 253]. To Clements, European-Americans represented an artificial, external disturbance that not only destroyed climax formations but that also disrupted and forestalled the process of succession back to climax. Thus, from this Clementsian point of view, there appears a sharp distinction between "natural" and "artificial" ecological conditions. The climax formation and the several successional seres leading up to it are natural. Anthropogenic landscapes created by European settlers are artificial.

Most ecologists in the first half of the twentieth century remained under the influence of Clements's theories [Tobey, 1981]. Many may have rejected his metaphysical idea that ecosystems were superorganisms, but few doubted his teleological concept of ecological succession, terminating in a climax community, which persisted, unless and until destroyed by some *external* disturbance. Few doubted Clements's hypothesis that after such a resetting of the ecological clock, the site would express the same successional series capped off by the same climax community, if only human beings armed with modern technology would leave it alone. Leopold too remained enthralled by this ecological myth. He wrote:

> The Wisconsin land was stable ... for a long period before 1840 [the year "settlement" began]. The pollens embedded in peat bogs show that the native plants comprising the prairie, the hardwood forest, and the coniferous forest are about the same now as they were at the end of glacial period, 20,000 years ago. Since that time these major plant communities were pushed alternately northward and southward several times by long climatic cycles, but their membership and organization remained intact. Thus in one northward push the prairie once reached nearly to Lake Superior; in one southward push the Canadian forest reached to Indiana. The bones of animals show that the fauna shifted with the flora, but its composition or membership likewise remained intact. [Leopold 1991, pp. 311–312]

THE MYTH OF CLEMENTSIAN EQUILIBRIUM ECOLOGY DEBUNKED

What's wrong with this picture? The most glaring thing is the putative interval between the present and "the end of the glacial period." When Leopold wrote this in 1944, the time back to the last glaciation was believed to be twice the actual interval [McIntosh, 1985]. On a clear day 20,000 years ago, from where Leopold stood at the dedication ceremony of the Wisconsin Arboretum and Wild Life Refuge, you would see a wall of ice on the northeastern horizon, and to the southwest you might see a herd of woolly mammoths. Also, faithfully reflecting the embryonic state of palynology in the 1940s, he tells us that the pollen record indicates that the Holocene biotic communities of Wisconsin—both prairies and forests—moved northward and southward *as units*. This, as Arthur Tansley notes, was also an idea expressly theorized by Clements:

> If a continental ice-sheet slowly and continuously advances or recedes over a considerable period of time all the zoned climaxes which are subjected to the decreasing or increasing temperature will, according to Clements's conception, move across the continent "as if they were strung on a string," much as the plant communities zoned around a lake will move towards the centre as the lake fills up. [Tansley, 1935, p. 302]

Contemporary palynology paints a very different picture. Plant species migrated from Pleistocene refugia from different directions at different rates [Davis, 1984; West, 1964]. This evidence supports the "individualistic" alternative to Clementsian holism, first championed by Henry Gleason [1926], a contemporary of Clements, in the first quarter of the twentieth century. According to Gleason [1926], what appears to be a tightly integrated ecological unit, is actually just an ill-defined, accidental assemblage of opportunistic organisms that are adapted to similar environmental gradients—such as soil pH, moisture, and temperature. Gleason was pretty much ignored during the first half of the twentieth century, but, beginning in the 1950s, his individualistic paradigm began to be vindicated [Curtis and McIntosh, 1951; Whittaker, 1951; 1967]. By the last quarter of the twentieth century it had triumphed over the Clemensian super-organism paradigm [McIntosh, 1975].

Presently, the modern more generally Clementsian "balance-of-nature" paradigm in ecology has been succeeded by a postmodern neo-Gleasonian "flux-of-nature" or "shifting" paradigm [Pickett and Ostfeld, 1995]. (I call it "post-modern" because, as Bryan G. Norton explains more fully in his chapter, putative ecological entities, such as biotic communities and ecosystems, if not socially constructed by ecologists, can no longer be regarded as having a robust ontological status independent of their investigation by ecologists.) What appeared to Clements and most of his contemporaries to be well-defined, self-regulating ecological units of various types, each with its tightly integrated complement of species, now appear to be ever-shifting mix-and-match collections or aggregates of species populations, interacting catch as catch can. Such assemblages or collections change gradually over time, stochastically, as new species chance to arrive and old ones leave. There is no fixed end-point or *telos*, no self-replicating climax community, which is the destination of successional change.

Inherently dynamic biotas are, moreover, subject to routine disturbances, each of which, depending on the spatial or temporal scale of reference, is incorporated into the system [Pickett and White, 1985]. For example, at a spatial scale of 1,000 hectares and a temporal scale of twenty years, fire in a mixed hardwood forest in the Upper Midwest is an abnormal and external event. But at a spatial scale of 100,000 hectares and a temporal scale of 200 years, fire in such a forest is "incorporated." With the postmodern shift in ecology from the balance-of-nature to the flux-of-nature paradigm, we have added disturbance regimes to energy flow and nutrient cycling as fundamental processes occurring in ecosystems. At appropriately chosen scales, some human disturbances—widely scattered shifting agriculture in moist tropical forests, for example—may also be regarded as incorporated [Sloan and Padoch, 1988]. Michael Soulé [1995, p. 143] sums up the current worldview in ecology quite bluntly:

> The idea that species live in biotic communities is a myth. So-called biotic communities, a misleading term, are constantly changing in membership. The species occurring in any given place are rarely convivial neighbors; their coexistence in certain places is better explained by individual physiological tolerances. . . . Current ecological thinking argues that nature at the level of local biotic assemblages has never been homeostatic. Therefore, any serious attempt to define the original state of a community or ecosystem leads to a logical and scientific maze.

What is the upshot for classical ecological restoration if there is no such thing as the "original" condition of a site? The condition that Dane County, Wisconsin was in at the moment European settlers saw it in the 1840s, to refer back to Leopold's classic articulation of the concept of ecological restoration, is but a snapshot in its ever-changing ecological odyssey. Why seize on that condition as the norm for restoration, rather than its condition at some earlier or, for that matter, later moment?

THE COLONIAL WILDERNESS MYTH DEBUNKED

Any earlier moment might be just as choice-worthy a norm, but any later moment, an apologist for classical ecological restoration might counter, would not be choice-worthy, because it would be an *artificial* condition. That invokes the wilderness myth, the core assumption of which is that the preColumbian inhabitants of North America were few in number and had no significant ecological impact. Demographers in the first third of the twentieth century, when Robert Marshall was waxing eloquent about the wilderness condition of the entire Western Hemisphere, had underestimated preColumbian American Indian populations by a factor of ten, because they failed to account for the disastrous effect of Old World diseases on New World peoples [Denevan, 1992]. If there were ten times more people here "when Columbus effected his immortal debarkation" than Marshall, Clements and their contemporaries supposed, the ecological effect of the indigenous peoples of the Western Hemisphere was proportionally greater than they supposed. Nor were American Indians as ecologically passive as Marshall and Clements represent them to have been [Kretch, 1999]. American Indian cultural fire, cultural predation, agriculture, and irrigation had significant and on-going effects on American ecosystems. Charles E. Kay [1994] argues that the ecological effects of cultural predation in North America have been seriously underestimated. So much so, that in the preColumbian period, elk were scarce in the Yellowstone, whereas until recently, protected from both human and wolf predation, the Yellowstone elk population grew to pestilential proportions.

The great unbroken forests in the East and great abundance of game everywhere that European explorers encountered in North America is attributable to the drastic reduction of Indian populations by Old World diseases, which spread from Indian to Indian well in advance of European conquest and settlement of the country [Dobyns, 1983]. Contemporary demographer William Denevan [1998] estimates that the total human population of North America (including European and African immigrants and Americans of European and African descent as well as remnant populations of American Indians) was thirty percent *smaller* in 1750 than it was in 1492. He concludes that because of the demographic debacle caused by Old World pathogens, North America was then in a state of "recovery" from the ecological effects of its indigenous human inhabitants [Denevan, 1998]. Less tendentiously, we might say simply that it was in a state of transition from one domain of ecological attraction to another [Holling, 1992]. Indeed, one might argue that, paradoxically, the wilderness condition encountered by European explorers and settlers of North America was itself artificial, created by the depopulation of the continent after its (re)discovery by Columbus.

Leopold's mention of the "end of the glacial period" raises another confounding question. What happened to the mammoths, mastodons, camels, horses, and all the more than thirty other genera of wildlife that were here 20,000 years ago and which all disappeared suddenly at the Pleistocene-Holocene boundary, about 10,000 years ago? Increasingly, the finger points to *Homo sapiens* as the dark angel of their extinction in the Western Hemisphere [Martin and Klein, 1984]. *Homo sapiens* may have been in the Western Hemisphere before eleven or twelve thousand years ago, but, as we well know from the European rediscovery of the Americas, groups of *Homo sapiens* differ significantly from one another culturally and, for that reason, also in their ecological impact. About eleven or twelve thousand years ago, a group of *Homo sapiens* culturally adapted to big game hunting, armed with Clovis spear points and atlatl throwing sticks, arrived in the hemisphere from Asia [Martin, 1973]. In a few centuries thereafter much of the big game they pursued was extinct. One alternative explanation of these extinctions is, of course, sudden climate change [Grayson, 1977]. But the species that went extinct this time had endured a series of glacial interstadials in which the climate had abruptly shifted from cold to warm. Another alternative explanation is the "hyperdisease theory": perhaps humans and/or their mammalian commensals brought a new highly lethal pathogen with them that jumped species and killed off the North American Pleistocene megafauna [McPhee, 1999]. That conjecture is analogous to the explanation of the decimation of American Indians by Old World diseases brought to the New World by Europeans. Nor is one explanation of an anthropogenic demographic debacle exclusive of another. In addition to disease, after all, American Indian populations were substantially reduced by genocidal warfare and ethnic cleansing. Analogously, probably all three factors offered to explain the mystery of the sudden Pleistocene megafauna extinctions in the Nearctic worked in combination. Climate change stressed them out, disease decimated their populations, and a new super-predator, the likes of which they had never experienced before, finished them off.

So how should we revise the picture of the ecological condition of the pre-Columbian Nearctic painted by Marshall and Leopold? And what are the implications of this revision for ecological restoration? First, the Nearctic was more dynamic than the ecologists of their day supposed. And for ten thousand years or so before the rediscovery of the Western Hemisphere by European peoples, *Homo sapiens* was not a negligible ecological force. Sudden climate change, cultural predation, and possibly pandemic disease suddenly and radically altered the composition of the fauna of the Nearctic, shortly after the arrival of the Siberian big game hunters at the Pleistocene-Holocene boundary. And by exerting unrelenting hunting pressure on the surviving fauna and setting fire to forests and grasslands, *Homo sapiens* became a keystone species in the Nearctic [Kay, 1995]. Therefore, the pre-settlement condition of an area appears to be a questionable target or norm for ecological restoration. Indeed, if, as Denevan [1998] notes, European settlers found the land in an abnormally fallow condition, such a condition would be an aberration in an ever-changing, and, for thousands of years, a largely anthropogenic landscape.

PLEISTOCENE PARKS?

Suppose we choose to think that ecological restoration should, indeed, aim to restore a site to its *natural* condition, and we choose to define its natural condition as relatively free from human influence, as Malcolm Hunter [1996] suggests we ought. But also suppose that we are persuaded by the "overkill" and "hyperdisease" hypotheses that hemispheric extinctions at the Pleistocene–Holocene boundary are anthropogenic. Then what? Two prominent thinkers register a bold answer: Back to the Pleistocene. First, Michael Soulé [1990, pp. 234–235, emphasis added], in his 1989 presidential address to the Society for Conservation Biology, commented that

> many of the genera of animals that most conservationists would consider alien in North America were actually part of that continent's biota *only moments ago in evolutionary time.* Thirty seven genera (57 species) of large mammals ... went extinct *just a few thousand years ago* in North America, whereas most of their plant prey survived. Some of these animal species still persist in the Old World, and many species of these genera could probably adapt to current North American conditions if they were allowed to "return." For many North American ecologists, the psychological adjustment to biogeographically recombined communities will be painful, but it might be facilitated by the realization that lions, cheetahlike cats, camels, elephants, horses, saiga antelope, yaks, and spectacled bears are native taxa to North America that disappeared *very recently.* The reintroduction of these large animals will be controversial, but I would not be surprised to read someday that cheetahs are helping to control deer and that mesquite is being "overbrowsed" by rhinoceroses. A cheerful way of viewing such faunal mixing is that it represents the *restoration* to the Nearctic of the great paleomammalian megafauna.

There is a hint of a tongue-in-cheek tone to Soulé's "modest proposal." Soulé I think considers what Aldo Leopold [1949, p. 217] lamented as a "world-wide pooling of faunas and floras" to be inevitable. To make it more palatable we can spin it as "the restoration to the Nearctic of the great paleomammalian megafauna." But Paul S. Martin, the leading exponent of the "overkill hypothesis," (writing with David A. Burney) expresses untempered enthusiasm for a back-to-the Pleistocene reintroduction program. According to Martin and Burney [1999, p. 59, emphasis added],

> In planning New World *restorations*, conservationists have endowed large mammals of *historic time* with the exclusive status of hallmarks, or flagships, overlooking the missing large mammals of the late Pleistocene. The animals that the first explorers and settlers saw and wrote about became incorporated in ideas of what constituted American wildness. The viewpoint imposed by a "Columbian Curtain" is unrealistic in *evolutionary time*. The historic fauna lacks the largest and most representative animals of the continent. Among the more common fossils of the late Pleistocene, which was dominated by equids, camelids, bovids, and especially bones, teeth, and tooth plates of proboscideans, only bison is represented.

Martin and Burney [1999] think big when they think about wildlife restoration. The title of their article is "Bring Back the Elephants," for that is how they suggest we *start* the restoration to the Nearctic of the great paleomammalian megafauna. They think big in another sense, in a temporal sense as well. The temporal scale on which both Soulé and Martin and Burney think is "evolutionary time." That's why their vision has something of a *Jurassic Park* feel to it. In fact, what Soulé muses about and Martin and Burney seriously propose is the creation of a system of Pleistocene parks in North America.

Soulé's whimsical suggestion and Martin and Burney's bold proposal were offered up in two relatively small and isolated intellectual barrios. As noted, Soulé's remarks were made in passing during his wide-ranging presidential address to the Society for Conservation Biology and subsequently published in *Conservation Biology* (the journal). Martin and Burney's proposal was published in *Wild Earth*, then a small, low-budget (and now defunct) journal, established by former affiliates of Earth First!, most notably, Dave Foreman. The Martin and Burney proposal, however, made its way out of the scientific backwaters and into the mainstream midway through the first decade of the 21st century. Dave Foreman, co-founder of Earth First!, later Executive Editor of *Wild Earth* and now head of the Rewilding Institute, and Michael Soulé, co-founder and past president of the Society for Conservation Biology, joined Paul S. Martin, David A. Burney, and eight others as co-authors of an article published in *Nature* seriously advocating a "Pleistocene re-wilding": "the restoration of large wild vertebrates into North America" [Donlan *et al.*, 2005]. That article was favourably noticed a month later on the op-ed page of the *New York Times* [Kristof, 2005].

In addition to "restoring" its extinct genera to North America, the Commentary piece in *Nature* offered, as a complementary rationale, the threat to elephants, cheetahs, camels, and the like in the places where they currently exist—Africa and Asia. In (benighted, it was implied) Africa and Asia—the authors allege, without evidence or argument—the prospects for these species to survive through the twenty-first century are dim; whereas in (presumably more enlightened) North America they might be protected in Pleistocene parks and thus saved from global extinction. Their 2005 *Nature* commentary was followed the next year by a fuller exposition in the *American Naturalist* by all the same authors [Donlan *et al.*,

2006]. Publication of that article was announced and summarized in a *Science News* cover story [Jaffe, 2006]. Which, along with coverage of the more condensed *Nature* article in the *New York Times*, took the idea out of the realm of closeted scientific musings into that of partisan public policy debate, as noted by Tim Caro [2007].

A QUESTION OF SCALE

A number of sceptics responded to the Pleistocene-parks proposal in subsequent issues of *Nature* [Chapron, 2005; Dinnerstein and Irvin, 2005; Schlaepfer, 2005; Shay, 2005; Smith, 2005], most worrying that it would be a distraction from more conventional conservation efforts, both in North America and in the potential donor countries of Africa and Asia. The most thorough brief against Pleistocene rewilding was filed by Dustin R. Rubenstein, Daniel I. Rubenstien, Paul W. Sherman, and Thomas A. Gavin. Their negative assessment is based on many considerations that range from the technical—such as the genetic similarity of extant "proxy" species to extinct species, of *Camelus* spp. to *Camelops* spp. and of *Elephus maximus* to *Mammuthus primigenius*—to the social and political: if residents in sparsely populated rural areas in the United States are hostile to reintroductions of grey wolves, how much more hostile are they likely to be to (re?)introductions of African lions and cheetahs, to say nothing of wild elephants? In my opinion, however, the deepest, and perhaps for that very reason, the least articulate matter of disagreement between proponents and critics of Pleistocene rewilding is a disagreement about the appropriate temporal and spatial scales of ecological restoration, especially the former. Soulé and those who have followed him frame their thinking on an evolutionary temporal scale, while their critics frame theirs on what might be called an ecological temporal scale (about which more shortly). These are biologically defined temporal scales. Another non-biological temporal scale creeps into the discourse in which this debate has been conducted, the historical temporal scale. Adding to the confusion, some writers conflate and confound the ecological and historical temporal scales because they are roughly coincidental—that is, the one maps coextensively fairly well on the other.

Temporal scales are defined by processes. The macroevolutionary temporal scale—which Soulé and Martin and Burney invoke—is defined by evolutionary processes, such as, most notably, speciation and the interval between speciation and extinction. Large mammals speciate slowly over tens of thousands of years and often endure as distinct species for several million years. The historical temporal scale is defined by historical processes, such as the migrations of peoples and the interval between the establishment and disestablishment of nations and systems of government—such as the rise and fall of the Roman Empire and the migration of Europeans to the Americas. The ebb and flow of historical processes is measured in decades and centuries. The Soviet Union lasted for approximately seven decades; the United States has been around for a little more than two centuries and a quarter; Christianity has been a historical phenomenon for a little more than two millennia. The ecological temporal scale is defined by ecological processes—such as, most notably, succession and disturbance regimes. Like the historical temporal scale, it is measured in decades and centuries—the interval between fires in a pine barrens or between floods on a river; the time it takes for an old-growth forest to replace an abandoned wheat field. Look again at the italicized words in the quotation from Martin and Burney. They in effect claim that the historical temporal scale ("historic time") is not appropriate for ecological restoration. Why? Because it is arbitrary; and it has nothing to do with biological processes. Instead, they suggest, the appropriate temporal scale is evolutionary ("evolutionary time"). And, as Soulé hyperbolically notes, the megafauna extinctions at the Pleistocene-Holocene boundary took place "*only moments ago in evolutionary time*"—that is, relatively recently on the evolutionary temporal scale, only eleven to thirteen thousand years ago.

But is the evolutionary temporal scale the appropriate scale for thinking about ecological restoration? I don't think so. We are, after all, struggling to make sense of the concept of *ecological* restoration—in an ever-changing, dynamic landscape, long influenced by our own species. Thus it would seem to make more sense to select the ecological temporal scale as the appropriate one for conceptualizing ecological restoration. This temporal framing discrepancy—proponents of Pleistocene rewilding framing the issue in evolutionary time, critics framing it in ecological time—is the crux of their scientific disagreement. Donlan *et al.* [2005] extol the successful rewilding of Przwalski's horse and the Asian ass as examples of what they are proposing to do with elephants, among other species, in North America. Rubenstein *et al.* [2006, p. 236] reply that

> Small-[spatial]scale reintroductions of these and other endangered equid species throughout Asia ... appear to be working. These are appropriate reintroductions and the sort of rewilding that makes evolutionary *and ecological* sense because *the time* between the species' extirpation and reintroduction has been short enough that neither *the native ecosystems* nor the animals themselves have changed [evolved] very much.

TEMPORAL SCALE AND THE PROBLEM OF SELECTING A REFERENCE SYSTEM FOR ECOLOGICAL RESTORATION

Eric Higgs [1997] is aware of the post-Clementsian ecology problem of identifying a reference system for ecological restoration, but seems to ignore it. By definition, ecological restoration aims at recreating a *past* ecological condition. Restoration should not be confused with another kind of ecological engineering: rehabilitation [Callicott *et al.*, 1999]. If an ecosystem has been radically and irreversibly altered, and is in a dysfunctional condition, it might be rehabilitated by creating a functional system of predator-prey dynamics more or less like of those in the past, but involving a set of species different from those of the past. After the extinction of several species of endemic deep-water ciscoes and the invasion of alewife and sea lamprey, the Great Lakes ecosystems were rehabilitated—how effectively is a matter of controversy—by lampreycide treatments and stocking Pacific salmon [Great Lakes Fishery Commission, 1992] to prey on the alewife. According to Higgs [1997, p. 343], "The goal of restoration is to reproduce by whatever means available a predetermined historic or indigenous ecosystem. This goal inscribes the concept of fidelity—that is, a quest to come as close as possible to restoring what existed on a specific site." By "what existed," Higgs refers to the components, the species, that existed on a site in the past. I understand "ecological restoration"—as do most ecological restorationists and restoration ecologists—in compositional terms. I understand "ecological rehabilitation" in functional terms—biomass production, tall trophic pyramids, lengthy food chains and complex food webs, efficient nutrient cycling, soil retention and soil building, hydrologic modulation and purification. If such past functions are recreated, but using a different set of species from "what existed" in the past, that's rehabilitation, not restoration. In addition to the thought experiment, another device employed by philosophers is the "stipulative definition." I stipulate that "ecological restoration" mean what it commonly does mean to restorationists and laypersons alike: reestablishing the species that once existed on a site; and I stipulate that "ecological rehabilitation" means recreating impaired functions that once existed on a site, if those functions are performed by a different set of species than those that once existed on a site. To return to the Great Lakes example; stocking Pacific salmon and chemically controlling the sea lamprey population is an attempt to rehabilitate not restore the Great Lakes.

But which ecological condition that existed on a specific site should be the target of true ecological restoration? There are many to choose from. In the quotation that follows, as his use of "so-called" indicates, Higgs [1997] is keenly aware of what he calls "postmodern" (*i.e.*, post-Clementsian) ecology and critiques of the wilderness myth. Nevertheless, he reverts to the classic norms:

> A completely faithful restoration, presumably, is one that exactly replicates *the* ecosystem (*i.e.*, the climax formation?). Hypothetically speaking, we could devise a test whereby ecologists were asked to view the *so-called original ecosystem* alongside the restored version. If no distinction could be made between them, this would be a perfect restoration.... A restored ecosystem must strongly resemble the structure and composition of the *so-called natural ecosystem* [Higgs, 1997, p. 343, emphasis added]

Higgs [1997, p. 343] admits that "There are several difficulties with this... definition of restoration, not the least of which is the idea of nature as a fixed, determinable entity. What vests us with the authority to make claims about the kind of ecosystem to be restored ...?" Here I try to overcome these difficulties and answer this question definitively. In short, the past norm for ecological restoration should be selected by reference to ecological, not evolutionary temporal scales. Ecological scales are also useful for accepting some anthropogenic ecological conditions as appropriate norms for ecological restoration and rejecting others.

Hierarchy theory in ecology identifies multiple temporal scales at which ecological processes occur [Pattee, 1973; Allen and Starr, 1982; O'Neill *et al.*, 1986; Allen and Hoekstra, 1992]. For example, nitrogen fixation by rhizobial bacteria occurs at a relatively rapid rate in comparison with ecological succession. Change occurs at all scales. However, we may regard the processes at the higher end of the hierarchy as relatively unchanging or stable if our interest focuses on processes at the middle or lower end. For example, if we are interested in the population cycles of North American arctic mammals, we may regard the latitude and elevation of the boreal biotic provinces of the North American continent to be stable, even though the North American plate is slowly drifting to the northwest and is still slowly rebounding from the weight of the retreating ice that once thickly covered its northern half.

So how does hierarchy theory help us think coherently about ecological restoration? It helps us at least to identify appropriate temporal horizons for locating restoration norms or targets.[1] Holling [1992, p. 480] identifies "three approximate scale ranges..., each defined by a broad class of processes that dominate over those ranges of scale. The microscales are dominated by vegetative processes, the mesoscales by disturbance and environmental processes, and the macroscale by geomorphological and evolutionary processes." The geological and evolutionary time scales, the scales on which continents migrate and species radiate, are too big. The diurnal, seasonal, and annual time scales on which individual organisms carry out their life processes, such as metabolism, growth, and development are too small.

Taking our clue from Holling [1992],[2] we might measure appropriate temporal mesoscales for norms of ecological restoration by disturbance regimes—the periodic intervals between disturbances of a particular and regularly occurring kind. For example, for coastal environments we might measure ecological time by the periodicity of disturbance by hurricane-force winds; for riparian environments by the periodicity of floods of various magnitudes, from seasonal fluctuation to the hundred-year flood cycle; for upland forests and grasslands, ecological time might be measured by the frequency of fire. Here, I am only trying to get a feel for what gross range of temporal intervals or units are ecologically meaningful. Let me make an analogy. In the course of a human life, some dynamic processes have little meaning or relevance because they are either too fast or two slow. The rate at which the Grand Canyon formed as the Colorado River's rate of erosion kept pace with the increased elevation of the plateau is too slow to register, and the speed of the Krebs cycle is too fast. A human lifetime might be meaningfully

[1]For further applications of hierarchy theory to issues of ecosystem health, integrity, management and restoration, see [Costanza *et al.*, 1992; Peterson and Parker, 1998; Norton, 2005; Falk *et al.*, 2006].

[2]For those familiar with Holling's work, my use of [Holling, 1992] in this context focuses on his characterization of micro, meso and macro scale ranges. In this paper I take no stand on Holling's more contentious theses concerning statistical evidence for scalar "lumping", or the so-called "adaptive cycle" model of ecosystem dynamics.

organized in half-decades and decades—a person's infancy, childhood, teen years, twenties, thirties, forties, and so on. Indeed, that is just the first scalar range that Holling [1992] characterizes as "vegetative"; it might more inclusively be termed the organismic scalar range. Now, what dynamic processes are meaningful and relevant for ecological restoration? By reference to disturbance regimes, I suggest we might meaningfully organize ecological time in terms of centuries.

THE CLASSIC NORMS OF ECOLOGICAL RESTORATION SCIENTIFICALLY JUSTIFIED

So that narrows the target window for ecological restoration to the Holocene—to between one and one hundred centuries ago. After the anthropogenic mass megafaunal extinction event in the New World at the Pleistocene–Holocene boundary, new ecological domains of attraction emerged which included the new primate super-predator as a factor in all, and a keystone species in many [Holling, 1992]. Other species—survivors of the ecological holocaust—adjusted to the new New-World order. Thus, we might justifiably select only Holocene, not Pleistocene, biotic communities at a given site as targets for *ecological* restoration. That selection would be narrowed further by what we might term "ecological drift," analogous to genetic drift in evolutionary biology. Ecosystems change over ecological time. They are, moreover, open to mutual influence from neighboring ecosystems. Selecting, as a target for restoration, a more recent past condition at a given site would auger better prospects for success and pose less risk of adversely affecting neighboring sites.

Risk of irreversible adverse ecological consequences is one of the main concerns of opponents of Pleistocene rewilding [Rubenstein *et al.*, 2006]. First, is it possible to reconstitute Pleistocene ecosystems by imposing proxies of Pleistocene fauna on late Holocene ecosystems, such as those now prevailing in the American Southwest and Great Plains, in late-Holocene (or, indeed, post-Holocene) climatic conditions? And were it possible, what would the effect of reconstituted Pleistocene ecosystems be on late-Holocene ecosystems neighboring Pleistocene parks? Proponents reply that carefully controlled small-scale experiments could be conducted to find out the answers to these questions [Donlan *et al.*, 2006]. But, counter the opponents, there is also unacceptable risk in scaling up from say a well-fenced ranch-sized site to a fenceless park-sized site [Rubenstein *et al.*, 2006]. Analogous, though less extreme, risks would be posed by trying to restore a site to its Holocene condition of say eight thousand years ago when many plants were still making their northerly way out of Pleistocene refugia.

Thus, more recent Holocene conditions seem to be the logical target for restoration, because the prospects of success are greater and the risks are lesser. I am, as you see, zeroing in on the conventional target and norm for ecological restoration, if not the condition of a site just prior to European-American "settlement," when it was in an abnormal state of "recovery," than in the condition it was in 1491. I am trying to do so, however, without invoking obsolete ecological assumptions about static equilibria: self-perpetuating, undisturbed climax communities (the "original" condition); or prattling about "pristine," "natural," "wilderness" conditions free of any significant human presence or influence.

Ecological restoration typically favors native species, so much so that to speak of a restoration project consisting of an indiscriminate mix of native and exotic species seems oxymoronic [Jordan *et al.*, 1987]. Indeed, restorationists would not only never think of deliberately employing exotic species in a restoration project, they constantly battle invasive exotics in the on-going management of restored sites [Egan and Glass, 1995]. But the distinction between native and exotic species is often unclear. Here again, proponents and opponents of Pleistocene rewilding differ about what species should be regarded as native and what exotic. And once again, the crux of their difference turns on the temporal framing of either party to the controversy. According to Donlan *et al.* [2006, p. 664], "Cultural conventions dictate which taxa are regarded as native and which are not." And Donlan and Martin [2004, p. 268] insist that "From a genetic, evolutionary, and ecological perspective, horses are native to North America." On the other hand, Rubenstein *et al.* [2006, p. 236] insist that "adding these *exotic* [horses among them] species to current ecological communities could potentially devastate populations of indigenous *native* animals and plants."

One sees bumper stickers in Florida proclaiming the vehicle's owner to be a native Floridian. It plainly signifies that the claimant was born in Florida, and is not one of the many immigrants to the state. If a native resident is a resident that resides in the state where he or she was born, then by a native species we might mean one that is found in the biological province where it was "born"—that is, where it evolved. For example, the several species of kangaroo are native to Australia and exotics elsewhere. And kangaroos evolved in Australia, but not elsewhere [Frith, 1969]. Donlan *et al.* [2005, p. 914] suggest that a species is native to its place of evolutionary origin when they call North America "the evolutionary homeland" of horses and camels. To insist, however, that a species is only a native in its place of evolutionary origin seems unduly restrictive. Armadillos evolved in South America and, when the Bolivar Trough disappeared and the Panamanian land bridge rose about three million years ago, they migrated to Central America and southern North America, where they are regarded as native [Marshall, 1988]. Some species, moreover, have evolved in one place, migrated to another, and gone extinct in their place of evolutionary origin. Camelids and equids are examples [Gauthier-Pilters, 1981]. They evolved in North America, but no wild populations of camels (or llamas) have existed on that continent for ten thousand years; nor had horses until only five hundred years ago. Few conservationists would argue that a species long residing in a place in which it did not evolve, but long extinct in its place of evolutionary origin, should either be exterminated altogether or exterminated in the place it is now found and reintroduced in its place of evolutionary origin. Would any sober conservationist advocate eradicating zebras from Africa in the name of removing an invasive exotic? Pace Donlan *et al.* [2005; 2006], as these considerations suggest, place of evolutionary origin, far from being a necessary condition of a species nativity, is not even a sufficient condition.

The concept of an exotic species is commonly delimited in terms of *natural* range and dispersal [Randall, 2000]. In an Office of Technology Assessment (OTA) report on harmful non-indigenous species in the United States, the following definition of exotic is provided: "the condition of a species being beyond its natural range or natural zone of dispersal" [U.S. Congress, Office of Technology Assessment, 1993, p. 53]. What "natural" means in this context is this: unaffected, directly or indirectly, intentionally or unintentionally, by human agency. As the OTA report makes clear, "natural range" means "the geographic area a species inhabits or would inhabit in the absence of significant human influence" [U.S. Congress, Office of Technology Assessment, 1993, p. 53]. Noss and Cooperrider [1994, p. 392] are equally explicit: "species that occur in a given place, area, or region as the result of direct or indirect, deliberate or accidental introduction of the species by humans, and for which introduction has permitted the species to cross a natural barrier to dispersal." This definition assumes the continued cogency of one important element of the obsolete Clementsian ecological paradigm, the sharp bifurcation of "man" and nature. The distinction between native and exotic species, however, is vitally important, not only to ecological restoration, but to the whole of conservation biology. How can we preserve the distinction, without invoking the scientifically indefensible segregation of human agency from all other kinds of causation?

Once more, considerations of appropriate temporal and spatial scale help us resolve the otherwise ambiguous and sometimes paradoxical native-exotic distinction. Take a specific example. Are horses and burros natives or exotics in North America? Again, Pace Donlan *et al.* [2005; 2006], most wildlife ecologists would classify them as exotics [Lodge and Shrader-Frechette, 2003; Rolston, 1998, Rubenstein *et al.*, 2006; Soulé, 1990]. But the genus *Equus* evolved in North America and spread from there into Asia, Africa, and Europe [Donlan *et al.*, 2006; Simpson, 1956]. It was extirpated from its place of evolutionary origin, in all probability anthropogenically, by the Clovis spearmen, along with the rest of the extinct Pleistocene megafauna of North America [Donlan *et al.*, 2006; Martin, 1973; Soulé, 1990]. The horse and burro were anthropogenically reintroduced as domestic beasts of burden by the Spanish in the late fifteenth century and soon thereafter established feral (or wild, depending on your point of view) populations in North America [Simpson, 1956]. If an evolutionary temporal scale is the only one of biological importance, as Donlan *et al.* [2005; 2006] appear to think, then we would have to agree with them and accept the horse and burro as restored native species. If presence in a place due to human agency were the defining characteristic of an exotic species, then because horses and burros are now in North America thanks to human agency, then, we must consider them to be exotics. That judgment must, however, by the same token, be immediately reversed, because *Equus* was, in all probability, absent in North America due to human agency, until reintroduced by the Spanish. However, if we reject human agency as a scientifically defensible way to distinguish native and exotic species *and* scale down temporally, and consider the horse in the context of reconfigured Holocene ecological relationships, then the conventional conservationist wisdom that the horse and burro are ecologically disruptive exotics in North America can be justified scientifically—and without ambiguity or equivocation. After the Pleistocene extinctions, which included *Equus* in North America, new ecological domains of attraction emerged. The sudden introduction of the horse and burro threw some of these into chaotic oscillations. In time, of course, *Equus* may be reincorporated in the ecosystems of western North America. But because ecological temporal scales are greater than the organismic scales on which we gauge changes meaningful to us, the horse and burro remain *personae non gratis* for contemporary conservationists and restorationists.

It should now be obvious that appropriate spatial as well as temporal scale is also crucial for distinguishing between native and exotic species. Every known species is native to some place on Earth. If our spatial scale of reference is global or planetary, then every earthly species is native to every earthly place. What more circumscribed spatial scale is appropriate for discriminating between native and exotic species? The back-to-the-Pleistocene advocates also think too big spatially, that is, they think in continental terms. Ecological spatial scales—patches, landscapes, biotic provinces—however are more appropriate, depending on the species in question. Some wide-ranging "cosmopolitan" species are native to many bioregions on several continents. The wolf is a good example [Harrington and Paquet, 1982]. At the opposite extreme, some species are endemic, that is, native to only a very restricted place. The Devil's Hole pupfish is a good example [Pister, 1974]. Considering intermediate spatial scales, the brown-headed cowbird, a nest parasite, is native to North America, but an alien in many North American bioregions [Brittingham and Temple, 1983]. Thus, for purposes of ecological restoration, it should be considered a noxious exotic to be eradicated in those areas outside its recent Holocene range. The southern magnolia is native to Texas, but not to all of Texas, a very large and ecologically diverse state [Wasowski, 1988]. Ecological restorationists in southeast Texas would do well to plant the species in restoration projects there, but not in those of other parts of the state.

The concept of a "naturalized" species seems to be a cross between the concepts of native and exotic species. According to Westman [1990, p. 252], "a naturalized species is defined as one that has been present so long among its associates that mutual coexistence (and dispersal) over a significant duration is demonstrated ... [but] it is unclear how long a species must be naturalized before it can be considered native." What is abundantly clear is that Westman regards a species' status as native, exotic, or naturalized to be determined not by reference to its place of evolutionary origin or vector of dispersal, but by reference to time. It is equally clear that both "present" (in a place) and "significant duration" in his definition of "naturalized" implicitly refer to ecological spatial and temporal scales. Besides being present in a place for a fairly long time (in ecological measures of time), but not long enough to be regarded as native, an additional ecological consideration is necessary, however, to distinguish a naturalized species from a persistent noxious exotic. As Westman here indicates without elaboration, for a species to qualify as naturalized also requires "mutual coexistence" with its adopted native (and fellow naturalized) associates. A naturalized species, in other words, is a well-established non-native that is also a well-behaved citizen of its adopted biotic community. That is, at the very least, to qualify as naturalized a non-native species must not displace or extirpate the native species in its adopted habitat, either by competitive exclusion or depredation and, more positively, if it turns out to be of use as habitat or food for the fellow citizens of its adopted biotic community, so much the better for its naturalized status. An example of naturalized species, so understood, provided by Westman [1990] are eucalypts in coastal California. Although ecological restorationists are unlikely to try proactively to establish naturalized species, they may be more tolerant of them, in their on-going management efforts, than they are of aggressive exotics [Westman, 1990]. Because of the conceptual morass that we are led into by the concepts or native and exotic (alien, non-indigenous) species, these concepts are gradually giving way to the concept of "invasive" species—species that competitively exclude other species and thus diminish biodiversity in the places they invade [Lodge and Shrader-Frechette, 2003].

Ecological restoration is an important component of the "transdiscipline" of conservation biology, the ultimate goal of which is the preservation of biodiversity [Groom *et al.*, 2006]. Without an acute sensitivity to considerations of spatial scale, however, management practices clothed in the mantle of biodiversity may be misguided. For some have argued—self-servingly, one suspects—that introducing exotic game species "enhances" the biodiversity of host communities [Tanner *et al.*, 1980]. But this is a specious argument when we consider biodiversity in respect to a hierarchy of spatial scales and levels of biological organization. Clear Lake in California, to take a case in point, had only twelve native fish species; it is now home to twenty-three [Moyle, 1989]. Thus its fish fauna is nearly twice as diverse as in its pre-Columbian Holocene condition. But the biota of Clear Lake is now compositionally similar to many other aquatic communities, reducing biodiversity at the community level of biological organization, that is, reducing biodiversity. More troubling, five of its native fishes were extirpated, of which two are now globally extinct, as a result of the deliberate anthropogenic introduction of what proved to be invasive non-indigenous fishes. According to Noss [1995, p. 35], "the global scale is the most critical scale for evaluating these kinds of changes." Of course, it must be remembered that sometimes the introduction—whether direct or indirect, intentional or unintentional—of particularly invasive or aggressive exotic species can dramatically decrease biodiversity at every scale [Coblenz, 1990]. A few such introductions are infamous: the aforementioned unintentional introduction of the sea lamprey and alewife in the upper Great Lakes; the unintentional introduction of the brown tree snake on Guam; the intentional introduction of kudzu to the southeastern United States; and the intentional introduction of the Nile Perch in Lake Victoria.

A SCALAR DISTINCTION BETWEEN PRE- AND POST-INDUSTRIAL HUMAN DISTURBANCE

The contemporary "flux of nature" paradigm in ecology, however, raises more fundamental and more challenging questions for restorationists in particular and conservationists in general. If human beings have been an ecological force on every continent, except Antarctica, throughout the Holocene; if disturbance, both anthropogenic and nonanthropogenic, has always been frequent, violent, and ubiquitous; and if, as a consequence, the landscape has always been a mosaic of ever-shifting patches, why should we be concerned with ecological restoration at all? The species composition of a given site has always been changing. What's wrong with the way things presently are? As Pickett and Ostfeld [1995, p. 273] note, "For all its scientific intrigue, the flux of nature is a dangerous metaphor. The metaphor and the underlying ecological paradigm may suggest to the thoughtless and greedy that since flux is a fundamental part of the natural world, any human-caused flux is justifiable." I have given reasons why targets selected in reference to the evolutionary time scale are inappropriate for ecological restoration; and I have given reasons why more distant points in the ecological time scale are also inappropriate targets for ecological restoration. But I haven't so far given any reasons why very recent points in the ecological time scale are inappropriate targets for ecological restoration. In Dane County, Wisconsin, for example, why not ecologically restore a retired farm, such as Leopold purchased in 1935, to its condition in the 1920s, rather than to its condition in 1830s [Meine, 1988]?

Eric Higgs [1996] hints at a scientifically defensible answer. What he calls "good" ecological restoration should exhibit "functional success." In general, according to Higgs [1996, p. 343], functional success is achieved when "biogeochemical processes" in restored ecosystems "operate normally." In other words, a target criterion for ecological restoration should be a condition that Aldo Leopold [1999b] called "land health" or a condition currently called "ecosystem health" [Costanza *et al.*, 1992]. Expressed in the terms stipulated above, a good restoration should also rehabilitate a site. A site might be rehabilitated without being restored, by establishing a suite of functional species that never existed there before. But a site should not be restored without also being rehabilitated. The examples of biogeochemical processes, which may be normal or abnormal, given by Higgs [1996, p. 343] are "flushing rates, ion exchanges, and decomposition." Leopold [1999b] stressed rates of soil erosion, loss or gain of soil fertility, amplitude of variation in stream flow (the "flashiness" of streams), length of food chains, complexity of food webs, and amplitude of variation in animal population cycles. The biogeochemical processes on unrestored sites affected by urban and suburban development, modern agriculture (especially industrial agriculture), and industrial forestry, unfortunately, do not function normally, that is, they do not manifest land or ecosystem health. Thus, "restoring" a retired farmstead to row crops and continuously grazed pastures would not be appropriate or "good" ecological restoration; it would not also rehabilitate it.

To counter the danger of the flux-of-nature metaphor and the underlying ecological paradigm, Pickett and Ostfeld [1995] identify three general ethical limitations on "human flux" in "the natural world"—physiological, historical, and evolutionary limitations. Industrial human beings challenge organisms with a suite of synthetic molecules that they are not adapted to handle. That's an example of the physiological limitation. A given site may not have the seed bank to respond to a historically unprecedented anthropogenic alteration, such as a strip mine or clear cut. That's an example of the historical limitation. Interrupting historical patterns of gene flow within populations of species, by isolation, or by artificially providing the opportunity for hybridization are examples of the evolutionary limitation.

Pickett and Ostfeld [1995] provide more general, scalar criteria for assessing anthropogenic changes imposed on nature. They identify "two characteristics of a human-induced flux [that] would suggest that it would be excessive: fast rate and large spatial extent" [Pickett and Ostfeld, 1995, p. 274]. For example, a bison herd passing over a patch of prairie denudes and tramples the grasses and forbs. The effect might be compared to plowing. But the same prairie patch might not be disturbed by a passing bison herd in the same way for a dozen years or more, while annual plowing would be an example of an anthropogenic disturbance or flux at an excessive rate—that is, of a temporal scale that exceeds the historical limitations of a site. To take another example, windfalls break up the continuity of forests. So do exurban real estate developments. If such patchy anthropogenic clearings were widely scattered in spatial distribution, they would not be ecologically problematic. But if their spatial distribution reduces an otherwise continuous forest to all edge, making it unfit habitat for interior obligates, then exurban real estate development becomes ethically reprehensible.

IS ECOLOGICAL RESTORATION HUBRISTIC?

Considerations of temporal and spatial scale, therefore, make it possible for us to distinguish between industrial and non-industrial human disturbance in a scientifically justifiable and non-arbitrary way, without invoking an evolutionarily suspect distinction between "man" and nature. Ecological restoration, however, must acknowledge the existence of preindustrial human disturbances and simulate at least some of them in recovery plans. Ecological restoration therefore presupposes ongoing site management, which might entail such activities as prescribed burns or regulated hunting, in addition to fighting off invasive species. Purists may charge that management is a form of human arrogance and hubris, because it assumes that human beings have more predictive knowledge about the workings of natural systems than can be legitimately claimed [Willers, 1992]. We should instead put at least a few favored places back the way they were before we mucked with them, and then leave them alone. Nature knows best. To this complaint I suggest two rejoinders.

First, who is "we"? We human beings or we *industrial* human beings whose disturbances have been so frequent and widespread that they have exceeded physiological, historical, and evolutionary limitations. Human beings have for a hundred centuries at least been part of terrestrial ecosystems everywhere except Antarctica. Therefore, on-going restoration management should aim not at controlling a landscape, but rather, as just noted, at simulating the well-integrated ecological effects of the ecologically incorporated indigenous *Homo sapiens*—to the extent that we can determine what they were, and to the extent that they did not exceed the ecological imitations specified by Pickett and Ostfeld [1995]. And, of course, restoration management should be adaptive, changing both its methods and goals in response to experience [Holling, 1978].

Second, one of the elements of ecological restoration most emphasized by restoration theorist William R. Jordan [1991] is the spiritual benefits it affords participants. The traditional wilderness idea either excludes people or relegates them to the role of voyeurs, attempting to move through the landscape with minimal effect [Plumwood, 1998]. The restoration idea provides a more active and meaningful role for human participants as enablers and co-creators [Jordan, 1991].

SUMMARY AND CONCLUSION

Let me sum up what I have tried to convey here. At first blush ecological restoration seems simple and easy in respect to ends, however complex and difficult it may be in respect to means. Ecological restoration should aim to recreate the original condition of a site—that is, the condition of the site at settlement. In what may be the first manifesto of ecological restoration, that is exactly what Aldo Leopold [1999a] said it should be about [Meine, 1988].

This simple and easy understanding of the appropriate norm for ecological restoration is premised on two myths that then prevailed—the wilderness myth and the ecological-equilibrium myth. Subsequent changes in cultural geography and ecology have made ecological restoration more problematic than in Leopold's day. *Homo sapiens* has been a ubiquitous and ecologically significant species on all continents except Antarctica throughout the Holocene. And the individualistic flux-of-nature paradigm in ecology has replaced the holistic balance-of-nature paradigm. If nature is but a series of human-influenced, ubiquitously disturbed, ever-changing landscapes, what moment—what snapshot from the past—should we attempt to restore?

Some prominent conservationists have suggested that the norm for ecological restoration in the Western Hemisphere should be the end of the Pleistocene period, because *Homo sapiens* was not a significant species in the Western Hemisphere until the advent of the Holocene. The end of the Pleistocene, that is, is the last time in which the Western Hemisphere was in a perfectly "natural" condition, a truly wilderness condition. That conclusion presupposes that the appropriate temporal scale for ecological restoration is evolutionary time. I suggest instead that the appropriate temporal scale for ecological restoration is ecological time, defined by the periodicity of ecological disturbances, by disturbance regimes. Correspondingly, the appropriate spatial scale for ecological restoration should also be defined ecologically—in terms of such units as landscapes and bioregions.

Ecological scales are more in accord with conventional intuitions about restoration, which make the condition of an area prior to disturbance and conversion by industrial *Homo sapiens* the target for restoration efforts. They are also useful in coherently distinguishing between native, exotic, and naturalized species. Disturbances wrought by industrial *Homo sapiens* exceed the limitations of ecological temporal and spatial scales. Finally, because *Homo sapiens* was a significant ecological force in the New World throughout the Holocene, to be successful, New World ecological restoration must simulate well-incorporated, preColumbian anthropogenic ecological disturbances, principally through prescribed burning and regulated hunting. Such activities provide contemporary people with an opportunity to interact meaningfully and positively with nature.

BIBLIOGRAPHY

[Allen and Starr, 1982] T. F. H. Allen and T. B. Starr. *Hierarchy: perspectives for ecological complexity.* University of Chicago Press, 1982.

[Allen and Hoekstra, 1992] T. F. H. Allen and T. Hoekstra. *Toward a unified ecology.* Columbia University Press, 1992.

[Brittingham and Temple, 1983] M. C. Brittingham and S. A. Temple. Have cowbirds caused forest songbirds to decline? *BioScience* 33: 31–35, 1983.

[Callicott *et al.*, 1999] J. B. Callicott, L. B. Crowder, and K. Mumford. Current normative concepts in conservation. *Conservation Biology* 13: 22–35, 1999.

[Caro, 2007] T. Caro. The Pleistocene re-wilding gambit. *Trends in Ecology and Evolution* 22: 281–283, 2007.

[Chapron, 2005] G. Chapron. Re-wilding: Other projects help carnivores stay wild. *Nature* 437: 318, 2005.

[Clements, 1916] F. E. Clements. Plant succession: An analysis of the development of vegetation. *Carnegie Institution Publication*, no. 42, Washington, DC, 1916.

[Clements, 1936] F. E. Clements. Nature and structure of the climax. *Journal of Ecology* 24: 252–284, 1936.

[Coblenz, 1990] B. E. Coblenz. Exotic organisms: a dilemma for conservation biology. *Conservation Biology* 4: 261–264, 1990.

[Costanza *et al.*, 1992] R. Costanza, B. G. Norton, and B. D. Haskell. *Ecosystem health: New goals for environmental management.* Washington, DC: Island Press, 1992.

[Curtis and McIntosh, 1951] J. T. Curtis and R. P. McIntosh. An upland forest continuum in the prairie-forest border region of Wisconisn. *Ecology* 32: 476–496, 1951.

[Davis, 1984] M. B. Davis. Climatic instability, time lags, and community disequilibrium. In J. Diamond and T. J. Case (eds.), *Community ecology*, pp. 269–284. New York: Harper and Row, 1984.

[Denevan, 1992] W. Denevan (ed.), *The native populations of the Americas in 1492*, 2nd Edition. Madison: University of Wisconsin Press, 1992.

[Denevan, 1998] W. Denevan. The pristine myth. In J. B. Callicott and M. P. Nelson (eds.), *The great new wilderness debate*, pp. 414–442. Athens: University of Georgia Press, 1998.

[Dinnerstein and Irivin, 2005] E. Dinnerstein and W. R. Irvin. Rewilding: No need for exotics as natives return. *Nature* 437: 476, 2005.

[Dobyns, 1983] H. F. Dobyns. *Their numbers became thinned: Native American population dynamics in eastern North America.* Knoxville: University of Tennessee Press, 1983.

[Donlan and Martin, 2004] C. J. Donlan and P. S. Martin. Role of ecological history in invasive species management and conservation. *Conservation Biology* 18: 267–269, 2004.

[Donlan *et al.*, 2005] J. Donlan, H. W. Green, J. Berger, C. E. Bock, J. H. Bock, D. A. Burney, J. A. Estes, D. Foreman, P. S. Martin, G. W. Roemer, F. A. Smith, and M. E. Soulé. Re-wilding North America. *Nature* 436: 913–914, 2005.

[Donlan *et al.*, 2006] J. Donlan H. J. Berger, C. E. Bock, J. H. Bock, D. A. Burney, J. A. Estes, D. Foreman, P. S. Martin, G. W. Roemer, F. A. Smith, M. E. Soulé, and H. W. Greene. Pleistocene rewilding: An optimistic agenda for Twenty-first century conservation. *American Naturalist* 168: 660–681, 2006.

[Egan, 1995] D. Egan and S. Glass. A new tool for killing woody species (Wisconsin). *Restoration and Management Notes* 13: 134–135, 1995.

[Falk *et al.*, 2006] D. A. Falk, M. A. Palmer and J. B. Zedler (eds.), *Foundations of restoration ecology: The science and practice of ecological restoration.* Washington, DC: Island Press, 2006.

[Frith, 1969] H. J. Frith. *Kangaroos.* Cheshire, Melbourne, 1969.

[Garrot *et al.*, 1993] R. A. Garrot, P. J. White, and C. A. Vanderbilt-White. Overabundance: An issue for conservation biologists? *Conservation Biology* 7: 946–949, 1993.

[Gauthier-Pilters, 1981] H. Gauthier-Pilters. *The camel: Its evolution, ecology, behavior, and relationship to man.* Chicago: University of Chicago Press, 1981.

[Grayson, 1977] D. K. Grayson. Pleistocene avifaunas and the overkill hypothesis. *Science* 183: 691–693, 1977.

[Great Lakes Fishery Commission, 1992] Great Lakes Fishery Commission. Strategic vision of the Great Lakes Fishery Commission for the decade of the 1990s. Great Lakes Fishery Commission, Ann Arbor, Michigan, 1992.

[Gleason, 1926] H. A. Gleason. The individualistic concept of the plant association. *Bulletin of the Torrey Botanical Club* 53: 1–20, 1926.

[Groom *et al.*, 2006] M. J. Groom, G. K. Meffe, and C. R. Carroll. *Principles of conservation biology*, 2nd Edition. Sunderland, Massachusetts: Sinauer Associates, 2006.

[Hagen, 1992] J. B. Hagen. *An entangled bank: The origins of ecosystem ecology.* New Brunswick, NJ: Rutgers University Press, 1992.

[Harrington and Paquet, 1982] F. H. Harrington and P. C. Paquet. *Wolves of the world: Perspectives of behavior, ecology, and conservation.* Park Ridge, NJ: Noyes Publications, 1982.

[Holling, 1978] C. S. Holling. *Adaptive environmental assessment and management.* New York: John Wiley & Sons, 1978.

[Holling, 1992] C. S. Holling. Cross-scale morphology, geometry, and dynamics of ecosystems. *Ecological Monographs* 62: 447–502, 1992.

[Hunter, 1996] M. L. Hunter, Jr. Benchmarks for managing ecosystems: Are human activities natural? *Conservation Biology* 10: 695–697, 1996.

[Jaffe, 2006] E. Jaffe. Brave old world: The debate over rewilding North America with ancient animals. *Science News* 170: 314–316, 2006.

[Jordan, 1991] W. R. Jordan. A new paradigm. *Restoration and Management Notes* 9: 64–65, 1991.

[Jordan *et al.*, 1987] W. R. Jordan, M. E. Gilpin, and J. D. Aber (eds.), *Restoration ecology: A synthetic approach to ecological research.* New York: Cambridge University Press, 1987.

[Kay, 1994] C. Kay. Aboriginal overkill: the role of Native Americans in structuring Western ecosystems. *Human Nature* 5: 359–398, 1994.

[Kay, 1995] C. Kay. Aboriginal overkill and native burning: Implications for modern ecosystem management. *Western Journal of Applied Forestry* 10: 121–126, 1995.

[Kretch, 1999] S. Kretch. *The ecological Indian: Myth and history.* New York: W.W. Norton, 1999.

[Kristof, 2005] N. Kristof. Where deer and lions play. *New York Times*, Late Edition (East Coast). December 13: A33, 2005.

[Leopold, 1949] A. Leopold. *A Sand County almanac: And sketches here and there.* New York: Oxford University Press, 1949.

[Leopold, 1991] A. Leopold. Conservation: In whole or in part? In S. L. Flader and J. B. Callicott (eds.), *The river of the mother of God and other essays by Aldo Leopold*, pp. 310–319. Madison: University of Wisconsin Press, 1991.

[Leopold, 1999a] A. Leopold. What is the University of Wisconsin arboretum, wild life refuge, and forest experiment preserve? *Transactions of the Wisconsin Academy of Sciences, Arts, and Letters* 87: 15–17, 1999.

[Leopold, 1999b] A. Leopold. *For the health of the land: Previously unpublished essays and other writings.* Washington, DC: Island Press, 1999.

[Lodge and Shrader-Frechette, 2003] Lodge, D. and K. Shrader-Frechette. Non-indigenous species: Ecological explanation, environmental ethics, and public policy. *Conservation Biology* 17: 31–37, 2003.

[McIntosh, 1975] R. P. McIntosh. H. A. Gleason, "individualistic ecologist," 1882–1975: His contribution to ecological theory. *Bulletin of the Torrey Botanical Club* 102: 253–273, 1975.

[McIntosh, 1985] R. P. McIntosh. *The background of ecology: Concept and theory.* New York: Cambridge University Press, 1985.

[MacPhee, 1999] R. D. E. MacPhee (ed.), *Extinctions in near time: Causes, contexts, and consequences.* New York: Kluwer Academic / Plenum Publishers, 1999.

[Marshall, 1988] L. G. Marshall. Land mammals and the great American exchange. *American Scientist* 76: 380–388, 1988.

[Martin, 1973] P. S. Martin. The discovery of America. *Science* 179: 969–974, 1973.

[Martin and Burney, 1999] P. S. Martin and D. Burney. Bring back the elephants! *Wild Earth* 9(1): 57–65, 1999.

[Martin and Klein, 1984] P. S. Martin and R. Klein (eds.), *Quarternary extinctions: A prehistoric revolution.* Tucson: University of Arizona Press, 1984.

[Marshall, 1998] R. Marshall. The problem of the wilderness. In J. B. Callicott and M. P. Nelson (eds.), *The great new wilderness debate*, pp. 85–96. Athens: University of Georgia Press, 1998.

[Meine, 1988] C. Meine. *Aldo Leopold: His life and work.* Madison: University of Wisconsin Press, 1988.

[Moyle, 1989] P. B. Moyle. Message from the president. *Newsletter of the Introduced Fish Section.* American Fisheries Society 9: 1–2, 1989.

[Norton, 2005] B. Norton. *Sustainability: a philosophy of adaptive ecosystem management.* Chicago University Press, 2005.

[Noss, 1995] R. F. Noss. *Maintaining ecological integrity in representative reserve networks.* World Wildlife Fund-Canada, World Wildlife Fund-United States, Toronto and Washington, DC, 1995.

[Noss and Cooperrider, 1994] R. F Noss and A. Y. Cooperrider. *Saving Nature's Legacy: protecting and restoring biodiversity.* Washington: Island Press, 1994.

[O'Neill *et al.*, 1986] R. V. O'Neill, D. L. DeAngelis, J. B. Wade, and T. F. H. Allen. *A hierarchical concept of ecosystems.* New Jersey: Princeton University Press, Princeton, 1986.

[Pattee, 1973] H. Pattee (ed.), *Hierarchy theory: the challenge of complex systems.* New York: Braziller, 1973.

[Peterson and Parker, 1998] D. L. Peterson and V. T. Parker (eds.), *Ecological scale: theory and applications.* New York: Columbia University Press, 1998.

[Pister, 1974] E. P. Pister. Desert fishes and their habitats. *Transactions of the American Fisheries Society* 103: 531–540, 1974.

[Pickett and Ostfeld, 1995] S. T. A. Pickett and R. S. Ostfeld. The shifting paradigm in ecology. In R. L. Knight and S. F. Bates (eds.), *A new century for natural resources management*, pp. 261–278. Washington, DC: Island Press, 1995

[Pickett and White, 1985] S. T. A. Pickett and P. S. White. *The ecology of natural disturbance and patch dynamics.* Orlando, Florida: Academic Press, 1985.

[Plumwood, 1998] V. Plumwood. Wilderness skepticism and wilderness dualism. In J. B. Callicott and M. P. Nelson (eds.), *The great new wilderness debate*, pp. 652–690. Athens: University of Georgia Press, 1998.

[Randall, 2000] J. M. Randall. Improving management of nonnative invasive plants in wilderness and other natural areas. In D. N. Cole, S. F. McCool, W. T. Borrie, and J. O'Loughlin (eds.), *Wilderness science in a time of change, Vol. 5: wilderness ecosystems, threats, and management*, pp. 64–73, 1999, May 23–27; Missoula MT, U.S. Department of Agriculture, Forest Service, Rocky Mountain Research Station, Ogden, Utah, 2000.

[Rolston, 1998] H. Rolston. The wilderness idea reaffirmed. In J. B. Callicott and M. P. Nelson (eds.), *The great new wilderness debate*, pp. 367–386. Athens: University of Georgia Press, 1998.

[Rubenstein *et al.*, 2006] D. R. Rubenstein, D. I. Rubenstein, P. W. Sherman, and T. A. Gavin. Pleistocene parks: Does re-wilding North America represent sound conservation for the 21st century? *Biological Conservation* 132: 232–238, 2006.

[Schlaepfer, 2005] M. A. Schlaepfer. Re-wilding: A bold plan that needs native megafauna. *Nature* 437: 951, 2005.

[Shay, 2005] S. Shay. Re-wilding: Don't overlook humans living on the plains. *Nature* 437: 476, 2005.

[Simpson, 1956] G. G. Simpson. *Horses: The story of the horse family in the modern world through six million years of evolution.* New York: Scribner, 1956.

[Sloan and Padoch, 1988] J. D. Sloan and C. P. Padoch (eds.), *People of the tropical rain forest.* Berkeley: University of California Press, 1988.

[Smith, 2005] C. I. Smith. Re-wilding: Introduction could reduce biodiversity. *Nature* 437: 318, 2005.

[Soulé, 1990] M. E. Soulé. The onslaught of alien species, and other challenges in the coming decades. *Conservation Biology* 4: 233–240, 1990.

[Soulé, 1995] M. E. Soulé. The social siege of nature. In M. E. Soulé and G. Lease (eds.), *Reinventing nature: Responses to postmodern deconstruction*, pp. 137–170. Washington, DC: Island Press, 1995.

[Tanner *et al.*, 1980] H. A. Tanner, M. H. Patriarche, W. J. Mullendore. *Shaping the world's finest freshwater fishery.* Lansing: Michigan Department of Natural Resources, 1980.

[Tansley, 1935] A. G. Tansley. The use and abuse of vegetational concepts and terms. *Ecology* 16: 284–341, 1935.

[Tobey, 1981] R. Tobey. *Saving the prairies: The life-cycle of the founding school of American plant ecology, 1895–1955.* Berkeley: University of California Press, 1981.

[US Congress, 1993] U.S. Congress, Office of Technology Assessment. Harmful Non-indigenous Species in the United States. Washington, DC: U.S. Government Printing Office, 1993.

[Wasowski, 1988] S. Wasowski. *Native Texas plants: Landscaping region by region.* Houston: Gulf Publishing Co., 1988.

[West, 1964] R G. West. Inter-relations of ecology and quarternary paleobotany. *Journal of Ecology* 52: 47–57, 1964.

[Westman, 1990] W. E. Westman. Park management of exotic species: problems and issues. *Conservation Biology* 4: 251–260, 1990.

[Whittaker, 1951] R. H. Whittaker. A criticism of the plant association and climatic climax concepts. *Northwest Science* 25: 18–31, 1951.

[Willers, 1992] B. Willers. Toward a science of letting things be. Conservation *Biology* 6: 605–607, 1992.

[Worster, 1994] D. Worster. *Nature's economy: A history of ecological ideas*, 2nd Edition. New York: Cambridge University Press, 1994.

HABITAT RECONSTRUCTION: MOVING BEYOND HISTORICAL FIDELITY

Sahotra Sarkar

1 INTRODUCTION

Today, only the very fortunate live and work in landscapes that do not bear the indelible mark of continuing highly intrusive anthropogenic transformation. Biota continue to be replaced by human artefacts or to be confined to small sub-optimal patches. Ecological and physical processes often manifest themselves only in forms dominated by the effects of human action. Besides the obvious harm to biota, the consequences of how land- and seascapes have been treated by human activities have typically included the depletion of natural resources, the deterioration of air and water quality, and our increased vulnerability to extreme weather events. We continue to alter climate in ways that may jeopardize the well-being of future generations.

All this is common knowledge, and often presented with much more rhetorical flourish than in the paragraph above. The standard twentieth-century response to these problems has been calls for increased reservation, setting aside land- and seascapes to preserve natural values. On land, at least superficially (that is, on paper if not in practice), this strategy has been quite successful during the last 25 years (1980–2005). By 2005, 17.1 million km^2 or 11.5% of Earth's terrestrial surface was supposed to be under some form of protection.[1] This is close to the 12% target, then considered visionary, that the Brundtland Commission recommended in 1987.[2] So far, the seas have not fared as well: only 2.35 million km^2 or 0.65% of the sea surface are under the aegis of a conservation plan.[3] However, efforts to designate more marine reserves continue.[4]

Nevertheless, if the long-term retention and enhancement of natural values is among our aspirations, reservation is not enough for at least six reasons.

1. The scale of the human impact on nature over the last few centuries has been so extreme that not enough areas with flourishing natural values remain.

[1]Naughton-Treves *et al.* [2005]. Most estimates are similar though the exact number depends on the criteria used to define protected areas.
[2]WCED [1987].
[3]Wood *et al.* [1987].
[4]See, for example, [Wood, 2007] for a global analysis aimed at protecting 1,038 fish and marine mammal species.

Handbook of the Philosophy of Science. Volume 11: Philosophy of Ecology.
Volume editors: Kevin deLaplante, Bryson Brown and Kent A. Peacock. General editors: Dov M. Gabbay, Paul Thagard and John Woods.

Even barely intact ecosystems (those with only a majority of their traditional biological components remaining viable) are becoming increasingly rare.

2. If genuine protection from human abuse is the relevant criterion, the numbers in the last paragraph are inflated. Some of these "protected" areas allow a wide range of human activities, including resource extraction so long as it is touted to be sustainable.[5]

3. Many of these protected areas are "paper parks," designated as such by some legal authority, but not implemented on the ground due to the lack or resources, political will, or both.

4. If biodiversity is one of the natural values that the areas are supposed to protect, almost 80% of the protected areas are not viable because they are less than 10,000 hectares in size[6] and thus not likely to ensure the long-term persistence of all their biota if they are isolated from other natural habitats.

5. Moreover, even if designation of new protected areas continues at the relatively high rate of the 1980–2005 period, which is unlikely, most new protected areas will continue to be small. Further transformation of the land will increasingly isolate these areas.

6. Perhaps most importantly, to maintain these individual protected areas in viable states (that is, to prevent habitat deterioration within them) will require active management: mere reservation will not be enough. There is no such thing as benign neglect any more: the human impact on Earth is now far too extensive and much too harsh.

For reasons such as these, it is by now widely recognized that we must move beyond reservation and actively manage habitats to enhance the natural values embodied in them.[7] Typically such efforts are viewed as ecological *restoration*. The aim of this paper is to argue that restoration as now understood by those who advocate it most systematically—for instance, the Society for Ecological Restoration International—is far too restrictive in the scope of its aims. The standard theoretical characterization of ecological restoration—though not always how it is practiced—is that it has two goals: ecological integrity and historical fidelity.[8] Both are normative goals that are supposed to guide policy. Ecological integrity is

[5]Naughton-Treves *et al.* [2005].

[6]This figure refers to any of the Types I–VI protected areas of the classification due to IUCN (the World Conservation Union, formerly known as the International Union for the Conservation of Nature and Natural Resources). If only strictly protected areas are considered the percentage would likely be even higher since these typically tend to be small in size.

[7]Hobbs & Norton [1996].

[8]See [Higgs, 2003, p. 95]. Hobbs and Norton [1996] also emphasized how restoration efforts (broadly construed, that is, what is being called habitat reconstruction here) should be guided by a dynamic criterion and a reference state criterion. Much earlier, Bradshaw and Chadwick [1980] had also implicitly embraced a dynamic criterion and a reference criterion (see the discussion below in Section 3).

what will be called here a *dynamic criterion*, that is, one specifying how a system should change over time.[9] Very roughly, it is a measure of how intact a system is in the way it behaves (see Section 4 for more detail). In contrast, *historical fidelity* is what will be called a *reference state* criterion, specifying, again roughly, what a system looks like as captured by its state variables (see Section 5).

We shall see, below, that ecological integrity may be questionable as a goal, or at least, as the only dynamic (or functional) goal of natural value enhancement programs. However, the real trouble with restoration is its insistence on historical fidelity as the (reference state) goal of such efforts. This paper will argue that there is rarely any justifiable normative ground for deifying historical fidelity, even without worrying about whether it is achievable in practice, or whether it is worth the resources it would consume.[10] Rather, habitat reconstruction should embrace the enhancement of a wide array of natural values through active intervention. Restoration, as the specific goal of habitat management, is a development of fairly recent vintage. The other natural values that are being advocated here formed part of many past habitat management efforts that have been called by a variety of names, especially before 1980—for instance, reclamation,[11] regeneration,[12] rehabilitation,[13] and revegetation,[14] besides, of course, reconstruction,[15] which is what this paper endorses.

To ground the later theoretical discussion in a tangible context, Section 2 presents an example in some detail: this is the case of the Balcones Canyonlands National Wildlife Refuge in central Texas, some 70 km west of Austin, Texas. Section 3 will note how the enhancement of degraded habitats, that is, their management for the enhancement of natural values, hereafter called *habitat reconstruction*, came to be conceptualized as restoration in the 1980s and 1990s. This is a complicated story, one that historians and philosophers have barely begun to explore,[16] and only a very few themes can be tracked here. Section 4 will note some problems with ecological integrity as the dynamic goal of reconstruction efforts but will then set that problem aside for more detailed exploration on some other

---

[9]This may initially appear strange because most definitions of ecological integrity explicitly refer to the state of an ecosystem (what species are in it, etc.) rather than how it is changing. However, when we probe deeper, the concern is typically over processes and, thus, about dynamics.

[10]Although this paper intends to give a fairly broad philosophical interpretation of what is called *ecological restoration* (and delves into some of its history), it does not address foundational issues about the science of *restoration ecology*. For an entry into that philosophical discussion, see Callicott [2002] and his similar contribution to this volume. Within restoration ecology, Higgs and Norton provide an influential pioneering discussion.

[11]See, for instance, [Kohnke, 1950; Costigan *et al.*, 1981; Schaller and Sutton, 1978]. Higgs [2003, p. 99] traces this term back to the late 1800s. In the United States, it was legally defined in 1977 by the Surface Mine Control and Reclamation Act [Jackson *et al.*, 1995].

[12]See, for instance, [Lewis, 1976].

[13]See, for instance, [US-NAS, 1974; Gudin and Syratt, 1975; Dibble and Bartha, 1979].

[14]See, for instance, [Gemmell, 1973; 1974; Johnson *et al.*, 1976].

[15]See, for instance, [Hall, 1941; Bradshaw, 1983].

[16]See, in particular, [Higgs, 2003; Hall, 2005]. A significant part of this paper is a response to [Higgs, 2003] which is a powerful statement of a different perspective.

occasion. Section 5 will argue the case against the pursuit of historical fidelity as a necessary normative goal of restoration. Section 6 will then explore other natural values that can also credibly (that is, with normative justification) serve as goals of habitat reconstruction. Section 7 will finally state the reconstructionist agenda in some detail. Section 8 will draw some conclusions. The most salient one will be noted here to prevent misunderstanding of the aims of this paper: much—though not all—of the *practice*, rather than the theory, of what is called ecological restoration remains untouched by this critique. That practice often does not adhere to the definition of restoration advocated by theorists—and that is how it should be. In fact, the reconstructionist agenda offers normative support for many of these broader practices. The analysis presented here should be viewed as an attempt to provide better foundations for those practices. Throughout, this paper draws heavily on examples from Texas because habitat reconstruction is typically best begun at home no matter where you happen to live.

2 THE BALCONES CANYONLANDS

The Hill Country of Texas is the eroded southeastern fringe of the semi-arid Edwards Plateau which is a large (> 70,000 km^2) oblong region of central Texas defined by its bedrock, thick flat layers of hard early Cretaceous limestone, sometimes 3,000 m deep. Its northern and southwestern limits are not geologically well-defined.[17] The Hill Country forms its southeastern border and gets its name from its many hills and valleys. Its northwest border is also not sharply defined, blending into the rest of the Edwards Plateau, but the southeast consists of the Balcones Escarpment, a now inactive fault line that resulted from a major upheaval in the Tertiary era. This is not a high cliff in spite of its name; rather, it is a band of canyon country also known as the Balcones Canyonlands. It marks a transition from the plateau to the plains of the south and east.

Around the city of Austin, in the north-eastern Balcones Canyonlands, natural habitat management is focused on two endangered migratory bird species, the Goldencheeked Warbler (*Dendroica chrysoparia*) and the Black-capped Vireo (*Vireo atricapillus*), which nest there during the summer.[18] Both winter in Mesoamerica and are endangered primarily because of the destruction of their nesting habitat in the United States. These are not the only species of conservation concern in the region. The limestone geology of the Edwards Plateau has led to the development of an intricate network of underground caves, fissures, and sinkholes. Erosion has led to the isolation of different underground karst cave systems from each other and resulted in the evolution of a suite of endemic cave invertebrates, several of which are endangered while the status of many others remain

[17] Fowler [2005]; Sarkar [forthcoming]. Most of the information about the Balcones Canyonlands National Wildlife Refuge used in this paper is from the latter source.

[18] Beatley *et al.* [1995]. During the discussion of the Balcones Canyonlands and other US examples, throughout this paper, the terms, "endangered," "threatened," etc., are being used in accordance with the US Endangered Species Act (1973).

to be properly investigated. These endangered troglobites include three insects, the Tooth Cave ground beetle (*Rhadine persephone*), the Kretschmarr Cave mold beetle (*Texamaurops reddelli*), and the Coffin Cave mold beetle (*Batrisodes texanus*), and four arachnids, the Bone Cave harvestman (*Texella reyesi*), the Bee Creek Cave harvestman (*Texella reddelli*), the Tooth Cave spider (*Neoleptoneta myopica*), and the Tooth Cave pseudoscorpion (*Tartarocreagris texana*).

Ultimately, what pushed all these species to the brink of extinction in the 1980s was the rapid urban expansion in and around Austin, and the concomitant transformation of natural habitat into residential and industrial developments. In response, during the early 1990s, the City of Austin, Travis County (which surrounds Austin), regional environmental organizations, and developers embarked on an ambitious attempt to formulate a habitat conservation plan (HCP) for ten at-risk species so as to permit continued development without violating the (United States) Endangered Species Act (ESA). A tortuous planning process[19] resulted in the creation of two protected area systems: the Balcones Canyonlands National Wildlife Refuge owned and administered by the United States Fish and Wildlife Service (USFWS) and the Balcones Canyonlands Preserve System belonging to the City of Austin and Travis County. Each system is administered separately and the focus of this paper will be on the former.

The Refuge acquisition area (that is, the geographical region within which USFWS can acquire land) is about 80,000 acres, of which 31,000 acres are planned for outright acquisition and a further 10,000 acres are envisioned to be managed under conservation easements.[20] Between 1992, when the Refuge was first established and now, about 20,000 acres have been acquired and another 2,000 acres have been put under conservation easements. Except for bird and surface vegetation surveys, the acquired areas have not been fully explored to date. So far, four aquatic and seven terrestrial troglobites have been found in the caves in the Refuge though none is yet listed as endangered or threatened. Consequently, management efforts are focused on the two endangered bird species (the Vireo and the Warbler). However, at least one karst troglobite, the ground beetle, *Rhadine russelli*, is endemic to the Post Oak Ridge which falls within the Refuge acquisition area. Four other species, a flatworm possibly of the genus *Sphalloplana*, a spider of the genus *Cicurina*, a centipede of the genus *Theatops*, and a millipede of the genus *Speodesmus*, are all believed to be unstudied species and probably also endemic to the Post Oak area. Finally four of the other species, an asselid isopod, *Caecidotea reddelli*, an amphipod, *Stygobromus bifurcatus*, a pseudoscorpion of the genus *Tartarocreagris*, and a mold beetle, *Batrisodes reyesi*, are recognized as species of concern in Texas. Since the cave habitats of these species are within the Refuge, they automatically have some legal protection even without active management.

Several rare plants are also present in the Refuge. The most interesting of these is Texabama croton, *Croton alabamensis* var. *texensis*, which is locally abundant

[19]See Sarkar [forthcoming].

[20]USFWS [2001]. We switch to the British system of measurement in dealing with the Refuge because all documentation connected with it uses this system.

in many canyons in the Refuge. Besides the Refuge population, which spills off into adjacent private lands, there are two other populations in Texas, one about 80 km north of the Refuge at Fort Hood, and one about 5 km south, at Pace Bend Park. These populations form a variety of their own (and probably should have the status of a subspecies) and the only other known populations are found in Alabama (*C. alabamensis* var. *alabamensis*) some 1,000 km away.[21] Locally rare plants include the microendemic Sycamore-leaf snowbell, *Styrax platanifolius* ssp. *platanifolius*, and the Shooting Star, *Dodecatheon meadia*. Two other species that probably merit endangered status but are yet to be listed are found in the neighborhood though, so far, have not been recorded within the acquisition area. These are the Canyon mock-orange, *Philadelphus ernestii*, and the Bracted twistflower, *Streptanthus bracteatus*. An experimental population of the latter was established within the Refuge in the 1990s but not allowed to persist at the conclusion of the experiment.

Most of the Refuge is limestone-dominated terrain. The plant community of the Refuge forms an ecotone with the mixing of species from different ecological communities. So far, over 600 plant species have been recorded in the Refuge. The historical vegetation pattern of the Refuge, as for the rest of Edwards Plateau, remains very poorly known.[22] Consequently, no habitat reconstruction effort can seriously claim to be guided by historical antecedent. Today, most of the Refuge is dominated by a variety of juniper-oak woodlands. Ashe juniper (*Juniperus asheii*) is ubiquitous as part of such woodlands and as an aggressively spreading shrub in open lands. Its spread must be managed to create sufficient desirable Black-capped Vireo habitat (see below). The oaks present vary with the terrain of the Refuge. Spanish oak (*Quercus buckleyi*) is mostly found on steep slopes and ravines; plateau live oak (*Q. fusiformis*) grows best on plateau tops and deep soil in valleys; shin oak (*Q. sinuata* var. *breviloba*) is present as thickets on very rocky plateau sites; and post oak (*Q. stellata*) is present on open savannahs on higher elevations of the Refuge—this region is called Post Oak Ridge.

Besides the juniper-oak woodlands, open grasslands are a major feature of the Refuge. Texas is famous for its wildflowers and a large diversity of these are conspicuous at the beginning of the spring and fall flowering seasons. A major problem is the spread of the invasive grass, King Ranch bluestem, *Bothriochloa ischaemum*, which has come to dominate many of the open grasslands and savannahs. Management of K. R. bluestem has proved to be particularly difficult throughout central Texas. Canyon bottoms and narrow valleys along creeks have riparian woodland corridors containing American elm (*Ulmus americana*) and sycamore (*Platanus occidentalis*) among many other species. Like the caves, the creeks have not been fully explored and may contain at-risk amphibians such as the Jollyville Plateau Salamander (*Eurycea tonkawae*) which is a candidate for listing under the ESA.

[21] See [Ginzbarg, 1992] on designating this population as a variety. Van Ee *et al.* [2006] raise the possibility of upgrading it to a sub-species. If that happens, the population would fall under the aegis of the ESA and could potentially be listed at least as threatened.

[22] Fowler [2005].

Habitat management in the Refuge is complicated because the two endangered birds have different habitat requirements and their needs must be balanced.[23] The Black-capped Vireo prefers to nest in mid-successional shrub oak or other shrub vegetation not dominated by Ashe juniper.[24] In sharp contrast to the Vireo, the Golden-cheeked Warbler prefers to nest in climax mixed oak-juniper woodlands with tree heights between 3 and 6 m.[25] Refuge staff *restore*,[26] that is, create and maintain such habitat by the active manipulation of vegetation in a variety of ways including prescribed burning. The spread of Ashe juniper requires active control. Throughout the Refuge—and, for that matter, the entire Balcones Canyonlands—expanding populations of white-tailed deer (*Odocoileus virginianus*) are a major problem for revegetation programs because of excessive browsing. Deer herd control includes hunting besides encouraging improvement of deer habitat in adjacent lands in order to decrease pressure on the Refuge. An additional serious problem for the Vireo is nest parasitism by Brown-headed Cowbirds (*Molothrus ater*). Cowbirds are trapped during the breeding season and removed.

Management plans include identification and evaluation of sub-climax habitat for monitoring and active protection.[27] White-tailed deer herd management is also required for the Warbler, as is control of Brown-headed Cowbirds though the latter do not present as significant a problem for this species as it does for the Vireo. Oak wilt management has emerged as an important issue in several areas of the Refuge.[28] Finally, excessive public use of the Refuge has had to be curtailed in some areas during the breeding season. An important decision for Refuge management is to decide how much habitat should be managed for each endangered bird species, and where these should be located.

3 DEFINING ECOLOGICAL RESTORATION

Though there is a long and rich history of antecedents (for instance, in the highly influential work of Aldo Leopold in North America),[29] *ecological restoration*, as we distinctively understand that practice today, emerged in the early 1970s amidst the variety of reconstructionist practices that were mentioned in Section 1. The idea of restoration as the goal of habitat management was articulately promoted by two major figures, A. D. Bradshaw in the United Kingdom and John Cairns Jr. in the United States along with many other proponents. A series of conferences during the 1970s highlighted the importance of restoration. For instance, on Friday, 16 April 1971, the (United States) Association of Southeastern Biologists (ASB)

[23] [USFWS, 2001, p. 7].
[24] Grzybowski [1995].
[25] Ladd and Gass [1999].
[26] USFWS [2001, p. 38].
[27] USFWS [2001, p. 37].
[28] USFWS [2001, p. 37].
[29] For the early history, see [Hall, 2005]. This book is an unusually valuable resource because it explores the history of reconstruction efforts beyond the customary North American context.

and the Center for Environmental Studies of the Virginia Polytechnic Institute co-sponsored a Symposium on the Recovery and Restoration of Damaged Ecosystems at the 32nd Annual Meeting of the ASB at the University of Richmond.[30] The conference had taken three years to organize. The program proclaimed:

> Much attention has been given to the prophets of doom who are bewailing the fate of the planet, and the terrible things that are happening to the environment. However, little attention has been given to the constructive and positive approaches to coping with environmental problems, either through deliberately restoring damaged ecosystems or relieving the pollutional stress sufficiently for them to recover by themselves. This symposium will deal with practical problems and case histories described and discussed by people who have worked with these real problems and case histories. [Anonymous, 1971]

The speakers included Cairns on damaged streams, Ronald D. Hill on mines, Robert R. Curry on forests, and Thomas L. Linton on estuaries.[31] Both Hill and Linton explicitly used restoration to describe their programs in what was one of the earliest modern uses of the term in an ecological context.

Another conference at the Virginia Polytechnic Institute in March 1975 led to an edited book, *Recovery and Restoration of Damaged Ecosystems*.[32] Across the Atlantic, a North Atlantic Treaty Organization (NATO) conference at Reykjavik, Iceland, in July 1976 produced *The Breakdown and Restoration of Ecosystems*.[33] By 1980, Bradshaw and M. J. Chadwick had produced *The Restoration of Land*, essentially a textbook waiting for the emergence of an academic discipline.[34] A theme that runs throughout these works is that "[l]and [and, presumably, also aquatic habitats] is not a resource which automatically renews itself like rainfall and sunlight.... [T]he profligate days are over. We have to be prepared to restore exploited land to a condition approaching its original biological potential."[35] In 1981 the University of Wisconsin—Madison Arboretum began publishing *Restoration and Management Notes*, the first periodical dedicated to the emerging field; in 1999 it changed its name to *Ecological Restoration*, which remains what it is called today. The Society for Ecological Restoration (SER) was founded in 1988; its journal, *Restoration Ecology*, began appearing in 1992. *Restoration ecology*, as the discipline dedicated to ecological restoration, thus emerges as an organized discipline at roughly the same period as conservation biology.[36] There was—and, to some extent, there remains—tension between the two fields with conservation biologists fearing that the acceptance of restoration as a goal would diminish sup-

---

[30] Anonymous [1971].

[31] See [Cairns *et al.*, 1971; Hill, 1971; Curry, 1971; Linton and Cooper, 1971].

[32] Cairns *et al.* [1977].

[33] Holdgate and Woodman [1978].

[34] Bradshaw and Chadwick [1980].

[35] Bradshaw and Chadwick [1980, p. 2].

[36] For the history of conservation biology, see [Sarkar, 2005]. The crucial date was 1986 when the Society for Conservation Biology and its journal, *Conservation Biology*, were both founded.

port for protecting whatever relatively intact ecosystems that remained.[37] By and large, during the late 1980s and early 1990s, the two disciplines evolved independently.[38]

However, in spite of its increasing popularity, the use of the term, *restoration*, remained fluid. In *The Breakdown and Restoration of Ecosystems*, the editors identified restoration with rehabilitation. In their textbook, Bradshaw and Chadwick explicitly chose a broad interpretation:

> *Reclamation* is often used where some new use of the land will be involved; *rehabilitation* is sometimes confined to improvements of a visual nature. *Restoration* may be used only where land is to be returned to its former use and *renewal* or *redemption* may be used in a form that enables flexibility in planning its re-use. The term *revegetation* is generally confined to situations where the original vegetation has been destroyed and its reappearance in some form is to be encouraged. In this book *restoration* is used as blanket term to describe all those activities which seek to upgrade damaged land that has been destroyed and to bring it back into beneficial use, in a form in which the biological potential is restored.[39]

Note that there is no insistence on achieving similarity to some past historical state or trajectory even though such a more restricted construal of *restoration* is noted as a possibility. Eight years later, Cairns produced a similarly general definition: "restoration ecology is the full or partial placement of structural or functional characteristics that have been extinguished or diminished and the substitution of alternative qualities or characteristics than the ones originally present with the proviso that they have more social, economic, or ecologic value than existed in the disturbed or displaced state."[40]

Once again there is no endorsement of historical fidelity. Nor is there much concern for forging a universally acceptable definition.[41] Nevertheless, by 1988, with the founding of the SER, definitional controversies began to get explicit attention. Between 1988 and 1994 the SER produced three official definitions of ecological restoration.[42] It was during these discussions that fidelity to a historical past began to become central to how restoration was conceived: restoration consisted of aiding and abetting a system's return to that past. In 1992, the Committee on Restoration of Aquatic Ecosystems appointed by the United States National

---

[37] Cairns [1988].

[38] See, however, [Cairns, 1986; Janzen, 1988] for early attempts to bridge the divide. An influential later attempt is [Hobbs and Norton, 1996].

[39] [Bradshaw and Chadwick, 1989, pp. 2–3]; italics as in the original. Note how this definition implicitly invokes a reference state criterion (upgrading degraded land) and a dynamic criterion (restoration of biological potential).

[40] Cairns [1988, p. 3].

[41] Hobbs and Norton [1996] even argued for the irrelevance of attempts to define *restoration* universally.

[42] Higgs [2003, p. 102].

Research Council produced a report on the state of aquatic restoration science.[43] During the course of its meetings, the committee canvassed the literature for suggestions on how the term should be defined. It came up with 60 definitions.[44]

Meanwhile, the SER produced two more official definitions based on recommendations of internal Science and Policy Working Groups, in 1996 and 2002.[45] The former effort arose from a symposium held during the 1994 annual SER conference at East Lansing, Michigan. The resulting definition was complicated: "Ecological restoration is the process of assisting the recovery and management of ecological integrity. Ecological integrity includes a critical range of variability in biodiversity, ecological processes and structures, regional and historical context, and sustainable cultural practices."[46] What the second sentence means is less than clear. It is oddly allusive which is particularly strange for a definition: we are told what ecological integrity includes but not what it is (which would require letting us know what it also excludes) or how it should be operationalized.

In any case, the explicit reference to ecological integrity was viewed by many SER members as involving unwarranted abstraction. Moreover, most SER members in the 1990s were North Americans and, perhaps expectedly, the explicit recourse to management and the reference to cultural practices both fell afoul of the authority that the wilderness model of nature protection has long had among North American environmentalists.[47] Consequently, the 2002 definition was shortened and side-stepped all three issues: "Ecological restoration is the process of assisting the recovery of an ecosystem that has been degraded, damaged, or destroyed."[48] Given the status of the SER (now SER International) as the only major collective organization in the field, presumably this is the definition most practitioners would endorse in principle, whether or not their practices followed its precepts carefully.

Note how this definition takes for granted that we know what it means for an ecosystem to be degraded, damaged, or destroyed. In practice, this is sometimes simple: in the Balcones Canyonlands National Wildlife Refuge we know when King Ranch bluestem has replaced native grasses, though we have no idea which grasses were originally present and when. But most decision scenarios are even worse.

Ecosystems are not static entities: they evolve over time and natural evolution may well result in changed, perhaps even simplified, communities (for instance, when a plant community proceeds towards a climax association through competitive exclusion). What counts as a degraded, damaged, or destroyed ecosystem

[43]MRC [1992].

[44]John Cairns Jr. (personal communication, 2007). Cairns Jr. chaired the NRC committee in question.

[45]Higgs [2003, p. 96].

[46]Higgs [2003, p. 109]. Note that this definition is quite different from the one that came out of the 1994 meeting itself. That definition is discussed by Jackson *et al.* [1995]: "Ecological restoration is the process of repairing damage caused by humans to the diversity and dynamics of indigenous ecosystems."

[47]For more on the wilderness model, see the papers collected in Callicott and Nelson [1998] and Sarkar [1998; 1999].

[48]Higgs [2003, p. 110].

can often only be determined only if a goal (typically involving what are called *reference conditions* in the literature) has been specified. The specification must be done with care: returning to the Refuge, by and large, we will never know the original vegetation cover for most areas. Moreover, the unwanted spread of Ashe juniper thickets, which must be controlled, does not obviously "look" like anything other than prospering forest recovery. The 2002 SER definition skirts the issue of specifying goals but it is unavoidable.[49]

Not requiring explicit attention to the goal of restoration also differentiates the 2002 SER definition from the earlier National Research Council definition which was, in glaring contrast, focused entirely on the goals: "*restoration* is defined as the return of an ecosystem to a close approximation of its condition prior to disturbance. In restoration, ecological damage to the resource is repaired. Both the structure and the function of the ecosystem are recreated.... The goal is to emulate a natural, functioning, self-regulating system that is integrated with the ecological landscape on which it occurs."[50]

In an obvious advance from purely goals-based and purely process-oriented definitions, in 1996, Cairns and Heckman included both process and goal in their influential review of the state of the field.[51] The 2002 SER definition was an unfortunate step backward. Even more importantly, cultural practice plays no role in the 2002 SER definition (unlike the 1996 SER definition). This is particularly surprising because of how restoration had been conceptualized by almost all practitioners. Unlike, for instance, biodiversity conservation, which was also emerging to the forefront of environmental movements during this period, restoration was widely perceived as socio-cultural practice. As early as 1988 Janzen had argued for the *biocultural* restoration of tropical habitats as a means to conserve biodiversity.[52] Within the SER, Cairns had long advocated restoration as including an *ecosocietal* process, arguing that "the field requires the input and cooperation of society to be successful"; restoration included "the process of reexamining human society's relationship with natural systems so that repair and destruction can be balanced and, perhaps, restoration practices ultimately exceed destructive practices."[53] Among other benefits, Cairns and Heckman argued that restoration practices may enable urban communities to establish worthwhile relationships with natural systems.[54]

For all its beguiling simplicity, the frailties of the 2002 SER definition are thus obvious and perhaps beyond easy repair. More recently, Higgs has returned to

[49]Somewhat strangely, in stark contrast to the definition itself, the SER primer, specifying how restoration should be practiced, paid full attention to the determination of reference conditions. See SERI [2004]. (Note that the definition being criticized in the text was also presented in this primer.)

[50]NRC [1992, p. 18]; italics as in the original.

[51]Cairns and Heckman [1996].

[52]Janzen [1988].

[53]Cairns [1995, p. 9]. A similar point is argued by Jackson *et al.* [1995] in their commentary on the 1994 SER definition.

[54]Cairns and Heckman [1996, p. 175]. A similar point was emphasized by Hobbs and Norton [1996, p. 95].

the 1996 SER definition and has insisted on both historical fidelity and ecological integrity as part of the definition of restoration.[55] Higgs is on solid ground in endorsing ecological integrity: some such criterion is invoked in most typical restoration efforts, at least in North America. For instance, even the Refuge Vision Statement of the Balcones Canyonlands National Wildlife Refuge claims that its management plan is designed to be "consistent with overall conservation of regional ecological integrity."[56] Cultural practice—more specifically, participatory or "focal" practice—is also part of what Higgs views as good restoration though it is not part of his definition. Higgs also endorses what he calls "wild" design, and we will return to wildness as a natural value in Section 6. Higgs has provided the most sustained philosophical discussion of ecological restoration to date and, in what follows, we will use his definition. Though the next section will note some problems with the concept of ecological integrity, the real target of this analysis is the reliance on historical fidelity, whether it has adequate normative justification as a goal of social (including ecological) policy. Note that, whereas ecological integrity disappears from the 2002 SER definition, that definition implicitly endorses historical fidelity since it requires *recovery* to the state of the habitat before it was degraded, damaged, or destroyed.[57] A critical evaluation of historical fidelity is thus not marginal to how ecological restoration is practiced today. The interesting question of *why* historical fidelity crystallized in the foreground of the mental maps of restorationists, especially in the 1990s, will be left for exploration on another occasion.

4 DOUBTS ABOUT INTEGRITY

Similarly, an adequately detailed evaluation of the concept of ecological integrity will also be left for another occasion. Recall how, in Section 1, we characterized ecological restoration as having two goals specified by a dynamic criterion and a reference state criterion. Note that this formulation is fully general: no matter how habitat reconstruction is construed, it will involve some dynamic criterion and some reference state criterion. What is at stake is which criteria should be chosen. Ecological restoration, as we have seen, uses historical fidelity to identify the desired reference state and we will examine that criterion in Section 5. Here we note a few worries about the use of ecological integrity as the dynamic criterion. Since these worries are fairly general, we will not distinguish between biological, ecosystem, and ecological integrity. The main problem has been that ecological integrity has never been fully satisfactorily defined.[58] Nor is there any compelling

[55]Higgs [2003, p. 122].

[56]USFWS [2001, p. 1]. This is just one of two references to ecological integrity in the two-paragraph statement.

[57]The SER primer also uses historical fidelity to define reference conditions and explicitly gives "prominence to the historically rich idea of 'recovery'" [SERI, 2004, p. 2].

[58]See, for instance, the desultory discussion of De Leo and Levin [1997], which remains one of the more elaborate attempts at providing a comprehensive account of integrity.

reason to expect that there is some "natural" definition of the concept. Like the concept of minimum viable population in conservation biology,[59] that of biological integrity was introduced by legislative fiat, when the United States Federal Water Pollution Control Act Amendments of 1972 stated that their objective was to "restore and maintain the chemical, physical, and biological integrity of the Nation's waters."[60] It is far from clear what the legislation meant by "integrity" beyond the quality of water and the reliability of its supply. However, that did not prevent the concept's scientific reification. In 1975, the United States Environmental Protection Agency's Office of Water and Hazardous Materials sponsored what it envisioned as a comprehensive symposium on the "integrity of water."[61] By this point integrity had already come to be reinterpreted as a property of any ecosystem, rather than of a water body. At the symposium, Cairns defined "[b]iological integrity ... as the maintenance of the community structure and function characteristic of a particular locale or deemed satisfactory to society"[62] and made an elaborate attempt to operationalize and quantify the concept. A few years later, Karr and Dudley produced a similar definition without the societal component: "the capability of supporting and maintaining a balanced, integrated, adaptive community of organisms having a species composition, diversity, and functional organization comparable to that of natural habitat of the region."[63]

There have been many variations on these themes[64] though, in recent years, there has been a tendency to produce multi-faceted measures of integrity without explicitly defining the concept.[65] Proponents have sometimes even argued for the replacement of biodiversity by biological integrity as a policy goal.[66] There have also been critics.[67] The point, though, is that there is no reason to suppose that a concept introduced by legislative fiat will necessarily be conceptually linked to (presumably more empirically-based) scientific concepts in such a way that it can be operationalized and quantified for scientific use. Perhaps the SER members had a point when they objected to the use of integrity in the 1996 SER definition of ecological restoration (see Section 3.) Given the relative lack of success with using ecological integrity, it may well behoove us to shift attention to what the concept was supposed to do for us to see if it may not be replaced by some other dynamic criterion.

For Higgs, ecological integrity is a metaphor.[68] It is related to ecological health but different from it and preferable because it is supposedly easier to define and more closely related to historical fidelity. Unfortunately, this is too vague to get

---

[59]For more on this example, see [Sarkar, 2005].

[60]See Agee [1977, p. 3].

[61]Ballentine and Guarraia [1977].

[62]Cairns [1977, p. 171].

[63]Karr and Dudley [1981, p. 56].

[64]See, for instance, [De Leo and Levin, 1997; Parrish *et al.*, 2003].

[65]See, for instance, [De Leo and Levin, 1997; Harwell *et al.*, 1999].

[66]Angermeier and Karr [1994].

[67]Sarkar [2005] has objected that the definitions of Karr and Dudley [1981] and Angermeier and Karr [1994] cannot be operationalized.

[68]Higgs [2003, pp. 122–124].

us very far. There is more help from those who have explicitly connected integrity to a variety of ways in which a potentially related concept, ecological stability, has been explicated within ecology. These include resistance, resilience, perturbation-tolerance, persistence, constancy, and reliability.[69] Of these, resistance (the inverse of the extent of change induced by a given perturbation on the system) and resilience (the rate at which a system recovers from a perturbation) have been most commonly invoked. For instance, Cairns took resistance to be an important component of integrity in his first attempt to quantify the concept.[70] Resilience is a goal that has repeatedly been emphasized in the literature[71] though Higgs has argued that resilience may be in conflict with what he takes to be integrity.[72] To the extent that each of these concepts measure the extent or way in which an ecosystem may avoid collapses, they are all singly or jointly appropriate as a dynamic criterion.

A theme that runs through many of these discussions is that, ultimately, the more self-sustaining a system is after reconstruction, the more successful the reconstruction has been.[73] In what follows we will take self-sustainability to be the dynamic criterion to judge success, and leave open the question of its relationship to integrity. The answer to that question will depend on how integrity is defined. We will also leave for another occasion the relation of self-sustainability to any of the ecological stability concepts mentioned above. Note, though, that self-sustainability will be a matter of degree. It can be quantitatively assessed through a variety of social and biological measures of intervention in the system: the cost of intervention, the amount of biomass introduced or removed, and so on, each of which provides an estimate of how much effort is required to maintain the system along the preferred trajectories.

Success does not necessarily require achieving a high degree of self-sustainability.[74] But it helps: if we acknowledge that success is a matter of degree, higher self-sustainability is associated with a higher degree of success. Returning, as usual, to the Balcones Canyonlands, the hope is that repeated prescribed burning will establish native grasses to such an extent that burning may eventually be reduced. Similarly, old growth Ashe juniper and oak forests, once established, will likely require little continued intervention to be maintained as good habitat for the Golden-cheeked Warbler. Both of these will constitute a high degree of self-

---

[69] For more on these concepts, see [Justus, 2007; Sarkar, 2007].

[70] Cairns [1977, p. 180].

[71] See, for instance, [Hobbs and Norton, 1996; Pavlik, 1996; White and Walker, 1997].

[72] Higgs [2001, p. 169]: "some integral ecosystems have relatively low levels of resilience, which explains their fragility and rarity."

[73] The significance of self-sustainability was noted by the United States National Academy of Science as early as 1974 in its attempt to characterize rehabilitation [US-NAS, 1974].

[74] Self-sustainability makes Higgs [2003, p. 169] uncomfortable because it seems to imply that restored systems are better if there is no human involvement in them. But this is only true if the notion of "self" excludes humans, that is, the systems are conceptualized without humans as components of them. No such assumption is being made in the text. Intervention, then, does not refer to all human interactions with the system; rather it refers to the additional interactions that are required beyond what was customarily achieved.

sustainability. In contrast, Black-capped Vireo habitat will probably have to be indefinitely created in order to prevent succession leading to mature Ashe juniper forests unsuitable for this species. In this case self-sustainability will be low but that is no indication of failure: rather, success or failure will depend on whether Vireo populations recover to safe levels.

5 PROBLEMS WITH FIDELITY

Let us now turn to the arguments for historical fidelity. Fidelity is not supposed to be interpreted as the exact replication of past ecosystems (which would be practically impossible in almost all cases). Rather, it means approximate achievement of past reference conditions with contextual factors, such as the information or resources available, determining what degree of approximation is sufficient. The strongest arguments for fidelity—perhaps not unexpectedly—are due to Higgs who has been developing a rationale for it for decades. For Higgs, historical fidelity or "historicity" is central to restoration. In his words: "In reviewing why historicity might be important, I have grown resolute in the belief that it is indispensable for theory and practice, no matter how social winds sculpt what we currently think ecological restoration is."[75] Since Higgs has produced the only sustained philosophical discussion of historical fidelity, this section focuses on his work.[76] He presents three explicit arguments for historicity: "nostalgia and the knowledge this brings of a better past; the capacity to create continuous stories that inform our understanding of a place, or what I call narrative continuity; and depth of time."[77] But behind these lurk what may be a much deeper and troubling worry: if we abandon history, "we will be giving in too much to the capricious nature of contemporary judgment."[78]

We will return to the worry about caprice later. Let us begin with nostalgia. Higgs is not naïve enough to think that the past was simply better than the present because it is the past. Rather:

> Why are we drawn to history in the first place? What is it about historical conditions that compels so much attention? ... The most obvious answer would be that the past is, or was, somehow better. But it is not better in any simple way, at least for most people. Past

[75]Higgs [2003, p. 157]. In its context in Higgs' original discussion, this quotation does not appear arrogant though it may seem that way out of context. This point is important to note because ecological restorationists (like biodiversity conservationists) have sometimes been accused of arrogance [Gobster, 2000].

[76]But Higgs is by no means alone, at least in North America, once we move beyond philosophers and include the broader restorationist community. For instance, the van Diggelen *et al.* [2001] embrace historical fidelity as the goal of "true" restoration though they are willing to use the term more broadly than Higgs. Allison [2004] at least partly endorsed historicity because it is supposed to define who we are.

[77]Higgs [2003, p. 132]. Allison [2004] also seems to rely on narrative continuity for the same reason.

[78]Higgs [2003, p. 131].

> landscapes ... represent a simple, less hurried time when fidelity to a more organic way of life was visible. But nostalgia ignores much of the difficulty of times past, and countervailing historical accounts are necessary for balance. In any case, the point is that the past shows an alternative sometimes better, model. [Higgs, 2003, p. 143]

But, if the past is only *sometimes* better, fidelity to the past should not be a *necessary* requirement imposed on all attempts to reconstruct habitats. We should only do so in those cases when we know that the past—some particular historical ecosystem—really is better. But what matters, then, are the criteria which make this historical ecosystem better. These criteria are what is normatively relevant as we decide what to do with habitats. Historical fidelity becomes merely a tool towards these other ends. Fidelity becomes relevant *only* if these other ends cannot be achieved without it or if these ends have some necessary connection with history.

In particular, if historical ecosystems are important because they provide a model of systems which are self-sustaining without intrusive human management, the relevant criterion is self-sustainability which, as we noted in Section 4, is reasonable as a dynamic criterion. Self-sustainability has no necessary connection with history. The past may even stand in the way of achieving the goal of self-sustainability. Recall, for instance, the much-studied case of the Mayan use of resources, an example that is compelling because of North American restorationists' infatuation with the pre-Columbian past. Assuming that the environmental resource overconsumption model of the classic Mayan collapse (roughly 800–900 C.E.) is at least partly correct,[79] the Mayan management of land hardly provides a model we should imitate. Moreover, there is much else in that Mayan past which we presumably would not want, for instance, the oppressive social hierarchy and the penchant for human sacrifice. In cultural contexts closer to us, presumably we would not be nostalgic over slavery, the domination of women, or colonialism; or for the Holocaust, even though the Nazis are supposed to have taken excellent care of nature. If we were designing a restoration plan for some place in the Deep South of the United States, historical fidelity to an environment (productive cotton fields) that was maintained through slavery hardly seems an ethically justifiable goal especially if, along with Higgs, we view our goals to include social processes along with ecological ones. Not everyone wants nostalgia.

A subsidiary theme, preference for simplicity, also runs through Higgs' defense of nostalgia. At one point, he writes: "most of us of a certain age can look back in one way or another to an era of greater integrity and simplicity";[80] shortly, thereafter, he explains his nostalgia: "the past [was] a simpler time with less fragmentation and a greater flow of ecological processes."; this is the "elegance" he longs for.[81] Attractive, though this may seem, it does not push the argument any further. If it is ecological simplicity that is at stake, we are faced with the problem

[79]See [Shaw, 2003].

[80]Higgs [2003, p. 144].

[81]Higgs [2003, p. 145]. Note that neither simplicity, nor the flow of ecological processes or elegance, has any necessary connection with history.

that, typically, damage to an ecosystem consists of the simplification of ecological communities. This cannot be what we are looking for, and the greater flow of ecological processes that Higgs calls for takes us back to self-sustainability and, perhaps, integrity, not historical fidelity through nostalgia. But Higgs, as noted earlier, is careful not to make a sharp distinction between ecological and cultural processes—restoration, for him, is as much a cultural process as an ecological one. But cultural simplicity also cannot be a reasonable goal in many, perhaps most, circumstances. Human well-being may well depend on the complexity of daily cultural interactions. Take the much-retold story of the Ik who were displaced from their traditional lands to create the Kidepo Valley National Park in Uganda.[82] Social life disintegrated and social processes were simplified as living became no more than a struggle for individual daily survival. There are many other such instances. Perhaps the only interpretation of simplicity that is defensible is one which refers to a situation in which human interactions with nature were less mediated by technology—recall Higgs' reference to "a more organic way of life." But, even then, what really matters is not the past, but rather a particular relation between humans and the rest of nature that may well be best achieved by imagining a new vision for the future. It does not *require* historical fidelity.

Turning to narrative continuity, Higgs argues that narratives associated with places make them meaningful to people when they come to value those places for what they are and, perhaps more importantly, what they may become through restoration efforts. Since human aspirations are central to the pursuit of all reconstructionist efforts, narratives may often provide important encouragement. But do narratives need historicity? It is easy to see that historical continuity helps narrative construction: a story about the past is a narrative about a place. But what does not follow is either that this is the only way in which a narrative can be associated with a place, or even that the past must dominate narrative construction. Nor does it follow that narratives, understood in the traditional sense of a story, are required to foster habitat reconstruction. People plant gardens, and this activity may well generate community and foster stewardship of habitats including the biota present on them. In Texas, the Parks and Wildlife Department (TPWD) encourages local residents to plant and maintain native vegetation, install birdbaths and ponds, and create other structures for use by wild animals.[83] The goal of this "Wildscapes Program" is to provide places for birds, small mammals, and other wildlife to feed and drink, escape from predators, and raise their young. Creating these "backyard habitats" is also easier than maintaining conventional lawns. Most native plants are hardy and drought-resistant and thus need little or no water or other care. These plants are also more tolerant of native insects and diseases. TPWD has a Texas Wildscapes certification program and, along with

[82] See [Turnbull, 1972]. However, some of Turnbull's conclusions have since been challenged—see [Heine, 1985].

[83] Damude *et al.* [1999]. When people create such non-historical natural habitats, they are often willing to defend them against restorationists—recall the controversy over attempted restoration in the Chicago metropolitan area in the mid-1990s when restorationists wanted to transform woodlands into shrublands and tall grass prairie guided by history [Gobster, 2000].

the National Wildlife Federation, also administers a Best of Texas Backyard Habitats certification program. To support the Program, biologists associated with the TPWD organize workshops and teach "wildscaping" courses in cooperation with local organizations. Texas Wildscapes is more than a backyard program. Besides homes, it embraces community parks, business offices, churches, and schools. Thousands of Texans participate in it and there is a tangible sense of community among those who attend the workshops and other events. Budgetary constraints have prevented TPWD from monitoring the program with quantitative evaluations of success though anecdotal reports suggest that such efforts are achieving some state-wide success.[84] It is a matter of taste whether we would want to call participation in such programs as embedding places in narratives. Either way, historical fidelity has nothing to do with the story.

The depth of time is supposed to be related to narrative continuity: "Continuity points to *time depth* ... [which] is the reach of history, the amount of time, and also the engagements that form between people and place over that interval."[85] But this claim does not yet provide a rationale for endorsing historical fidelity. Rather, establishing a relevant connection between fidelity and time depth is based on a claim about rarity: "The older an ecosystem is, judging by the length of time without major human simplification of processes and patterns, typically the rarer it is. Depth depends on rarity: they are really two sides of the same coin."[86] But, now the crucial normative criterion is rarity. Higgs explicitly notes that there is no necessary connection between rarity and history: "it is possible for something to be rare without being historical.... Rarity depends often on depth of history, but it can stand alone, too."[87] What we should conclude is that historical fidelity may sometimes guide us towards what is really important: increasingly rare ecosystems. Fidelity, once again, is a means, not an end. There is no reason to suggest that historical fidelity can never be important. In some circumstances trying to recapture a past use of the land may be an important social value: those are the situations when reconstruction may reduce to restoration provided that ecological integrity, rather than some other goal, is also taken as the dynamic criterion (see Section 7). But there is no reason to accept that achieving historical fidelity should be a *necessary* component of every habitat reconstruction plan. However, before we fully accept this conclusion, let us turn to what may be the most compelling argument for taking history seriously: if we do not, Higgs says, we may fall prey to the caprices of the present.

There are two responses to this argument. First, what we do with habitats is guided by our values including our dynamic criteria and other natural values that may serve as reference state criteria (see Section 6). As we shall see in Section 7, not much room is left for caprice after we negotiate through these values.

---

[84] Alford [2005]. Note that many restorationists have emphasized the role of explicit monitoring in restoration efforts [Hobbs and Norton, 1996]. The TPWD can be faulted on this ground [Margules and Sarkar, 2007].

[85] Higgs [2003, p. 154]; italics as in the original.

[86] Higgs [2003, pp. 154–155].

[87] Higgs [2003, p. 155].

Rather, the real problem is that there may be few truly different alternatives left once we implement all these values. Suppose, for instance, that habitats must be managed for biodiversity, productivity, and ecosystem services. Our choices are more likely to be overdetermined rather than left to the whims of contemporary caprice. Second, concern for historical fidelity hardly provides relief from possible caprice. There is no single past state for any ecosystem. Which period of time should we aspire to? In the so-called Old World (essentially, Africa, Asia, and Europe) this question cannot conceivably have a non-controversial answer. In Australasia and the Americas, the state of the land before European contact is often taken to be the ideal. In the Americas, at least, the authority of this view has largely come from the now antiquated belief that the First Nations did not actively manage much of the land.[88] Once we recognize the extent to which the First Nations constructed the ecosystems that were invaded by the Europeans, 1492, besides the introduced alien species, has no compelling ecological significance in spite of all its devastating moral, cultural, political, and social consequences.

Here comes space for caprice. In August 2005, with much aplomb, a group of scientists and a right-wing political activist announced a new *restoration* plan for North America.[89] They proposed to introduce and promote species closely related to North American megafauna from the late Pleistocene era, some 13,000 years ago. Partly the goal was to prevent Asian and African species, such as the Asian ass (*Equus hemionus*), Przewalski's horse (*Equus przewalskii*), Bactrian camel (*Camelus bactrianus*), African cheetah (*Acinonyx jubatus*), African elephant (*Loxodonta africana*), Asian elephant (*Elephas maximus*), and lion (*Panthera leo*), from becoming globally extinct. But the alleged real purpose was to restore the evolutionary potential of North American biota, and to assuage the guilt that the authors felt for the allegedly anthropogenic extinction of North American megafauna: "humans were probably at least partly responsible for the Late Pleistocene extinctions in North America, and our subsequent activities have curtailed the evolutionary potential of most remaining large vertebrates. We therefore bear an ethical responsibility to redress these problems."[90]

Within the conservation biology community, the proposal was largely greeted with derision mainly because it was designed (probably unintentionally) to turn conservation attention and resources away from countries of the South in which the megafauna were under serious threat.[91] The ethics of reallocating such resources away from the South, particularly from sub-Saharan Africa, had escaped the attention of the plan's proponents while they explored their moral responsibilities to long-extinct species. Biological critics pointed out that we really do not know

[88] The antiquated view has been repudiated by a vast body of historical research during the last 25 years, starting with the pioneering work of Cronon [1983]. Mann [2005] provides an engaging synthesis of what has recently been learnt of pre-Columbian America.

[89] Donlan *et al.* [2005; 2006]. The recognizable right-wing political activist was Dave Foreman—for more on Foreman's politics, see [Lee, 1995]; see, also, [Sarkar, 2005].

[90] Donlan *et al.* [2005, p. 913].

[91] Chaptron [2005].

what such massive introductions of exotic species would do to habitats.[92]

Nor do we know enough about the Pleistocene era to judge whether we can recover the ecological and evolutionary potential of that distant era.[93] Critics also pointed out that there were plenty of native North American species that were under threat and would benefit from "re-wilding" through their reintroduction to parts of their historical habitat from which they had been extirpated. These include the bison (*Bison bison*), pronghorn (*Antilcapra americana*), elk (*Cervus elaphus*), jack rabbit (*Lepus townsendii*), black-footed ferret (*Mustela nigripes*), bobcat (*Lynx rufus*), badger (*Taxidea taxus*), swift fox (*Vulpes velox*), various ground-dwelling squirrel species (*Spermophilus* spp.), and prairie dog species (*Cynomys* spp.). Moreover, the puma (*Puma concolor*) is a much closer relative of the long-extinct American cheetah (*Miracinonyx trumani*) than the African cheetah, and to prefer the latter to the former for re-introduction is biologically bizarre. Its proponents had never calculated the cost of the plan, and the resources that would be necessary even to initiate it would probably be much better deployed elsewhere. The Pleistocene rewilding proposal is a superb exercise of caprice, motivated in part by its proponents' fascination with megafauna which reflects Northern values of the moment. The point is that a deep concern for historical fidelity provided no guard against caprice.

We should also not forget that attempts to establish historical fidelity are typically inordinately expensive and, in most cases, impossible to achieve because of incomplete information and the uncertainty of ecological predictions.[94] It is expensive because data acquisition and the reconstruction of what the past looked like usually takes extraordinary effort. Sifting through historical archives for old photographs and records takes time. For many parts of Earth no such records exist even within historical time. Reconstructing the past from palynological records is even more difficult and expensive. Efforts spent in these direction may well be better spent in acquiring land and reconstructing it into something better. Moreover, even if we manage an accurate reconstruction of what some habitat was at a specified time in the past, we have no way to reliably predict how it would have changed under different management conditions. Let us also not forget that establishing historical fidelity would likely harm species that have come to benefit from the changes in the land—recall the discussion of the Goldencheeked Warbler in Section 2. We need a broader conception of natural values. Finally, in an era in which largely irreversible global change, in particular, climate change is beginning to dominate ecological changes in habitats, returning to some historical conditions may have little correlation with what can survive into the future.

[92]For a review of the criticisms, see [Rubenstein *et al.*, 2006].

[93]Leave aside the problem that *ecological potential* and *evolutionary potential* are rather vague terms, never clearly defined, let alone operationized for scientific use. There is little that is of positive value in Donlan *et al.*'s [2005; 2006] proposal.

[94]Note that Higgs [2003, p. 129] explicitly acknowledges and emphasizes these problems.

6 NATURAL VALUES

Natural values are those that promote the welfare of non-human biota or enhance aspects of the physical environment that are not the result of continuing human action.[95] They are human values, just like historical fidelity. There are two false contrasts that may potentially be read into this characterization and it will be useful to reject them explicitly. (1) No *type*-distinction is being endorsed between humans and the rest of nature. Rather, an operational distinction is being made between values on the basis of the extent to which they promote the flourishing of non-human features of the environment. The pursuit of ecologically destructive mineral extraction, for instance, oil extraction in the Niger Delta by Shell and other multi-national companies, falls at one end of the spectrum, in this case accompanied by gross human rights violations.[96] Planting native wildflowers along highways as, for instance, promoted by the Texas Department of Transportation since 1932,[97] falls at the other end. (2) It is not being suggested that reconstruction should be directed towards "pristine" aspects of the environment bereft of anthropogenic influence. That would be self-contradictory: reconstruction is human activity. Rather, this formulation is designed to reflect the view that anthropogenically modified parts of the physical environment—Mount Rushmore, for example—embody natural values to diminishing extents as they are increasingly modified. Finally, natural values are not the only ones that are normatively well-grounded. Reconstruction efforts must negotiate with other, sometimes even more ethically compelling, demands on habitats, for instance, even potentially ecologically destructive resource extraction if that is the only way in which (human) individuals can survive.[98]

Environmental ethicists have unfortunately spent far more time on how natural values should be founded than on finding out what they are, and how they interact with each other. Consequently, any attempt to produce a list must be exploratory and subject to critical future refinement. We have already encountered ecosystem integrity, resistance, resilience, perturbation-tolerance, persistence, constancy, and reliability as possible goals of habitat reconstruction in Section 4. Each of these individually, or in combination with any of the others, can serve as the dynamic criterion in reconstruction efforts. Historical fidelity, however, may be replaced by a reference state criterion reflecting other natural values. Even a perfunctory

---

[95]The qualification "continuing" is needed because some physical features we now value may once have been partly created or influenced by human actions. Presumably the actions of even our recent evolutionary ancestors should be regarded as not different from, say, those of beavers (*Castor canadensis*) building dams across landscapes.

[96]See [Okanta and Douglas, 2001] for a book-length discussion of the environmental and social problems created by oil extraction from the Niger Delta of Nigeria.

[97]See [Markwardt, 2005]. This program has been remarkable successful, at least at the sociocultural level, by increasing appreciation of wildflowers in Texas.

[98]Luckily, from this ethical perspective, especially in the South, where most poverty is concentrated, except in some urban slums, the most economically dispossessed groups are those that rely most heavily on continuing resource production from natural ecosystems—the groups from the Niger Delta provide a good example [Okanta and Douglas, 2001].

examination of reconstruction efforts produces five broad categories.[99]

1. *Contextualized Biodiversity*: That biodiversity is a desirable natural value probably no longer needs any defense—it has become part of our cultural landscape, at least in Northern societies. But not all components of biodiversity are equally appropriately promoted in all contexts. Rather, the biodiversity that it is appropriate to pursue on a given land- or seascape must be locally contextualized. At least five criteria have been used for this purpose and each has a sound normative grounding.

   1.1. *Vulnerability*: In the United States, where much of biodiversity conservation is practiced under the aegis of the ESA, it requires no further argument to defend the use of the vulnerability of taxa as a criterion for contextualizing biodiversity. Beyond the United States, the Red List of at-risk species maintained by the World Conservation Union (IUCN) is widely used in a similar way.[100] The normative argument for focusing on vulnerable taxa is straightforward: if preventing the disappearance of biota is one of our goals, we have no option other than to focus on vulnerable biota (and the depth of focus should track the degree of vulnerability) because these are the ones most likely to disappear without our attention. There is room for argument as to whether it is wise social polity to allocate limited resources on biota that may disappear even in spite of our best efforts, as opposed to allocating them to biota with better prognoses. But that does not affect the appropriateness of using vulnerability as a criterion for contextualizing biodiversity in habitat reconstruction efforts. In the Balcones Canyonlands National Wildlife Refuge, much of management focuses on the Golden-cheeked Warbler and Black-capped Vireo precisely because they are endangered. Management consists as much of creating new suitable habitat as protecting existing such habitat. In the rest of the Canyonlands, attention is equally appropriately focused on the other at-risk species discussed in Section 2.

   1.2. *Rarity*: Rare biota deserve our attention. Most often rarity is taken to be an indicator of potential vulnerability and, in such cases, this criterion is not entirely independent of the last one and has the same normative grounding. But rare biota may also deserve attention because they draw our attention to unusual features of our habitats (including the complexity of interactions between biota) and encourage us to think harder about them. Rarity is also widely used in habitat management efforts. Returning to the Refuge, rare endemic invertebrates (even though they are not recognized to be endangered or threatened) are protected in the cave systems; similarly rare plants (including the

[99]Note that these categories are not supposed to be independent of each other. The order in which they are presented is not intended to suggest any judgment about their relative importance.
[100]Margules and Sarkar [2007].

Shooting Star, Sycamore-leaf snowbell, and Texabama croton) are also protected to the extent possible.

1.3. *Environmental Suitability*: It makes sense to support a species at a locale if that locale, in terms of its environmental features, is particularly suitable for that species. Moreover, environmental suitability makes success more likely when biota have to be established (or re-established) on any habitat. These may appear to be prudential claims but prudence is not entirely normatively irrelevant. Consequently, this criterion is an obvious one to use when determining what is appropriate biodiversity to promote in any context. Finally note that this criterion can be used by itself or in conjunction with any of the others to prioritize biota for attention at any locale.

1.4. *Proximity to Native Range*: By and large, it should be fairly uncontroversial that local species should get preference over non-local ones when all other things (for example, vulnerability or rarity) are equal. Nevertheless, this criterion has generated a fair amount of controversy during the last few decades, among other reasons because on many occasions it may reflect socio-cultural prejudices such as nativism rather than any genuinely natural (or ecological) value. Arguably "alien," "invasive," and so on, are human terms which become metaphors when introduced in ecological contexts. Chew and Laubichler have emphasized the problems with interpreting nature in such human terms.[101] Nevertheless, there is ample evidence that non-native taxa often pose serious threats to the persistence of native taxa and, on that ground alone, they should be discouraged.

1.5. *Cultural Role*: In many societies, biota play significant cultural roles. Species have totemic value, religious significance, are used for hunting or other forms of recreation, and so on.[102] In many regions of Earth, entire biologically rich habitats have been preserved or even created as sacred groves.[103] In the North, charismatic species have long been used to identify which habitats to protect or enhance. Sometimes large amounts of resources have been dedicated to a single charismatic species. For instance, in Texas, the Aransas National Wildlife Refuge was set up in 1937 to protect just one species, the Whooping Crane (*Grus americanus*).[104] Managing that Refuge has involved modifying thousands of hectares of habitat to make it suitable for just this species which, for some, became North America's "symbol of conservation."[105]

---

[101] Chew and Laubichler [2003a; 2003b]; for a response, see [Perry and Schueler, 2003].
[102] The literature on this topic is vast. Jardine *et al.* [1996] provides a useful entry. Berkes [2008] is an extensive discussion of sacred ecology.
[103] For African examples, see [Sheridan and Nyamweru, 2008]; for India, see [Malhotra *et al.*, 2007].
[104] McNulty [1966].
[105] Pratt [1966].

Those who see a sharp nature-culture divide may be uncomfortable with cultural role as a natural value but, then, they would presumably be equally uncomfortable with conscious anthropogenic habitat modification in the first place. The point is that these cultural choices satisfy the definition of natural value with which we started.

2. *Productivity*: Enhancing the productivity of ecosystems (typically measured in biomass produced) has often been a goal for the reconstruction of habitats for a variety of reasons including improving ecosystem services and enhancing environmental security, both of which we discuss below.[106] Here we focus on agriculture and aquaculture. Reclaiming land for agriculture has traditionally been an important rationale for reconstructing habitats particularly after they had reached the end of their useful "non-natural" lives, for instance, as mines or quarries.[107] In the United Kingdom in the 1940s, Hall even went as far as to identify land reconstruction largely with "farming in the national interest."[108] Similarly, aqueous bodies can be managed for fish and other useful taxa, often ultimately intended for human consumption. Increasing productivity includes plantation forestry to ensure or increase supplies of fiber, fuelwood, and other wood products. These are all valid natural values to be pursued, once again "natural" for those who do not demand a sharp nature-culture divide.

3. *Ecosystem Services*: Agriculture and aquaculture do not exhaust the services that natural habitats provide. In general, forests may serve as sinks for carbon and other products of natural or anthropogenic processes. Constructed wetlands can be used to remove pollutants from water.[109] The Millennium Ecosystem Assessment listed more than twenty ecosystem services besides productivity (of food, fuelwood and fiber) and environmental security (see below).[110] These include other provision resources (fresh water, biochemicals [including natural medicines and pharmaceuticals], ornamental resources, and genetic resources), regulating services (air quality maintenance, climate regulation, disease regulation, water regulation, water purification, pollination, and biological control), cultural services (including spiritual and religious services, recreation, aesthetic services, inspirational services, education, conveying a sense of place, cultural heritage, and social relations), and supporting services (soil formation and nutrient services besides primary production). Once again, these are all valid natural values, at

[106]Productivity has sometimes been controversial as a goal of restoration (*sensu stricto*)—see, for example, [Jackson *et al.*, 1995]. However, this does not detract from its being a natural value. Hobbs and Norton [1996] defended its use as a goal of restoration broadly construed, in effect, reconstruction.

[107]See, for example, [Konke, 1950; Smith *et al.*, 1971; Bradshaw *et al.*, 1982].

[108]Hall [1942].

[109]See, for example, [Samecka-Cymerman *et al.*, 2004].

[110]MEA [2003, pp. 53–60]. The list that follows excludes those services that are also classified here under primary productivity and environmental security.

least for those who do not demand a sharp nature-culture divide.

4. *Environmental Security*: A particularly important type of service provided by natural habitats is security against weather-related harm.[111] In general, wetland creation can be used for flood reduction. In particular, reconstructing mangroves along coasts can guard against the effects of storms.[112] These are included by the Millennium Ecosystem Assessment as regulating services.[113] Reconstructing habitat for wildlife so as to minimize human-wildlife contact may reduce the risk of zoonotic disease transmission to humans which has increased significantly since 1940.[114] Once again, these are all valid natural values for those who do not demand a sharp nature-culture divide.

5. *Wild Nature*: Finally, we turn to wild nature, a constant preoccupation of Northern environmentalists, and often equally strongly criticized by environmentalists from the South.[115] Two positions should be distinguished, (i) the pursuit of wild nature as a goal, along with other natural and socio-cultural goals, and (ii) the situation when wild nature trumps these other goals because of the presumed intrinsic value of wild nature (or some other such reason). While the second position may well be ethically bankrupt,[116] the first suffers from no such problem unless we assume that there is some necessarily problematic aspect to any appreciation of wild nature. To the best of our knowledge, such a case has never been successfully made. Note that the pursuit of wild nature as a natural value has been interpreted in two strikingly different ways.

   5.1. *Wildness*: Wildness refers to unpredictability, in nature it refers to the ability of a system to generate surprises and, perhaps, evade human control. It promotes humility in humans, demanding caution in how they modify the rest of nature. In the context of the United States, this may well be what Thoreau had in mind when he promoted wildness. Higgs is also one of those who embraces wildness as a goal of restoration.[117] Normative justifications for the pursuit of wildness range from attempts to attribute intrinsic value or internal agency to natural systems[118] to forging anthropocentric arguments for the value of wildness. The latter arguments often involve the power of wildness to transform felt preferences of individuals.[119]

---

[111] Mea [2003, p. 58].
[112] Ewel *et al.* [1998].
[113] Hey and Phillipi [1995].
[114] Jones *et al.* [2008].
[115] For a spectrum of views, see [Callicott and Nelson, 1998; Sarkar, 1998; 1999].
[116] Sarkar [1999; 2005] makes this argument.
[117] Higgs [2003].
[118] Sarkar [2005] surveys this literature.
[119] For more on such transformative values see [Norton, 1987; Sarkar, 2005].

5.2. *Wilderness*: Finally, we turn to what is perhaps the most contested goal in environmental philosophy, wilderness, typically construed as habitats which were not molded by human actions and which do not admit permanent human presence.[120] If we take seriously the requirement that wilderness habitats be ones that are not molded by human actions, then the idea of reconstructing wilderness *by our actions* is self-contradictory. But, by now, we know that the typical exemplars of wilderness almost anywhere in the world are habitats which were often actively created by traditional inhabitants whose memory has been erased, typically by Northern colonists.[121] Consequently, we may reject the "no human molding" requirement, and may argue for the creation of wilderness defined by no permanent human presence. The normative justification of such a goal will be similar to that of wildness, either based on intrinsic value or on transformative value, in the latter case typically relying on the sublime power of solitude to transform human minds and preferences.

We now have a tentative catalog of natural values which may be used to replace historical fidelity as the reference state criterion in attempts to reconstruct habitats. It is time to lay down the reconstructionist agenda.

7 THE RECONSTRUCTIONIST AGENDA

After all this preparation, framing the reconstructionist agenda is rather straightforward. As in Higgs' definition of ecological restoration, there are two criteria to be satisfied: a dynamic criterion and a reference state criterion. Earlier we suggested self-sustainability as the dynamic criterion and left open its relation to various concepts of ecological integrity and stability. We will not pursue this issue further here though it merits more discussion. Suffice it to note that some such criterion of success is necessary in order to ensure that the end point of a habitat reconstruction effort is not ephemeral, that is, it will have some degree of permanence. It is not being assumed that the reconstructed habitats will require no future human intervention at all. But it is being assumed that the less intervention that is required to maintain a reconstructed habitat, the greater the success of a reconstruction plan.

Turning to the reference state criterion, for habitat reconstruction, any of the natural values listed in the last section may be used singly or jointly to specify what that criterion is. For the reasons discussed in Section 5, what we aspire to make in a habitat should not be constrained by a deification of history. The

[120]The classic critique is Guha [1989]. Callicott and Nelson [1998] collect many of the more important contributions from both sides of the debate. Sarkar [1999; 2005] provides a scientific critique of wilderness preservationism. Woods [2001] attempts a systematic overview.

[121]For the United States, see, in particular, [Cronon, 1983; 1996]. For rainforests, see [Willis *et al.*, 2004]. Sanderson *et al.* [2002] provide a global assessment.

broad spectrum of natural values that reconstruction embraces is also consistent with the practice of much of what is called restoration today. Historical fidelity is not adopted as a goal in some circumstances because the historical past of a locale cannot be reconstructed with the data that are available.[122] In the Balcones Canyonlands and, in fact, throughout the Edwards Plateau, we know little of the vegetation cover even a hundred years ago, let alone earlier.[123] There is obviously nothing normatively suspect in using other natural values in such cases. But, in many circumstances, the natural values of Section 6 are pursued because they are deemed worthwhile themselves even when ensuring historic fidelity is feasible and what is being practiced is called *restoration*. These efforts are still called *restoration* presumably because of an assumption that what is really at stake is that some dynamic (functional) or reference state attribute of the system is being *restored* to a more desirable state. In the Balcones Canyonlands National Wildlife Refuge, prescribed fire is used to *restore* grasslands and shrublands in many areas that may or may not have been such communities in the first place. *Restoration* is also used to describe the creation and maintenance of habitat for the endangered birds (the Black-capped Vireo and the Golden-cheeked Warbler) wherever it is appropriate (that is, there is suitable environment, embodying yet another of the natural values of Section 6) whether or not these areas were historically occupied by these species.[124]

Is there room for any choice? Suppose we worry about only one natural value: vulnerability. In the Balcones Canyonlands, as noted in Section 2, ensuring enough adequate habitat for just these two endangered bird species requires a careful balancing act because they have radically different habitat requirements. What about caprice? It is hard enough even to ensure that just two endangered species recover. Our choice is over-determined—mostly we wish we had more leeway in what often turns out to be a frustrating search for a reasonably optimal habitat management plan. If more than one natural value from our list is used to devise a reference state criterion, there is scope for even more conflict: values may be incompatible with each other.

Enhancing productivity may conflict with promoting native species. Promoting native species may conflict with promoting rare or vulnerable species. Miscellaneous ecosystem services may be incompatible with each other, for instance, recreation with water purification.[125] And so on. Interpreting each value as a criterion to be satisfied, negotiating such potential conflicts will typically require formal multi-criteria analysis (MCA) for complex decisions. MCA techniques, mainly developed by the decision analysis community over the last few decades, are already routinely used in systematic conservation planning for biodiversity

[122]Higgs [2003, p. 129] also notes this difficulty.

[123]Fowler [2005]; Sarkar [forthcoming]. The historical record is so sparse that the prospect of ever reconstructing this history is negligible.

[124]USFWS [2001, p. 6].

[125]Kareiva *et al.* [2007] have also emphasized the importance of explicitly analyzing tradeoffs between ecosystem services.

conservation.[126] They are yet to be systematically used in restoration ecology but their availability indicates that there is no fundamental problem with using a wide spectrum of potentially incompatible values for habitat reconstruction.

From this perspective, ecological restoration (*sensu stricto*, that is, with historical fidelity as the reference state criterion) is just one special form that habitat reconstruction can take. So is each of reclamation, regeneration, rehabilitation, revegetation, and so on. What is being embraced here is a broadening of the goals of what we should do with habitats to enhance the natural values embodied in them. But, obviously, as was emphasized earlier, this does not amounts to any rejection of ecological restoration as one possible goal for habitat management. It depends primarily on whether historical fidelity is deemed appropriate as a social goal. But it also depends on whether adequate data are available to reconstruct the past with any degree of certainty, whether it is even possible to return a system at least approximately to that past state, and whether it is worth the cost. There is probably no general answer to these questions. Any answer will depend on the particular ecological and social context of formulating policy for a habitat.

Note that the discussion of Section 6 did not explicitly list historical fidelity as a natural value even though that is the reference state criterion for ecological restoration. However, since historical fidelity was treated as an alternative to the natural values of Section 6, there is an implicit suggestion that it, too, is a natural value. However, there is an equally compelling sense in which the pursuit of any history is a cultural value. We will leave open the question as to whether historical fidelity is a natural or cultural value, or both. For the argument of this paper, it does not matter. We have not assumed a sharp nature-culture distinction and, in any case, the practice of reconstruction (or even restoration *sensu stricto*) does not require an answer to this question.

Finally, let us return for one last time to the question whether jettisoning historical fidelity will lead to caprice. After all, this was the deepest worry raised by Higgs' defense of historical fidelity (see Section 5). We saw in Section 5 that requiring fidelity is no guard against caprice. Moreover, if we take seriously both whatever dynamic criterion we adopt, and a reference state criterion that incorporates the natural values of Section 6, habitat reconstruction may well turn out to be as, if not more, restrictive than many traditional forms of ecological restoration (*sensu stricto*). Caprice requires underdetermination of choice by criteria. In the case of restoration, underdetermination occurs, among other ways, because of the freedom to decide what historical period should provide the reference state. In contrast, our earlier discussion of caprice in this Section shows that the natural values of Section 6 will typically leave no such freedom.

In reconstruction we may have more reason to worry about too many constraints rather than the possibility of caprice.

---

[126] Moffett and Sarkar [2005] review the literature; for illustrations, see [Moffett *et al.*, 2006; Margules and Sarkar, 2007].

8 CONCLUSIONS

Where does all this leave us? We have two options.[127] If we construe ecological restoration narrowly, requiring both ecological integrity and historical fidelity, then restoration will very often not be the most normatively salient goal for how we choose to treat our habitats. If we construe restoration more broadly, as originally championed by Bradshaw and Chadwick,[128] then restoration would both describe much more of what is practiced as such. But we may then confine historical fidelity to those cases in which it is deemed appropriate as a cultural value. Meanwhile our goals can be established using a wide spectrum of natural values, including those discussed in Section 6. What we do at the Balcones Canyonlands National Wildlife Refuge when we create and maintain habitat for the Black-capped Vireo or the Golden-cheeked Warbler does count as restoration. If we choose to encourage the spread of rare plants on the Refuge, we are still doing restoration.[129] In other words, *restoration*, broadly construed is no different from what we have been defending as *reconstruction*.

By itself, the terminological issue is probably not important.[130] But whether we endorse historical fidelity as our sole reference state criterion does matter, both for practice and in theory. It matters for practice because of two reasons. (1) It restricts the range of natural values that can be used to guide what we do with our habitats. None of the values discussed in Section 6 would be legitimate goals to pursue. (2) It makes the reconstruction projects expensive, probably quite often impossible. It requires us to expend resources to determine what the past condition of a system was. In many cases, for instance, in the Balcones Canyonlands, this is virtually impossible. Perhaps even more importantly, it matters in theory because of four reasons. (1) It deifies nostalgia. Ecologically this may be illegitimate if the past state of a system has little relevance to what it would have become today even without anthropogenic change. Given the largely irreversible global changes that have lately been taking place, including climate change, it is difficult to see how the past, especially the distant past, can have any relevance except in very rare cases. Socially and culturally, nostalgia is equally problematic because the past may well not be something we wish to encounter again. (2) It is arbitrary because any past period that we choose to replicate is just one of an indefinite number of possibilities. For instance, the much-used pre-Columbian past in the Americas is arbitrary and based on a demonstrably false assumption that the First Nations of the continent did not actively manage their natural environments. (3) It is also arbitrary because we have no reliable method of knowing what any past state would have evolved into under a different interaction regime with humans. Theoretical ecology simply does not have the tools to make accurate predictions of

[127] These are essentially the same options also demarcated by [Higgs, 2003, p. 129].

[128] Bradshaw and Chadwick [1980].

[129] This expanded definition is also endorsed elsewhere, for instance, by Hobbs and Norton [1996] and Gobster [2000].

[130] This point was forcefully argued by Hobbs and Norton [1996]; for a response, see [Higgs, 2003, pp. 94–95].

that kind. (4) It assumes without argument that the other natural values discussed in Section 6 are not important, or at least they are trumped by historical fidelity. As we saw in Section 6, there is a multiplicity of natural values and each of these values is normatively well-grounded. Individuals and groups will disagree about these values. Public discussions to plan for the future are inevitable, and desirable in democratic societies. Invoking nostalgia (and historical fidelity) amounts to foreclosing such debates about nature by fiat, ruling against a continued public cultural discussion on what is appropriate in human interactions with the rest of nature.

In Section 5 we saw that every argument for historical fidelity turned out ultimately to be based on some other end: simplicity, an "organic" relation to habitat, rarity, and so on. Fidelity was at best a means to an end, and this end had no necessary connection to history. What this paper as a whole is intended to have shown is that we should move beyond historical fidelity and, as a result, beyond ecological restoration construed narrowly. Rather we should embrace the full spectrum of reconstructionist practices that have a much longer history than what ecological restoration became when historical fidelity emerged as its sole reference state goal. The practical problems with historical fidelity give us prudential reasons for rejecting its authority. Even more importantly, while framing policies for the management of habitats, the theoretical problems show that the pursuit of historical fidelity *alone* as such a goal is normatively misguided. Luckily, in practice, much of what is called ecological restoration is not construed narrowly and this critique leaves it untouched. Perhaps it even helps establish good normative foundations for these practices by delving into all the natural values we can justifiably embrace as we manage our land- and seascapes.

ACKNOWLEDGEMENTS

The material presented here formed the basis of a talk at Ohio State University (Spring 2007); discussions there with Kevin deLaplante and Clark Wolf encouraged the development of these ideas. For information on the Balcones Canyonlands National Wildlife Refuge, thanks are due to Deborah Holle, Kelly McDowell, and, especially, Chuck Sexton. Many students have been involved in the research at the Refuge in our attempts to enhance natural values there: (in alphabetical order) thanks are particularly due to Michael Ciarleglio, Trevon Fuller, Tiffany Hollon, Keith Kaden, Matthew Moskwik, Nancy Shackelford, Blake Sissel, and Stavana Strutz. Thanks are also due to John Cairns Jr., J. Baird Callicott, Marc Hall, Eric Higgs, Curt Meine, and Alexander Moffett for help with sources and for many critical discussions. Needless to say, none of these individuals should be presumed to agree with the conclusions reached here which remain the sole responsibility of the author. This work was supported by NSF Grant No. SES-0645884, 2007–2009.

BIBLIOGRAPHY

[Agee, 1977] J. L. Agee. The Problem. In Ballentine, R. K. and Guarraia, L. J. (eds.), *The Integrity of Water: A Symposium*. Washington, DC: U.S. Environmental Protection Agency Office of Water and Hazardous Materials, pp. 3–4, 1977.

[Alford, 2005] A. Alford. Wildlife Tax Breaks Lack State Oversight. *Austin-American Statesman*, May 29, 2005.

[Allison, 2004] S. K. Allison. What Do We Mean When We Talk about Ecological Restoration? *Ecological Restoration* 22: 281–286, 2004.

[Angermeier and Karr, 1994] P. L. Angermeier and J. R. Karr. Biological Integrity versus Biological Diversity as Policy Directives. *BioScience* 44: 690–697, 1994.

[Anonymous, 1971] Anonymous. About the Richmond Meeting. *Association of Southeastern Biologists Bulletin* 18: 3–7, 1971.

[Ballentine and Guarraia, 1977] R. K. Ballentine and L. J. Guarraia (eds.), *The Integrity of Water: A Symposium*. Washington, DC: U.S. Environmental Protection Agency Office of Water and Hazardous Materials. 1977.

[Beatley *et al.*, 1995] T. Beatley, T. J. Fries, and D. Braun. The Baclones Canyonlands Conservation Plan: A Regional, Multi-Species Approach. In Porter, D. R. and Salvesen, D. A. (eds.), *Collaborative Planning for Wetlands and Wildlife: Issues and Examples*, pp. 75–92. Washington, DC: Island Press, 1995.

[Berkes, 2008] F. Berkes. *Sacred Ecology*, 2nd Edition. New York: Routledge, 2008.

[Bradshaw, 1983] A. D. Bradshaw. The Reconstruction of Ecosystems. *Journal of Applied Ecology* 20: 1–7, 1983.

[Bradshaw and Chadwick, 1980] A. D. Bradshaw and M. J. Chadwick. *The Restoration of Land: The Ecology and Reclamation of Derelict and Degraded Land.* Oxford: Blackwell, 1980.

[Bradshaw *et al.*, 1982] A. D. Bradshaw, R. H. Marrs, R. D. Roberts, and R. A. Skeffington. The Creation of Nitrogen Cycles in Derelict Land. *Philosophical Transactions of the Royal Society (London) B* 296: 557–561, 1982.

[Cairns, 1977] J. Cairns, Jr. Quantification of Biological Integrity. In Ballentine, R. K. and Guarraia, L. J. (eds.), *The Integrity of Water: A Symposium*, pp. 171–187. Washington, DC: U.S. Environmental Protection Agency Office of Water and Hazardous Materials, 1977.

[Cairns, 1986] J. Cairns, Jr. Restoration, Reclamation, and Regeneration of Degraded or Damaged Ecosystems. In Soulé, M. E. (ed.), *Conservation Biology: The Science of Scarcity and Diversity*, pp. 465–483. Sunderland, MA: Sinauer Associates, 1986.

[Cairns, 1988] J. Cairns, Jr. Restoration Ecology: The New Frontier. In J. Cairns Jr. (ed.), *Rehabilitating Damaged Ecosystems* 1: 1–11. Baton Rouge: CRC Press, 1988.

[Cairns, 1995] J. Cairns, Jr. Ecosocietal Restoration: Reestablishing Humanity's Relationship with Natural Systems. *Environment* 37(5): 4–9, 30–33, 1995.

[Cairns *et al.*, 1971] J. Cairns, Jr., J. S. Crossman, K. L. Dickson, and E. E. Herricks. The Recovery of Damaged Streams. *Association of Southeastern Biologists Bulletin* 18: 79–106, 1971.

[Cairns *et al.*, 1977] J. Cairns, Jr., K. L. Dickson, and E. E. Herricks (eds.), *Recovery and Restoration of Damaged Ecosystems.* Charlottesville: University Press of Virginia, 1977.

[Cairns and Heckman, 1996] J. Cairns, Jr. and J. R. Heckman. Restoration Ecology: The State of an Emerging Field. *Annual Review of Energy and the Environment* 21: 167–189, 1996.

[Callicott, 2002] J. B. Callicott. Choosing Appropriate Temporal and Spatial Scales for Ecological Restoration. *Journal of Biosciences* 27(S2): 409–420, 2002.

[Callicott and Nelson, 1998] J. B. Callicott and M. P. Nelson (eds.), *The Great New Wilderness Debate.* Athens, GA: University of Georgia Press, 1998.

[Chapron, 2005] G. Chapron. Re-Wilding: Other Projects Help Carnivores Stay Wild. *Nature* 437: 318, 2005.

[Chew and Laubichler, 2003a] M. K. Chew and M. Laubichler. Natural Enemies–Metaphor or Misconception? *Science* 301: 52–53, 2003.

[Chew and Laubichler, 2003b] M. K. Chew and M. Laubichler. Metaphors, Misuse, and Misconceptions. *Science* 301: 1481–1482, 2003.

[Costigan *et al.*, 1981] P. A. Costigan, A. D. Bradshaw, and R. P. Gemmell. The Reclamation of Acidic Colliery Soil. I. Acid Production Potential. *Journal of Applied Ecology* 18: 865–878, 1981.

[Cronon, 1983] W. Cronon. *Changes in the Land: Indians, Colonists, and the Ecology of New England.* New York: Hill & Wang, 1983.

[Cronon, 1996] W. Cronon. The Trouble with Wilderness; or, Getting Back to the Wrong Nature. In Cronon, W. (ed.), *Uncommon Ground: Rethinking the Human Place in Nature*, pp. 69–90. New York: W.W. Norton, 1996.

[Curry, 1971] R. R. Curry. Soil Destruction Associated with Forest Management and Prospects for Recovery in Geologic Time. *Association of Southeastern Biologists Bulletin* 18: 117–128, 1971.

[Damude *et al.*, 1999] N. Damude, K. Bender, D. Foss, and J. Gowen. *Texas Wildscapes: Gardening for Wildlife.* Austin: Texas Parks and Wildlife, 1999.

[De Leo and Levin, 1997] G. A. De Leo and S. Levin. The Multifaceted Aspects of Ecosystem Integrity. *Conservation Ecology* 1(1): 3. 1997. `http://www.consecol.org/vol1/iss1/art3/`; last accessed 31-Mar-08.

[Dibble and Bartha, 1979] J. T. Dibble and R. Bartha. Rehabilitation of Oil-Inundated Agricultural Land: A Case History. *Soil Science*: 56–60, 1979.

[Donlan *et al.*, 2005] C. J. Donlan, H. W. Greene, J. Berger, C. E. Bock, J. H. Bock, D. A. Burney, J. A. Estes, D. Foreman, P. A. Martin, G. W. Roemer, F. A. Smith, and M. E. Boulé. Re-Wilding North America. *Nature* 436: 913–914, 2005.

[Donlan *et al.*, 2006] C. J. Donlan, J. Berger, C. E. Bock, J. H. Bock, D. A. Burney, J. A. Estes, D. Foreman, P. S. Martin, G. W. Roemer, F. A. Smith, M. E. Soulé, and H. W. Greene. Pleistocene Rewilding: An Optimistic Agenda for Twenty-First Century Conservation. *American Naturalist* 168: 660–681, 2006.

[Ewel *et al.*, 1998] K. C. Ewel, R. R. Twilley, and J. E. Ong. Different Kinds of Mangrove Forests Provide Different Goods and Services. *Global Ecology and Biogeography Letters* 7: 83–94, 1998.

[Fowler, 2005] N. Fowler. An Introduction to the Vegetation and Ecology of the Eastern Edwards Plateau (Hill Country) of Texas, 2005. `http://www.sbs.utexas.edu/fowler/epveg/epmainpage.htm`; last accessed 09-Mar-08.

[Gemmell, 1973] R. P. Gemmell. Revegetation of Derelict Land Polluted by a Chromate Smelter. Part 1. Chemical Factors Causing Substrate Toxicity in Chromate Smelter Waste. *Environmental Pollution* 5: 181–197, 1973.

[Gemmell, 1974] R. P. Gemmell. Revegetation of Derelict Land Polluted by a Chromate Smelter. Part 2. Techniques of Revegetation of Chromate Smelter Waste. *Environmental Pollution* 6: 31–37, 1974.

[Gobster, 2000] P. H. Gobster. Restoring Nature: Human Actions, Interactions. In Gobster, P. H. and Hull, R. B. (eds.), *Restoring Nature: Perspectives from the Humanities and the Social Sciences*, pp. 1–19. Washington, DC: Island Press, 2000.

[Ginzbarg, 1992] S. Ginzbarg. A New Disjunct Variety of *Croton alabamensis* (Euphorbiaceae) from Texas. *Sida* 15: 41–52, 1992.

[Graves, 2003] J. Graves. *Texas Hill Country.* Austin: University of Texas Press, 2003.

[Grzybowski, 1995] J. A. Grzybowski. Black-capped Vireo. *Birds of North America* 181: 1–24, 1995.

[Gudin and Syratt, 1975] C. Gudin and W. J. Syratt. Biological Aspects of Land Rehabilitation Following Hydrocarbon Contamination. *Environmental Pollution* 8: 107–112, 1975.

[Guha, 1989] R. Guha. Radical American Environmentalism and Wilderness Preservation: A Third World Critique. *Environmental Ethics* 11: 71–83, 1989.

[Hall, 1941] A. D. Hall. *Reconstruction and the Land: An Approach to Farming in the National Interest.* London: Macmillan, 1941.

[Hall, 2005] M. Hall. *Earth Repair: A Transatlantic History of Environmental Restoration.* Charlottesville: University of Virginia Press, 2005.

[Harwell *et al.*, 1999] M. A. Harwell, V. Myers, T. P. Young, A. Bartuska, N. Gassman, J. H. Gentile, C. C. Harwell, S. Appelbaum, J. Barko, B. Causey, C. Johnson, A. McLean, R. Smola, P. Templet, and S. Tosini. A Framework for an Ecosystem Integrity Report Card. *BioScience* 49: 543–556, 1999.

[Hey and Phillipi, 1995] D. L. Hey and N. S. Phillipi. Flood Reduction Through Wetland Restoration: The Upper Mississippi River Basin as a Case History. *Restoration Ecology*: 2–17, 1995.

[Higgs, 2003] E. Higgs. *Nature by Design: People, Natural Process, and Ecological Restoration.* Cambridge, MA: MIT Press, 2003.

[Hill, 1971] R. D. Hill. Restoration of a Terrestrial Environment—The Surface Mine. *Association of Southeastern Biologists Bulletin* 18: 107–116, 1971.

[Hobbs and Norton, 1996] R. J. Hobbs and D. A. Norton. Towards a Conceptual Framework for Restoration Ecology. *Restoration Ecology* 4: 93–110, 1996.

[Holdgate and Woodman, 1978] M. W. Holdgate and M. J. Woodman (eds.), *The Breakdown and Restoration of Ecosystems*. New York: Plenum, 1978.

[Jackson *et al.*, 1995] L. L. Jackson, N. Lapoukhine, and D. Hillyard. Ecological Restoration: A Definition and Comments. *Restoration Ecology* 3: 71–75, 1995.

[Janzen, 1988] D. H. Janzen. Tropical Ecological and Biocultural Restoration. *Science* 239: 243–244, 1988.

[Jardine *et al.*, 1996] N. Jardine, J. A. Secord, and E. C. Spary (eds.), *Cultures of Natural History*. Cambridge, UK: Cambridge University Press, 1996.

[Johnson *et al.*, 1976] M. S. Johnson, A. D. Bradshaw, and J. F. Handley. Revegetation of Metalliferous Fluor Mine Tailings. *Transactions of the Institution of Mining and Metallurgy* 85A: 32–37, 1976.

[Jones *et al.*, 2008] K. E. Jones, N. G. Patel, M. A. Levy, A. Storeygard, D. Balk, J. L. Gittleman, and P. Daszak. Global Trends in Emerging Infectious Diseases. *Nature* 451: 990–993, 2008.

[Justus, 2007] J. Justus. The Stability-Diversity-Complexity Debate of Community Ecology: A Philosophical Analysis. PhD Dissertation, University of Texas, 2007.

[Kareiva *et al.*, 2007] P. Kareiva, S. Watts, R. McDonald, and T. Boucher. Domesticated Nature: Shaping Landscapes and Ecosystems for Human Welfare. *Science* 316: 1866–1869, 2007.

[Karr and Dudley, 1981] J. R. Karr and D. R. Dudley. Ecological Perspective on Water Quality Goals. *Environmental Management* 5: 55–68, 1981.

[Kohnke, 1950] H. Kohnke. The Reclamation of Coal Mine Spoils. *Advances in Agronomics* 2: 317–349, 1950.

[Ladd and Gass, 1999] C. Ladd and L. Gass. Golden-cheeked Warbler. *Birds of North America* 420: 1–23, 1999.

[Lee, 1995] M. F. Lee. *Earth First! Environmental Apocalypse*. Syracuse: Syracuse University Press, 1995.

[Lewis, 1976] J. W. Lewis. Regeneration of Coastal Ecosystems after Mineral Sand Mining. *Australian Mining*, July, pp. 1–3, 1976.

[Linton and Cooper, 1971] T. L. Linton and A. W. Cooper. Damaged Estuarine Ecosystems, Their Restoration and Recovery. *Association of Southeastern Biologists Bulletin* 18: 129–136, 1971.

[Malhotra *et al.*, 2007] K. C. Malhotra, Y. Gokhale, S. Chatterjee, and S. Srivastava. *Sacred Groves in India: An Overview*. New Delhi: Aryan International, 2007.

[Mann, 2005] C. C. Mann. *1491: New Revelations of the Americas Before Columbus*. New York: Knopf, 2005.

[Margules and Pressey, 2000] C. R. Margules and R. L. Pressey. Systematic Conservation Planning. *Nature* 405: 242–253, 2000.

[Margules and Sarkar, 2007] C. R. Margules and S. Sarkar. *Systematic Conservation Planning*. Cambridge, UK: Cambridge University Press, 2007.

[Markwardt, 2005] D. Markwardt. Texas Roadside Wildflowers. *Native Plants Journal* 6: 69–71, 2005.

[McNulty, 1966] F. McNulty. *The Whooping Crane: The Bird that Defies Extinction*. New York: E. P. Dutton & Co., 1966.

[MEA, 2003] Millennium Ecosystem Assessment. *Ecosystems and Human Well-being: A Framework for Assessment*. Washington, DC: Island Press, 2003.

[Moffett *et al.*, 2006] A. Moffett, J. S. Dyer, and S. Sarkar. Integrating Biodiversity Representation with Multiple Criteria in North-Central Namibia Using Non-Dominated Alternatives and a Modified Analytic Hierarchy Process. *Biological Conservation* 129: 181–191, 2006.

[Moffett and Sarkar, 2005] A. Moffett and S. Sarkar. Incorporating Multiple Criteria into the Design of Conservation Area Networks: A Minireview with Recommendations. *Diversity and Distributions* 12: 125–137.

[Naughton-Treves *et al.*, 2005] L. Naughton-Treves, M. Buck Holland, and K. Brandon. The Role of Protected Areas in Conserving Biodiversity and Sustaining Local Livelihoods. *Annual Review of Environment and Resources* 30: 219–252, 2005.

[Norton, 1987] B. G. Norton. *Why Preserve Natural Variety?* Princeton: Princeton University Press, 1987.

[NRC, 1992] National Research Council. *Restoration of Aquatic Ecosystems.* Washington, DC: National Academy Press, 1992.

[Okanta and Douglas, 2001] I. Okanta and O. Douglas. *Where Vultures Feast: Shell, Human Rights, and Oil in the Niger Delta.* San Francisco: Sierra Club Books, 2001.

[Parrish *et al.*, 2003] J. D. Parrish, D. P. Braun, and R. S. Unnasch. Are We Conserving What We Say We Are? Measuring Ecological Integrity within Protected Areas. *BioScience* 9: 851–860, 2003.

[Pavlik, 1996] B. M. Pavlik. Defining and Measuring Success. In Falk, D. A., Millar, C. I., and Olwell, M. (eds.), *Restoring Diversity*, pp. 127–155. Washington, DC: Island Press, 1996.

[Perry and Scheueler, 2003] G. Perry and F. W. Scheueler. Metaphors, Misuse, and Misconceptions. *Science* 301: 1480–1481, 2003.

[Pratt, 1996] J. J. Pratt. *The Whooping Crane: North America's Symbol of Conservation.* Prescott, AZ: Castle Rock Publishing, 1996.

[Rubenstein *et al.*, 2006] D. R. Rubenstein, D. I. Rubenstein, P. W. Sherman, and T. A. Gavin. Pleistocene Park: Does Re-Wilding North America Represent Sound Conservation for the 21st Century? *Biological Conservation* 132: 232–238, 2006.

[Samecka-Cymerman *et al.*, 2004] A. Samecka-Cymerman, D. Stepien, and A. J. Kempers. Efficiency in Removing Pollutants by Constructed Wetland Purification Systems in Poland. *Journal of Toxicology and Environmental Health A* 67: 265–275, 2004.

[Sanderson *et al.*, 2002] E. W. Sanderson, M. Jaiteh, M. A. Levy, K. H. Redford, A. V. Wannebo, and G. Woolmer. The Human Footprint and the Last of the Wild. *BioScience* 52: 891–904, 2002.

[Sarkar, 1998] S. Sarkar. Restoring Wilderness or Reclaiming Forests? *Terra Nova* 3(3): 35–52, 1998.

[Sarkar, 1999] S. Sarkar. Wilderness Preservation and Biodiversity Conservation–Keeping Divergent Goals Distinct. *BioScience* 49: 405–412, 1999.

[Sarkar, 2005] S. Sarkar. *Biodiversity and Environmental Philosophy: An Introduction.* New York: Cambridge University Press, 2005.

[Sarkar, 2007] S. Sarkar. From Ecological Diversity to Biodiversity. In Hull, D. L. and Ruse, M. (eds.), *The Cambridge Companion to the Philosophy of Biology*, pp. 388–409. Cambridge, UK: Cambridge University Press, 2007.

[Sarkar, forthcoming] S. Sarkar. *The Balcones Canyonlands National Wildlife Refuge.* Austin: University of Texas Press, forthcoming.

[Schaller and Sutton, 1978] F. W. Schaller and P. Sutton. *Reclamation of Drastically Disturbed Lands.* Madison, WI: American Society for Agronomy, 1978.

[SERI, 2004] Society for Ecological Restoration International Science and Policy Working Group. 2004. *The SER International Primer on Ecological Restoration.* Tucson: Society for Ecological Restoration International, 2004. `http://www.ser.org/content/ecological{\_}restoration{\_}primer.asp`; last accessed 29-Mar-08.

[Shaw, 2003] J. M. Shaw. Climate Change and Deforestation: Implications for the Maya Collapse. *Ancient Mesoamerica* 14: 157–167, 2003.

[Sheridan an dNyamweru, 2008] M. J. Sheridan and C. Nyamweru (eds.), *African Sacred Groves: Ecological Dynamics and Social Change.* Oxford, UK: James Currey, 2008.

[Smith *et al.*, 1971] R. M. Smith, E. H. Tryon, and E. H. Tyner. *Soil Development on Mine Spoil.* Morgantown: West Virginia University Agricultural Experimental Station, 1971.

[Turnbull, 1972] C. M. Turnbull. *The Mountain People.* New York: Simon and Schuster, 1972.

[US-NAS, 1974] National Academy of Science. *Rehabilitation Potential of Western Coal Lands.* Cambridge, MA: Ballinger, 1974.

[USFWS, 2001] United States Fish and Wildlife Service. *Balcones Canyonlands National Wildlife Refuge: Comprehensive Conservation Plan and Environmental Assessment.* U.S. Fish and Wildlife Service, 2001.

[van Ee *et al.*, 2006] B. W. van Ee, N. Jelinski, P. E. Berry, and A. L. Hipps. Phylogeny and Biogeography of Croton alabamensis (Euphobiaceae), a Rare Shrub from Texas and Alabama, Using DNA Sequence and AFLP Data. *Molecular Ecology* 15: 2735–2751, 2006.

[van Diggelen *et al.*, 2001] R. van Diggelen, A. P. Grootjans, and J. A. Harris. Ecological Restoration: State of the Art or State of the Science. *Restoration Ecology* 9: 115–118, 2001.

[WCED, 1987] World Commission on Environment and Development, 1987. *Our Common Future*. Oxford: Oxford University Press, 1987.

[White and Walker, 1997] P. S. White and J. L. Walker. Approximating Nature's Variation: Selecting and Using Reference Information in Restoration Ecology. *Restoration Ecology*: 338–349, 1997.

[Willis *et al.*, 2004] K. J. Willis, L. Gillson, and T. M. Brncic. How 'Virgin' Is Virgin Rainforest? *Science* 304: 402–403, 2004.

[Wood, 2007] L. J. Wood. The Global Network of Marine Protected Areas: Developing Baselines and Identifying Priorities. PhD Dissertation. Department of Geography, University of British Columbia, Vancouver, Canada, 2007.

[Wood *et al*, 2007] L. J. Wood, L. Fish, J. Laughren, and D. Pauly. Assessing Progress Towards Global Marine Protection Targets: Shortfalls in Information and Action. *Oryx*: in press, 2007.

[Woods, 2001] M. Woods. Wilderness. In Jamieson, D. (ed.), *A Companion to Environmental Philosophy*, pp. 349–361. Malden: Blackwell, 2001.

MODELING SUSTAINABILITY IN ECONOMICS AND ECOLOGY

Bryan G. Norton

1 INTRODUCTION: THE CHALLENGE

A little over ten years ago, Michael Toman and I worked together on a grant from Environmental Protection Agency to examine the relationship between the sciences of economics and ecology in the context of articulating, choosing and implementing policies that are more "sustainable".[1] In the culminating essay of that research, we identified what we thought were two of the most challenging—if not *the* two *most* challenging—obstacles to achieving a unified theory and a standardized measure of sustainable economic growth [Toman, 1994; Norton and Toman, 1997]. We explored two areas where conceptual and theoretical disagreements between ecologists and economists have been most conflicted, and we gave them the shorthand names: "reversibility and substitutability" and "the accounting problem".

Put briefly, the *reversibility and substitutability problem* refers to a cluster of problems about whether the path to sustainability will involve the successful substitution of one type of resource for others in the effort to expand human welfare, or whether there are essential natural features of natural systems which, if lost, will lead to irreversible damages. Economists, who engage in analysis of valuation for goods and services "at the margin", focus their attention on overall welfare, manifest as the average ability of consumers to pay for goods and services they desire. Ecologists, however, are suspicious of marginal analysis of the goods derived from nature, because they believe ecological systems are complex systems that contain many thresholds and multiple "regimes of functioning." Many ecologists believe that it is possible for human actions to cause "flips", irreversible transformations in the functional regimes of productive systems, into a less productive state. Further, they believe that ecological theory, resilience theory in particular, provides a scientific explanation for the resilience and the fragility of landscapes under intensive (usually economic) use.

We also noted that economists and ecologists have very different ways of thinking about the value of natural products and the processes that create them, and that while this conflict can be referred to conveniently as a *problem of accounting*, the theoretical and conceptual issues go deeper than usual accounting differences.

[1]Grant to Resources for the Future, US Environmental Protection Agency, Office of Exploratory Research ("Clarifying the Requirements of Sustainability").

Handbook of the Philosophy of Science. Volume 11: Philosophy of Ecology.
Volume editors: Kevin deLaplante, Bryson Brown and Kent A. Peacock. General editors: Dov M. Gabbay, Paul Thagard and John Woods.

Economists and ecologists do not just disagree *how much* value various objects have; they also disagree about *what things* have value and *in what terms those values are to be measured or counted.* Despite these deep disagreements about the nature of natural value, ecologists, who are often anxious to remain value neutral, have largely ceded discussion of social values to economists and environmental ethicists.

Looking back at that essay today, I see little to disagree with: the emphasis on problems of substitutability and accounting, which are all tied up with differing conceptions of scale, are no less daunting today. It is still the case that mainstream economists conceptualize value in nature as the value nature has in creating human welfare, and it is still the case that mainstream economists generally advocate partial equilibrium models and marginal analysis. Ecologists, on the other hand—while often ceding most of the evaluative task to economists—have worried that valuing nature only as a contributor in increments to the aggregated welfare effects on humans will miss the many and dissimilar ways that natural systems are of value to humans. While we are disappointed in the extent to which our pluralistic but integrated, "two-tier" approach based on hierarchical models of scale has been explicitly adopted by policy analysts, ideas similar to ours are emerging in a number of fields and in interdisciplinary collaborations (see, for example, [Schelhas *et al.*, 2001; Taylor, 2005].

Since I do not disagree with our earlier assessments of those two areas, in Section 2 I will briefly reprise the situation with respect to environmental accounting, first, and then proceed to show the durability of the set of problems about substitution and irreversibility. Indeed, this latter set of problems has become entwined with a growing recognition of the importance and complexity of problems of scale. Problems of substitutability and irreversibility must now be addressed within a more complex context, which include a changing view (originating in the philosophy of science) of disciplinary models. These contextual changes result from other and related issues at the intersection of economics and ecology, especially issues having to do with decisions regarding the scale at which environmental policies and problems are conceptualized. Can the scale at which substitutions take place and the scale at which benefits are aggregated be justified on descriptive grounds? Apparently not. If, however, our scaling decisions have no scientific basis, the question arises: How *do* scientists and resource managers choose the scale at which they will model a system in order to illustrate and analyze an "environmental problem"? In Section 3, I will focus on problems of scale and scaling which, it turns out, confound simple understanding of the substitutability and irreversibility problem. These problems also raise questions about the realism attributed to ecological models as objective and independent of human purposes. Ecologists make liberal use of hierarchical reasoning, but many advocates of hierarchy theory insist these models are simply conventions for understanding; so a serious discussion of scalar aspects of problems requires clarification as to the degree to which posited hierarchies are real or conventional. If developing a hierarchical model requires *choices* regarding the scale and boundaries of system-representations, on what basis are these choices to be made? In Section 4, I survey recent developments at the intersection of economics and ecology, and discuss several sources of new insights that promise to help economists and ecologists to heal the rift between their fields. This work, by self-described "ecological economists," cognitive psychologists, and decision analysts, as well as philosophers, has undertaken to correct economists' assumptions about the nature of preferences and values. These developments, far from healing the rift, show that there is no shared way to express economic facts and values that seem central to ecology. I will conclude that there is no common discourse which allows an integrated discussion of normative and scientific/descriptive aspects of environmental problems.

Finally, in Section 5, I will explore recent developments in the field of ecology, as important ecological theorists have suggested a new role for ecology in the process of evaluating environmental change. This new line of reasoning is consonant with recent thinking about the essentially analogical and metaphorical nature of basic scientific reasoning [Gentner, *et al.*, 1997; Hallyn, 2000; Nersessian, 1992; 1999]. Leading ecologists now argue that the very act of choosing and scaling the "generic" ecosystem model to apply to a particular problem almost always involves the introduction of metaphors, metaphors which express deep values and deep cultural attitudes as well as scientific observation and theory [Pickett *et al.*, 2004]. This work is particularly illuminating because, when understood within emerging views of science and scientific modeling in the philosophy of science, these references to metaphors *explain* the underlying conflicts between ecology and economics: economists have adopted a single metaphor for natural productivity and its role in economics: ecosystems as machines productive of human welfare. Ecologists, who treat ecological systems as complex and capable of changing from one functional regime to another, use diverse and varied metaphors to understand some aspects of those complex systems. This latter view on models and metaphors, however, departs from the positivistic views of descriptive science as existing independently of values. Resonant metaphors, redolent with value, shape both economic and ecological models, according to the emergent view of models and their association with analogical cognitive processes. The models are tools of human understanding, not ontological elements of the universe, and they reflect and express the values of those who embrace the metaphors.

2 ACCOUNTING AND REVERSIBILITY AND SUBSTITUTABILITY: STILL CONTROVERSIAL

2.1 Environmental Accounting

Since the mid-1980s, a major contribution of one school of environmental economists, especially in Europe, but also with some contributors in North America and elsewhere, has been to develop methods of accounting for environmental damages and resource depletion, offering these as corrections to normal computation in the System of National Accounts. Much as accountants depreciate manufacturing equipment so as to give an accurate accounting of the actual assets of a company as plants and equipment age and become obsolete, they recommend "satellite accounts" [Lange, 2007] to measure and compute losses to national wealth due to depletion of resources. This kind of accounting is very useful in bringing attention to important and often missed losses to national wealth due to over-exploitation of resources. Environmental accounting is a work in progress, and it is subject to criticism both internally, by economists, and externally, by members of other disciplines who object to its overly "economistic" view of value. Even its advocates agree that environmental accounting practice is still plagued with problems in that most national accounts don't attribute value to nonmarket uses of resources, and many benefits from renewable resources are assigned to other economic sectors. These problems are well summarized in [Lange 2007], and will not be discussed further here, where it is more important to examine the deep theoretical differences between economists, ecologists, and environmental philosophers over how to evaluate environmental change.

The practical problems just mentioned pale by comparison to the very large issues associated with what could be referred to as the "ontological" problems that afflict the discourse about environmental values and valuation. Different disciplines, of course, develop independent theoretical and conceptual frameworks to explain their respective areas of study, but in the area of environmental valuation the turf wars and incommensurabilities across disciplinary and theoretical lines have settled into trench warfare, with economists and philosophers as the main antagonists, and with ecologists standing by with crucial insights, but reluctant to enter the values fray for fear of losing their status as "objective, value-neutral, scientists."[2]

Writing in the mid-nineties, Toman and I said:

> How should we assign values, and what values, however measured, to changes in states of the world that degrade or protect the environment? Advocates of more stringent environmental policies, and those who oppose them as sometimes going too far, often cannot agree on what really counts. And you can't *ac*count until you know what to count! If the disputants could agree on what to count, the problem of evaluation would be, if not easy, at least significantly advanced. [Norton and Toman, 1997]

As I said above, I have no reason to alter this assessment of the situation. First and foremost, the disciplinary disagreements between environmental economists and environmental ethicists who have adopted a nonanthropocentric stance rage on. This debate, very unlikely to be resolved any time soon, has led to a deep rift between the two fields [Norton and Minteer, 2002]. In the context of this

[2]These generalizations, of course, are subject to many counter-examples; they are offered as very rough guides to the interdisciplinary argumentative terrain [Norton, 1998]. In fact, much has changed in this area, as the Society for Conservation Biology has become increasingly activist, and increasingly policy oriented, and more international in scope.

volume, the anthropocentrism/nonanthropocentrism debate is not directly, but indirectly, involved. Sustainability is most basically an intertemporal concept, not an interspecific concept. While it seems at least possible to advocate a nonanthropocentric approach to sustainability—arguing that earth should be treated sustainably for all creatures—such a theory would remain at odds with economic theory, and nonanthropocentrists have not adopted this strategy; instead, they have generally treated sustainability as a human-centered concept and attacked it as such. Unfortunately, this concentration on interspecific ethics, and the monistic interpretation of these ethical concerns by nonanthropocentrists, has tended to divert the attention of most environmental ethicists away from intertemporal ethics, and cross-temporal moral relations have not been explored in as much detail as cross-species moral relations.

Intertemporal ethical arguments—and often assumptions—apply more directly to sustainability discussions, and there is no shortage of controversial issues. Here, however, the debate does not rage neatly across the ethicist/economist lines, as issues like discounting future costs and benefits of environmental changes are controversial both within economics and within the broader cross-disciplinary debate about intergenerational morality. There is simply not much agreement across disciplines, or even within disciplines, regarding how to formulate and evaluate our impacts on the future, which seems like a crucial aspect of any sustainability program. As noted in Norton and Toman [1997], while there is agreement that sustainability involves moral considerations of equity and fairness over and above considerations of economic efficiency, the plurality of moral conceptualizations—utilitarianism, rights theory, and contract theory—have all been invoked as interpretations of that sense of fairness. In utilitarian language, economists have articulated "weak sustainability," which requires that we maintain the stock of total capital of the society, so that individuals in the future will have the opportunity to enjoy levels of welfare equal or greater than ours. There have also been explorations of "rights" views, whereby future individuals are accorded rights [Howarth and Norgaard, 1990]. Another approach is to argue that current generations hold essential natural resources and beautiful places "in trust" for future generations [Brown, 1994; Weiss, 1986]; others suggest a "constitutional" model, arguing that protection of, for example, biodiversity, should be mandated in national constitutions [Schlickheisen, 1994], or at least understood as being understood as exhibiting a logic that is analogous to the special place accorded constitutional rights [Norton, 2005].

As has been true since the Brundtland Commission [Brundtland, 1987] published its call to make sustainable development an international commitment, no discipline has been able to offer a coherent and convincing conceptualization of intergenerational obligations, much less a comprehensive accounting system for measuring progress toward sustainability. For the foreseeable future, this situation will continue and some philosophers and others have advocated giving up on the process of articulating a monistic, one-size-fits-all moral theory that will capture the many aspects of intergenerational moral relations [Norton, 2005; O'Neill *et al.*, 2008]. The alternative sometimes suggested is to seek a "post-metaphysical" conception of public deliberation and discourse in which varied formulations of value compete in attempts to achieve cooperative solutions without resolving underlying "ontological" questions about value [Dietz and Stern, 1998; Dryzek, 1997; 2002; Habermas, 1984; 1987; Norton, 1991]. This process-oriented move does not of course resolve the deep disagreements about how to characterize environmental value. The search for definitive value positions, however, gives way to a more open-ended, action-oriented discourse [Norton, 1991; 1995; 2005; O'Neill *et al.*, 2008]; by focusing on finding policies and acts that are widely acceptable to those with varied positions. Philosophical advice to the policy process now emphasizes the offering of heuristics that help communities involved in policy decisions to focus on the right questions. Discursive pluralism has the advantage that one can seek policies that serve multiple values, without becoming bogged down in arguments about what has ultimate value.

The problem of environmental accounting is very difficult to separate from the problem of defining—or, at least, measuring—sustainability [Norton and Toman, 1997]. Economics, ethics, and ecology all weigh in with important insights into the problem of specifying a sustainable pathway into the future. It also seems, however, that none of these disciplines has anything like a universal perspective from which to define and identify sustainable pathways, so it seems inevitable that discussions of value issues affecting sustainability discourse will be chaotic and confusing for some time. Two central controversies will continue to divide the discourse about sustainability into apparently incommensurable dialects. First, there is the debate among those, on the one hand, who see "sustainability" as a matter of science and use it as a descriptive term [Costanza, *et al.*, 1981] and those who, on the other hand, see the choice to live sustainably as requiring that we determine what has value in order to know what we should save. Second, there is the related but even deeper disagreement about whether factual and evaluative discourse should be "segregated", so that science is done independent of values, which contrasts with the emergent, pragmatist viewpoint that values and science are all mixed together in the maelstrom of public deliberation about what to do about an environmental problem [Norton, 2005].

One new controversy has broken out at the edges of economics and ecology and it has to do with how to place value on the "ecosystem services" humans derive from other species and the functioning of natural processes. Many ecologists, and some economists have endorsed the calculation of dollar values to represent such services [Costanza, *et al.*, 1997; Costanza and Patten, 1995; Daly, *et al.*, 1997]. This has led, however, to further controversy between ecologists and "ecological" economists, who favor such representations, and more traditional environmental economists who think such representations are methodologically suspect. Traditionalists, noting that service estimates are often based on "replacement cost" and other methods that do not consider alternative replacements of the service, cannot be considered as, or aggregated with, economic values, which are established as prices at the margin. Marginal analysis seeks to determine the "willingness-to-pay" (wtp) of consumers—the price at which the consumer will purchase one more of an item as the price rises. If the price rises too far, the consumer will shift to an alternative or wait for the price of the item to decline [Bockstael *et al.*, 2000]. This is an ongoing controversy, but the trend toward more and more use of dollar estimates for ecosystem services [Millenium Ecosystem Assessment, 2005], continues. Despite disparagement by traditional economists, the concept of ecosystem services, while perhaps not an accurate way to determine natural values at a highly aggregated level, may nevertheless be useful in particular contexts when, for example, downstream users of rivers are willing to compensate upstream users for protection of the riparian zone. An estimate of key ecosystem services might prove useful in the future as a way to identify a fair price when some resource users pay others to maintain favorable environmental conditions for the former. For example, in some partnerships between upstream land users and downstream land users, downstream users pay upstream users to maintain riparian buffers and in other ways protect the downstream water quality. In contexts such as this, estimates of the monetary value of services may be very useful, even though there is reason for skepticism that the ecosystem services strategy of counting and aggregating toward a "total value of nature" is feasible.

2.2 The Cluster of Problems around Reversibility, Substitutability

At the center of our examination of conflicts between economists and ecologists, for Toman and I, was the economists' idea of substitutability of resources for each other, and ecologists' insistence that some losses are irreversible and not compensable. We argued that, at smaller scales, substitutability is relatively high. If a public park is closed, picnickers find another, perhaps less nice, perhaps further away, park in which to recreate. One might reasonably estimate the lost benefits in lost enjoyment or travel costs to reach an alternative park. If, however, some important resource is being lost across a whole area—such as a distinctive ecosystem or an important species, it seems less plausible to speak of "substitutes." Instead, a more appropriate characterization may be "reversible-or-not"; but these latter judgments cannot, in turn, be smoothly integrated with the marginal analysis of preferences-as-prices of partial equilibrium economics.

Economists and ecologists also differ in their approaches to the problem of spatio-temporal scaling. Economists, who generally engage in discounting when calculating economic impacts over multiple years to reflect time preference for the present in their calculations of costs and benefits, in effect, collapse time into the present, ignoring duration and temporal scale. This tendency is embodied in their choice of partial equilibrium models and in their counting values in present dollars. This approach, whatever else its merits and faults, clearly raises challenges for understanding ecological aspects of sustainability because ecological change is believed by ecologists to be prevalent, but episodic and somewhat unpredictable. As will be discussed further, below, from the beginning of the emergent field, ecological economics, one of its advocates' strongest complaints was that mainstream economics "ignores the *scale*" of an economy with respect to the ecological base on which the economy relies. Ignoring scale, they argued, is tantamount to assuming there are no limits to the resource-producing and the pollution-absorbing capacity of ecosystems [Daly and Cobb, 1989] and to giving unlimited growth of "throughput" a free range.

Economists' models seem to assume that the usefulness of ecological systems is assessed economically, and "natural resources" appear in their models only as one "sector" of the economy, "natural resource extraction." Even if one sides with the ecological economists and insists that the economy must "fit" into the ecosystem in which it is embedded (rather than the other way around), scalar problems continue to provoke difficulties. Toman and I proposed a conceptual device to explicate the "meta-logic" of a "two-tier" or "two-phase" approach to environmental action and goal-setting, which we called "risk decision squares" (RDS) [Norton and Toman, 1997]. We proposed that we adopt, based on thinking independent of the means-oriented logic of the action sphere, a more reflective sphere in which we articulate and defend a plurality of action rules, such as Cost-Benefit Analysis (BCA), Safe Minimum Standard of Preservation (SMS) (The SMS rule says: "Save the resource, provided the social costs are bearable."), and the Precautionary Principle (PP). We then engage in a discourse about which of the various criteria are appropriate in particular problematic situations. We proposed that the choice of an appropriate decision rule should be guided by a careful analysis of the "scale" of a problem.

As in Figure 1, we plotted environmental problems on a decision space defined by a temporal and a spatial continuum. On one axis, we plotted the temporal scale of a problem—how long would it take to repair damage if the action has worst-case impacts, ranging from quickly reversible to essentially irreversible—against the spatial extent of the area under risk. We hypothesized that, in the high-reversibility outcomes, threats over smaller spatial scales, and threats with both of these characteristics, it seems reasonable to apply a rule such as CBA, which ignores scale and which collapses time to the present. If, however, there is a large area or a whole system at risk in situations with low reversibility, then one should apply the SMS criterion. The SMS is a very useful rule in situations of great uncertainty, but where there is risk of irreversible loss—such as a species extinction. It thus occupies a place somewhere between CBA and PP, recommending caution but recognizing some conservation goals—even goals to avoid irreversible losses—can be over-ridden if the costs of protection of the resource conflict with other social priorities. We still believe that this two-tier approach is the only way that one can begin to integrate the marginalist analyses of CBA with the recognition of ecologists that economic systems that produce important resources are complex systems and can "flip" into alternate, less desirable, states of functioning.

Our work on scalar reasoning may help to choose appropriate decision rules in particular situations; it was associated with related work on hierarchy theory in ecology. A central axiom of hierarchy theory is that ecological systems are to be envisioned as smaller, nested subsystems within larger systems, with smaller subsystems—having their own dynamics and their own "integrity"—that cycle at

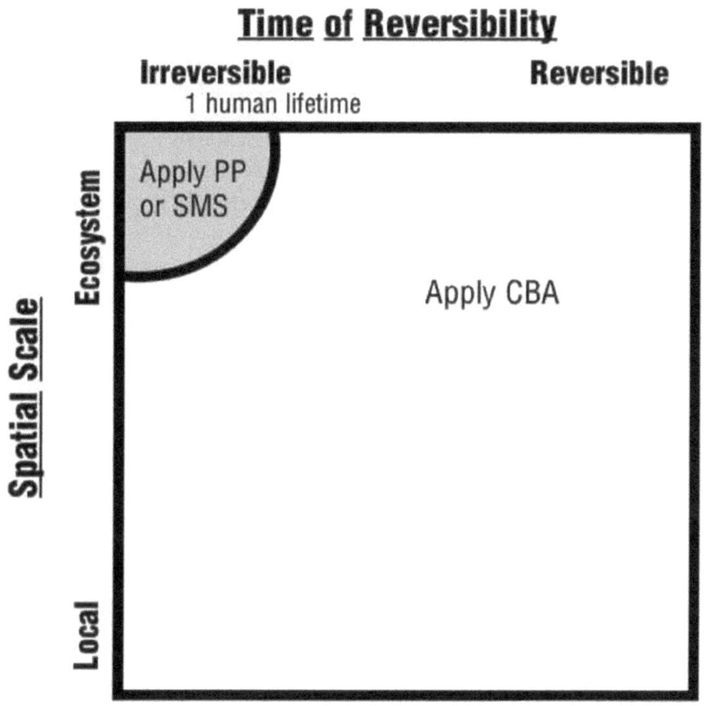

Figure 1. Risk Decision Square, showing how plotting time scale against spatial scale of damage risks from a policy can guide choices of decision rule. "CBA" = Cost-Benefit Analysis; "SMS" = Safe Minimum Standard of Conservation and "PP" = Precautionary Principle.

a more rapid pace than do the larger, slower, "environmental" systems of which they function as parts. This systems theory model corresponds to the top and the left side of the RDS: temporal scale of a problem can be measured, across the horizontal dimension, as the time/pace of recovery from disturbances (degrees of reversibility); the vertical axis of the RDS can be calibrated, from bottom to top, in measures of increasing scale (extent). As time of recovery and scale of impact increase, longer-term concerns—sustainability concerns—come into play. Thus, [Norton and Ulanowicz, 1992] were able to exploit this correlation of size with slowness, and proposed that hierarchical models could be scaled in ways useful to management planning and boundary-setting by first setting the temporal horizon of a management goal and associated regimen, and then using this temporal clue to infer the appropriate spatial scale and the relevant physical dynamics associated with processes of social importance (also see [Giampietro, 2004; Waltner-Toews *et al.*, 2008]).

3 SCALE DETERMINATION: THE GREATER CHALLENGE TODAY

3.1 Scale and Decision Making.

We thought that these decision squares may help us to sort problems according to "spatial" and "temporal" scale and we still think it is a useful device that allows at least rough quantitative measures that might help managers to choose the appropriate decision rule for the problem at hand. While useful, however, this device might be criticized for hiding deeper problems about decisions as to what scale on which to model emergent environmental problems. The problem of scale has become even more complex because, as the descriptivist, value-neutral and positivist view of science continues to lose plausibility as a philosophy of science, it apparently seems that human values deeply affect how we "scale" problems both physically and temporally. For example, if an activist cares deeply about protecting a local river system from pollution she may, in the course of discussing the problem with others, recognize that the river system is part of a larger watershed and that many of the problems of the river she emphasizes cannot be addressed without addressing the problems of the whole watershed. As her attention shifts from river to watershed, one might say that, originally, her efforts to save the watershed are just a means to protect her original, local valuing of the river; but as she attends watershed-wide meetings, she finds herself expanding her mental attachment and soon values progress in saving the watershed at least as much as she cared for her original river. This example represents a personal example of a broader phenomenon known as "scale-jumping" [Adger *et al.*, 2005; Hajer, 2003]) or "macroscoping". The point of the example is that decisions as to how to "bound" a problem—what aspects of the physical system to model—and what scale at which to address a problem ethically, economically, and politically are simultaneously open for choice [Lasswell, 1970; Ulrich, 1998, 2000, 2003].

So, while we stand behind our scalar approach to sorting problems according to scale as a way of choosing appropriate decision criteria, the risk decision squares only become operable once the appropriate scale for addressing a problem is already determined. In recent research, it has been recognized that this prior question is itself a thorny problem deserving careful thought and deliberation, because our (sometimes unconscious) choices of a scale at which to address an environmental "problem" are tied up with both scientific, descriptive information about dynamics driving change and also with value assumptions often hidden in choices of how to formulate the problem in question (see Section 5, below).

On the hypothesis that most environmental problems are "wicked problems"—problems that have no single, uncontested formulation because different individuals and different groups come to the situation with differing values and perspectives—it follows that most contentious environmental decisions involve situations where questions of scale are themselves unanswered and contentious [Funtowicz *et al.*, 1999; Funtowicz and Ravetz, 1994; 1995; Norton, 2005; Rittel and Webber, 1973; Taylor, 2005]. To take a stylized but usually accurate example, in disputes about water quality and quantity, the upstream advocates usually favor smaller models of local water quality and movements, because they want autonomy in managing their local water. Downstream advocates, on the other hand, insist on a more "holistic" model encompassing local situations in a larger picture [Hirsch, 2009]. Thus, while we continue to think that scalar determinations are useful in helping to choose the correct decision rule, it must be admitted that there is a prior question of the appropriate scale at which to address a given problem. Thus, while Norton and Ulanowicz [1992] were able, given a temporal horizon over which biological diversity is to be managed, to narrow the range of appropriate spatial scales for management plans considerably, it remains true that no purely descriptive account of scalar relationships can determine the appropriate scale at which to model a problem. Setting sustainability goals, according to this view, is a decision regarding what we will care about and, hence, requires important decisions about what is valued and what counts as one pays attention to particular scales and dynamics in natural systems. Acknowledging that this viewpoint is controversial, I turn now to a discussion of the epistemological and evaluative status of ecological models.

3.2 Scale and Hierarchy: Realism vs. Pragmatism in Ecological Science

A number of authors have suggested that hierarchy theory, an emergent ecological theory that is designed to make problems of spatio-temporal scale more manageable conceptually, can be employed to help sort out scalar problems and clarify policy options [Batie and Shugart, 1991; Gunderson and Holling, 2002; Gunderson *et al.*, 1995; Holling, 1996; O'Neill, 1988]. As an early and strong advocate of this approach in work with ecologists and economists [Norton, 1991; 1995; 2005; Norton and Ulanowicz, 1992b; Norton and Toman, 1997], I am heartened to see that this idea has now been discovered in environmental ethics as Callicott, in this volume, proposes that a form of hierarchy theory—he cites [Holling, 1992]—be used to choose an appropriate temporal scale for restoration projects. He argues that we should choose species that were present in the area in question in an "ecological" time frame, rather than going back into the Pleistocene to find species that were lost up to 15,000 years ago. He says, "*ecological* restoration should be framed in ecological spatio-temporal scales, which may be defined temporally in reference to ecological processes such as disturbance regimes and recovery intervals and spatially in reference to ecological units such as landscapes, ecosystems, and biological provinces" [Callicott, this volume]. Callicott argues for reintroducing species lost more recently, while rejecting the proposal to undertake evolutionary restorations of animals once present in earlier ecological ages.

As far as it goes, this kind of analysis and the recommendations that go with it are obviously correct; attempts at reintroductions of wooly mammoths and saber-toothed tigers would detract attention from real ecological problems in today's systems. We should focus on species that have some chance of ecological integration into future landscapes. I accept Callicott's argument and have, likewise, taken it as given that ecological management and restoration operates over the "mesoscale" as defined by Holling [1992]. Callicott proposes that we identify "disturbance regimes," such as the periodicity of hurricanes hitting beaches or, perhaps, frequency of wildfires, and use these to guide us toward appropriate temporal scales for planning and implementing on-the-ground restoration projects. I am convinced, then, that, if the question is between the Pleistocene re-wilding proposal and the more commonly advocated view that we are protecting and restoring ecological processes, hierarchy theory may be helpful. So far, however, Callicott has simply used it to refute a highly aberrant position—he even admits that Pleistocene re-wilding was proposed partly in jest, and carried on by advocates whose arguments and proposals are simply not taken seriously by responsible restorationists as possible policy. That argument, in other words, is "theoretical" in the worst sense: it discusses the theoretical value of a policy that could never happen in the foreseeable future. Aside from the highly speculative proposals Callicott quotes and dismisses, 99% of environmental managers and activists are not even discussing the Pleistocene restorations as serious *policy* options.

The analysis employed in this argumentative strategy against Pleistocene re-wilding, however effective in ruling out that already-implausible proposal, seems much less likely to help us make the real-life and more fine-grained decisions about reintroductions and restorations that arise every day when restorationists have to develop a scalar model and choose appropriate species to effect a restoration. Settling on ecological frames of time, a decision that is unquestioned by virtually all managers and restorationists, does not even begin to identify the scale at which to model a particular environmental problem or an environmental management project such as an ecological restoration. The range of scales that are "ecological" (which Callicott defines as measured "in terms of centuries"), which Callicott says are "biologically defined temporal scales," is enormous: Some fire-regulated communities function on a roughly 20-year cycle of fires, while other forests cycle over centuries. The ecological processes necessary to replace old-growth forests with their characteristic soils and undergrowth is hundreds to perhaps ten thousand years. What does Callicott's tell us that is useful in complex decisions? How does it help us to restore a wetland damaged by an oil spill, for example? And how does it help us to choose among recently extirpated species as the most appropriate to reintroduce?

Practicalities aside, it seems reasonable to explore whether Callicott's proposed method, based on appeals to "hierarchy theory," has promise to help us to choose, from this vast variety of possibilities, the appropriate ecological scale to apply with respect to a particular restoration problem or project? Can hierarchy theory establish a "correct," or at least scientifically based, "appropriate" ecological scale for restoration projects? At this point, two key ambiguities arise, an ambiguity related to two different aspects of hierarchical thinking in ecology, and an ambiguity in how to understand hierarchy theory "philosophically" (epistemologically).

Holling incorporates hierarchy theory into both his scientific work and into his practical work on adaptive management. The first ambiguity is between *hierarchy theory*, and particular, concrete *hierarchical models*, which are constructed in order to clarify particular relationships and particular processes either for scientific or managerial—or both—motives. Hierarchy theory proposes that we observe and understand ecological processes as actors in a "hierarchically" organized system, and develop models on which smaller subsystems change at a pace faster than the larger, slower systems of which they are parts. As such, HT is best thought of as a heuristic by which to choose models that are useful in understanding multiscalar processes. Hierarchical models themselves are specific, illustrative and predictive models that, by focusing on certain, well-chosen dynamics, gain better understanding of a system's functioning. Advocates for these models believe that, by simplifying and organizing the overwhelming amount of data coming from all the different scales of complex dynamic systems, they can provide models useful for describing ecological processes under study. Sometimes these hierarchical models are used to clarify what is going on in processes of environmental degradation, other times they are used to understand how to re-establish ecological processes once they are damaged. They are, that is, models that share certain formal characteristics that qualify them as "hierarchical," but the models of this type are countless and can be constructed for any purpose, from "understanding" some ecological process (by identifying key driving variables in a system and modeling them hierarchically), to constructing management goals that correspond to multiple scales in a proposed hierarchical model. "Hierarchical models", so understood, simply refers to a set of models that share certain formal characteristics; the set is, in turn, defined by "axioms" which specify those formal qualities. So understood, it is difficult to see how Callicott thinks a theory, made up of formal axioms, or a defined set of models that share formal characteristics, could guide restorationists in choosing an "appropriate" scale. Perhaps Callicott's argument can be better understood by clarifying the epistemological ambiguity.

This second ambiguity, which has to do with how we interpret these hierarchical models, epistemologically, goes to the heart of the ongoing debates between many "nonanthropocentrists" such as Callicott, and the philosophical approach known as "environmental pragmatism." Recognition of the importance of hierarchical organization in understanding systems emerged from systems theory in the last half of the Twentieth Century. Herbert Simon [Simon, 1947; 1982] argued for the importance of hierarchical organization and Arthur Koestler [1967] developed these ideas into a theory of the nature of life and self-organizing systems. Basic ideas in the area were developed in [Pattee, 1973]. There exists a tradition, then, that sees the hierarchies found in nature as real and given in nature [Simon, 1947; Wimsatt, 1994], but most ecologists have used hierarchy theory as a conventional tool rather an as a description of the structure of real systems. Hierarchy theory was explicitly applied to ecological systems by Allen and Starr, who said that they did not "mean to imply that reality, independent of our cognizance is in its nature hierarchical" [Allen and Starr, 1982, p. 6]. They describe the choices of scales and their construction into a hierarchy as based only on an "epistemological stance in a utilitarian philosophy," and questions of "ontology" were not addressed.

Writing with a different co-author ten years later, Allen went beyond this careful agnosticism, arguing that:

> It is essential to resist the temptation to base ecological understanding on a belief that ecological systems are an ultimate reality beyond observation. Ignoring human subjectivity will not make it go away. Since one makes arbitrary decisions anyway, ignoring them abdicates responsibility needlessly. All decisions come at the price of not having made some other decision. By acknowledging subjectivity one can make it reasoned instead of capricious. A real danger in suppressing the ecologist in ecology is to be bound by unnecessarily costly decisions. These could be exchanged for a more cost-efficient intellectual device if only the subjectivity of the enterprise were acknowledged. [Allen and Hoekstra, 1992, pp. 51–52]

Building on Allen's two characterizations, and turning the straw man they argue against into a form of scientific realism, we can apparently identify three positions, which might be described as three points on the "conventionalism" to "realism" continuum in the philosophy of science.

1. Choices of hierarchical models are "arbitrary", but still rationally made by seeking a more "cost-efficient intellectual device" [Allen and Hoekstra, 1992].

2. Choices of hierarchy have an "epistemological" basis; we choose particular categories because they allow human beings as conscious creatures to better understand ecological relationships [Allen and Starr, 1982].

3. We can observe patterns in ecological systems that allow us to identify hierarchical patterns and these patterns can be described as "real" characteristics of ecological systems just as the relationships between atoms and molecules represent observable relationships that exist in reality (Allen's straw man, [Wimsatt, 1994]).

Early advocates of hierarchy theory, happy to see it as a useful heuristic, avoided the strong ontological claims of 3, contenting themselves with either 1 or 2, or ignored the issue altogether. It turns out, however, that both 1 and 2 are themselves philosophically ambiguous:

1. a. Rational choice of a particular hierarchy to understand an ecological system is based on "linguistic convenience" (simplicity of expression, clarity of communication, etc.).

   b. Rational choice of a hierarchical model is justified because the model is more successful than others in allowing human beings, given our cognitive equipment, to understand and make predictions regarding ecological systems.

2. a. Rational choice of a particular hierarchy to understand an ecological system is based on epistemological and cognitive grounds: given human cognitive abilities, ecological systems will be more systematically and more effectively described as exhibiting hierarchical organization.[3]

   b. Rational choice of a hierarchy to understand an ecological system is a matter of choosing carefully a hierarchy that will illuminate processes that have important implications for human values. (For example, a particular hierarchy may illuminate how human activities threaten systems and lead to loss of ecological services or other values better than any other known model). This is the position favored by environmental pragmatists, who construct hierarchical models as *tools* for both understanding and action, and readily admit that such models reflect social values as well as reality.

Most ecologists who discuss hierarchy theory, I believe, have had views that cluster around 1b and 2a. An exception is C. S. Holling, who has proposed (as noted by Callicott, this volume) that *nature is indeed organized hierarchically* (also see [Wimsatt, 1994]). One would assume, given that Callicott cites Holling—who has speculated that 3, above, may be true, that Callicott also endorses 3. This bold position, which asserts that "animals living in specific landscapes should demonstrate the existence of ... lumpy architecture by showing gaps in the distribution of their sizes," faces some formidable hurdles, however, as Holling himself notes. He is careful to present the idea as a speculative hypothesis more than as accepted theory [Holling, 1992, in Samson, 1996, p. 352]. Careful examination of body masses of animals in Western Australia and the United States, however, has not supported the expectation of gaps in body mass distribution [Siemann and Brown, 1999]. Furthermore, advocates of this hypothesis, while implying it is an *empirical hypothesis* consistent with hierarchy theory, as a theory, do *not* claim that recognition of discontinuities in the distribution of body sizes will guide a scientist to a uniquely appropriate, specific hierarchical model [Allen and Starr, 1982, p. 161].

While I do not intend to impute to Callicott any particular position on this philosophical continuum, it seems fair to ask for clarification of this ambiguity. He states the goal of his chapter as demonstrating that: "the appropriate temporal scale for ecological restoration is ecological time, defined by the periodicity of ecological disturbances, by disturbance regimes. Correspondingly, the appropriate spatial scale for ecological restoration should also be defined ecologically—in terms of such units as landscapes and bioregions." What exactly is the meaning of "appropriate" in this context? Remember that, as noted above, Callicott describes ecological scales as "biologically defined temporal scales." But he also, at other points in his discussion, seems to grant that humans often pay special attention to some dynamics and associated hierarchies because those dynamics affect what we value: "We may regard the processes at the higher end of the hierarchy as

[3]It may lead to unnecessary quibbling to insist that 1b and 2a represent independent positions, but since we are examining a continuum, that probably will not matter to our further discussion.

relatively unchanging or stable if our interest focuses on processes at the middle or lower end". At another point, Callicott suggests we avoid larger-scale models because they are greater than the scales "on which we gauge changes meaningful to us." When he emphasizes the "biological definition" of scale, he seems to be endorsing a view on the continuum near 3. When emphasizing that our interests and what we find meaningful determine the scale on which we focus, it seems as if we construct hierarchical models which are useful given whatever human "problem"—epistemological or managerial—we are trying to understand or resolve. This would apparently be a more conventionalist or pragmatic view of how hierarchies are chosen and justified.

Callicott, at the least, owes an explanation of where he stands on these ambiguities and how, exactly, he expects that invoking Holling's speculatively descriptive version of hierarchical model will determine whether water quality efforts are best tackled river-by-river or at the watershed scale. Can descriptive HT, by itself, point the way toward "appropriate" hierarchies of scale by which to understand the natural world? Is Callicott asserting that there exists one or more actual, biologically describable hierarchies in nature that exist independently of our human concerns and values? Is he, that is, endorsing interpretation 3, above? If so, is he also suggesting that, once we determine descriptively the appropriate HT for an ecological system, then we should use that scale as a guide to analyzing all (some?) environmental problems that occur in that system? If so, he is placing an enormous burden upon descriptive HT, an interesting, but highly speculative interpretation of HT, supported by few ecologists and hierarchy theorists.

I can now explain my remark that the uses of HT in environmental modeling, ecological understanding, and environmental management, brings into sharp relief the contrast between environmental pragmatists and Callicott. If Callicott endorses interpretation 3, the realist interpretation of hierarchies, then his realistic approach differs sharply from more pragmatic applications which articulate scales appropriate to human concerns, rather than arraying all human concerns on a single model claiming to be "real".

Pragmatists would not search for a single, empirically correct model of ecological systems; a pragmatist would begin, not at the theoretical level of analysis, but with a specific problem, and then construct models that are useful representations of the dynamics involved in systems that are considered problematic [Light and Katz, 1996]. For pragmatists, models are tools of the understanding, but those tools must be orientated by reference to a concrete problem demanding action. One does not, first, create a "model" of an ecological system, complete with a number of featured scales organized hierarchically, and then array all environmental problems on that single hierarchy. One has no idea which of the indeterminate number of possible scalar hierarchies that could be represented in models until one's "interest" is engaged. Interest is engaged, according to the pragmatists, when something valued is at stake. Problems are situations in which something of human value is underprovided or unprotected. Norton and Ulanowicz [1992], for example, provide a detailed explanation of how, given a combination of value determinations, reflective use of hierarchical modeling can help to choose an appropriate scale for endangered species management. This approach presupposes not interpretation 3, but interpretation 2b.

If choices regarding scalar calibration must be made in describing natural systems, according to pragmatists and adaptive managers, it is human values and interests that allow us to focus on appropriate time frames *for the production or protection of a human good.* To assume that there is one "real" scalar system for ecological systems would be to take us back to the modern period in which the world was assumed to be a mechanical, reversible machine arrayed across "absolute space," where objects existing at larger scales are assumed to be reducible to smaller and smaller components, ending with atoms. If Callicott presupposes all of these modernist assumptions about space and time, and if he were to claim that all interesting ecological change can be assigned "biologically defined temporal scales", he might claim to be offering a "non-anthropocentric" theory of ecological scale. But of course, the world we live in is today's post-modern world in which it is difficult to defend the idea of models that are "correct" as descriptions of nature without some reference to human perspectives and values and, while biological data and theory certainly constrain the range of possibly useful models, it would require discredited modernist assumptions about scientific realism to claim that the appropriate scale can be determined *solely* by data and theory.

Further confusing the matter, Callicott also recognizes a shorter, "historical" horizon, the pace at which human interactions with natural systems change the latter: "Another non-biological temporal scale creeps into the discourse in which this debate has been conducted, the historical temporal scale. Adding to the confusion, some writers conflate and confound the ecological and historical temporal scales because they are roughly coincidental—that is, the one maps coextensively fairly well on the other."

While I'm not sure what to make of this remark, which is not developed, the pragmatist viewpoint would reject Callicott's apparent assumption that ecological scales and human, historical scales are similar by some form of coincidence that only confuses us; the pragmatist would instead argue that we understand ecology—especially when we bring it to bear upon practical problems and choices that are involved in restoration policy and actions—in a way that makes it commensurate with our conception of human history. This would seem to be the implication of any interpretation of hierarchical models other than the realist version 3, as described above; and, it seems also to be a consequence of HT itself, since a key axiom of HT is that all observation and interpretation of the system takes place from within the system, which amounts to a repudiation of the general applicability of Newtonian, modernist ideas of "absolute" space and "real" time.

4 REFORMING OR REPLACING ECONOMICS: WHAT IS NEEDED?

Since the early 1960s and 1970s, at least, there has been a growing "opposition" within the ranks of economists to many aspects of the "rational choice" model for conceptualizing environmental valuation and for analyzing decision options. This group is varied and eclectic, but its many advocates, while playing different instruments, are playing the same song: the methods of mainstream economics are not up to the task of evaluating environmental change, especially not up to evaluating systemic changes that emerge slowly and over large spaces. These critics have included, of course, E. F. Schumacher, Nicholas Georgescue-Roegen; Kenneth Bolding, Herman Daly, Daniel Bromley, and many others. While big ideas about overshoot and steady-state economy created a great stir from the 1960s forward, more detailed evidence has come from many fields showing how, both in principle and detail, the methodological assumptions necessary to quantify human preferences are theoretically indefensible, unstable over different contexts, and lead to clearly inaccurate or meaningless empirical results. Perhaps the most telling criticisms come from cognitive psychology and the cognitive sciences, including decision theory, and there are several families of strong challenges to the economic theory, to its methods of analysis, and to the use of CBA's to guide environmental policy.

It would be beyond the scope of this paper to survey and analyze each and every one of these critiques, because many of them address the question of sustainability only indirectly: criticisms of the methods used by mainstream neoclassical economists to measure and aggregate social values, while important, apply very broadly to all attempts to guide environmental policy by economic analysis. I will mention briefly several of these critiques, but will reserve a full-fledged discussion for the emerging, trans-disciplinary field of study, "Ecological Economics," which has sometimes been referred to as "the science of sustainability" [Costanza, 1981]. While ecological economists use all or most of these broader critiques in drawing contrasts between their "ecological" approach and the more traditional methods of mainstream economists, I will concentrate in detail only on the progress of ecological economists toward more ecologically conscious decision making and toward a more integrated way of understanding environmental and, especially, ecological values. In this section, then, I will briefly mention a variety of these critics and their concerns about economism, and then discuss in detail the criticisms coming from five areas. These are A. Ethics, B. Cognitive Science, C. Decision Theory, D. Process Approaches; and E. Ecological Economics. Some of these criticisms reflect deep issues in the philosophy of science, and range from methodological and definitional issues to deep disagreements about the relationship between scientific and value discourse, and will be revisited in Section 5.

4.1 *The Critique from Ethics*

It is widely accepted that neoclassical economics incorporates a specific form of utilitarian ethics, sometimes called "preference utilitarianism," into its conceptual foundations. According to this view, "happiness"—as the key variable in utilitarian ethics dictum, "do that which results in the Greatest Happiness for the Greatest Number"—can be operationalized by treating willingness-to-pay (wtp) as a measure of preferences. A person's willingness to pay for a good or service is thus understood as a quantifiable measure of its impact on their personal welfare. Social choice, then, can be analyzed in terms of aggregations of wtp across relevant affected groups. Many ethicists, perhaps most prominent among them, Mark Sagoff, have criticized this analytic tool because it conflates actions taken according to principle or based on one's deeper-than-consumptive values with simple preferences. Sagoff goes on to attack economistic reasoning as vacuous, because "preferences" are only observable by inference from choices, and cannot be thought of as data about human satisfaction or happiness: "The term 'preference' as it occurs in microeconomics ... combines the idea of a stipulated value of a variable in an epistemological program and the idea of a psychological spring, motive, or trigger for action." Sagoff describes this combination as a "mix-up of epistemological and psychological 'givens'" and says that this mix-up is what makes the theory impossible to test and ultimately even to understand" [Sagoff, 2004]. Philosophers and others have also criticized economists' version of preference utilitarianism by arguing against the methodologically based assumptions (a) that individuals are always the best judges of their own welfare, (b) that individual preferences can be treated as "stable" across any temporal decision frame, and (c) that preferences pre-exist choice situations [Norton, 1991; 1995; Norton *et al.*, 1998]. The last of these criticisms has been developed and argued with empirical evidence by cognitive psychologists, cognitive scientists, and decision theorists, and these critiques are summarized briefly in the next sub-section [Gregory *et al.*, 1993; Slovic, 1993].

4.2 *The Critique from Cognitive Science and Decision Theory*

A group of cognitive psychologists and decision theorists have empirically studied how people actually make decisions and they have compared these results to the assumptions (especially (b) and (c), above) of mainstream economists. Some of the most powerful results, sometimes referred to as evidence of "preference reversals," showed that, as one varies the choice situation and the way one provides information to an individual facing a choice, very different preferences can be expressed by the same person (for a review, see [Kahneman and Knetsch, 2005; Tversky and Kahneman, 1981; 1990]). This work has been supplemented by decision theorists who have argued that it is more accurate to describe preferences as "constructed" within particular choice situations [Gregory, *et al.*, 1993; Slovic, 1993]. This concept, which further undermines the economists' assumption that preferences are pre-existing and fixed, seems to encourage a more pluralistic and fluid approach to value formation and reformation.

Perhaps the most profound challenge to rational choice and neoclassical economics has come from decision theory, where it has been shown that most people do not act as a "rational calculator of expected utilities." Limitations on time and energy lead human individuals to follow rules of thumb, rather than rationally calculating expected utilities. Rather than always calculating utilities, most people engage in self-binding behaviors which thwart our day-to-day preferences; and act in many ways that reduce our utility all the time [Simon, 1982]. These limitations, when applied to complex and recalcitrant negotiations regarding policy with many winners and many losers, the Pareto Improvement rule is inapplicable. In these situations, however, the most that can be achieved is to "satisfice"—to choose among the achievable policies, given constraints of interest groups and other political realities, and the most that can be achieved is a temporary compromise [Lindblom, 1959; Simon, 1947; 1982].

This rather pessimistic view of the limits of rationality in policy formation is consonant with another formative finding in decision theory, Rittel and Webber's [1973] convincing argument that most remaining environmental and planning problems are "wicked problems." Unlike "benign problems," which have definitive answers, wicked problems invoke different "frames" [Gray, 2003] and "mental models" of a given problem [Bostrom *et al.*, 1994; Doyle and Ford, 1998; Johnson-Laird, 2000], and a shared sense of success and the ability to articulate widely shared goals is elusive. Rittel and Webber show that wicked problems are susceptible to multiple problem formulations because of varied interests and competing perspectives. They provide a strong and very general argument that the most difficult aspects of a decision which is treated as a problem of "optimizing" one or a few variables must be resolved or assumed before the problem can be addressed with the powerful algorithms of optimality.

All of these criticisms can be assembled under the label of "contextual" environmental policy analysis, which eschews computational approaches that seek algorithmic calculations of best outcomes, and favors a more process-based, iterative process by which decisions are always tentative, problem formulation may have to be revisited, and no final solutions emerge. Good outcomes are temporary resolutions gained by negotiation and partial agreements. Cooperative action can be achieved, but it is usually partial and incremental. On the more positive side, however, these theories of limited reason and complex, wicked problems have opened up a dialogue, mentioned above, about how collaboration, deliberation, and social learning can be a reasonable response to the waning of computational and algorithmic decision models. These new models, which can all provide guidance within a process of adaptive management, locate good decision making in fair and intelligent processes seeking cooperative solutions, rather than in algorithms for the assessment and ranking of policies according to their production of maximal utility. It turns out that there is a sort of "middle ground" between adaptive management processes and the search for algorithms. Conservationists and conservation biologists have developed algorithms for choosing reserves using decision analysis within a practical process of conservation planning. While algorithms are prominently discussed in this approach, the process is an open-ended one of developing more and more effective reserves [Caughley, 1994; Pressey, *et al.*, 1997; Sarkar, 2005], so this approach is ultimately a process-based approach.

4.3 Process Approaches: Discourse Ethics, Deliberation, and Collaboration

The work on the construction of preferences, cited above, suggests that one should perhaps think less of "individual, fixed preferences that pre-exist policy discourse," and further emphasize the processes by which people communicate, and this approach encourages and supports a more process-based, collaborative approach to management in which communities "construct" goals and objectives in a deliberative process that encourages "social learning" [Dryzek, 2002; Hajer and Wagenaar, 2003; Lee, 1993; Norton, 2005; Sagoff, 1998; Torgerson, 1985; Wondolleck and Yaffee, 2000]. Social learning, in turn, has become very important in the theory of adaptive management [Gunderson and Holling, 2002; Gunderson, *et al.*, 1995; Lee, 1993; Norton, 2005], which seeks to manage ecological systems in a process of "learning by doing," with the goal of gradually learning to live sustainably through a combination of experimental management and collaborative public processes.

An important, remaining question is exactly how adaptive collaborative management will incorporate systematic thinking about values into adaptive management processes. In some cases, it seems adaptive managers—many of them trained as ecological scientists and mathematical bioeconomists—simply capitulate to economics, insisting only that ecosystem services be counted broadly in aggregative calculations. Others have explored "double-loop learning," which can involve actual transfiguration of the way a problem is formulated and values are expressed [Lee, 1993]. Norton [2005] is an extended attempt, based on insights from the philosophy of language, to encourage transformative learning about sustainability goals by employing certain "heuristics" that will help communities to re-frame problems and find cooperative solutions to them [Fisher *et al.*, 1991; Sagoff, 2004; Wondolleck and Yaffee, 2000]. Other scholars have studied successful public-private partnerships in managing watersheds and other lands and waters, and this literature seems to show that, when communities organize and work together, initially chilly relations between members of different interest groups thaw and, even partial agreements can build trust and encourage increasing cooperation [Leach and Pelkey, 2001; Leach *et al.*, 2002; Lubell, 2004; Moore and Koontz, 2003; Sabatier *et al.*, 2005; Sagoff, 2004]. The economist Richard Howarth has recently published important papers showing how communities committed to cooperative behavior in spite of conflicting interests, can be seen to follow a rational decision rule that is based more in process than in determining "rational outcomes" in advance and striving to bring them about by any means [Howarth and Wilson, 2006].

This work on cognitive processes and collaborative decision making has important applications to understanding sustainability because, whereas mainstream economists have tended to see sustainability as a purely descriptive measure of impacts of resource use and substitution on the ability of future generations to achieve per capita welfare equal or greater than earlier generations, these approaches plunge the discussion of sustainability right into public debates about what is valuable and what goals should be set, which creates a much more dynamic and value-laden search for sustainability [Norton and Toman, 1997].

One other possible source of guidance from within economics may at last be forthcoming from the area of economics called "institutional economics," and should be mentioned briefly here, see [Bromley, 2006; Hodgson, 1997; Schmid, 2002]. Advocates of this approach, which sees economics as inevitably tied up with politics, practice a kind of political economy characteristic of thinkers prior to the divorce, forced by economists as they sought to become more positivistic, between economics and political theory. Institutional economists criticize mainstream, utilitarian economists for failing to recognize that markets often described as "free" are in fact shaped and controlled by political factors, and the current constraints and arrangements are designed to favor those with wealth and political power. Given that creating sustainable lifestyles will require appropriate institutions that incentivize ecologically benign activities and punish those that are ecologically harmful, it could be argued that institutional economics should replace the microeconomics of most environmental economists as the main source of environmental policy insights. This approach has its roots in pragmatist thought, and this connection has recently been developed in an outstanding and useful book by Bromley [2006].

4.4 A More "Ecological" Economics?

Following upon the attacks by heretical economists upon neoclassical economics and its understanding of environmental problems, led by Schumacher, Georgescue-Rogen, Boulding, and by the early work of Herman Daly, an oppositional force emerged in the mid-1980s, a transdisciplinary area of study called "Ecological Economics." Its early advocates, including Daly, John Cobb and, especially, the systems ecologist Robert Costanza, spoke of this study as "the science of sustainability" [Costanza, 1981]. A vibrant group formed that was united by shared, and sometimes highly varied, objections against mainstream economism and by a shared, but not entirely articulate and far from systematic ideas about how to move toward a more "ecological" understanding of environmental problems.

Perhaps it is possible to identify 5 key ideas that often appear in the journal *Ecological Economics* and in other publications by Ecological Economists, and thereby define the core ideas of the emerging field. Some ecological economists may not accept all of these ideas, but by adopting some elements of this cluster of ideas, advocates of this emerging field are expressing their independence from the methodological and theoretical assumptions of mainstream economics.

1. Mainstream economics does not pay enough attention to the *scale* of economic systems vis-à-vis the *scale* of ecological systems. Indeed, economists

construct models on which natural systems simply appear as one "sector" (resource extraction) in the economy. Ecological economists reverse this relationship, and place economic systems and their activities *inside* ecological systems. They insist that economic systems must be trimmed to fit ecological systems without overwhelming the natural processes supporting the latter.

2. More generally, ecological economists pay more attention to matters of scale, as noted above in Section 3, because they model the natural systems encompassing economic systems of production as complex, dynamic systems, rather than as partial equilibrium systems. This encourages ecologists to abandon "mechanical," "deterministic," and reversible systems, and to model environmental systems as multi-scaled [Norton, 2005]

3. Building on 1 and 2, ecological economists, and ecologists working with economists, have developed numerous models that mix human and nonhuman elements, and they have explored ways to make ecological and economic models interactive [Gunderson and Holling, 2002; Gunderson, *et al.*, 1995].

4. Ecological economists express, in various vocabularies and in various venues, the concern that the methods of economists—or the current means of employing them in practice—are too narrow and that counting only values that can be expressed and measured as willingness-to-pay will greatly restrict the kind of values we see and we count. Advocates of this line of reasoning can be divided into at least two primary sub-groups,

   A. There are critics who tend to accept economic methodologies, which measure values as aggregated wtp, but who try to expand the use of economic methods such as contingent valuation and shadow pricing in order to include a much wider range of "benefits" and "resources" that people will be willing-to-pay for, thus increasing the proportion of total value derived by societies from natural systems, thus increasing incentives to protect systems. This may, for example, include insistence on "environmental accounting" [Lange, 2007], attempts to establish a wtp for saving holistic systems like "wetlands" and "landscapes," and liberal use of the idea of "existence values," the value people place on simply knowing something exists and is protected. This first group, then, tries to make economics more ecological without breaking the standard rules by which economic activity is measured.[4]

   B. There are critics who believe that the wtp means of evaluation, which is necessarily a marginalist analysis, is inadequate to specify the full range of values humans derive from nature. This group has introduced the

[4]See [Bockstael, *et al.*, 2000] for a careful explanation of why the difference between A and B, below, is so important and unavoidable.

concept of "ecological services," the utilitarian and sometimes less tangible values one derives from ecosystems, and used liberalized means of measuring these values, such as "replacement cost," or "avoided costs," which cannot be aggregated meaningfully with wtp values measured by mainstream economists [Bockstael, *et al.*, 2000]. The ecological services literature, which is large and growing [Costanza, *et al.*, 1997; Daily, 1997; Millenium Ecosystem Assessment, 2005], is successful in creating publicity and such values surely have some rhetorical value, but they occupy an odd, betwixt-and-between position in that they *appear to be economic values* because expressed in dollars, but they are *not economic values* in that they are not measured with the normal tools of economic value estimation.

Some advocates of this ecological services accounting especially emphasize one specific service, the maintenance of "ecological resilience," the features of ecosystems that maintain them within their current regime of ecological functioning. This "service" can be thought of as "overhead": a complex, dynamic, and self-organizing system must take some high-energy resources from its environment and expend those in maintaining internal structure and an external "membrane" that keeps the system from collapsing and flipping into another domain of attraction. According to this line of reasoning, resilience—a descriptive characteristic of ecosystems in temporary equilibrium—has instrumental values to humans because it represents the overhead necessary to maintain the system in its desired and productive state.

5. A more radical criticism of economic valuation argues that there are some aspects of nature such that, if they were lost, people in the future will be worse off, even if they are as wealthy as prior generations. This position, which leads to the development of lists of elements or features of natural systems that should be protected—ecological capital—these items should be protected even if this means significant sacrifice. Norton, 2005, contains a careful analysis of these various nuances in the definition of sustainability.

One particular manifestation of strong sustainability theory is the method, quite antithetical to the usual methods of economics, of calculating the costs of environmental over-use and degradation as a function of an individual's, a community's, or a country's "ecological footprint" [Wackernagel and Rees, 1996]. This technique, which is still under development, may provide an alternative to economic valuation as a way to measure, quantify, and evaluate human impacts on the natural world. This method makes one painfully aware of the impacts of actions on ecological systems, and thus provides an antidote to mainstream economists' predilection for counting welfare effects and ignoring impacts on ecological systems; it remains to be seen, however, whether ecological footprinting can provide a comprehensive alternative to CBA and other welfare-based calculations of benefits and costs of human activities, or whether this technique might acquire a status as one of several important criteria for judging sustainability of policies.

Assessing the progress of ecological economics as a transdiscipline must be an ongoing process and only a snapshot evaluation can be provided here. It has been argued that, if the criterion of success for this transdiscipline is the integration of scientific knowledge and ecological knowledge in the analysis and management of environmental problems, ecological economics can be considered a reasonable success, with more successes to be expected. On the other hand, if one hoped from ecological economics the development of a new, comprehensive method to identify and measure environmental values, one should be disappointed. The prospects for an early breakthrough in this area remains dim, largely because of the deep, theoretical issues, referred to above as the problem of substitutability and reversibility and the problem of accounting. I fear that these theoretical differences are hardly closer to resolution than in 1997. This assessment may, however, demand too much; what may emerge is the enthusiastic endorsement by ecological economists of "institutional" economics and/or the development of a more process-oriented approaches in which both economics and ecology provides empirical assistance within many place-based, process-oriented, open-ended, collaborative partnerships [Howarth and Wilson, 1996].

5 REFLEXIVE PHILOSOPHY OF SCIENCE: MODELS, METAPHORS, AND THE DISCERNMENT OF SOCIAL VALUES

So far, I have concentrated on criticism of the rational choice, economic model from empirical and normative disciplines that compete with it for space in the public discourse about environmental policy. In this Section, I will explore briefly a very new and perhaps revolutionary understanding of the interaction of ecological science and evaluation of human-induced environmental changes, an understanding that originates in two important strains of thought in contemporary philosophy of science which, when combined with developments in cognitive psychology and cognitive science, have led several ecologists to propose that we engage in "reflexive" ecology, and that we do so, not by becoming better fact-finders, but by becoming more self-critical—and more creative–about the modeling decisions that must be made [Taylor, 2005].

Note that this new direction is especially appropriate if one accepts the arguments of most hierarchy theorists that specific hierarchical models are chosen for their usefulness in the pursuit of human endeavors in specific situations. As we saw in Section 4, above, the rationales for such choices could be limited to *descriptive* ones—as in interpretation 2a. On this interpretation, a hierarchy is chosen as the best model to *describe* a given ecological system. One might construct hierarchies as a means to better understand and protect *human values* derived from a natural system. If one can make a convincing argument for 2b as the proper interpretation of choices regarding hierarchical models, then it follows that choosing a specific hierarchical model in many given situations will be based, at least indirectly, on value considerations. On this view, one cannot know what model to choose in a given situation unless one understands *both* the social value driving the research *and* the structure and dynamic functioning of the ecological systems involved in the production of those values. In other words, endorsing 2b will necessarily involve reasoning described in both 2a *and* 2b.

To borrow some language from John Dewey's logic, a position limited to 2a would describe "situations" simply in terms of uncertainty about how to *describe* what is happening in an ecological dynamic, while embracing 2b, *in addition to* addressing scientific uncertainty in the situation, would also require an irreducible evaluative element. A situation, that is, demands enough resolution of the factual situation to allow action—but all actions gain their justification from the *desire* to obtain or protect some human value. So, any reasonable interpretation of 2b will *include* information described in 2a; to defend 2a, as against 2b, however, apparently requires the positivistic assumption that facts can be separated from values, and that specific hierarchical models can be chosen on grounds of "pure science". Once again, we see that the interpretation of hierarchical modeling choices goes to the heart of the arguments between environmental pragmatists and others, who insist upon a sharp separation of factual and evaluative content, and who thereby see it as possible to choose "correct" models based on observation and scientific theory alone.[5] Further, we can now see why pragmatists, unlike scientific realists who would favor interpretation 3, see ecological modeling as essentially involving human values, human perspectives, and human goals as well as involving a struggle to understand which natural dynamics are drivers of the ecological factors important to the particular situation in which the model is being constructed. To fully explore the territory surrounding the extent to which values—and which type of values—intrude into science would be beyond the scope of this review paper.

Reflexive ecology is consonant with two important strains in contemporary philosophy of science. First, reflexive ecology adopts the view that scientific models do not provide *the* single, correct account and explanation of the functioning of a physical system; rather, for many philosophers of science today, the scientist constructs multiple, partial models of systems, and uses these models either alone or in tandem to understand systems that are too complex to be modeled deterministically and uniquely. Nancy Cartwright captures this viewpoint when she says: "We live in a world rich in different things with different natures, behaving in different ways. The laws that describe this world are a patchwork, not a pyramid" [Cartwright, 1999]. William James put the point elegantly when he claimed that we live in a "pluriverse." A second, related, strain in contemporary philosophy of science has to do with how we manage and get along with patchworks that can never be complete or definitively associated with a complete and deterministic reality. Philosophers who emphasize the role of "model-based reasoning" recognize, also, the important role of analogy and metaphor in the management

[5]As we saw above, [Callicott, this volume] seems to assume the realist assumption 3, even as some of his comments seem to belie this as he refers to human meaning and other apparently evaluative aspects of the problem, which seems to suggest that 2a will tend to collapse into 2b, unless he can sharply separate purely descriptive from evaluative aspects of the situation.

of this patchwork, and at least some of them recognize the key role of values in such management [Lakoff and Johnson, 1980; Johnson, 1987; 2007; Nersessian, 1992; 1999]. The choice of specific models in specific situations cannot be based on simple observation, on pure ecological theory, or strong forms of realism. As we face situations, seeking to fulfill important, but gradually emerging goals in situations involving "wicked problems", we are at best armed with many partial models that can be deployed in multiple ways to increase understanding and enable action, even if that action is only "experimental," and may require—if unexpected or undesired outcomes result—revisiting the modeling choice and, in some cases, reconsidering the problem formulation itself.

This broad approach has emerged within ecology over the past few years and promises to move the discussion of the interaction of scientific modeling with social values in a new and promising direction consistent with the view that all science is guided by the search for models that will be useful in particular situations and, in order to choose an appropriate model, one must "experiment" with varied models and metaphors until one develops an appropriate model for such situations. These ecologists do not propose that one first *describe* changing systems and then *evaluate* the changes according to a single computation of the effects on human welfare. Instead, they embrace an open-ended search for many partial, but complementary, models that tell stories from multiple points of view, recognizing that this search will be guided by our diverse values and purposes. Taylor [2005, pp. 226–227] refers to such an approach to research as "reflexive" ("applying one's method to one's own work") and as involving "practical reflexivity" ... "that takes into account the range of practical conditions that enable researchers to build and gain support for their *representations*."[6]

Steward Pickett, in a series of papers with several co-authors, has explored how the creation and development of models can be integrated into management process, moving back and forth between concerns for managing complex systems with natural and human components and various possible models that can illuminate choices and guide management toward *both* better understanding of the dynamics of problems *and* how better to protect and sustain human values [Pickett and Cadenasso, 2006; Pickett *et al.*, 2004]. Advocating the use of the ecosystem concept as a useful tool for communication among scientists and among scientists and the interested public, including stakeholders and government agencies [Pickett and Cadenasso, 2002, p. 5; Pickett *et al.*, 2004], Pickett and colleagues argue that analogies and metaphors connect modeling decisions with our values and human concerns. These authors frame the question as one of choosing a model appropriate to one's purpose, arguing that "The richness of topics, complexity of model domains, and range of behaviors that models can exhibit suggest that ecosystem models can be used for diverse purposes" [Pickett and Cadenasso, 2002, p. 5; Pickett and Cadenasso, 2006; Kolasa and Pickett, 2005]. This approach to models is

[6]We have, with my own set of colleagues, developed what we call a "two-phased process" of policy formation and evaluation, which incorporates reflexive thinking into adaptive management (Norton *et al.*, 1998; Norton, 2005).

linked by Pickett and co-authors with an explicit endorsement of the importance of metaphors associated with ecosystems, seeing them as having a creative and generative role in science; and as valuable in communicating ecological ideas to the public and policy makers in public discourse. Pickett and Cadenasso describe a reciprocal relationship in which use of the scientific concept of "ecosystem" is applied in public discourse where metaphorical reasoning and value implications abound. Metaphors, when engaged by participants in a management process, allow public discussion of values and useful metaphors; these metaphors and the values associated with them must be "stripped off" so that scientific concepts can again be applied non-metaphorically [Pickett and Cadenasso, 2002, p. 8]).

What is significant, and directly relevant to the topic of economics and ecology of sustainability in this work is that it is based on a recognition that human purposes—goals, values, priorities—are integral to ecological model-building. Pickett and Cadenasso [2002, p. 6] say, "This area of communication includes education, the media, policy making, and management. In such public uses, the precision and narrow focus of technical terms is eschewed in favor of richness of connotation and in support of societally important, if sometimes controversial, values." Substantively, Pickett and Cadenasso also advocate the identification of ecological systems with spatially defined areas, and also advocate encouragement of recognition of systems as "places" with social meaning and endowed with "responsibility and empowerment" [Pickett and Cadenasso, 2002, p. 6]. This work is relevant here because, referring to recent thought in the philosophy of science, Pickett and colleagues seek an integrated dialogue about environmental policy and scientific research that is post-positivist and self-reflexive about the choices that are made in building models and framing environmental problems.

To see the relevance of this approach to ecological and economic approaches to sustainability, consider a discussion, by the respected economist Myrick Freeman, III, who discusses the extent to which mainstream economic analysis can comprehend and measure ecological changes in terms of expected changes in human welfare [Freeman, 1993; 2003]. Freeman, in the 1993 edition of his highly respected and widely cited book, suggested that most aspects of natural systems valued by environmentalists, such as species, diverse ecosystems, etc., cannot be measured by mainstream economic methods [1993, p. 485]. More recently, in the second edition, Freeman has suggested that, if certain conditions are fulfilled, one might be able to estimate welfare values for changes in ecological states. His explanation, in discussing the role of models and the explosion of new economic models to apply, is very revealing: "Because the true model cannot be known, we must add model uncertainty to the list of sources of uncertainty in welfare measures." "Model Uncertainty" is actually a misnomer. Economists and other modelers *choose* the scale at which they build models; no observations can tell them which scale to choose. As Freeman himself says, "There has been a virtual explosion in the number of models available to exploit," clearly implying that what modelers face is a choice among models, and that researchers and managers will choose those models that are best suited to achieve a goal, either scientific, managerial, or both.

In the Second Edition, Freeman is much more optimistic regarding how (and whether) to evaluate ecosystem services, especially those that provide support for human welfare only indirectly (such as the maintenance of "ecosystem function)." Here, he seems more hopeful, suggesting that one might be able to use "stated preference" (contingent valuation) methods to estimate what consumers would be willing to pay if they were able to "know the link between that function and the ecosystem service flows that it supports." If that were known, "the value of a bird species is the sum of the values of all its services." In an understatement, he follows this precondition for placing any welfare (wtp) value upon indirect ecosystem services by saying: "This will not always be easy to uncover". Indeed, one of the most vexing problems of ecological modeling is to even identify all the interdependencies that, as he admits, must be identified, measured, and summed. In other words, while sketching a strategy for placing economic values (prices) on indirect services, he sets conditions that require knowledge not available to ecologists today or in the foreseeable future. Methodological barriers seem to block any estimates of the value of ecological functions for as long as can be anticipated; nevertheless, Freeman offers advice as to how one might begin to design a scenario that would allow stated preference estimates of the economic value of ecological functions. This advice is also revealing: "One approach to establishing this link is to think of the relevant components of the ecosystem as being involved in a production process..." [2003, pp. 458–459]. In other words, if one wants to measure the contribution of all ecosystem services to human welfare, *imagine by analogy that ecological systems are modeled after human productive systems.* The lesson to be learned from Freeman's implicit imposition of a mechanical, productive metaphor upon natural systems is that economists, no less than ecologists "choose" their models, by applying analogies and metaphors to natural systems, and those models therefore "express" what economists value and think important to understand. Ecologists, I am suggesting, are ill advised to grant economists the right to choose the guiding metaphors when trying to model and evaluate changes to ecological functions. Ecologists might prefer "ecosystem as *web* of life," "ecosystem as *fragile* system," or "ecosystem as garden," while Muirians might prefer "ecosystems as *cathedrals*". Values will ultimately guide us in the choice of metaphors. But shouldn't such an important aspect of how we evaluate ecological change at least be brought out into the open and discussed? Practically speaking, this realization should encourage a more public, richer, pluralistic, and open-ended discussion of the many ways nature is conceived and valued. Choosing appropriate metaphors to express our deepest values is thus as much a part of the process of adaptive management as the process of making predictions based upon "models".

It is important for ecologists and adaptive environmental managers to respond to this opportunity, implicit in the new, more metaphorical approaches to identifying models for managing ecological systems, to re-think the relationship between the models used to describe natural processes and the models used to evaluate changes in their processes. Pickett and the others cited above are advocating no less than an inversion of the usual thinking about science, values, and policy.

The old positivist model advocated first gathering descriptive information and data, and then predicting impacts of actions, followed by a microeconomic estimate of the dollars-worth of impacts on the welfare of consumers. Norton [2005] calls this "the Serial View of Science and Policy" and criticizes it in more detail. What is suggested here is that, at its deepest level, the ecosystem concept rests on metaphors, and these metaphors connect values and emotions with choices of models. In order to be applied to real-world situations, the ecosystem concept demands experimentation with new analogies and interpretations, and this level of experiment is deep enough to connect to our values, fears, and aspirations. These ecologists embrace an open-ended search for many partial, but complementary, models that tell stories from multiple points of view, recognizing that this search will be guided by diverse values and purposes. O'Neil and colleagues [2008] provide a provocative explanation of a pluralistic, layered decision process consistent with the idea of multiple perspectives and multiple narratives.

The new direction in ecological thought discussed in the last Section involves work that meshes nicely with the work of adaptive managers who take a broad look at iterative management and the possibilities of social learning (see above, Section 4.3), especially with work that recognizes that interaction with wild species and natural systems can have "transformative" value [Norton, 1991; Sarkar, 2005]. Indeed, one might follow this new direction in ecology, combining it with work by philosophers on the formation and reformation of human values, and advocate an emerging position that might be referred to as "cognitive transformationalism." Cognitive transformationalism, rejecting realistic interpretations of theories and models and accepting that models are conventional tools of understanding, may open the way toward a deeper understanding of how factual information and social values interpenetrate and how they encourage cognitive change at more holistic, systemic levels through the interaction of information, values, and metaphorical thinking. Although there is not space here for the full development of this promising theory, it can be noted that such a theory would go a long way toward understanding one of the great iconic transformations in environmental history, which Aldo Leopold described as his "learning to think like a mountain" [Leopold, 1949; 1966].

In telling the story of how he had worked, first, to extirpate wolves and mountain lions from the Southwest Territories to improve hunter's success and tourism expenditures, Leopold was employing, as Freeman suggests, a "model" of nature as a system productive of human welfare. When the wolves were gone and the deer population exploded, Leopold—who was a careful observer who gathered facts about the deer and the system that supports their populations—learned that the productivity model is too simple to take into account systemic variables such as the amount of vegetative cover and the amount of erosion of topsoil on the mountain. At some point, Leopold described the transformation as triggered by seeing "the fierce green fire" dying in an old wolf's eyes, and he adopted a new "model" that was more complex (included vegetation and soils), that was "hierarchical". Wolves were considered to have value on the scale of the mountain, in that they ensure the health of vegetation by trimming the deer herd, and that clearly implied he and his colleagues were responsible for the negative outcome of a huge die-off of deer in a hard winter. As expected by philosophers of science who emphasize the role of metaphor in the development of scientific theories [Brown, 1993; Lakoff and Johnson, 1980; Nersessian, 2005]. Leopold's transformation was not the result of deductive reasoning from first principles or a corollary of scientific realism; the transformation, rather, was driven metaphorically (technically, by a *simile*) as he "learned to think like a mountain."

6 CONCLUSION: PROSPECTS FOR A TRULY ECOLOGICAL APPROACH TO SUSTAINABILITY

This survey of current thought on the relationship between ecological and economic approaches to understanding sustainability began by noting two long-standing theoretical differences that make the models of economists and ecologists incommensurable and impossible to integrate: The Accounting Problem and The Substitutability/Reversibility Problem. It was noted that these longstanding differences still stand in the way of effective communication and integration in public discussions and in interdisciplinary discourses, but it was also noted that the growing recognition of the complexities of scale and of model-building and the gradual erosion of positivistic commitments to sharply separating descriptive and prescriptive discourse, have created a new context in which these problems and incommensurabilities are being contested.

Emphasis on context, accordingly, has become important in the discourse about policies and actions to protect natural systems and the species they support. More contextual analyses, analyses that tailor systems thinking to particular situations, in turn require a more reflective and reflexive approach to model-building. An important consensus is emerging in academic and public discourse that "getting the scale right" is crucial in understanding and addressing environmental problems. As noted in Section 4, however, there remains disagreement about the philosophical interpretation of scalar models, with at least some philosophers apparently defending the view that the hierarchical models of nature are based on natural biological and ecological cycles, allowing a realistic interpretation of scale choices in ecology. This minority view seems strongly at odds with another, more promising approach, which is to accept that scale choices in ecological modeling are human choices based on human purposes. Given varied human purposes and varied expressions of human value, it seems more accurate and enlightening to problematize implicit choices of scale, to ask what guiding metaphors and underlying values encourage us to accept some models and to make choices regarding which models to use in many different particular situations.

The overlapping and interactive fields of economics and ecology, and the transdisciplinary science of ecological economics, face a serious challenge because the two fields use very different models and metaphors to understand the processes they attempt to describe and analyze. Environmental philosophy, unfortunately, has concentrated too heavily on axiological and meta-ethical analyses and has not contributed to the very interesting and important problems in the philosophy of modeling science: how, in public discourse, infused with values, can our discourse *express* values, even as it *describes* a system in ways that help us to understand and to sustain it for future generations.

BIBLIOGRAPHY

[Adger *et al.*, 2005] W. N. Adger, K. Brown, and E. L. Tompkins. The political economy of cross-scale networks in resource co-management. *Ecology and Society* 10(2): 9, 2005.

[Allen and Hoekstra, 1992] T. F. H. Allen and T. W. Hoekstra. *Toward a Unified Ecology*. New York: Columbia University Press, 1992.

[Allen and Starr, 1982] T. F. H. Allen and T. B. Starr. *Hierarchy: Perspectives for Ecological Complexity*. Chicago: University of Chigago Press, 1982.

[Batie and Shugart, 1991] S. Batie and H. Shugart. The biological consequences of climate changes: An ecological and economic assessment. In Resources for the Future, Inc., *Greenhouse Warming: Abatement and Adaptation*, pp. 121–131. (SEE N 92-10228 01-45), 1991.

[Bockstael *et al.*, 2000] N. E. Bockstael, I. A. Myrick Freeman, R. J. Kopp, P. R. Portney, and V. K. Smith. On Measuring Economic Values for Nature. *Environmental Science and Technology* 34: 1384–1389, 2000.

[Bostrom *et al.*, 1994] A. Bostrom, M. G. Morgan, B. Fischhoff, and D. Read. What do people know about global climate change? 1. Mental Models. *Risk Analysis* 14(6): 959–970, 1994.

[Bromley, 2006] D. W. Bromley. *Sufficient Reason*. Princeton, NJ: Princeton University Press, 2006.

[Brown, 1993] G. Brown. The Mother of Necessity: Technology Policy and Social Equity. *Science and Public Policy* (December), 411–416, 1993.

[Brown, 1994] P. Brown. *Restoring the Public Trust: A Fresh Vision for Progressive Government in America*. Boston, MA: Beacon Press, 1994.

[Brundtland, 1987] G. H. Brundtland. *Our Common Future: Report of the World Commission on Environment and Development*. Oxford: Oxford University Press, 1987.

[Cartwright, 1999] N. Cartwright. *This Dappled World: A Study of the Boundaries of Science*. Cambridge, UK and New York, NY: Cambridge University Press, 1999.

[Constanza, 1991] R. Costanza (ed.), *Ecological Economics: The Science and Management of Sustainability*. New York, NY: Columbia University Press, 1991.

[Constanza *et al.*, 1997] R. Costanza, R. d'Arge, R. de Groot, S. Farber, M. Grasso, B. Hannon, K. Limburg, S. Naeem, R. V. O'Neill, J. Paruelo, R. G. Raskin, P. Sutton and M. van den Belt. The Value of the World's Ecosystem Services and Natural Capital. *Nature* 387: 253–260, 1997.

[Constanza and Patten, 1995] R. Costanza and B. C. Patten. Defining and predicting sustainability. *Ecological Economics* 15(3): 193–196, 1995.

[Constanza *et al.*, 1996] R. Costanza, L. Wainger, and N. Bockstael. Integrating Spatially Explicit Ecological and Economic Models: Theory and Application in the Patuxent River Watershed, in R. Costanza, O. Segura, and J. Martinez-Alier (eds.), *Getting Down to Earth: Practical Applications of Ecological Economics*. International Society for Ecological Economics, 1996.

[Daily, 1997] G. C. Daily. *Nature's Services: Societal Dependence on Natural Ecosystems*. Washington, DC: Island Press, 1997.

[Daly and Cobb, 1989] H. Daly and J. Cobb. *For the common good: Redirecting the economy toward community, the environment, and a sustainable future*. Boston: Beacon Press, 1989.

[Dietz and Stern, 1998] T. Dietz and P. C. Stern. Science, Values, and Biodiversity. *Bioscience* 48(6): 441–444, 1998.

[Doyle and Ford, 1998] J. K. Doyle and D. N. Ford. Mental models concepts for system dynamics research. *System Dynamics Review* 14(1): 3–29, 1998.

[Dryzek, 1997] J. Dryzek. *The Politics of the Earth: Environmental Discourses*. New York: Oxford University Press, 1997.

[Dryzek, 2002] J. S. Dryzek. *Deliberative Democracy and Beyond: Liberals, Critics, Contestations.* New York: Oxford University Press, 2002.

[Fisher *et al.*, 1991] R. Fisher, W. Ury, and B. Patton. *Getting to Yes: Negotiating Agreement Without Giving In.* New York: Houghton Mifflin, 1991.

[Freeman, 1993] A. M. Freeman. Resource Evaluation and Public Policy. In A. M. Freeman (ed.), *The Measurement of Environmental and Resource Values*, pp. 1–18. Washington, DC: Resources for the Future, 1993.

[Funtowicz *et al.*, 1999] S. Funtowicz, J. Martinez-Alier, G. Munda, and J. R. Ravetz. *Information tools for environmental policy under conditions of complexity.* Luxembourg: European Communities, 1999.

[Funtowicz and Ravetz, 1994] S. Funtowicz and J. R. Ravetz. Emergent Complex Systems. *Futures* 26(6): 568–582, 1994.

[Funtowicz and Ravetz, 1995] S. Funtowicz and J. R. Ravetz. Science for the Post Normal Age. In L. Westra and J. Lemons (eds.), *Perspectives on Ecological Integrity*, pp. 146–161. Kluwer Academic Publishers, 1995.

[Gentner *et al.*, 1997] D. Gentner, S. Brem, R. W. Ferguson, A. B. Markman, B. B. Levidow, P. Wolff and K. D. Forbus. Analogical Reasoning and Conceptual Change: A Case Study of Johannes Kepler. *Journal of the Learning Sciences* 6(1): 3–40, 1997.

[Giampietro, 2004] M. Giampietro. *Multiscale Integrated Analysis of Agroecosystems.* Boca Raton, FL: CRC Books, 2004.

[Gray, 2003] B. Gray. Framing of environmental disputes. In R. Lewicki, B. Gray and M. Elliot (eds.), *Making sense of intractable environmental conflicts*, pp. 11–34. Washington: Island Press, 2003.

[Gregory *et al.*, 1993] R. Gregory, S. Lichtenstein, and P. Slovic. Valuing Environmental Resources: A Constructive Approach. *Journal of Risk and Uncertainty* 7: 177–197, 1993.

[Gunderson and Holling, 2002] L. Gunderson and C. S. Holling. *Panarchy: Understanding Transformations in Human and Natural Systems.* Washington: Island Press, 2002.

[Gunderson *et al.*, 1995] L. Gunderson, C. S. Holling, and S. S. Light. Barriers broken and bridges built: a synthesis. *Barriers and bridges to the renewal of ecosystems and institutions*, pp. 490–532. New York: Columbia University Press, 1995.

[Habermas, 1984] J. Habermas. *Theory of Communicative Action, Vol I: Reason and the Rationalization of Society.* Boston: Beacon Press, 1984.

[Habermas, 1987] J. Habermas. *Theory of Communicative Action, Vol. II: Reason and the Rationalization of Society.* Boston: Beacon Press, 1987.

[Hajer, 2003] M. Hajer. Policy without polity? Policy analysis and the institutional void. *Policy Sciences* 36(2): 175–195, 2003.

[Hajer and Wagenaar, 2003] M. Hajer and H. Wagenaar. Introduction. In M. Hajer and H. Wagenaar (eds.), *Deliberative Policy Analysis: Understanding Governance in the Network Society*, pp. 1–32. Cambridge: Cambridge University Press, 2003.

[Hallyn, 2000] F. Hallyn. *Metaphor and Analogy in the Sciences*: Kluwer Academic Pub, 2000.

[Hodgson, 1997] G. Hodgson. Economics, Environmental Policy and the Transcendence of Utilitarianism. In J. Foster (ed.), *Valuing Nature? Economics, Ethics and Environment*, pp. 48–63. Routledge Publishing, 1997.

[Holling, 1992] C. S. Holling. Cross-Scale Morphology, Geometry, and Dynamics of Ecosystems. *Ecological Monographs* 62(4): 447–502, 1992.

[Holling, 1996] C. S. Holling. Engineering Resilience versus Ecological Resilience. In Schulze, Peter C., *Engineering within Ecological Constraints*, pp. 31–43. National Academy Press, 1996.

[Howarth and Norgaard, 1990] R. B. Howarth and R. B. Norgaard. Intergenerational Resource Rights, Efficiency, and Social Optimality. *Land Economics* 66(1): 1–11, 1990.

[Howarth and Wilson, 2006] R. B. Howarth and M. A. Wilson. A Theoretical Approach to Deliberative Valuation: Aggregation by Mutual Consent. *Land Economics* 82(1): 1–16, 2006.

[Johnson-Laird, 2000] P. N. Johnson-Laird. The current state of the mental model theory, in J. Garcia-Madruga, M. Carriedo and M. J. Gonzalez-Labra (eds.), *Mental models in reasoning*, pp. 17–40. Madrid: UNED, 2000.

[Kahneman and Knetsch, 2005] D. Kahneman and J. L. Knetsch. Valuing Public Goods: The Purchase of Moral Satisfaction. *The Earthscan Reader in Environmental Values*, pp. 229–243, 2005.

[Koestler, 1967] A. Koestler. *The Ghost in the Machine.* London, UK: Penguin Books, 1967.

[Kolasa and Pickett, 2005] J. Kolasa and S. T. A. Pickett. Changing Academic Perspectives of Ecology: A View from within. In Johnson, E. A. and M. J. Mappin (eds.), *Environmental Education and Advocacy: Changing Perspectives of Ecology and Education.* Cambridge: Cambridge University Press, 2005.

[Lakoff and Johnson, 1980] G. Lakoff and M. Johnson. *Metaphors We Live By.* Chicago: University of Chicago Press, 1980.

[Lange, 2007] G.-M. Lange. Environmental and Resource Accounting. In G. Atkinson, S. Dietz and E. Neumayer (eds.), *Handbook of Sustainable Development.* Cheltenham, UK: Edward Elgar, 2007.

[Lasswell, 1970] H. Lasswell. The Emerging Conception of the Policy Sciences. *Policy Sciences* 1: 3–14, 1970.

[Leach and Pelkey, 2001] W. D. Leach and N. W. Pelkey. Making Watershed Partnerships Work: A Review of the Empirical Literature. *Journal of Water Resources Planning and Management* 127(6): 378–385, 2001.

[Leach *et al.*, 2002] W. D. Leach, N. W. Pelkey, and P. A. Sabatier. Stakeholder partnerships as collaborative policymaking: Evaluation criteria applied to watershed management in California and Washington. *Journal of Policy Analysis and Management* 21(4): 645–670, 2002.

[Lee, 1993] K. N. Lee. *Compass and Gyroscope: Integrating Science and Politics for the Environment.* Washington, DC: Island Press, 1993.

[Leopold, 1949] A. Leopold. Thinking like a mountain. *A Sand County Almanac*, pp. 137–140. New York: Ballantine Books, 1949, 1966.

[Light and Katz, 1996] A. Light and E. Katz. *Environmental Pragmatism.* London and New York: Routledge Publishers, 1996.

[Lindblom, 1959] C. Lindblom. The Science of Muddling Through. *Public Administrative Review* 19: 79–88, 1959.

[Lubell, 2004] M. Lubell. Collaborative Watershed Management: A View from the Grassroots. *Policy Studies Journal* 32(3): 341–361, 2004.

[Millenium Ecosystem Assessment, 2005] Millenium Ecosystem Assessment. from `http://www.millenniumassessment.org/en/index.aspx`, 2005.

[Moore and Koontz, 2003] E. A. Moore and T. M. Koontz. Research Note A Typology of Collaborative Watershed Groups: Citizen-Based, Agency-Based, and Mixed Partnerships. *Society and Natural Resources* 16(5): 451–460, 2003.

[Nersessian, 2005] N. Nersessian. *Interpreting Scientific and Engineering Practices: Integrating the Cognitive, Social, and Cultural Dimensions.* New Jersey: Lawrence Erlbaum Associates, 2005.

[Nersessian, 1992] N. Nersessian. How do scientists think? Capturing the dynamics of conceptual change in science. *Cognitive Models of Science* 15: 3–44, 1992.

[Nersessian, 1999] N. Nersessian. Model-Based Reasoning in Conceptual Change. In L. Magnani, N. J. Nersessian and P. Thagard (eds.), *Model-Based Reasoning in Scientific Discovery.* New York: Kluwer Academic/Plenum Publishers, 1999.

[Norton, 1991] B. Norton. *Toward Unity Among Environmentalists.* New York: Oxford University Press, 1991.

[Norton, 1995] B. Norton. Ecological integrity and social values: at what scale? *Ecosystem Health* 1: 228–241, 1995.

[Norton, 2005] B. Norton. *Sustainability: A Philosophy of Adaptive Ecosystem Management.* Chicago: University of Chicago Press, 2005.

[Norton *et al.*, 1998] B. Norton, R. Costanza, and R. C. Bishop. The evolution of preferences: Why sovereign preferences may not lead to sustainable policies and what to do about it. *Ecological Economics* 24(2–3): 193–211, 1998.

[Norton and Minteer, 2002] B. Norton and B. A. Minteer. From environmental ethics to environmental public philosophy: ethicists and economists, 1973–future. *The International Yearbook of Environmental and Resource Economics, 2003*, pp. 373–407, 2002.

[Norton and Toman, 1997] B. Norton and M. Toman. Sustainability: Ecological and Economic Perspectives. *Land Economics* 4: 553–568, 1997.

[Norton and Ulanowicz, 1992] B. Norton and R. Ulanowicz. Scale and Biodiversity Policy: A Hierarchical Approach. *Ambio* 21(3): 244–249, 1992.

[Norton, 1998] B. Norton. Improving Ecological Communication: The Role of Ecologists in Environmental Policy Formation. *Ecological Applications* 8(2): 350–364, 1998.

[O'Neill, 1988] R. O'Neill. Hierarchy Theory and Global Change. In T. Rosswall, R. G. Woodmansee and P. G. Risser (eds.), *Scales and Global Change*: John Wiley & Sons, 1988.

[O'Neill *et al.*, 1983] R. V. O'Neill, D. L. Deangelis, J. B. Waide, and G. E. Allen, *A Hierarchical Concept of Ecosystems*. Princeton, NJ: Princeton University Press, 1983.

[O'Neill *et al.*, 2008] J. O'Neill, A. Holland, and A. Light, *Environmental Values*. London: Routledge Publishers, 2008.

[Pattee, 1973] H. H. Pattee. *Hierarchy Theory: The Challenge of Complex Systems*. New York: G. Braziller, 1973.

[Pickett *et al.*, 2004] S. Pickett, M. L. Cadenasso, and J. M. Grove. Resilient cities: meaning, models, and metaphor for integrating the ecological, socio-economic, and planning realms. *Landscape and Urban Planning* 69: 369–384, 2004.

[Rittel and Webber, 1973] H. W. J. Rittel and M. M. Webber. Dilemmas in a General Theory of Planning. *Policy Sciences* 4: 155–169, 1973.

[Sabatier *et al.*, 2005] P. A. Sabatier, C. Weible, and J. Ficker. Eras of Water Management in the United States: Implications for Collaborative Watershed Approaches. In P. A. Sabatier, W. Focht, M. Lubell, Z. Trachtenberg, A. Vedlitz and M. Matlock (eds.), *Swimming Upstream: Collaborative Approaches to Watershed Management*, pp. 23–52. Boston: The MIT Press, 2005.

[Sagoff, 1998] M. Sagoff. Aggregation and deliberation in valuing environmental public goods: A look beyond contingent pricing. *Ecological Economics* 24: 213–230, 1998.

[Sagoff, 2004] M. Sagoff. *Price, Principle, and the Environment*: New York: Cambridge University Press, 2004.

[Samson and Knopf, 1996] F. B. Samson and F. L. Knopf. *Ecosystem Management: Selected Readings*. New York, NY: Springer-Verlag, 1996.

[Sarkar, 2005] S. Sarkar. *Biodiversity and Environmental Philosophy: An Introduction to the Issues*. New York: Cambridge University Press, 2005.

[Schelhas *et al.*, 2001] J. Schelhas, L. E. Buck, and C. C. Geisler (eds.), Introduction: The Challenge of Adaptive Collaborative Management. In *Biological Diversity: Balancing Interests through Adaptive Collaborative Management*, pp. xxii–xxxv. Boca Raton, FL: CRC Press, 2001.

[Schlickeisen, 1994] R. Schlickeisen. Protecting Biodiversity for Future Generations. An Argument for a Constitutional Amendment. *Tulane Environmental Law Journal* 8: 181–221, 1994.

[Schmid, 2002] A. A. Schmid. All Environmental Policy Instruments Require Moral Choices as Whose Interests Count. In D. W. Bromley and J. Paavlova (eds.), *Economics, Ethics, and Environmental Policy: Contested Choices*, pp. 133–147. Oxford, UK: Blackwell Publishing, 2002.

[Siemann and Brown, 1999] E. Siemann and J. H. Brown. Gaps in Mammalian Body Size Distributions Reexamined. *Ecology* 80: 2788–2792, 1999.

[Simon, 1947] H. A. Simon. *Administrative Behavior: A Study of Decision-Making Processes in Administrative Organization.*. The Free Press, 1947.

[Simon, 1982] H. Simon. *Models of Bounded Rationality* (Vol. 1 and 2). Boston: MIT Press, 1982.

[Slovic, 1993] P. Slovic. Perceived risk, trust, and democracy. *Risk Analysis* 13(6): 675–682, 1993.

[Taylor, 2005] P. J. Taylor. *Unruly Complexity: Ecology, Interpretation, Engagement*. Chicago: University Of Chicago Press, 2005.

[Toman, 1994] M. A. Toman. Economics and 'Sustainability': Balancing Trade-Offs and Imperatives. *Land Economics* 70(4): 399–413, 1994.

[Torgerson, 1985] D. Torgerson. Contextual Orientation in Policy Analysis: The Contribution of Harold D. Lasswell. *Policy Sciences* 18(3): 241–261, 1985.

[Tversky and Kahneman, 1981] A. Tversky and D. Kahneman. The framing of decisions and the psychology of choice. *Science* 211(4481): 453–458, 1981.

[Tversky and Kahneman, 1990] A. Tversky and D. Kahneman. Rational Choice and the Framing of Decisions. In K. S. Cook and M. Levi (eds.), *The Limits of Rationality*, pp. 60–89. Chicago: University of Chicago Press, 1990.

[Ulrich, 1998] W. Ulrich. *Systems Thinking as if People Mattered: Critical Systems Thinking for Citizens and Managers*. Lincoln, UK: Lincoln School of Management, 1998.

[Ulrich, 2000] W. Ulrich. Reflective practice in the civil society. *Reflective Practice* 1(2): 247–268, 2000.

[Ulrich, 2003] W. Ulrich. Beyond methodology choice: critical systems thinking as critically systemic discourse. *The Journal of the Operational Research Society* 54: 325–342, 2003.

[Wackernagel and Rees, 1996] M. Wackernagel and W. Rees. *Our Ecological Footprint.* Philadelphia: New Society Press, 1996.

[Waltner-Toews *et al.*, 2008] D. Waltner-Toews, N.-M. E. Lister, and S. Bocking. *The Ecosystem Approach: Complexity, Uncertainty, and Managing for Sustainability.* New York: Columbia University Press, 2008.

[Weiss, 1986] E. B. Weiss. *In Fairness to Future Generations.* Tokyo: United Nations University, Transnational Publishers, 1986.

[Wimsatt, 1994] W. Wimsatt. The Ontology of Complex Systems. *Canadian Journal of Philosophy* 20: 207–274, 1994.

[Wondolleck and Yaffee, 2000] J. Wondolleck and S. Yaffee. *Making Collaboration Work.* Washington, DC: Island Press, 2000.

DIVERSITY AND THE GOOD

Gregory M. Mikkelson

"The diverse shall be no less diverse, but they shall flow and unite..."
"Great is liberty! Great is equality! I am their follower..."
[Whitman, 1855]

1 INTRODUCTION

As a species, we humans are consuming more than our fair share of resources, and producing more than our fair share of waste. This problem is getting worse every day, with increasingly catastrophic results for life on Earth. We are launching the first mass extinction since a meteor destroyed the dinosaurs and most other life forms 65 million years ago [Leakey and Lewin, 1995; Rosenzweig 2003]. Not only are we arrogating a large disproportion of the biomass produced globally each year, but we are actually suppressing the total amount of photosynthesis taking place on land [Haberl *et al.*, 2007]. In other words, as we are snuffing out species, it is not the case that those remaining are making up the difference through increased population size. Instead, there is simply 10% less terrestrial life happening now than there was as recently as 60 years ago—despite the fact that we are putting as much nitrogen into the biosphere as do all natural processes combined, in order to boost the productivity of our preferred species (crops, the livestock they feed, and of course ourselves) [Vitousek *et al.*, 1997]. Excessive nitrogen and phosphorous are in turn wreaking havoc on aquatic systems, as for example dead zones are spreading further each year from the mouths of rivers disgorging agricultural runoff [Diaz and Rosenberg, 2008].

Nor can these and other such damages to the living world be justified in terms of improvements to the human condition. For example, the massive economic growth fueled by increasing consumption of the planet's resources over the past several decades does not appear to have had any overall effect on human health. Among countries across the globe there is no significant correlation whatsoever between changes in the gross domestic product per capita and changes in life expectancy between 1960 and 2005 [the temporal range of the data available from the World Bank, 2007 and United Nations, 2005]. Nor has growth improved human happiness. For example, despite increases in per capita income of up to six-fold, self-reported happiness and satisfaction have not increased at all in the United States, the United Kingdom, or Japan over the past 30 to 50 years [Speth, 2008]. At the same time, the gap between the rich and poor has widened in most countries

Handbook of the Philosophy of Science. Volume 11: Philosophy of Ecology.
Volume editors: Kevin deLaplante, Bryson Brown and Kent A. Peacock. General editors: Dov M. Gabbay, Paul Thagard and John Woods.

[data from Pitt Inequality Project, 2006]. And of course, political freedoms have declined since September 11th, 2001.

Science has played a role in creating the snarl of interacting travesties sampled above. Individual sciences, like ecology, have also helped reveal the nature and extent of these problems, as well as some partial solutions. However, many now recognize that disciplinary science alone is radically insufficient, and have thus initiated programs of interdisciplinary research. These initiatives hark back to early-20th-Century ambitions for the unity of science. I submit, however, that the unity of science—as it is typically conceived—is not enough either. Solving the ecological and social problems sketched above will require vigorous, broad-based, long-term political struggle. It will also require an unprecedented integration of science with ethics—in other words, a unity of not just "science" but of knowledge in general. In this chapter, I therefore propose a unified perspective on a subject that crosses not just the purported divide between natural and social science, but also the supposed boundary between science and ethics.

That subject is diversity. I will anchor my discussion of its nature and value in the science of ecology. However, I will argue that a theoretical perspective developed in health-care economics offers the most perspicuous way to understand a key consequence of biological diversity, namely ecosystem productivity. The two central concepts involved in this perspective are diminishing returns and higher-level interaction. The theory also sheds light on the benefits of economic equality; and helps to clarify the intrinsic value of diversity, equality, and related properties; as well as having many other applications.

2 SPECIES RICHNESS AND BIOLOGICAL PRODUCTIVITY

Biodiversity has always fascinated ecologists. Thoreau [1854], Darwin [1859], Elton [1927], Leopold [1949], MacArthur [1955], and Carson [1962] all celebrated its importance in their own ways. Darwin noted that species-rich areas have more life in them than species-poor areas do:

> It has been experimentally proved, that if a plot of ground be sown with one species of grass, and a similar plot be sown with several distinct genera of grasses, a greater number of plants and a greater weight of dry herbage can thus be raised. [Darwin, 1859, p. 113]

Present-day ecologists have now begun elucidating this and other effects of species richness (the number of species), which is one component of biodiversity at the species level. The current siege of extinctions caused by one species in particular—*Homo "sapiens"*—renders interest in such topics urgent.

Besides species richness, the other component of species diversity is the degree of equality or evenness in the number of organisms among the different species. The most widely-used measure of species diversity (besides species richness itself)—the Shannon-Weaver index—reflects both of these components. For a given level of evenness, the more species there are, the greater the Shannon-Weaver index of diversity. And for a given number of species, the closer to equality they are, the greater the diversity. Perhaps because it is often easier to estimate species richness than evenness, studies of biodiversity loss and its consequences have so far focused on the former rather than the latter. However, some data allow for both components of diversity to be estimated. Richness and evenness turn out to have similar links to some variables of interest, but contrasting relationships with others [Mikkelson *et al.*, in review].

The overall pattern

Experiment 120 at the Cedar Creek Long Term Ecological Research station in Minnesota is similar to the studies cited by Darwin in the quotation above. In 1994, researchers planted different numbers of grassland species in 168 previously-cleared nine-by-nine-meter plots. Ever since, they have periodically measured a host of population- and ecosystem-level response variables in those plots. One of these variables is total productivity—the amount of new biomass created through photosynthesis in a given year. As Figure 1 shows, species richness enhances the biological productivity of these plots.

Since results like these began to appear in the mid-1990s, ecologists have discussed and debated different mechanisms to explain them. Here I propose a new perspective that offers a way to disentangle two kinds of mechanism that have heretofore proven to be "difficult to distinguish in practice" [Loreau and Hector, 2001]. This perspective also generalizes to other contexts both within and outside of ecology. Loreau and Hector [2001] refer to the two mechanisms in question as "niche differentiation" vs. "facilitation". In order to emphasize the analogies between the diversity-productivity case and others discussed later in this paper, I will refer to the two mechanisms as "compositional" vs. "contextual" effects of diversity on productivity. As the name suggests, the compositional effect has more to do with the properties of the individual species populations that compose an ecosystem, and less to do with interactions between those populations and/or between that ecosystem and others. Contextual effects have more to do with such interactions.

The compositional effect

If the compositional effect were the only one in operation, then any given species would interact with others only by excluding them from the areas within the same ecosystem (*i.e.*, the same experimental plot) to which it was better adapted. The more species that are present, the more likely it is that any given part of the habitat is occupied by a species particularly well suited to it. The total productivity of species-rich plots is thus higher than if only one or a few species had to cover the whole area. Figure 2 depicts the compositional effect. The curves represent three hypothetical species, and moving from left to right along the x-axis represents more and more habitat being taken up by a given species. Presumably, if a species takes up only a small area, it is most likely to occupy the portion of the plot to which it is best adapted. Therefore, as a given species takes up more and more area, its total productivity will increase, but at a decreasing rate, because every additional unit of habitat will be less suitable for it than the previous unit. In other words, we will observe a pattern of diminishing returns.

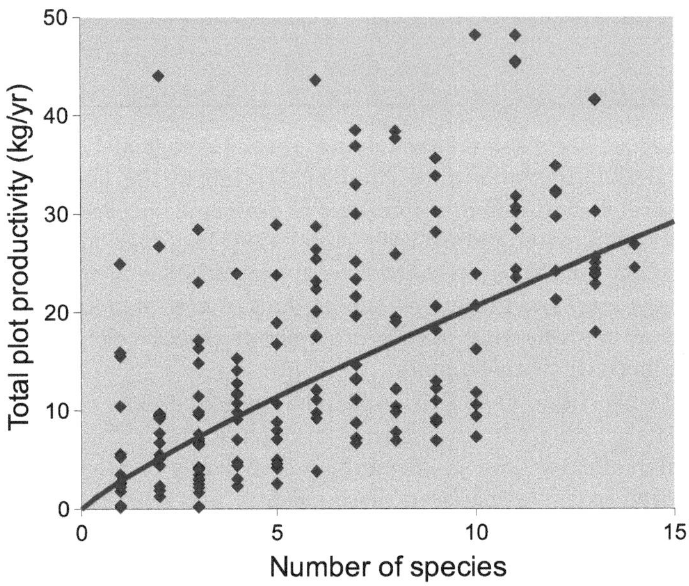

Figure 1. Relationship between species richness and total plot productivity in Cedar Creek Experiment 120 (www.lter.umn.edu/research/data.html). The data used here concern the species present in each plot in 2003, that were among the 18 species originally planted in 1994.[1] The vertical axis represents the amount of new aboveground biomass produced collectively in 2003 by the combination of these species in a given plot. One outlier (a plot with eight such species and a productivity of 85 kg/yr) is not shown in the graph, but was nevertheless used in calculating the best-fit power curve that is shown ($y = 2.8x^{0.86}$).

If the diminishing-returns relationships for different species are similar enough, then having more species will generally result in greater total plot productivity. For example, if an entire 81 m^2 plot were taken up by any one of the species shown in Figure 2, the productivity of the plot would be around seven kilograms per year. But if two of those species each took up roughly half of the plot, then plot productivity would be over $5.5 + 5.5 = 11.0$ kg/yr. And if all three species were present in roughly equal proportions (27 m^2 each), plot productivity would

[1]For each plot, this includes any and all of those 18 species that were either planted or colonized naturally, but it does not include any other species that colonized naturally.

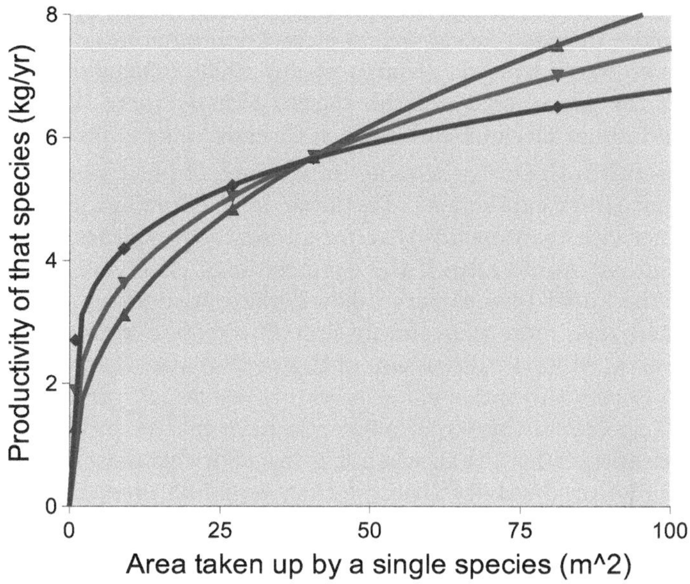

Figure 2. Hypothetical relationships between the area taken up by single species within grassland plots the same size as those in Cedar Creek Experiment 120 (81 m^2), and the productivity of those species within those plots.

exceed $5 \times 3 = 15$ kg/yr. As we shall see, while the real Cedar Creek species do show diminishing returns, they are not nearly as dramatic as those depicted for the sake of illustration in Figure 2. For this reason, the compositional effect alone does not result in productivity gains that are as great as those described in this hypothetical example.

Contextual effects

Contextual effects of species richness on biological productivity are ecosystem-level (inter-populational) or landscape-level (inter-ecosystemic) processes that go beyond the simple sorting of species into suitable portions of the available habitat. Contextual effects either boost, or reduce, the productivity of a given species occupying a given amount of habitat. Loreau and Hector [2001] call positive inter-specific contextual effects "facilitation". These include the buffering of one species by another against disease and predation, and the fertilization of soil by legumes (members of the pea family) that benefits other grassland plants [Lambers *et al.*, 2003; Jactel and Brockerhoff, 2007]. "Allelopathy"—the poisoning of one plant species by another—is an example of a negative contextual effect. In addition to these processes occurring within ecosystems, interactions between ecosystems (landscape-level processes) conceivably also affect individual species' productivity. For example, more diverse ecosystems resist invasion by non-native species better than less diverse ecosystems do [Knops *et al.*, 1999]. Depending on how such invaders affect the productivity of the species already there, keeping them out could be an additional mechanism by which diversity affects productivity.

Figure 3 illustrates the overwhelming importance of positive contextual effects within the Cedar Creek experiment. The lower (red) curve in Figure 3A estimates the area-productivity relationship that a representative species (*Panicum virgatum*, or switchgrass) would have if the compositional effect were the only one in operation. As the upper (green) curve shows, however, putting additional species into an ecosystem does more than simply shift *P. virgatum* leftward along its area-productivity curve, as in the discussion of Figure 2 above. Increasing the number of species also boosts the entire curve upward. Besides *P. virgatum*, all but one of the other 17 species in this experiment also have greater productivity per area when species richness is high than when it is low. For this reason, species richness enhances total plot productivity far more than would be predicted on the basis of the compositional effect alone. In other words, the upper (green) curve in Figure 3B approximates the true curve shown in Figure 1 much more closely than the lower (red) curve in Figure 3B does.

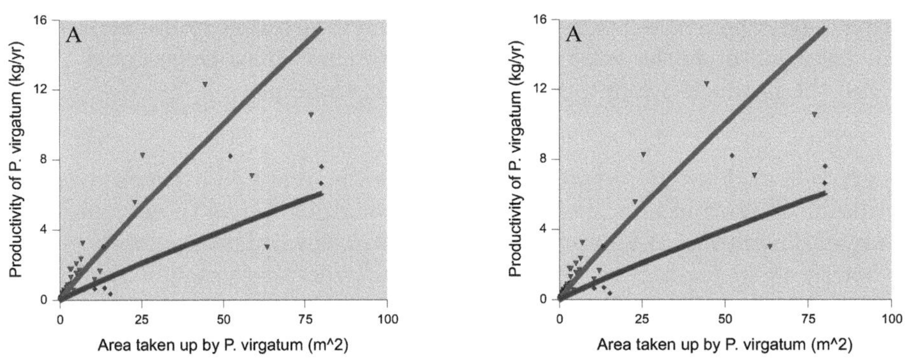

Figure 3. (A) Estimated area-productivity relationships of *P. virgatum*. The proportion of total plot productivity contributed by this species was used as a proxy for the area taken up by it. Lower (red) curve: low species richness (1–5 species)—$y = 0.11x^{0.91}$. Upper (green) curve: high species richness (6–14 species)—$y = 0.27x^{0.92}$. (B) Predicted species-productivity relationships of entire plots. The lower (red) curve ($y = 5.1x^{0.20}$) was estimated based on the low-richness single-species area-productivity relationship for each species. The upper (green) curve ($y = 3.4x^{0.78}$) was estimated based on the low-richness single-species area-productivity relationships for plots containing 1–5 species, and the high-richness relationships for plot containing 6–14 species. Both estimates take the area proxy for each species in each plot as given.

Interactions between populations and/or ecosystems, rather than the individualistic properties of those populations, thus play the dominant role in the richness-productivity relationship at Cedar Creek. This goes some way toward refuting an idea with a long history of adherents and detractors, namely that ecosystems can be reduced to the populations that compose them [Odenbaugh, 2006; Mikkelson, 2007 and in press; deLaplante, in press]. When it comes to the effects of species richness on biological productivity, these ecosystems are far more than the sums of their parts.

However, I predict that the relative importance of the compositional effect should expand with the spatial scale of investigation. The compositional effect of species richness on biological productivity depends on spatial differences in habitat characteristics. While these characteristics vary somewhat across an 81 m^2 plot, they obviously vary more across larger scales. This, in turn, gives the compositional effect more room to operate. Also, contextual effects require interaction between species, but the species in a large region do not generally interact as intimately as do those living together within a small plot. For these reasons, the compositional effect should explain progressively more of the diversity-productivity relationship at larger and larger scales.[2]

3 ECONOMIC INEQUALITY AND BIODIVERSITY LOSS

Intriguing parallels exist between the distribution of ecological resources—and therefore the number of organisms—among species, and the distribution of economic resources among members of our own species. For example, just as abundance tends to be distributed according to a certain pattern, called the log-normal distribution; so also do wealth and income tend to be distributed log-normally [May, 1975]. In this section, I will focus on analogies between the consequences, rather than the statistical nature, of biological diversity and economic equality. The same general theoretical framework as that applied above to the ecological realm—involving compositional and contextual effects—clarifies the mechanisms by which equality has its effects in the economic sphere.

Just as ecologists conceive of species diversity as having two components—species richness and species evenness (see above)—we can conceive of social "diversity" in an analogous way. Typically, a society with more people in it is more diverse than an economy with fewer people. Larger human populations can accommodate greater diversity of languages, ethnicities, worldviews, etc. By analogy with ecological diversity, we can also consider more economically egalitarian societies to be more diverse. Because of the close connections between economic wealth and various kinds of power, a more egalitarian society will allow a greater number of people to succeed in expressing their different perspectives.[3]

[2]Changing the spatial scale also changes the diversity-productivity relationship itself. At progressively larger scales, more and more species are required to maintain any given average level of productivity per unit area (cf. [Mikkelson, 2007]).

[3]Alternatively, one could conceive of economic inequality (*i.e.*, the existence of very different

I noted above that in ecology, less work has been done on the effects of species evenness than on species richness. In contrast, a growing body of literature has focused on the social analogue of species evenness, namely economic equality. Below, I will concentrate on this component of social "diversity".

History of the general framework employed in this paper

Far from being foreign to economics, the theoretical framework introduced above actually originated there. In the 1920s, Pigou spelled out an implication of the diminishing marginal utility of money. One's first $10,000 worth of monetary wealth, or yearly income, increases one's well-being immensely—by protecting one from starvation, exposure, etc. The second $10,000 also comes in very handy. However, with each successive increment, the additional utility declines. For instance, it is hard to imagine a billionaire even noticing an additional $10,000, let alone being made much happier or healthier by it. Intuitively, that same $10,000 would do a pauper far more good. These considerations provide one rationale for ensuring a reasonable degree of economic equality [Putnam, 2002].

Seventy years later, public-health researchers began finding empirical evidence supporting, but also pointing beyond, this kind of rationale. Average lifespans are higher, and health is generally better, in more egalitarian societies. This pattern is partly—but only partly—explicable in terms of a diminishing-returns relationship between individual income and health. Other factors also come into play. For example, levels of trust between members of a society, the functioning of public institutions (*e.g.*, public-health agencies), and a number of other social properties and processes are enhanced by economic equality. These factors mediate social-level contextual effects of equality on health, that operate in addition to the compositional effect brought about by diminishing returns at the individual level [Kawachi and Kennedy, 2002].

Negative contextual effects of equality on health are also conceivable—*e.g.*, because more egalitarian societies experience slower economic growth [Partridge, 2005]. However, the strong overall equality-health relationship proves that these are outweighed by the positive compositional and contextual effects of equality. The importance of positive contextual effects indicates that, were income re-distributed from the rich to the poor, even the rich would end up healthier despite their drop in income, because of improvements to the society in which they live [Wilkinson and Pickett, 2009].

The distribution of resources among people and species

My colleagues and I recently discovered a link between economic inequality and biodiversity loss [Mikkelson *et al.*, 2007]. What we did not report in that article,

economic classes) as a kind of diversity. However, at least with respect to the "goods" considered in this section (utility, health, and biodiversity), inequality as diversity lies outside the domain of the current theory. This is because inequality has negative, rather than positive, effects on those goods.

but I shall now discuss, is a compositional effect that might partly explain that link. I will also outline some of the contextual effects that may help to mediate the equality-biodiversity connection. In this case, however, we are not yet in a position to quantify the relative importance of composition vs. context.

Societies that distribute their economic resources more equitably among members of the human species also tend to share ecological resources more equitably among different species. The Gini ratio of income inequality in a society is strongly and positively correlated with the number of species that are threatened or declining within its territory. This pattern holds among countries across the world, and among states across the US. And it persists after controlling for the most likely confounding variables: a society's total number of species, human population size, and per capita gross domestic product or income [Mikkelson *et al.*, 2007].

In this case, the compositional effect is based on the fact that individuals tend to treat biodiversity conservation as a "necessary good". The amount of money people spend on a necessary good has a diminishing-returns relationship to income. In other words, while the rich may spend more in absolute terms, they spend less in proportion to their incomes, on necessities than do the poor (see Figure 4A).

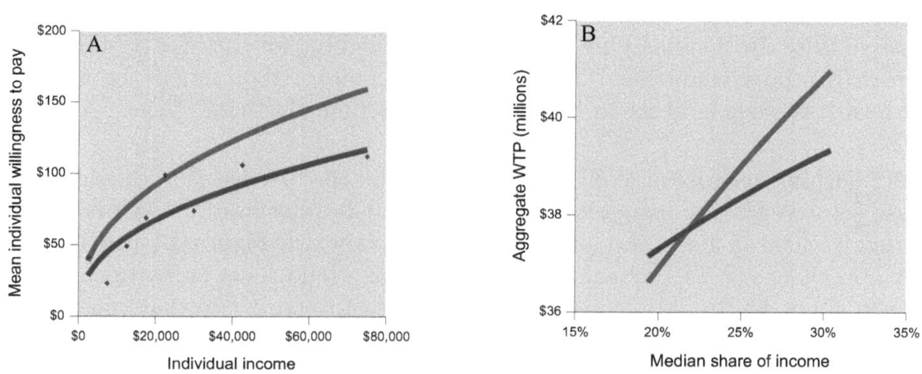

Figure 4. Conservation as a necessary good. (A) Lower (red) curve: Power model fitted to actual data on Utah citizens' income and willingness to pay (WTP) for setting aside 15% of their state as wilderness [Pope and Jones 1990, Table 5]. Upper (green) curve: How that relationship might look were Utah to achieve a more equitable distribution of income, and assuming positive contextual effects. (B) Power models fitted to predictions made using the Utah income-WTP model (lower, red curve) and hypothetical contextual effects (upper, green curve); taking into account the actual number of households, and mean WTP, in Utah in 1986 [*ibid.*, p. 169 and Table 2, respectively]; and assuming constant per capita income, log-normally distributed. The x-axis (the median share of income) is the proportion of Utah's total income that is collectively earned by the poorest 50% of the population.

In all, we found 17 "contingent valuation" studies that report how people's "willingness to pay" (WTP) for habitat protection varies with individual income. The human populations surveyed in these studies range from single towns to entire countries, and the biodiversity "targets" range from a single shrimp species to tropical rainforests worldwide. In all cases, the "income elasticity" of WTP is less than one. This means that a 1% rise in income is associated with less than a 1% rise in WTP. In other words, there is a diminishing-returns relationship between income and WTP [Aldy *et al.*, 1999; Bateman *et al.*, 1995; Bateman *et al.*, 1996; Garrod and Willis, 1997; Hammitt *et al.*, 2001; Kramer and Mercer, 1997; Muriithi and Kenyon, 2002; Pope and Jones, 1990; Stanley, 2005; Toivonen, 2004; Turpie, 2003; Willis and Garrod, 1993]. This result appears to be quite robust, characterizing not just conservation measures in particular, but charitable giving in general: while the rich tend to contribute more in absolute per capita terms, they contribute substantially less as a proportion of their income, than do the poor and middle class [Aldy *et al.* 1999].

These findings fly in the face of the widespread assumption that the environment is a "luxury good" that only rich people can afford to care about [Kriström and Riera, 1996; Jacobsen and Hanley, 2009]. Larry Summers—former chief economist at the World Bank, Secretary of the Treasury under Bill Clinton, and president of Harvard University; and now a member of Barack Obama's administration—parroted this false dogma in his infamous memo arguing for the export of pollution from rich to poor countries: "The demand for a clean environment for aesthetic and health reasons is likely to have very high income-elasticity" [*The Economist*, 1992].

Diminishing returns of WTP to income imply that re-distribution should increase total WTP. For example, if we took $10,000 away from a Utah household making $75,000 in 1986, the lower (red) curve shown in Figure 4A predicts that that household's WTP for wilderness protection would have fallen from around $120 to about $110. This would represent a loss of roughly $10 that could otherwise have gone into the conservation "pot". But if we then gave the $10,000 to a household making only $2,500, it would have boosted this second household's WTP from roughly $30 to almost $60—a gain of around $30 for wilderness protection. On balance, this re-distribution would have raised the WTP summed over the two households from about $150 to around $170—a net gain of roughly $20.

Again, we can use the individual-level relationship, this time between income and WTP, to predict the results of re-distributing income in the absence of contextual effects (the lower, red curve in Figure 4B). And we can speculate that contextual effects might enhance the effect of equality on aggregate WTP for conservation efforts (the upper, green curve in Figure 4B). At this point, we are not in a position to assess the relative importance of compositional vs. contextual effects on WTP, or even the degree to which WTP results in positive conservation outcomes (*i.e.*, species and their habitats actually being preserved and/or restored). Nevertheless, the congruence between the observed micro-level income-WTP relationship (the lower, red curve in Figure 4A) and the macro-level equality-biodiversity relationship [Mikkelson *et al.*, 2007] is intriguing.

What contextual factors might reinforce, or undermine, the equality-biodiversity connection? The positive effects of equality on social cooperation, mentioned above, will also likely help to explain the equality-biodiversity connection. After all, conservation efforts, like public health, require such cooperation. What about economic growth? One study found that more egalitarian US states have experienced slower increases in per-capita income over the past several decades than have more inequitable states [Partridge, 2005]. *Ceteris paribus*, this should weaken or reverse the positive effect of equality on aggregate WTP. However, given the environmental damage wrought by economic production and consumption in the first place, slowing growth arguably benefits wild species more than enough to offset a possible reduction in funds available for alleviating the effects of unabated growth [Czech *et al.*, 2005]. In addition to social cooperation and economic growth, it would also be worth exploring the role of inter-societal processes. For example, perhaps more unequal societies are more susceptible to environmental exploitation by corporations owned by shareholders living elsewhere.

As a final note about WTP, I would like to offer a caveat. In this section, I have considered individuals' WTP as one potential *means* for achieving positive conservation outcomes. However, aggregate WTP is usually treated in a very different way, namely as a measure of the "value" of wild species and habitats considered as *ends*. The latter approach to "valuation" has fundamental flaws that in my opinion render it far inferior to non-monetary evaluation through ethically- and scientifically-informed deliberative democracy. As Daly and Farley [2004] put it, "Attempting to estimate social values by aggregating individual tastes... is a category mistake..."

4 THE INTRINSIC VALUE OF DIVERSITY AND EQUALITY

Speaking of ends, Kelly [2003] has recently articulated an ambitious theory of what makes any goal or end worthy of pursuit. His theory has intriguing parallels with the above discussion of diversity and equality, except that Kelly's is about intrinsic or final value, whereas the examples above involve instrumental value. Intrinsic value is the value of something for its own sake, while instrumental value is a tendency for the thing to augment other items of value. In this section, I will propose that something like Kelly's theory might fit naturally into the framework developed in this paper. According to that theory,

> there is only one property of value and it is the property of being *rich*, of being a unified variety. Examples of richness are living creatures, artworks, ecosystems, societies and human relationships. [Kelly, 2003, p. iii]

Miller [1982] (see also [Bradley, 2001]) proposed a similar idea early in the development of environmental ethics, but his proposal has so far not received much attention by mainstream practitioners of that field.

Kelly posits a diminishing-returns relationship between the number of parts of a given type, and the value added to the whole by those parts. For example,

> Considered alone, a reproduction of a Rembrandt has some value; it is a unified variety. Call this V. But now consider a booklet made of three hundred of the same Rembrandt. What is the value of this whole? It is certainly not 300V. The repetition of the same painting (in the absence of some other rich reason for this repetition) clearly reduces the value of each individual copy. [p. 25]

A booklet with greater diversity—*e.g.*, 300 different paintings—would presumably be richer and therefore more valuable. But diversity is not the only determinant of value. Unity can increase (and dis-unity can decrease) the value of a whole comprising a diverse set of parts:

> [I]magine putting together the two halves of Shakespeare's sonnet. It is clear that the value of the whole sonnet is greater than the value of the two halves by themselves. [p. 26]

These reflections on the good cohere remarkably with the perspective espoused above on certain ecological and economic issues. We can therefore schematize the claims just quoted in graphical form (Figure 5). To emphasize the parallels between ecology, economics, and ethics, I have depicted the kind of situation in which the diversity of the parts enhances the unity of the whole. It should be kept in mind, though, that in some cases a diverse set of parts might interact (or fail to interact) in such a way as to undermine the unity or harmony of the whole—perhaps even more than enough to offset the value added to the whole through the compositional effect of diversity.

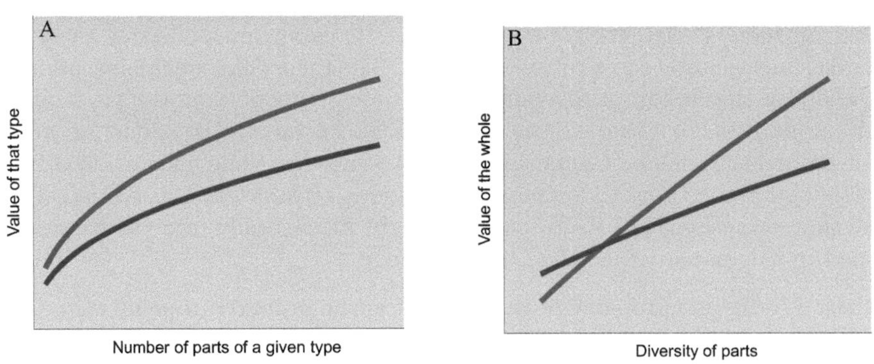

Figure 5. The good as richness, *i.e.*, unified variety.

Kelly's theory, like Miller's, implies that biological diversity and economic equality have intrinsic value, as well as contributing to other valuable properties like biological productivity and human health. Many if not most ecologists do in fact care about biodiversity for its own sake, as well as what it does for humans [Soulé, 1985; Ehrenfeld, 1988]. And yet they have all but completely effaced any arguments based on the intrinsic value of nature, *e.g.*, in their attempts to put dollar values on "ecosystem services" (see, *e.g.*, [Costanza *et al.*, 1997; Millennium Ecosystem Assessment, 2004]).

This, I submit, is a tragically misguided attempt to "sell" biodiversity to the "public" on the basis of selfish interests alone. This approach obviously appeals to some elite Western corporate, governmental, and academic audiences. But burgeoning popular movements on behalf of animals, species, and ecosystems indicate that professional ecologists are out of touch with what it will take to build the grassroots support necessary for effective conservation [Rollin, 2006; Cullinan, 2008; Revkin, 2008]. In other words, in this respect ecologists actually represent a regressive, rather than progressive, force relative to the rest of society. In a different context, Levins [1974] issued a prescient warning that "[a]cceptance of the constraints of the present, unless these are also subjected to criticism, results in their reinforcement". Unless ecologists "come out of the closet" with regard to the intrinsic value of nature, they will arguably reinforce the same selfish, economistic attitudes that continue to devastate the natural world.

5 OTHER POTENTIAL APPLICATIONS

In addition to the ecological, economic, and ethical applications described above, the compositional-contextual framework propounded in this paper can also shed light on a host of other important issues. In ecology, the relationships between habitat heterogeneity and species richness [Kohn and Walsh, 1994] and between species evenness and ecological stability [Mikkelson *et al.*, unpublished manuscript] are amenable to analysis in these terms.[4] The concomitant reduction in global species diversity and terrestrial productivity cited in Section 1 above is a large-scale example of the richness-productivity pattern examined in Section 2 [Haberl *et al.*, 2007], and therefore also explicable by such effects.

To name one example outside of ecology, the Condorcet Jury Theorem concerns the epistemological value of democracy [Christiano, 2006]. Democracy boils down to giving each of many diverse people, and therefore perspectives, equal political power. And its epistemic advantages can be viewed as another manifestation of compositional and/or contextual effects. In short, any case of diminishing returns, of some valuable property or value itself, along with the possibility of contextual influences on that property, might potentially yield the kinds of insights offered above.

[4]Thanks to Denis Walsh for suggesting this application.

6 CONCLUSION

Philosophers and scientists have long recognized the importance of diversity and equality, in both ecological and social settings (cf., *e.g.*, [Mill, 1859; Rawls, 1971; Naess and Sessions, 1984; Wilson, 1988]). In this paper, I have focused on their value as contributors to biological productivity and species diversity; and as basic constituents of the good. To elucidate the consequences of diversity and equality, I have discussed potential mechanisms at three different levels of organization: below the level of the relationship between diversity/equality and its consequence(s), at the same level, and above it.

Figure 6 portrays this multi-level perspective. For the diversity-productivity case, the lower level is that of the individual population, while same-level pathways involve other properties at the ecosystem level, and higher-level mechanisms involve interactions between the ecosystem and the landscape in which it is embedded. For the equality-biodiversity case, the lower, same, and higher levels are that of an individual person, a society, and a group of societies, respectively. And for the ethics case, they involve an individual part of a valuable whole, the whole itself, and some larger system in which the whole is embedded. In the sections above, I followed the convention developed by equality-health researchers, in calling a particular kind of lower-level pathway the compositional effect, and same- and higher-level pathways contextual effects.

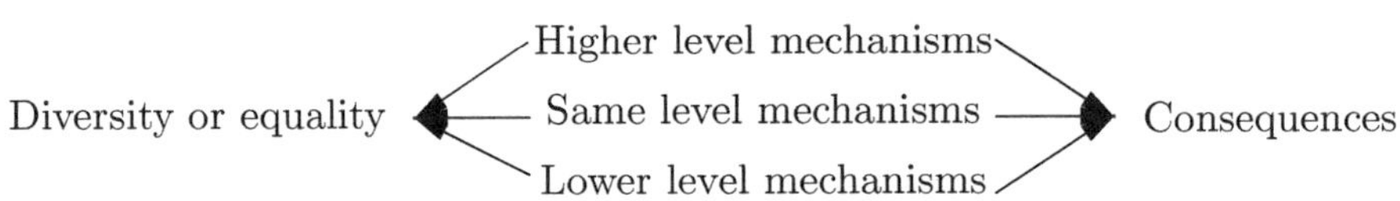

Figure 6. A multi-level perspective on the effects of diversity and equality

We live in environmentally and politically fraught times—but also heady times, when it comes to intellectual inquiry. I follow Dewey's [1938] lead in asserting that to avoid social and ecological disaster, we sorely need to re-connect our knowledge of different aspects of reality—knowledge that the past few hundred years of intellectual history have fragmented into separate disciplines. The emphasis placed by many universities on inter-disciplinarity suggests a re-kindling of the "unity of science" ideal. Various efforts to put ethics back on objective foundations connote an even older (and in my opinion, even more important) ideal, namely the unity of all knowledge (cf., *e.g.*, [Kelly, 2003; Oddie, 2006]). I hope and intend that the synthesis offered in this paper; of various findings in ecology, economics, and ethics; will make some contribution toward the fulfillment of these ideals.

ACKNOWLEDGMENTS

I thank Kevin deLaplante for the opportunity to present these ideas in writing. Olivier Bellefleur, David Goodin, Genevieve Gore, and audiences at the Second Queensland Biohumanities Conference: Philosophy of Ecology (2006), the Dubrovnik philosophy of science conference (2008), and the Ecological Society of America meeting (2008) provided helpful feedback on them. Concerning the biodiversity-productivity relationship, I thank Michel Loreau and Jon Shurin for fruitful discussion. With regard to the equality-biodiversity relationship, I thank Steph Palmer for excellent research assistance; Andy Gonzalez, Tim Holland, and Garry Peterson for many stimulating discussions; and the McGill School of Environment for a collaborative research grant.

BIBLIOGRAPHY

[Aldy *et al.*, 1999] J. E. Aldy, R. A. Kramer, and T. P. Holmes. Environmental equity and the conservation of unique ecosystems: An analysis of the distribution of benefits for protecting southern spruce-fir forests. *Society and Natural Resources* 12: 93–106, 1999.

[Bateman *et al.*, 1995] I. J. Bateman, I. H. Langford, R. K. Turner, K. G. Willis, and G. D. Garrod. Elicitation and truncation effects in contingent valuation studies. *Ecological Economics* 12: 161–179, 1995.

[Bateman *et al.*, 1996] I. J. Bateman, E. Diamand, I. H. Langford, and A. Jones. Household willingness to pay and farmers' willingness to accept compensation for establishing a recreational woodland. *Journal of Environmental Planning and Management* 39: 21–43, 1996.

[Bradley, 2001] B. Bradley. The value of endangered species. *The Journal of Value Enquiry* 35: 43–58, 2001.

[Carson, 1962] R. Carson. *Silent Spring*. Boston, MA: Houghton Mifflin, 1962.

[Christiano, 2006] T. Christiano. Democracy. In Zalta, E. N. (ed.), *Stanford Encyclopedia of Philosophy*. `http://plato.stanford.edu/entries/democracy`, 2006.

[Costanza *et al.*, 1997] R. Costanza, R. d'Arge, R. de Groot, S. Farber, M. Grasso, R. Hannon, K. Limburg, S. Naeem, R. V. O'Neill, J. Paruelo, R. G. Raskin, P. Sutton, and M. van den Belt. The value of the world's ecosystem services and natural capital. *Nature* 387: 253–260, 1997.

[Cullinan, 2008] C. Cullinan. If nature had rights. *Orion*, January/February, 2008.

[Czech *et al.*, 2005] B. Czech, D. L. Trauger, J. Farley, R. Costanza, H. E. Daly, C. A. S. Hall, R. F. Noss, L. Krall, and P. R. Krausman. Establishing indicators for biodiversity. *Science* 308: 791–792, 2005.

[Daly and Farley, 2004] H. E. Daly and J. Farley. *Ecological Economics: Principles and Applications*. Washington, DC: Island, 2004.

[Darwin, 1859] C. Darwin. *On the Origin of Species by Means of Natural Selection*. London: John Murray, 1859.

[deLaplante, in press] K. deLaplante. Philosophy of ecology: An overview. In Jorgensen, S. E. and B. Fath (eds.), *Encyclopedia of Ecology*. Elsevier, in press.

[Dewey, 1938] J. Dewey. Unity of science as a social problem. *Encyclopedia of Unified Science* 1: 29–38, 1938.

[Diaz and Rosenberg, 2008] R. J. Diaz and R. Rosenberg. Spreading dead zones and consequences for marine ecosystems. *Science* 321: 926–929, 2008.

[Ehrenfeld, 1988] D. Ehrenfeld. Why put a value on biodiversity? In Wilson, E. O. (ed.), *Biodiversity*, pp. 212–216. Washington, DC: National Academy, 1988.

[Elton, 1927] C. S. Elton. *Animal Ecology*. New York, NY: Macmillan, 1927.

[Garrod and Willis, 1997] G. D. Garrod and K. G. Willis. The non-use benefits of enhancing forest biodiversity: A contingent ranking study. *Ecological Economics* 21: 45–61, 1997.

[Haberl *et al.*, 2007] H. Haberl, K. H. Erb, F. Krausmann, V. Gaube, A. Bondeau, C. Plutzar, S. Gingrich, W. Lucht, and M. Fischer-Kowalski. Quantifying and mapping the human appropriation of net primary production in earth's terrestrial ecosystems. *Proceedings of the National Academy of Sciences, USA* 104: 12942–12947, 2007.

[Hammitt *et al.*, 2001] J. K. Hammitt, J.-T. Liu, and J.-L. Liu. Contingent valuation of a Taiwanese wetland. *Environment and Development Economics* 6: 259–268, 2001.

[Jacobsen and Hanley, 2009] J. B. Jacobsen and N. Hanley. Are there income effects on global willingness to pay for biodiversity conservation? *Environmental and Resource Economies* 43: 137–160, 2009.

[Jactel and Brockerhoff, 2007] H. Jactel and E. G. Brockerhoff. Tree diversity reduces herbivory by forest insects. *Ecology Letters* 10: 835–848, 2007.

[Kawachi and Kennedy, 2002] I. Kawachi and B. P. Kennedy. *The Health of Nations: Why Inequality Is Harmful to Your Health.* New York, NY: The New Press, 2002.

[Kelly, 2003] C. Kelly. *A Theory of the Good.* PhD dissertation, University of Colorado. Boulder, CO, 2003.

[Knops *et al.*, 1999] J. M. H. Knops, D. Tilman, N. M. Haddad, S. Naeem, C. E. Mitchell, J. Haarstad, M. E. Ritchie, K. M. Howe, P. B. Reich, E. Siemann, and J. Groth. Effects of plant species richness on invasion dynamics, disease outbreaks, insect abundances and diversity. *Ecology Letters* 2: 286–293, 1999.

[Kohn and Walsh, 1994] D. D. Kohn and D. M. Walsh. Plant species richness: The effect of island size and habitat diversity. *The Journal of Ecology* 82: 367–377, 1994.

[Kramer and Mercer, 1997] R. A. Kramer and D. E. Mercer. Valuing a global environmental good: U.S. residents' willingness to pay to protect tropical rain forests. *Land Economics* 73: 196–210, 1997.

[Kriström and Riera, 1996] B. Kriström and P. Riera. Is the income elasticity of environmental improvements less than one? *Environmental and Resource Economics* 7: 45–55, 1996.

[Lambers *et al.*, 2004] J. H. R. Lambers, W. S. Harpole, D. Tilman, J. Knops, and P. B. Reich. Mechanisms responsible for the positive diversity–productivity relationship in Minnesota grasslands. *Ecology Letters* 7: 661–668, 2004.

[Leakey and Lewin, 1996] R. Leakey and R. Lewin. *The Sixth Extinction: Biodiversity and Its Survival.* London, England: Weidenfeld and Nicolson, 1996.

[Leopold, 1949] A. Leopold. *A Sand County Almanac.* New York, NY: Oxford University, 1949.

[Levins, 1974] R. Levins. Genetics and hunger. *Genetics* 8: 67–78, 1974.

[Loreau and Hector, 2001] M. Loreau and A. Hector. Partitioning selection and complementarity in biodiversity experiments. *Nature* 412: 72–76, 2001.

[MacArthur, 1955] R. H. MacArthur. Fluctuations of animal populations, and a measure of community stability. *Ecology* 36: 533–536, 1955.

[May, 1975] R. M. May. Patterns of species abundance and diversity. In Cody, M. L. and J. M. Diamond (eds.), *Ecology and Evolution of Communities*, pp. 81–120. Cambridge, MA: Harvard University, 1975.

[Mikkelson, 2007] G. M. Mikkelson. Ecology. In Hull, D. L. and M. Ruse (eds.), *Cambridge Companion to the Philosophy of Biology*, pp. 372–387. New York, NY: Cambridge University, 2007.

[Mikkelson, in press] G. M. Mikkelson. Part-whole relationships and the unity of ecology. In Skipper, R. A., Jr., C. Allen, R. Ankeny, C. F. Craver, L. Darden, G. M. Mikkelson, and R. C. Richardson (eds.), *Philosophy Across the Life Sciences.* Cambridge, MA: MIT Press, in press.

[Mikkelson *et al.*, 2007] G. M. Mikkelson, A. Gonzalez, and G. D. Peterson. Economic inequality predicts biodiversity loss. *Public Library of Science (PLoS)* ONE 2:e444, 2007.

[Mikkelson *et al.*, in review] G. M. Mikkelson, B. J. McGill, S. Beaulieu, and P. L. Beukema. Multiple links between the bioological diversity andd ecological stability of bird communities across North America. In review.

[Mill, 1859] J. S. Mill. *On Liberty.* London, England: John W. Parker, 1859.

[Millennium Ecosystem Assessment, 2005] Millennium Ecosystem Assessment. *Ecosystems and Human Well-being: Biodiversity Synthesis.* Washington, DC: World Resources Institute, 2005.

[Miller, 1982] P. Miller. Value as richness: Toward a value theory for an expanded naturalism in environmental ethics. *Environmental Ethics* 4: 101–114, 1982.

[Muriithi and Kenyon, 2002] S. Muriithi and W. Kenyon. Conservation of biodiversity in the Arabuko Sokoke Forest, Kenya. *Biodiversity and Conservation* 11: 1437–1450, 2002.

[Naess and Sessions, 1984] A. Naess and G. Sessions. The deep ecology platform. Available at `www.deepecology.org/platform.htm`, 1984.

[Oddie, 2006] G. Oddie. *Value, Reality, and Desire.* New York, NY: Oxford University Press, 2006.

[Odenbaugh, 2006] J. Odenbaugh. Ecology. In Sarkar, S. and J. Pfeifer (eds.), *The Philosophy of Science: An Encyclopedia*, pp. 215–224. New York, NY: Routledge, 2006.

[Partridge, 2005] M. D. Partridge. Does income distribution affect U.S. state economic growth? *Journal of Regional Science* 45: 363–394, 2005.

[Pitt Inequality Project, 2006] Pitt Inequality Project. Standardized Income Distribution Database. Version 1. `www.pitt.edu/~\inequal/2006SIDD-1-Aug21.xls`, 2006.

[Pope and Joes, 1990] C. A. Pope III and J. W. Jones. Value of wilderness designations in Utah. *Journal of Environmental Management* 30: 157–174, 1990.

[Putnam, 2002] H. Putnam. *The Collapse of the Fact/Value Dichotomy, and Other Essays*. Cambridge, MA: Harvard University Press, 2002.

[Rawls, 1971] J. Rawls. *A Theory of Justice*. Cambridge, MA: Harvard University, 1971.

[Revkin, 2008] A. Revkin. Ecuador constitution grants rights to nature. Dot Earth, *The New York Times* 2008.

[Rollin, 2006] B. E. Rollin. *Science and Ethics*. New York, NY: Cambridge University, 2006.

[Rosenzweig, 2003] M. L. Rosenzweig. Reconciliation ecology and the future of species diversity. *Oryx* 37: 194–205, 2003.

[Soulé, 1985] M. E. Soulé. What is conservation biology? *BioScience* 35: 727–734, 1985.

[Speth, 2008] J. G. Speth. *The Bridge at the End of the World: Capitalism, the Environment, and Crossing from Crisis to Sustainability*. New Haven, CT: Yale University, 2008.

[Stanley, 2005] D. L. Stanley. Local perception of public goods: Recent assessments of willingness-to-pay for endangered species. *Contemporary Economic Policy* 23: 165–179, 2005.

[Summers, 1992] L. Summers. Let them eat pollution. *The Economist*: 322 (February 8), 66, 1992.

[Thoreau, 1854] H. D. Thoreau *Walden; or, Life in the Woods*. Boston, MA: Ticknor and Fields, 1854.

[Toivonen, 2004] A. L. Toivonen. The economic value of recreational fisheries in Nordic countries. *Fisheries Management and Ecology* 11: 1–14, 2004.

[Turpie, 2003] J. K. Turpie. The existence value of biodiversity in South Africa: How interest, experience, knowledge, income and perceived level of threat influence local willingness to pay. *Ecological Economics* 46: 199–216, 2003.

[United Nations, 2005] United Nations. *World Population Prospects: The 2004 Revision*. New York, NY: United Nations, 2005.

[Vitousek *et al.*, 1997] P. M. Vitousek, H. A. Mooney, J. Lubchenco, and J. M. Melillo. Human domination of the Earth's ecosystems. *Science* 277: 494–499, 1997.

[Whitman, 1855] W. Whitman. *Leaves of Grass*. New York, NY: Rome Brothers, 1855.

[Wilkinson and Pickett, 2009] R. G. Wilkinson and K. E. Pickett. *The Spirit Level: Why More Equal Societies Almost Always Do Better*. New York, NY: Penguin, 2009.

[Willis and Garrod, 1993] K. G. Willis and D. G. Garrod. Valuing landscape: A contingent valuation approach. *Journal of Environmental Management* 37: 1–22, 1993.

[Wilson, 1988] E. O. Wilson (ed.), *Biodiversity*. Washington, DC: National Academy, 1988.

[World Bank, 2007] World Bank. *2007 World Development Indicators Online*. Washington, DC: World Bank, 2007.

INDEX

abduction, 3, 4, 25
Abrams, P., 120
accidental assemblage, 17
accounting, 19, 364, 366, 385, 387
  problem of, 363, 393
acrasin, 242
adaptation, 4, 14, 202
adaptation-distribution, 35
adaptations and distributions, 29
adaptive capacity, 213
adaptive collaborative management, 383
adaptive cycle, 212
adaptive environmental managers, 391
adaptive management, 383
adaptive restoration, 19
adequacy conditions, 147
aesthetics, 97
agency, 14
agriculture, 350
Ahmadjian, V., 219
alewife, 312
algorithms, 382
alien species, *see* invasive species
American Indians, 303
anagenesis, 215
analogy, 42, 365
Anderson, H., 303
Anker, P., 96
anthropocentrism/nonanthropocetrism, 367
anthropocentrists, 367
anthropogenic, 347
anti-cancer virus as symbiotic, 243
approximations, 269
aquaculture, 350
area effect, 111
Aristotle, 181
artifactual ecologies, 16
Asian ass, 312
assortative equilibrium, 115
atomic physics, 267
Austin, 330, 331
authority, 251, 253, 255–257
autotroph
  as energy valve, 240
axiological criteria, 17

bacteria, 222
balance of nature, 181
balance of nature paradigm, 306
balance of selection, 15
Balcones Canyonlands, 329, 330, 336, 338, 340, 348, 353, 355
Barbour, M. G., 66
barrier zones, 137
Bateson, W., 222
Beaupre, Steven, 125
Begon, M., 93
Behe, M., 232
Bessey, C. E., 52, 87
biocönose (biocenose), 31
biochemistry, 267
biodiversity, 12, 39, 169, 177, 186, 197, 289, 293, 319, 328, 337, 339, 345, 348, 400
biodiversity–ecosystem function, 189
biodiversity–ecosystem function debate, 169, 188
biodiversity–ecosystem function experiments, 178
biodiversity-ecosystem function debate, 196
biogeography, 124
biological diversity, 373
biological integrity, 178

biome, 207
bioregions, 323
biotic resistance, 138, 139, 143
Bismarck Archipelago, 120
Black-capped Vireo, 333, 341, 348, 353, 355
Blackman, F. F., 59
Bolding, K., 380, 384
Bonevac, D., 266
Bonner, J. T., 238
Botkin, D., 60
boundaries, 49, 52, 54, 56–59, 61, 64, 67, 68, 73, 74, 76, 77, 79, 83, 91, 93–95, 102
boundary conditions, 271
Boyle, M., 216
Bradshaw, A. D., 333–335, 355
Braun, E. L., 49, 95
British Ecological Society, 40
brittleness, 257
"broken stick" model, 110
Broome, H., 303
Brown, B., 16
Brundtland Commission, 327, 367
Buffon, C. de, 252, 253

Cadenasso, M. L., 207, 212
Cairns, J. Jr., 333–335, 337, 339
Callicott, J. B., 17, 18, 20, 55, 95
camels, 301, 308
cancer, 219, 241–244
capital
    depreciation of, 19
caprice, 341, 344–346, 354
carbon offsets, 17
carbon sequestration, 286, 291, 292
carbon trading, 17, 286–288, 293, 295, 296
carrying capacities, 124
carrying capacity, 123, 241
Cartwright, N., 269
causation, 57, 58, 70–76, 81–83, 90, 94–103
cause
    final, 6
CBA, 370, 380
celestial mechanics, 260
cell theory, 221
centrifugal, 21
centripetal, 21
challenge
    data, 270
    model complexity, 270
    natural complexity, 270
    openness, 270
Chamberlin, T. C., 256
chance, 50, 62, 93
changes
    anthropogenic, 19
chaos, 262
chaos, mathematical, 65
checkerboard pattern, 121
cheetahs, 301, 311
Cheke, A., 264
choice making, 17
chronosequence, 34
Churchill, W. S., 230
cladogenesis, 215, 216
Clements, F., 303
Clements, F. E., 4–7, 17, 27, 51–62, 66–100, 102–103
Clementsian, 183
Clementsian equilibrium ecology, 302
Clementsian view, 12
climate, 274
climate change, 19, 288, 294, 346, 355
climax, 6
climax ecosystem, 17
climax formation, 34
climax plant formation, 304
closed, 261
cognitive linguistics
    and image schemata, 133
cognitive lingusitics
    and science, 133
cognitive science, 381
cohesion, 205, 206
Cold War, 54

Collier, J., 13, 14, 21
Colyvan, M., 17
commensalism, 15, 226
communism and socialism, 5
community, 4, 49–51, 54–57, 62, 64, 66–69, 72–103, 148
  climax, 53, 54, 60, 61, 69–76, 78, 80, 81, 84–86, 88–90, 92–94, 96, 104, 106
community ecology, 109
community equilibrium, 32
community organisation, 31
competition, 7, 71, 74, 101
  interspecific, 110, 111
competition coefficients, 120
competitive exclusion principle, 117
complex adaptive system, 201, 203, 204
complex dynamical system, 210, 211
complexity, 11, 262, 266, 268
compositional, 21
compositional effect, 401
Comte, A., 37
conclusion, 266
conditional premises, 271
confirmation, 259
Connor, E. F., 120, 121
consequence relation, 269
conservation, 50–56, 66–69, 85, 86, 97–102
conservation biology, 111, 177, 319, 334, 335, 345
conservation of mass and energy, 268
constancy, 340, 347
constitutions, 367
constitutive metaphors, 8
constraints, 273
constructivism, 142
constructivist philosophy of nature
  arguments against, 131–133
CONTAINER image schemata, 133–137, 142
  and invasive species, 136
contextual, 21, 22
contextual effects, 403
contextualised, 348
continental drift, 269
contingency hypothesis, 125
contingent valuation, 391
continuity, 19
convergence, 4, 25
Cooper, W. S., 90
cooperation, 15
cooperative interaction, 14
Corixa
  affinis, 116
  macrocephala, 116
  punctata, 116
cost-benefit analysis, *see* CBA
Cowles, H. C., 34, 57, 74, 87
cultural, 337, 338
cultural practices, 336
cultural role, 349
culture and nature, 131
Cumming, G., 13, 14, 21
Cummins, R., 12

Darwin, C., 26, 231, 272
Darwinian evolution, 215
Dawkins, R., 219, 233, 237
de Bary, A., 220, 224, 226, 245
De Vries, H., 59
decision theory, 17, 20, 285, 289, 294, 296, 297, 381, 382
decision-making, 20
Deconstructivism, 55
deep ecology, 289
deep time, 16
deep-water ciscoes, 312
defining hypotheses, 35, 44
deforestation, 202
Delage, Y., 221
deLaplante, K., 11, 12, 17
Demodex, 226
deontic absolution, 17
deontological criteria, 17
deontological rule, 17
dependence, 7

depth of time, 341, 344
Descartes, R., 63, 225
descriptive, 21, 22
Desert Laboratory, 31
determinism, 61, 64, 68, 69, 72, 80, 84, 88, 90, 92, 93
Dewey, J., 388
Diamond, Jared, 119–121
diminishing returns, 400
discounting, 367, 369
discursive pluralism, 368
displacement, 13, 14
dissipation, 216
dissipative structure, 14, 229
distance effect, 111
distance metric, 153
distributed genome, 14
disturbance, 18, 42
disturbance regime, 306, 314
diversity, 9–11, 21, 22, 147, 194, 400
diversity and stability, 11
diversity–stability debate, 170, 196
Dobzhansky, T., 229
Dodo, 264
domesticates, 14
Doolittle, F., 219
Douglas, A., 232
Drude, O., 37, 56, 87
Dunham, A., 125
Dureau de la Malle, A. J. C. A., , 33, 69
Dust Bowl, 53, 86
dynamic, 13
dynamic criterion, 328, 329, 335, 338–340, 342, 344, 348, 352
dynamic equilibrium, 31
dynamical system, 205, 216
dynamical unity, 210

Earman, J., 272
ecesis, 71, 72, 81, 96
ecological, 387
  community, 4, 5
  diversity, 22
  economics, 365, 380, 384, 387
  economists, 365, 385
  equilibrium myth, 303
  footprint, 386
  function, 185
  integrity, 328, 336, 338, 344, 355
  models, 254, 271
  process, 185
  processes, 17
  reconstruction, 338, 352, 355
  rehabilitation, 313
  resilience, 386
  restoration, 17, 22, 301, 328–330, 354, 356, 377
  services, 22, 386
  societies, 40
  spatial scale, 318
  subgroups, 44
  temporal scale, 318
  triage, 286
  values, 17
Ecological Society of America, 40, 95
Ecology, 40
ecology, 219, 254, 261, 262, 268
  community, 110, 120, 124
  evolutionary, 110
  human, 220
  mathematical, 117
  plant, 12
  population, 110, 124
  reflexive, 21
  spatial, 116
economic equality, 400
economic inequality, 405
ecosystem, 201–203, 207–209, 211, 215, 227
  complexity, 210
  ecology, 169
  function, 11, 12, 169, 173, 178, 179, 186, 197
  functioning, 169, 197
  health, 178, 210, 320
  identity, 13, 208, 209
  individuation, 202–204, 207, 210

integrity, 178, 347
more than the sums of parts, 405
productivity, 21
reflexive, 20, 388
restoration, 17
services, 13, 345, 350, 353, 368
stability, 39
ecosystem unity
network account, 210
ecosystems, 5, 55, 56
ectosymbiosis, 224
Edwards Plateau, 332, 353
Einstein, A., 225
Eldredge, N., 228, 231
electron microscope, 222
elephants, 301, 311
eliminative induction, 265
Eliot, C., 4–7
Elton, C., 27, 138, 170
embodied realism
and invasive species, 142–143
emerald ash borer (EAB), 129
emergence, 17, 67, 68, 95, 96
endangered species, 379
Endangered Species Act, 101, 331
endosymbiont, 15, 221, 222
endosymbiosis, 224
entropy, 229
environmental
accounting, 366, 368
ethicists, 347
impact statement, 294
policy, 173
pragmatism, 375
resistance, 139
security, 351
suitability, 349
triage, 17, 285, 287, 289, 293, 295
Environmental Protection Agency, 339, 363
epistemology, 254
equation, 266
equilibrium, 4, 56, 67, 68, 82, 95
equilibrium biogeography, 123
equilibrium or balance, 12
equilibrium or extinction, 11
equity, 367
erosion, 262
ESA, 348
ethics, 16, 17, 297, 381, 383
environmental, 55, 100, 101, 285, 293, 297
eucaryotic cell, 222
Euclidean distance, 153
European settlement, 18
evenness, 10, 150
evolution, 259, 260
cultural, 16
internal vs. external, 236
evolutionary biology, 208, 219
evolutionary limitation, 321
evolutionary temporal scale, 317
evolutionary tie, 309
exergy, 216
exosymbiosis, 224
exotic species, 316
expected utility, 288, 290
explanation, 64, 65, 70–83, 87, 93–96, 99, 103, 252–254, 261
independence of, 268
explanatory narrative, 17
external disturbance, 304
external relations, 210
extinction
anthropogenic, 19

facilitation, 401
factors limiting growth, 28
fairness, 367
Famintsyn, A. S., 221
Fausto-Sterling, A., 221
fecundity, 270
feedback, 229, 262
fidelity, 328
First Nations, 345
Fisher, R. A., 116, 122
fitness, 122, 239

competitive, 240
constructive, 240
cooperative, 240
Lotkan, 240, 246
Fitzsimmons, A. K., 55, 56
flipping, 386
flips, 363
flood geology, 256
Florida Keys, 114
flux of nature paradigm, 306
food chains, 32
Forbes, E., 27
Forbes, S., 32
FORCE dynamic schema, 137–139
FORCE dynamics, 133–137, 141
Forel, F.-A., 27
Frank, B., 220, 245, 303
Frank, S. A., 229
free energy, 229, 238
function, 12, 13, 74–76
function talk, 182
function vs. purpose, 234
functional diversity, 173, 178
fundamental principles, 268
fusion, 14
future values, 17

Gödel, K., 211
Gaia hypothesis, 14, 219, 227, 233, 236
minimal, 238
Galileo, 264
game theory, 295, 296
genes, 50
genetic redundancy, 225
genetics
Mendelian, 59
genome, 222
distributed, 235
geological column, 269
geological narratives, 269
geology, 254, 259, 260
Georgescue-Rogen, N., 384
germination, 264, 265
Ghiselin, M., 13, 208
Gilpin, M., 121
Gleason, H. A., 5–7, 17, 35, 49, 51–56, 60–62, 65, 66, 68–70, 77–86, 89–91, 93–95, 98, 102, 305
Gleasonian, 11, 184
Godfrey-Smith, P., 50
Golden-cheeked Warbler, 333, 341, 346, 348, 353, 355
good, 22
Goodin, R. E., 289, 292, 293, 296
Gould, S. J., 231
Great Plains, 53
greenhouse emissions, 288
Grime, J. P., 38
Grisebach, A., 29
group selection, 184, 215, 219, 238
guild, 121
Gunderson, L. H., 215

Haberlant, G., 29
habitat, 5
habitat conservation plan, 331
habitat reconstruction, 329, 338, 344, 347, 352
Hacking, I., 50, 132, 140
haecceity, 205
Haekel, E., 39
Hagen, J. B., 74, 83, 92
Haldane, J. B. S., 122
Hall, M., 350
Hardin, G., 219, 237, 239
Hempel, C., 71
Heraclitean flux, 6
Heraclitus, 62–64, 68
Hesse, M., 9
heterarchies, 206
heterotroph, 15, 241
heuristics, 368, 376, 383
hierarchical, 392
hierarchical model, 364, 375, 376, 379
hierarchy, 373, 377, 378, 387
hierarchy theory, 364, 373, 374

Higgs, E. S., 19, 338, 340–343, 346, 351, 352, 354
higher-level interaction, 400
higher-level mechanisms, 412
historical fidelity, 19, 329, 330, 335, 338, 341–347, 353–356
historical individuals, 14
historical limitation, 321
historical narrative, 259
historical science, 5, 16, 252, 255, 259–261, 268, 269
historicity of restoration, 19
historiography, 83, 94, 103
holism, 6, 62, 67, 68, 74, 93, 95–98, 102, 174, 179, 196, 219
holistic, 4, 11, 12, 183
Holland, J. H., 216
Holling, C. S., 212, 214, 215
hominin, 259
homology, 73, 79
Horn, Henry, 117, 120
horses, 308, 309
HT, 378
Hull, D., 13, 208
human impact, 327
human societies, 37
human well-being, 202
humanity as evolutionary experiment, 244
humans, 14
    indigenous, 18
Humboldt, A. von, 29
Hume, D., 7
Husley, T. H., 240
Hutchinson, G. E., 98–100, 110, 116, 117, 124
Huxley, T., 220
hyperdisease theory, 308
hypotheses
    initial defining, 4
hypothesis, 25

Idealism, 76
    German, 89
identity, 204
    dynamical, 204
identity criteria, 13
ideology, 179
Ik, 343
image schemata, 133
    and invasion biology, 133, 136
    and metaphors, 133
income equality, 21
incommensurability, 17
incommensurate value, 289, 291
inconsistency, 5
individual age, 270
Individualistic Hypothesis, 54, 77–83, 85, 86, 90–93
inefficiencies in energy transfer, 33
inference, 17, 257, 267, 273
    predictive, 259
    reliability, 21
    selective, 269
inferential links, 268
inferential reduction, 266–268
inferentialism, 271
infinite value, 289, 290
information theory, 159
initial condition ensemble, 274
institutional economics, 384, 387
integration, 8
integrity, 338, 339, 342, 343
intelligent design, 235
interaction, 22
interdependence, 22
internal relations, 209
intervention, 272
intrinsic value, 290, 351, 400
introduced species, 18
invasion biology
    and image schemata, 133, 136
invasional meltdown, 139
invasive species, 8, 9, 14, 139, 319
    alternative views of, 142–143
    and container image schema, 136
    and cultural contexts, 140–141
    and embodied realism, 142–143

and path image schema, 136
and propagule pressure, 139
and resistance, 139
as metaphor, 141–143
definition, 130
functions of, 142
harmful, 142
impact debate, 130–132
impact on native communities, 132
superabundant, 142
irreducible complexity, 211
irreversible, 370
island biogeography, 39, 109, 124
equilibrium model, 111
isocline analysis, 111

James, W., 388
Jax, K., 185
Jeon, K. W., 227
Jeon, M. S., 227
Johnson, M., 8, 9
Jordan, V., 59
Journal of Ecology, 40
Justus, J., 9, 10

karst caves, 330
Kay, J. J., 216
Kelly, C., 22
Kelvin, *see* Thompson, W., 257
Kepler, J., 270
Kidepo Valley National Park, Uganda, 343
kin selection, 238
Kingsland, S. E., 94
Koestler, A., 375
Krakatoa, 125
Kropotkin, P., 220, 240
Kuhn, T. S., 52, 53, 56, 60, 63, 64, 86
Kyburg, H., 271, 273

laboratory, 259
Lack, David, 109, 117
Lakoff, G., 8, 9
land ethic, 55
land health, 320
Lande, H., 148
landscapes, 323
language, 62, 64, 66, 78, 79, 83, 88, 90, 93, 94
Laraudagoitia. P., 272
Larson, B., 8, 9
law of the minimum, 28
laws
ceteris paribus, 92
of probability, 62
of science, 71, 94
of succession, 56, 61, 69–72, 79, 92
laws of biochemistry, 267
laws of physics, 267
legislative fiat, 339
Leibniz' Law, 204, 205
Leigh, Egbert, 110
Lenton, T., 233, 237
Leopold, A., 16, 55, 245, 302, 333
less sharp or uniform, 6
levels of organisation, 412
Levins, R., 75, 110, 117
Lewontin, R. C., 75, 110, 115
lichen, 220
Liebig, J. von, 28
limiting similarity, 109, 111, 116, 123
Linnaeus, 251
lions, 311
Lotka, A. J., 240
Lotka-Volterra equations, 119
Lovejoy, A. O., 91
Lovelock, J., 219, 227
lower-level mechanisms, 412
Lyell, C., 260

Möbius, K. A., 31
MacArthur, R. H., 7, 8, 11, 109–113, 115–126, 170
MacArthur-Wilson equilibrium model, 112–116
Mackie, J. L., 83

MacLeish, A., 86
macroparasites, humans as, 244
Macropygia
  mackinlayi, 120
  nigriostris, 120
macroscoping, 372
magnetic surveys, 269
Malthus, T. R., 16, 26, 221, 226, 272
mammoths, 308
management, 288
mangroves, 351
Margulis, L., 14, 219, 222, 223, 227
market efficiency, 288
Marshall, R., 303
marsupials, 9
mastodons, 308
material inference, 267
mathematical model, 8
Mauritius, 264, 265
May, R., 11, 117–120
Mayan, 342
McCoy, E. D., 5, 56, 95
McIntosh, R. P., 83, 84, 90–93
McKay, B., 303
McKenzie, R. D., 38
McNeill, W., 244
mechanical philosophy, 181
mechanisms, 412
Medawar, P. B., 225
Meine, C., 302
Merezhkovskii, K. S., 221
meta-model, 211–213, 216
  dynamic limitation, 215
  random walk, 213
  replacement, 213
  succession, 214
  system Evolution, 215
metagenomy, 235
metaphor, 9, 54, 62, 73, 75, 87–89, 94, 103, 365, 390, 392
metaphoric resonance, 143
metaphors
  and image schemata, 133
  examples of, 138
  invasive species as, 141–143
  pressure as a, 138
  resistance as a, 138
  role in scientific practice, 133–135
  roles of, 133
metaphors in science
  use of, 138
metaphysical, 255, 263, 266
metaphysics, 253, 254
metapopulation models, 111, 116
metazoa, 242
metazoan, 15
mid ocean ridges, 269
migration, 14
Mikkelson, G., 21
Millennium Ecosystem Assessment, 350, 351
minimum standard of preservation (SMS), 370
Mitman, G., 50
mitochondria, 221, 222
Mixotricha paradoxa, 235
model, 258, 264, 271, 273
  climate, 274
  Club of Rome, 273
  global circulation (GCM), 274
  limitations, 272
  of stock markets, 273
  phenomenological, 273
  sequence of, 274
  weather, 274
model complexity, 270
model uncertainty, 294, 390
models, 65, 82, 83, 87, 88, 257, 261, 263, 268, 269
  community, 270
  deterministic, 270
  ecological, 270
  ecosystem, 270
  population, 270
  semantic, 269
  stochastic, 270
monism, 367

monoclimax theory, 88, 90, 92–94
Moore, G. E., 22
Moran, N., 225
Morgan, M., 272, 274
Morrison, M., 123, 269, 272, 274
Moss, C. E., 5, 56–60, 94
Mousetrap Problem, 232
multi-criteria analysis (MCA), 353
multi-model ensembles, 274
mutation
    directed, 233
    point, 222
mutual information, 210
mutualism, 14, 22, 220, 226, 229
    facultative, 226
    obligate, 226
    of clones, 241
mutualistic, 15

narrative, 254, 261, 265, 268, 344
narrative continuity, 341, 343
narrative links, 19
native communities
    and image schema, 136
    resisting invasion, 139
native range, 349
native species, 316, 353
nativism, 349
natural functions, 181
natural history, 58, 251
natural individuals, 208
natural kinds, 208
natural philosophy, 251
natural range, 317
natural sciences, 257, 259–263, 268, 269
natural selection, 12, 13, 230
natural selection as feedback process, 239
natural unity, 206
natural values, 327–329, 347
naturalized species, 318
Nature, 302
nature and culture, 131
Nazis, 342
necessary good, 407
neo-Darwinian synthesis, 181
neo-Darwinism, 229
New York Times, 302
Newton, I., 379
Neyman-Pearson testing, 121
niche
    fundamental, 117
    realized, 117
niche complementarity, 190
niche differentiation, 401
niche overlap, 117–119
Nicolson, M., 83
Niger Delta, 347
non-indigenous species, *see* invasive species
non-native species, *see* invasive species
nonanthropocentrists, 367
norm for ecological restoration, 322
normative, 21, 22, 328, 330
norms, 13, 267
Norton, B. G., 19–21, 272
nostalgia, 341–343, 355
novel hypotheses, 26
nucleocentrism, 221
null hypotheses, 120, 121

Oberholtzer, E., 303
obligate mutualism, 264
observation, 17
Odenbaugh, J., 5, 7, 8, 17, 51, 55, 72, 95, 98–103
Odum, E. P., 11, 16, 39, 55, 67, 170, 227
oncogene, 242
ontogeny, 4
ontological, 366, 368, 376
ontological elements, 365
ontology, 60, 66, 67, 89, 100, 101, 266, 375
openness, 252, 263, 280
operations research, 286
Oreskes, N., 269, 271, 272

organic analogy, 42
original condition, 302
origins of ecology, 27
overkill hypothesis, 309

Pacific salmon, 313
paleontology, 254
Paley, W., 235
Paracer, S., 219
paradigms, 52, 56, 60, 63, 64, 66, 86, 89
parametric indices, 148
parasitism, 14, 15, 220
    chronic, 226
    pathogenic, 226
Pareto improvement, 382
parthood, 13
PATH image schemata, 133–137
    and invasive species, 136
patterns, 268
Peacock, K. A., 14, 15
Peirce, C. S., 3, 25
Perry, G., 13, 256
persistence, 340, 347
perturbation-tolerance, 340
pessimism, 271
pests, 14
Phillips, J., 95, 96
philosophy of modeling science, 394
philosophy of science, 255
    reflexive, 387
physical anthropology, 254
physics, 268
physiological limitation, 321
Picasso, V., 11, 12, 17
Pickett, S. T. A., 207, 212
Pielou, E. C., 150
Pimm, S., 11
plant geography, 29
plant succession, 5
plastids, 221
plate tectonics, 269
Plato, 64, 68, 96, 181
Pleistocene, 345, 346
    megafauna, 301
    parks, 310
    re-wilding, 374
Pleistocene-Holocene boundary, 18, 301, 308
plurality, 368
Poirier, 14
policy, 13, 20, 328
    public, 50
pollinator to pollen-producer, 14
pollution, 288
Popper, K., 255
population, 5
    biology, 8, 109, 110, 123
    density, 122
    genetics, 124
    regulation, 26
    structure, 265
Portier, P., 221
positivism, 365, 372, 392
Possingham, H. P., 296
post-metaphysical, 368
post-positivism, 390
postmodern science, 203
Poststructuralism, 55
practical limits, 268
pragmatic, 377, 378
pragmatism, 368, 373
pragmatist, 379, 384, 388
precautionary principle (PP), 370
predation, 110
predator-prey relations, 210
prediction, 53, 65, 66, 68, 69, 72, 80, 83, 97, 99, 102, 258, 263, 272
    retrospective, 258
predictive power, 258
preference reversals, 381
preferences, 20
preservation, 50, 51, 54–56, 67, 68, 95–99, 101–103
primitive, 9
Principles of Geology, 260
probabilistic assemblage, 17

probability, 62, 81, 287, 294
procaryote, 222
processes, 13, 253–255, 257–262, 268, 272
productivity, 21, 194, 345, 350, 353
productivity-diversity relation, 21
progress, scientific, 50, 52, 58, 60
propagule pressure, 138, 139, 143
 and invasive species, 139
Przwalski's horse, 312
punctuated equilibrium, 219
purpose, 12

quadrat method, 93
quantum mechanics, 225, 228

radiological dating, 269
rarity, 19, 348
rational choice, 382
rational choice model, 380
Ray, J., 251
re-wilding, 346
realism, 142
 view of invasive species, 132
realism, scientific, 51, 56, 98, 100–102
realist philosophy of nature
 arguments vs. realist view, 131–133
realistic, 273
reallocation costs, 296
reconstruction, 340, 343, 346, 347, 350, 352, 354
reconstructionist, 356
recursion, 229, 246
reduction, 4, 11, 17, 174, 183, 196, 225, 253, 266, 267
 of scientific theories, 266–267
 practical limits, 268
reference state, 329, 352, 354
reference state criterion, 352
rehabilitation, 312
Reichenbachian 'forks', 258
Reinheimer, H., 221
reliability, 340, 347
replacement cost, 368
reports, 266
reproductive potential, 273
rescue effect, 115
reservation, 19, 327
reserve design, 111
resilience, 340, 347
resistance, 340, 347
 and invasive species, 139
resource reallocation, 295
resource spectrum, 117, 118
restoration, 18, 19, 330, 333, 335, 337, 344, 345, 350, 353, 374
 ecological, 301, 328, 329, 374
retrodictive, 259
retrodictive testing, 274
reversibility, 363, 370
reversible, 369, 385
rewilding, 19
richer or more formal, 11
richness, 22, 409
Ricklefs, R. E., 93, 94
risk aversion, 292
risk decision square (RDS), 370, 372
robustness, 14
Roegen, N. G., 380
Rosen, R., 210
Ross, E. A., 37
Rutherford, E., 254, 255, 266

saiga antelope, 309
Saint-Simon, H., 37
same-level mechanisms, 412
sampling effect, 192
Sapp, J., 220, 224
Sarkar, S., 19, 270
scale, 18, 20, 21, 369, 370, 372, 373, 375, 377, 379, 384
 temporal, 18
scale-jumping, 372
Schimper, A. F. W., 30
Schrader-Fechette, K. S., 5
Schumacher, E. F., 380, 384
Schwendener, S., 29, 220
Science News, 302

science
  and cognitive linguistics, 133
  unity of, 268
scientific explanation, 183
sea lamprey, 312
selection, 12, 14
  density-dependent, 111, 122–124
selection effects, 194
selective advantage, 221
selective pressure, 272
self-organization, 201, 202, 209, 211, 213
self-regulating systems, 214
Sellars, W., 9, 266
Semper, C. G., 33
sensitivity analysis, 294
SER, 335–339
sere, 6
serial endosymbiosis theory (SET), 14, 222, 227
set theory, 117
settlement, 322
Shannon, C. E., 9, 10, 147
Shannon–Weiner index, 178
Shelford's law of tolerance, 29
Shelford, V. E., 29, 61
Shell, 347
shifting paradigm, 306
Shrader-Frechette, K. S., 56, 95
Sideroxylon grandiflorum, 264
Simberloff, Daniel, 114, 115, 120, 121
simplicity, 342, 343
simplification, 269
Simpson, E. H., 9, 10, 147
Simpson, G. G., 230, 234
simulation, 274
sinks for carbon, 350
slime molds, 238, 242
Smith, D. C., 228
Smith, W., 253
SMS, 370
Smuts, J. C., 96
Sober, E., 239
social constructivism, 131, 132
social learning, 392
social welfare function, 294
Society for Ecological Restoration, 43, 334
sociology, 37
species, 50, 208
species abundance, 148
species as individuals, 14
species equilibrium, 115
species evenness, 9
species richness, 9, 112, 124, 149
species selection, 239
species-area effect, 111, 113, 114
spectacled bears, 309
Spencer, H., 36, 84, 87, 240
spiritual benefit, 322
stability, 340
stable community, 31
stated preference, 391
Steele, K., 17
Stegmüller, W., 89
Sterelny, K., 5, 95, 125
Stevens, N., 61
stratigraphy, 253
substitutability, 363, 364
Substitutability/Reversibility problem, 393
substitution, 19
succession, 33, 49, 51, 53, 56, 61, 69–72, 75–82, 84–93, 95–98, 304
  Clementsian, 4
succession/super-organism hypothesis, 35
super-organism, 34, 55, 93, 95, 184
supervenience, 253, 266
sustainability, 19, 220
  ecosystem, 202
  management, 202
symbiogenesis, 221, 227, 235
symbiome, 224, 233
  thermodynamic interpretation, 229
symbionts, 220
symbiosis, 14, 15, 219
  causal link interpretation, 223

complexity of, 225
contact interpretation, 223
definition, 220, 223
mutualistic, 220
scale of, 226
varieties of, 220
symbiotic transition, 227, 243
for humans, 245
system individuation, 216
systematics, 58

Tambalacoque, 264, 265
Tansley, A. G., 56, 59, 88, 89, 201, 207, 227
target effect, 115
taxis, 222
taxonomy, 254
technology, 85
teleology, 12, 182, 234
telescoping unification, 267
Temple, S. A., 264
tension zone, 7
Terbough, John, 119
Terra nullius, 9
Texabama croton, 332
Texas, 329, 343
Texas Wildscapes, 344
theoretical ecology, 109
theoretical principles, 261
theoretical purity, 261
theoretical sciences, 255
theory, 273
logical, 269
thermodynamics, 14, 268
thermodynamics and evolution, 238
Thompson, W., 256
Thoreau, H. D., 33, 351
Tilman, D., 11, 101, 163, 172
Tobey, R. C., 60, 86–90, 92, 94
traces, 258, 268
tragedy of the commons, 219
TransAmazon highway, 202, 203
transformative value, 392
tRNA, 222
troglobites, 331
trophic relations, 210
trophic-dynamic hypothesis, 33
trophic-limitation hypothesis, 35
tropical antbird, 119
Turing, A. M., 211
turnover rate, 115
Twain, Mark, 257

Ulanowicz, R. E., 210, 216
uncertainty, 271, 292–294, 297
understanding, 258
unification, 8, 22
in ecology, 109
unified ecology, 43
unified variety, 22, 409
unit of selection, 219, 228
unity, 13, 205
unity of all knowledge, 412
unity of science, 266
universality, 131
University of Wisconsin Arboretum and Wild Life Refuge, 302
utilitarian, 381, 384
utilitarianism, 375
utilization function, 118, 120

Valk, A. van der, 3–5
value, 13, 294
value-neutral, 19
van Beneden, P.-J., 220
van Valen, L., 110
Verhulst, P.-F., 26
Vireo, 331
vocabulary, 266
Volkens, G., 30
von Humboldt, A., 87
vulnerability, 348

Wallace, A. R., 272
Wallin, I., 14, 221, 231
Warbler, 331
warbler
Black-throated green, 117
Myrtle, 117

Ward, L. F., 37
Warming, E., 30, 57, 87
Watase, S., 14, 221
water, shade, etc.
of, 6
web of life, 391
Wegener, A., 269
wetlands, 350
what if, 272
Whatmough, G. A., 16, 245
Whittaker, R. H., 39, 77, 94
wicked problem, 20, 372, 382, 389
Wiggins, David, 208
wild nature, 351
wilderness, 336, 352
wilderness myth, 303
Wilderness Society, 303
wildness, 338, 351, 352
Wildscapes Program, 343
Wilson, E. O., 110, 112–116, 124
Witmer, M., 264
wolves, 311
World Conservation Union (IUCN), 348
Worster, D., 55, 60, 64, 66–68, 84–86, 88–90, 95
Wright, S., 122

yaks, 309
Yard, R. S., 303

Kolibri 360 GmbH
Ferdinand-Jühlke-Straße 7
99095 Erfurt, Deutschland
produktsicherheit@kolibri360.de